Bioelectrochemical Systems

Integrated Environmental Technology Series

The *Integrated Environmental Technology Series* addresses key themes and issues in the field of environmental technology from a multidisciplinary and integrated perspective.

An integrated approach is potentially the most viable solution to the major pollution issues that face the globe in the 21st century.

World experts are brought together to contribute to each volume, presenting a comprehensive blend of fundamental principles and applied technologies for each topic. Current practices and the state-of-the-art are reviewed, new developments in analytics, science and biotechnology are presented and, crucially, the theme of each volume is presented in relation to adjacent scientific, social and economic fields to provide solutions from a truly integrated perspective.

The *Integrated Environmental Technology Series* will form an invaluable and definitive resource in this rapidly evolving discipline.

Series Editor

Dr Ir Piet Lens, Sub-department of Environmental Technology, IHE-UNESCO, P.O. Box 3015, 2601 DA Delft, The Netherlands (p.lens@unesco-ihe.org).

Published titles

Biofilms in Medicine, Industry and Environmental Biotechnology: *Characteristics, analysis and control*

Decentralised Sanitation and Reuse: *Concepts, systems and implementation*

Environmental Technologies to Treat Sulfur Pollution: *Principles and engineering*

Phosphorus in Environmental Technology: *Principles and applications*

Soil and Sediment Remediation: *Mechanisms, techniques and applications*

Water Recycling and Resource Recovery in Industries: *Analysis, technologies and implementation*

Environmental Technologies to treat Nitrogen Pollution

Bioelectrochemical Systems: From extracellular electron transfer to biotechnological application

Forthcoming title

Treatment of Micropollutants in Water and Wastewater

Bioelectrochemical Systems

From Extracellular Electron Transfer to Biotechnological Application

Edited by
Korneel Rabaey, Largus Angenent,
Uwe Schröder and Jürg Keller

Publishing
London • New York

Published by **IWA Publishing**
Alliance House
12 Caxton Street
London SW1H 0QS, UK
Telephone: +44 (0)20 7654 5500
Fax: +44 (0)20 7654 5555
Email: publications@iwap.co.uk
Web: www.iwapublishing.com

First published 2010

British Library Cataloguing in Publication Data
A CIP catalogue record for this book is available from the British Library

Library of Congress Cataloging-in-Publication Data
A catalog record for this book is available from the Library of Congress

ISBN: 184339233X
ISBN 13: 9781843392330

Contents

Foreword

Energy and water supply are two of the biggest challenges facing humanity in the coming decades. The present efforts to develop strategies for the recovery or efficient usage of resources are, therefore, highly justified. Groundbreaking technology is needed to allow novel means of converting and conserving resources. Bioelectrochemical Systems (BESs) fit within this endeavour.

Within the context of this book, we will discuss microbial BESs. These systems use whole cell biocatalysts to drive oxidation and reduction reactions at solid-state electrodes. Due to the separation between the oxidation and the reduction reaction, a whole range of applications is possible. The most widespread application is presently the microbial fuel cell, which aims to generate power – or at least decrease the usage of power associated with wastewater treatment. In the slipstream of microbial fuel cells, microbial electrolysis cells have more recently emerged. The versatility of the latter has notably expanded the range of applications of BESs. In this book, the reader will find applications in wastewater treatment, sediment-based electrical power generation, product generation, and bioremediation.

A key advantage of BESs is that electric power or products can be generated with lower greenhouse gas emission rates compared to conventional technologies. The unique control of BESs enables real-time monitoring of the biocatalytic process. Mathematical modelling will increasingly allow interpreting electrochemical data relative to microbial conversions. As this also allows studying microbial metabolism and biocatalysis in far better controlled conditions than possible in conventional lab experiments, many researchers are using BESs for basic microbial and microbial ecology research. The results

© 2010 IWA Publishing. *Bioelectrochemical Systems: From Extracellular Electron Transfer to Biotechnological Application*. Edited by Korneel Rabaey, Largus Angenent, Uwe Schröder and Jürg Keller. ISBN: 9781843392330. Published by IWA Publishing, London, UK.

of this research are increasingly coupled back to other fields, such as geomicrobiology. In addition, researchers have been using BESs as biosensors.

The development of BESs requires expertise in electrochemistry, materials science, microbiology, engineering, and other fields. For this reason, a number of chapters in this book are dedicated to each of these aspects, including an overview of the different measurement techniques available. This versatility of techniques is reflected in the versatility of the process, and we hope that the reader will enjoy the multidisciplinary approach many authors use when describing the different aspects of BESs. We also hope that the broad platform described here stimulates new thoughts and ideas, which can then develop into a new generation of BES applications.

Korneel Rabaey,
Uwe Schröder,
Largus Angenent and
Jürg Keller

List of Contributors

Chapter 1

Korneel Rabaey
Advanced Water Management Centre,
Gehrmann Building, Research Rd,
The University of Queensland,
Brisbane QLD4072, Australia

Chapter 2

Caroline M. Plugge
Laboratory of Microbiology,
Dreijenplein 10, 6703 HB,
Wageningen, The Netherlands

Jules B. van Lier
Sanitary Engineering Section,
Department of Water Management,
Faculty of Civil Engineering and Geosciences,
Delft University of Technology, Stevinweg 1,
2628 CN Delft, The Netherlands

Alfons J.M. Stams
Laboratory of Microbiology, Dreijenplein 10,
6703 HB, Wageningen, The Netherlands

David Jeison
Universidad de La Frontera,
Faculty of Engineering Sciences and Administration,
Department of Chemical Engineering,
Av. Francisco Salazar 01145,
4811230 Temuco, Chile

Chapter 3

Zana Zulic
Saint Louis University,
Department of Chemistry,
3501 Laclede Ave.,
St. Louis, MO 63103

Shelley D. Minteer
Saint Louis University,
Department of Chemistry,
3501 Laclede Ave.,
St. Louis, MO 63103

Chapter 4

Enrico Marsili
School of Biotechnology,
Collins Avenue, Glasnevin,
Dublin 9, Ireland

Xiaoming Zhang
Room 601, Building 510,
3-21-1 Takezono, Tsukuba,
Ibaraki 305-0032, Japan

Chapter 5

Orianna Bretschger
J. Craig Venter Institute,
10355 Science Center Drive,
San Diego, CA 92121

Yuri A. Gorby
J. Craig Venter Institute,
10355 Science Center Drive,
San Diego, CA 92121

Kenneth H. Nealson
J. Craig Venter Institute,
10355 Science Center Drive,
San Diego, CA 92121

Chapter 6

Miriam Rosenbaum
Department of Biological and Environmental Engineering,
Cornell University,
214 Riley-Robb Hall,
Ithaca, NY 14850

Largus T. Angenent
Department of Biological and Environmental Engineering,
Cornell University,
214 Riley-Robb Hall,
Ithaca, NY 14850

Chapter 7

Uwe Schröder
Sustainable Chemistry and Energy Research,
Institute of Ecological Chemistry,
Technische Universität Braunschweig,
Hagenring 30, 38106 Braunschweig, Germany

Falk Harnisch
Sustainable Chemistry and Energy Research,
Institute of Ecological Chemistry,
Technische Universität Braunschweig,
Hagenring 30, 38106 Braunschweig, Germany

Chapter 8

Edward LaBelle
Department of Microbiology and BioTechnology,
Institute University of Minnesota,
140 Gortner Laboratory,
1479 Gortner Avenue, St. Paul, MN 55108

Daniel R. Bond
Department of Microbiology and BioTechnology,
Institute University of Minnesota,
140 Gortner Laboratory,
1479 Gortner Avenue, St. Paul, MN 55108

Daniel A. Lowy
DAVIA Corporation, 13255 Revilo Loop,
Woodbridge, VA 22191

Aswin K. Manohar
Mork Family Department of Chemical Engineering
and Materials Science,
University of Southern California,
Los Angeles, CA 90089

Zhen He
Department of Civil Engineering & Mechanics
University of Wisconsin – Milwaukee
3200 N Cramer St.
Milwaukee, WI, 53211

Florian Mansfeld
Mork Family Department of Chemical Engineering
and Materials Science,
University of Southern California,
Los Angeles, CA 90089

Chapter 9

Bruce E. Logan
H2E Center & Engineering Environmental Institute,
231Q Sackett Bldg, Penn State University,
University Park, PA 16802

Chapter 10

Bert Hamelers
Sub-department of Environmental Biotechnology,
Bomenweg 2, 6703HD, Wageningen,
The Netherlands

Tom H.J.A. Sleutels
Sub-department of Environmental Biotechnology,
Bomenweg 2, 6703HD, Wageningen,
The Netherlands

Adriaan W. Jeremiasse
Sub-department of Environmental Biotechnology,
Bomenweg 2, 6703HD,
Wageningen,
The Netherlands

Jan W. Post
Sub-department of Environmental Biotechnology,
Bomenweg 2, 6703HD,
Wageningen,
The Netherlands

David P.B.T.B. Strik
Sub-department of Environmental Biotechnology,
Bomenweg 2, 6703HD,
Wageningen,
The Netherlands

René A. Rozendal
Advanced Water Management Centre,
Gehrmann Building, Research Rd,
The University of Queensland,
Brisbane QLD4072, Australia

Chapter 11

Stefano Freguia
Bio-analytical and Physical Chemistry Laboratory,
Division of Applied Life Sciences,
Graduate School of Agriculture,
Kyoto University, Sakyo,
Kyoto 606-8205, Japan

Chapter 12

Paritam K. Dutta
Advanced Water Management Centre,
Gehrmann Building, Research Rd,
The University of Queensland,
Brisbane QLD4072, Australia

Jürg Keller
Advanced Water Management Centre,
Gehrmann Building, Research Rd,
The University of Queensland,
Brisbane QLD4072, Australia

Zhiguo Yuan
Advanced Water Management Centre,
Gehrmann Building, Research Rd,
The University of Queensland,
Brisbane QLD4072, Australia

René A. Rozendal
Advanced Water Management Centre,
Gehrmann Building, Research Rd,
The University of Queensland,
Brisbane QLD4072, Australia

Korneel Rabaey
Advanced Water Management Centre,
Gehrmann Building, Research Rd,
The University of Queensland,
Brisbane QLD4072, Australia

Chapter 13

René A. Rozendal
Advanced Water Management Centre,
Gehrmann Building, Research Rd,
The University of Queensland,
Brisbane QLD4072, Australia

Falk Harnisch
Sustainable Chemistry and Energy Research,
Research Institute of Ecological Chemistry,
Technische Universität Braunschweig,
Hagenring 30, 38106 Braunschweig, Germany

Adriaan W. Jeremiasse
Sub-department of Environmental Biotechnology,
Bomenweg 2, 6703HD,
Wageningen,
The Netherlands

Uwe Schröder
Sustainable Chemistry and Energy Research,
Research Institute of Ecological Chemistry,
Technische Universität Braunschweig,
Hagenring 30, 38106 Braunschweig, Germany

Chapter 14

Peter Clauwaert
Laboratory for Microbial Ecology and Technology (LabMET),
Ghent University, Coupure Links 653, 9000 Ghent, Belgium

Chapter 15

Federico Aulenta
Department of Chemistry,
Sapienza University of Rome,
Piazzale Aldo Moro 5,
00185, Rome, Italy

Mauro Majone
Department of Chemistry,
Sapienza University of Rome,
Piazzale Aldo Moro 5,
00185, Rome, Italy

Chapter 16

Peter R. Girguis
Harvard University Biological Laboratories,
Rm. 3085, 16 Divinity Avenue,
Cambridge, MA 02138-2020

Mark E. Nielsen
Harvard University Biological Laboratories,
Rm. 3085, 16 Divinity Avenue,
Cambridge, MA 02138-2020

Clare E. Reimers
College of Oceanic and Atmospheric Sciences,
Oregon State University,
104 COAS Administration Building,
Corvallis, OR 97331-5503

Chapter 17

Byung Hong Kim
Water Environment & Remediation Research Center,
Korea Institute of Science and Technology,
39-1, Hawolgok-dong, Sungpook-ku,
136-791, Seoul, South Korea

I.N. Seop Chang
Department of Environmental Science and Engineering,
Gwangju Institute of Science and Technology (GIST),
261 Cheomdan-gwagiro, Buk-gu,
Gwangju 500-712, South Korea

Geoffrey M. Gadd
Division of Molecular and Environmental Microbiology,
College of Life Sciences,
University of Dundee,
DD1 5EH, Scotland

Chapter 18

Freda R. Hawkes
Sustainable Environment Research Centre,
University of Glamorgan,
Pontypridd, RCT CF37 1DL, UK

Jung Rae Kim
Sustainable Environment Research Centre,
University of Glamorgan,
Pontypridd, RCT CF37 1DL, UK

Godfrey Kyazze
Sustainable Environment Research Centre,
University of Glamorgan,
Pontypridd, RCT CF37 1DL, UK

Giuliano C. Premier
Sustainable Environment Research Centre,
University of Glamorgan,
Pontypridd, RCT CF37 1DL, UK

Chapter 19

Miriam Rosenbaum
Department of Biological and Environmental Engineering,
Cornell University,
214 Riley-Robb Hall,
Ithaca, NY 14850

Matthew T. Agler
Department of Biological and Environmental Engineering,
Cornell University,
214 Riley-Robb Hall,
Ithaca, NY 14850

Jeffrey J. Fornero
Department of Energy, Environmental & Chemical Engineering,
Washington University in St. Louis,
St. Louis, MO 63130

Arvind Venkataraman
Department of Biological and Environmental Engineering,
Cornell University,
214 Riley-Robb Hall,
Ithaca, NY 14850

Largus T. Angenent
Department of Biological and Environmental Engineering,
Cornell University,
214 Riley-Robb Hall,
Ithaca, NY 14850

Chapter 20

Ioannis Ieropoulos
Bristol Robotics Laboratory,
University of Bristol and
University of the West of England,
DuPont Building, Bristol Business Park,
Coldharbour Lane,
Bristol, BS16 1QD, UK

John Greenman
Bristol Robotics Laboratory,
University of Bristol and
University of the West of England,
DuPont Building, Bristol Business Park,
Coldharbour Lane,
Bristol, BS16 1QD, UK

Chris Melhuish
Bristol Robotics Laboratory,
University of Bristol and
University of the West of England,
DuPont Building, Bristol Business Park,
Coldharbour Lane,
Bristol, BS16 1QD, UK

Ian Horsfield
Bristol Robotics Laboratory,
University of Bristol and
University of the West of England,
DuPont Building, Bristol Business Park,
Coldharbour Lane,
Bristol, BS16 1QD, UK

Chapter 21

Jorge Rodriguez
Sustainable Environment Research Centre,
University of Glamorgan,
2 Forest Grove (2FG),
CF37 1UB Pontypridd, Wales, UK

Giuliano C. Premier
Sustainable Environment Research Centre,
University of Glamorgan,
2 Forest Grove (2FG),
CF37 1UB Pontypridd, Wales, UK

Chapter 22

Jürg Keller
Advanced Water Management Centre,
Gehrmann Building, Research Rd,
The University of Queensland,
Brisbane QLD4072, Australia

René A. Rozendal
Advanced Water Management Centre,
Gehrmann Building, Research Rd,
The University of Queensland,
Brisbane QLD4072, Australia

Largus T. Angenent
Department of Biological and Environmental Engineering,
Cornell University,
214 Riley-Robb Hall,
Ithaca, NY 14850

Uwe Schröder
Sustainable Chemistry and Energy Research,
Institute of Ecological Chemistry,
Technische Universität Braunschweig,
Hagenring 30, 38106 Braunschweig, Germany

Piet Lens
Sub-department of Environmental Technology,
IHE-UNESCO,
P.O. Box 3015, 2601 DA Delft,
The Netherlands

Korneel Rabaey
Advanced Water Management Centre,
Gehrmann Building, Research Rd,
The University of Queensland,
Brisbane QLD4072, Australia

1

Bioelectrochemical Systems: a new approach towards environmental and industrial biotechnology

Korneel Rabaey

1.1 FUEL CELLS AND BIO-ELECTRICITY

In 1839, William Grove realized that if electrolysis, using electricity, could split water into hydrogen and oxygen, then the opposite should also be true (Grove 1839). Thus, he built the world's first gas battery, later renamed the fuel cell. While his invention was an initial success, the interest in this technology dwindled rapidly through the emergence of fossil fuels. More than 70 years later, Potter described how microbial conversions could create reducing power, and

© 2010 IWA Publishing. *Bioelectrochemical Systems: From Extracellular Electron Transfer to Biotechnological Application*. Edited by Korneel Rabaey, Largus Angenent, Uwe Schröder and Jürg Keller. ISBN: 9781843392330. Published by IWA Publishing, London, UK.

thus electrical current (Potter 1911). Cohen (1931) investigated the capacity of bacteria to create reducing power, and noted that the main limitation of the current generation was the small capacity that microorganisms could deliver. It took until the 1960s before this microbial capacity was truly combined with fuel cells and the first microbial fuel cells (MFCs) emerged (Berk and Canfield 1964; van Hees 1965).

A MFC can be defined as a system in which microorganisms function as catalysts to convert chemical energy into electrical energy. MFCs are a type of biofuel cell, the latter more broadly employing a biocatalyst (from whole cell to enzyme) to achieve oxidation and/or reduction reactions. Biofuel cells using enzymes will be discussed in Chapter 3. Microbial fuel cells have recently attracted considerable interest, which is indicated by the number of citations in the field for the past twenty years in Figure 1.1 The upsurge of MFC research in the past decade is directly linked to the concerns with respect to energy and water supply, aside from a number of microbial and engineering breakthroughs.

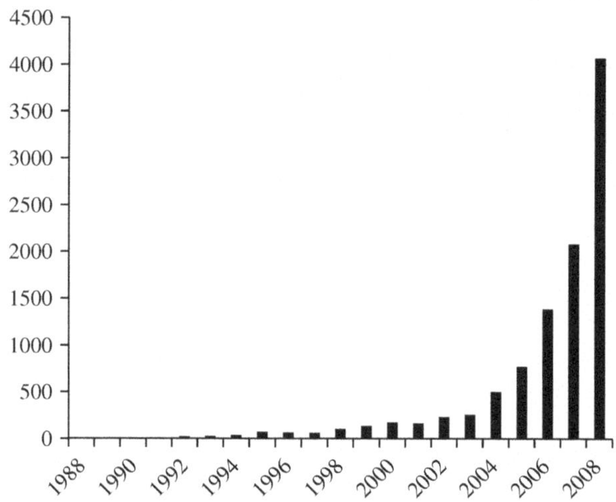

Figure 1.1. Number of citations on microbial fuel cells for the past 20 years (Web of Science, data collected on January 5th 2009)

Biomass is increasingly considered as an energy source. One kilogram of biomass chemical oxygen demand represents about 4 kWh worth of energy. The main waste biomass types used for energy purposes are wood chips/shavings/ pellets, crop residues, and household vegetable, fruit and garden waste. However, much biomass is also present in dissolved or suspended forms in wastewater. Chapter 18 reviews these liquid or semi-solid feedstocks.

Table 1.1. Some common terminology for the technology, processes and microorganisms relevant for this book

	Name	Abbrev.	Definition	Reference
Technology	Bioelectrochemical system	BES	An electrochemical system in which biocatalysts perform oxidation and/or reduction at electrodes (in the context of this book, the biocatalysts are whole microorganisms)	(Rabaey *et al.* 2007)
	Microbial fuel cell	MFC	A BES that produces net electrical power	(Roller *et al.* 1984)
	Microbial electrolysis cell	MEC	A BES to which net electrical power is provided to achieve a certain process or product formation	(Logan *et al.* 2008)
	Bioelectrochemically assisted microbial reactor	BEAMR	A BES to which net electrical power is provided to achieve a certain process or product formation	(Liu *et al.* 2005)
	Bio-electrical reactor	BER	A reactor in which current is provided to microorganisms to stimulate their metabolism	(Thrash and Coates 2008)
	Biocatalyzed electrolysis cell	BEC	A BES to which net electrical power is provided to achieve a certain process or product formation	(Rozendal *et al.* 2006)
	Biochemical fuel cell	BFC	An electrochemical system in which biocatalysts function as catalysts for oxidation and/or reduction reactions at electrodes	
	Biofuel cell		An electrochemical system in which biocatalysts function as catalysts for oxidation and/or reduction reactions at electrodes	(Katz *et al.* 2003)
	Sediment microbial fuel cell	SMFC	Microbial fuel cell operated at underwater sediment interface	
	Benthic unattended generator	BUG	Microbial fuel cell operated at underwater sediment interface	(Tender 2007)

(*continued*)

Table 1.1. *Continued*

	Name	Abbrev.	Definition	Reference
Process	Electrogenesis		Creation of electrical power using a BES (note that electrogenesis also has different meanings in other contexts)	(Ishii *et al.* 2008a)
	Electro-hydrogenesis		Creation of hydrogen gas by supplying power to a BES	(Logan *et al.* 2008)
	Microbial electrolysis		Creation of a product or realization of a process through providing net power to an MEC	
	Biocatalyzed electrolysis		Creation of a product or realization of a process through providing net power to an MEC	(Rozendal *et al.* 2006)
Organisms	Anodophiles		Microorganisms capable of reducing electrodes	(Park and Zeikus 2000)
	Electrochemically active bacteria	EAB	Microorganisms capable of transporting electrons to and from their extracellular environment	(Chang *et al.* 2006)
	Electroactive bacteria	EB	Microorganisms capable of transporting electrons to and from their extracellular environment	(Dulon *et al.* 2007)
	Electricigens		Microorganisms capable of reducing electrodes	(Lovley 2006)
	Exo-electrogens		Microorganisms capable of transporting electrons to and from their extracellular environment	(Logan 2008)
	Electrode reducing bacteria	ERB	Bacteria using an anode as electron acceptor	(Lovley 2008)
	Electrode oxidizing bacteria	EOB	Bacteria using a cathode as electron donor	(Lovley 2008)
	Anode respiring bacteria	ARB	Microorganisms capable of reducing electrodes	(Torres *et al.* 2007)

More recently, it was found that the production of electrical power is not necessarily the only outcome of MFCs. By adding power to the system, it was shown that hydrogen could be produced at the cathode (see also Chapter 14). In this case, the system is referred to as a microbial electrolysis cell (MEC) (Logan *et al.* 2008). Recently, MFCs and MECs were grouped together as Bioelectrochemical Systems (BESs) (Rabaey *et al.* 2007), here almost exclusively denoting BESs using whole microorganisms. The rapid evolution of this research domain has led to a plethora of terms describing the technology, the underlying processes and the microorganisms involved. Table 1.1 overviews the terms commonly found. The wide variety of possible applications of BESs correlates with a wide variety of novel products and processes in the emerging biorefinery industry and wastewater treatment field. An overview of such products and processes is provided in Chapter 2.

1.2 UNDERLYING PRINCIPLES

1.2.1 Microorganisms and current

Electron flows are inherent to the microbial metabolism. Microorganisms transfer electrons from an electron donor (lower potential) to an electron acceptor (higher potential). The electron acceptor can be external, in which case the metabolism is called respiration. If the electron acceptor is internal, the metabolism is called fermentation. Not all electron donors can be fermented, mainly due to thermodynamic constraints.

The theoretical energy gain ΔG (kJ/mol) for microorganisms relates directly to the potential difference between the donor and the acceptor: $\Delta G = -n\, FE_{\mathrm{emf}}$ with n the amount of electrons exchanged in the reaction, F Faraday's number (96485 C/mol) and E_{emf} the potential difference between the donor and acceptor (V). Evidently, inefficiencies will decrease this energy gain, for more in-depth information on energy conservation by microorganisms, we refer to Heijnen's review (Heijnen 1999).

Microorganisms will, within their capacity, attempt to maximize their energy gain by selecting the electron acceptor with the highest potential available. Often soluble electron acceptors are depleted in the microbial environment. Microorganisms can then either turn to fermentation or use non-soluble electron acceptors. In the latter case, microorganisms have to transport the electrons outside the cell to achieve the reduction. This process is called extracellular electron transfer (EET). It is ubiquitous in nature, where mainly minerals containing iron and manganese oxides are being reduced. In BESs, the anodes function as insoluble electron acceptors.

Likewise, microorganisms will also, within their capacity, maximize the energy gain by selecting the electron donor with the lowest potential available which can be easily metabolized. Similarly to electron acceptors, soluble electron donors may become depleted, in which case microorganisms can oxidize insoluble electron donors via EET. Thus, EET can relate to electron transport both in and out of the cell.

Several mechanisms have been described for EET in the past years, they can be categorized as follows (Rabaey *et al.* 2007):

- Direct electron transfer, which involves membrane bound or associated enzyme complexes, and may involve conductive pili or pilus-like structures. The latter are also referred to as nanowires. Direct electron transfer is the main focus of Chapter 5.
- Indirect electron transfer, where an organic or inorganic, soluble compound is reduced or oxidized at the cell and subsequently diffuses towards the insoluble electron acceptor/donor. Examples of organic shuttles are pyocyanin and humic acids. Examples of inorganic shuttles are sulfur compounds and hydrogen. These will be discussed in depth in Chapter 4.

Extensive research has been done to elucidate both pathways using type organisms such as *Shewanella* and *Geobacter* sp., and attempts are made to engineer this metabolism (as discussed in Chapter 6).

In many studies, mixed populations of microorganisms were used at anodes and/or cathodes (see also Chapter 4). The analysis of these communities revealed a high degree of complexity, and hinted at microbial interactions driving the electron flow (Pham *et al.* 2008; Rabaey *et al.* 2004). Further research into these interactions is required to elucidate whether an effective difference exists between pure and mixed population capacities for electron flow (Ishii *et al.* 2008b; Nevin *et al.* 2008). Mixed populations will also give rise to a foodweb involving several groups of microorganisms for a single oxidation process. Examples of this are described in Chapter 12.

1.2.2 Microbial communities in BESs

Anodic communities. Analyses of the microbial communities on anodes and cathodes of BESs has revealed a high diversity. While the capacity to perform or at least participate in extracellular electron transfer appears ubiquitous among all the major phyla (Aelterman *et al.* 2008), notably Proteobacteria and Firmicutes are found at high abundancy in BES anodes (Lovley 2006; Rabaey *et al.* 2007).

The most widely described type species are *Geobacter sulfurreducens* and *Shewanella oneidensis* MR-1. However, an abundancy of isolates has been obtained in recent years, and increasingly metabolic and/or genomic information has become available. Key examples are *Pseudomonas aeruginosa* (Rabaey *et al.* 2004), *Rhodopseudomonas palustris* (Xing *et al.* 2008), *Clostridium acetobutylicum* (Park *et al.* 2001) and *Thermincola* sp. (Wrighton *et al.* 2008). Interestingly, while at first glance anodic reduction appears identical to metal reduction, recent studies hint that this may not be the case. For example, in one study it was demonstrated that *Pelobacter carbinolicus* cannot reduce an anode, while being capable of reducing ferric iron (Richter *et al.* 2007). Whether this is due to a physiological difference from the microorganism perspective, or due to different thermodynamic or chemical conditions (e.g. presence/absence of sulfide) needs to be further evaluated.

In the anodic biofilm, a complex foodweb may establish, leading to associations between EAB, fermenting organisms and *Archaea*, notably methanogens (Freguia *et al.* 2008). Further analysis of the communities, notably with respect to the presence and activity of *Archaea*, is certainly a necessity.

Cathodic communities. While anodic communities have been extensively studied, only in recent years has cathodic biocatalysis emerged. Only a limited amount of studies have analyzed the microbial communities from cathodic biofilms. Key players in oxygen biocatalysis at the cathode belong to the *Bacteroidetes* (e.g. *Sphingobacterium*) and the Proteobacteria (e.g. *Acinetobacter*) (Rabaey *et al.* 2008). Isolates of the aforementioned species could cause modest improvement to the catalysis of the cathode. *Geobacter metallireducens* was shown to use a cathode as electron donor and link this to the reduction of nitrate to nitrite, or fumarate to succinate (Gregory *et al.* 2004). Likewise, analyses of the microbial communities reducing nitrate, perchlorate, etc. have been performed, the results of which are discussed further in this book.

1.2.3 From microbial metabolism to electrical current

By linking up the microbial metabolism to electrodes via EET, electrical current can be generated or consumed. In an MFC mode, the anode is reduced to a low potential by microorganisms, while a high(er) potential is maintained at the cathode. Overall, the voltage over the MFC can be described as (Larminie and Dicks 2000):

$$E_{\text{cell}} = E_{\text{emf}} - (\Sigma\eta_{\text{a}} + |\Sigma\eta_{\text{c}}| + IR_{\Omega})$$

with E_{cell} the effective cell voltage, $\Sigma\eta_{\text{a}} + |\Sigma\eta_{\text{c}}|$ electrode related losses and IR_{Ω} losses related to the internal resistance of the electrochemical system. Chapter 8 will discuss these losses in depth. There are a number of methods to define the individual contributions of each loss (Chapter 8).

The end result of linking up an oxidation process at the anode and a reduction process at the cathode is a BES that has a high versatility for both the electron donors and acceptors (Figure 1.2).

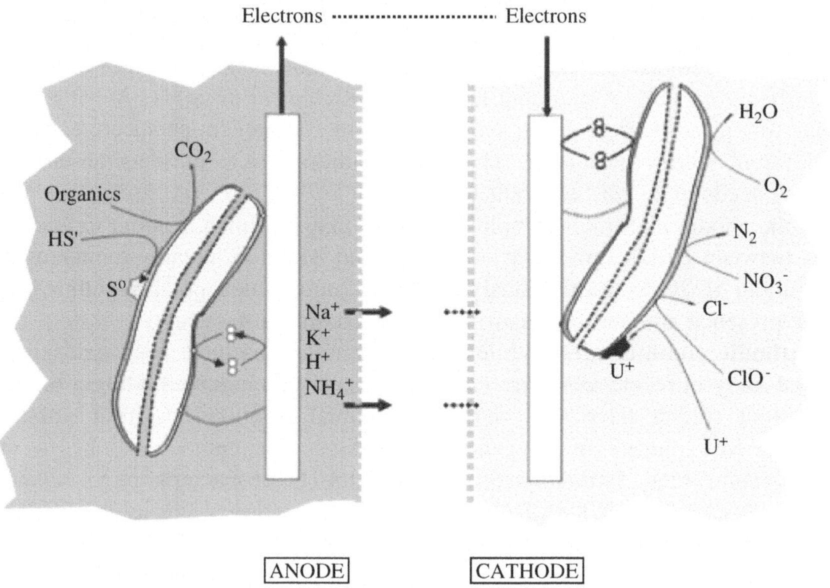

Figure 1.2. A Bioelectrochemical System allows separation of an oxidation reaction at the anode from a reduction reaction at the cathode. As a result, a high versatility exists in both distinguished processes (Rabaey *et al,* 2007)

The effectiveness of BESs will be highly dependent on the design choice. A number of critical bottlenecks have been identified in the past years; Chapter 10 provides an overview of these technical factors. The influence of the materials selection (electrodes and membranes) on these losses will also be highlighted in Chapter 9. In addition, a range of peripheral equipment and process controls will need development (Chapter 20).

1.3 MEASURING AND DEFINING PERFORMANCE

There exists a wide variety of measures to describe the performance of BESs (Logan *et al.* 2006; Logan *et al.* 2008). Lack of appropriate information has led to considerable confusion about the results achieved in a number of studies. We can categorize the performance indicators as either efficiency based or rate based. The assessment of these indicators can occur through a range of (electro) chemical techniques (Chapter 10).

1.3.1 Measuring potentials

The potential of a redox couple can be predicted via the Nernst equation, which highlights the dependence of the potential on the activity of the oxidized and reduced species, and the solution pH. However, the solution potential is not necessarily the corresponding electrode potential – there are losses between the solution and the electrode, depending on the kinetics. Also, several redox active species can be simultaneously present in the solution, making potential calculations complex. More in-depth information on these losses can be found in Chapters 8 and 21. In most cases, the electrode potential cannot be accurately predicted and must be measured.

Generally, the potentials of the anode and/or cathode are measured against a reference electrode. The normal (or standard) hydrogen electrode (NHE or SHE) is used as universal reference ($E_{ref} = 0$ V). Due to the complexity of this electrode, a wide variety of other reference electrodes exists (Table 1.2). Note that the potential values of reference electrodes vary according to temperature and salinity, often necessitating a correction of the reference potential.

Table 1.2. Common reference electrodes and their potential (Volt) ($T = 25°C$)

Common Name	Electrode	E_{ref} (V)
Normal Hydrogen Electrode (NHE)*	H_2/H^+	0.0
Standard Calomel Electrode (SCE)	Hg/Hg_2Cl_2/sat. KCl	0.241
Calomel	Hg/ Hg_2Cl_2/1M KCl	0.28
Mercurous sulphate	Hg/Hg_2SO_4/sat. K_2SO_4	0.64
	Hg/Hg_2SO_4/0.5M H_2SO_4	0.68
Mercurous oxide	Hg/HgO/1M NaOH	0.098
Silver chloride	Ag^+/AgCl/sat. KCl	0.197
Copper sulphate	Cu^{2+}/sat. $CuSO_4$	0.316
Zinc/seawater	Zn/seawater	−0.800

*Also called *Standard Hydrogen Electrode.*

Once the potentials are clearly defined and the current is measured, the necessary data is available to calculate efficiencies and rates. One important aspect in reporting performance is the distinction between volumetric and surface based performance. From an engineering perspective, it is desirable to report power and current on a volumetric basis. Both compartment based reporting (e.g. power per anode volume) or total volume reporting (e.g. power per volume reactor) are widely used, depending on the priority of the researcher. From an electrochemical and electrocatalysis point of view, reporting values versus electrode or membrane surface area are preferred, because this provides information on the catalysis itself. The main point with reporting performance values is supplying sufficient information to allow the reader to calculate to any other representation (Logan *et al.* 2006).

1.3.2 Rate based performance indicators

Power. Power is perhaps the most used indicator, certainly when MFCs are described. As mentioned earlier, power can be expressed both on a volume and surface basis. Net power indicates the power output of the MFC minus the energy invested to operate the MFC. In the case of MECs, power is added to the system, thus net power is the sum of the power input for driving the reactions and for operating the MEC.

Current. Current is a key parameter to describe conversion rates. Certainly for BESs where products are formed, the current directly relates to the quantity of product formed. Current also directly relates to the organic loading rate (see below). From an engineering perspective, current per unit volume of reactor is perhaps the most informative parameter, although not widely used at the moment.

Organic loading rate. In environmental engineering, the organic loading rate expresses the amount of organics converted per unit reactor over a given time. Certainly for BESs operated in the wastewater field, reaching a sufficient organic loading rate is crucial, as this is directly related to size and thus footprint of the reactor. There are two ways of expressing this parameter, being the *volumetric* loading rate (kilogram organics per volume reactor per day) or *sludge* loading rate (kilogram organics per kilogram biocatalyst per day).

Product formation. The rate of product formation at for example a cathode, not only reflects the rate of electron transfer throughout the BES, but also takes the conversion efficiency towards the product into account.

1.3.3 Efficiency based performance indicators

Coulombic efficiency: also Charge Transfer Efficiency or wrongly denominated 'columbic' efficiency. For an anode, this value represents the relative amount of electrons recovered from the electron donor as current. For a cathode, this value represents the amount of electrons ending up in reduction products relative to the amount of electrons supplied by the cathode. One can define an overall Coulombic efficiency as the quantity of reduction products generated relative to the amount of electron donor consumed. For calculations and more background information, we refer to a Chapter 19 and a number of existing manuscripts (Logan *et al.* 2006; Logan 2008; Rabaey *et al.* 2005a).

Energetic efficiency: similar to Coulombic efficiency, energetic efficiency compares the energy output of the system to the energy input(s) of the system. It includes both the quantity of electrons produced/spent in the process and the energy associated with them.

Removal efficiency: in cases where wastewater treatment or bioremediation are the objectives, the overall removal efficiency of the target compounds is a key parameter. This will define the system effluent concentrations, which are often legal drivers for the treatment. The relationship between the removal efficiency and the Coulombic efficiency is a proxy of how efficient electrons are transferred between the electrodes to and from the electron donor/acceptor. Typically, the higher the desired removal efficiency, the lower the removal rate will become due to the slower kinetics at lower substrate concentrations.

1.4 A PLETHORA OF APPLICATIONS

The possibility to control oxidation and reduction reactions at electrodes has created a vast array of application niches, performance expectations, and targeted products. Table 1.3 highlights some of the major niches in which BESs have been deployed.

Table 1.3. Most common BES deployments and target products or bioremediation purposes. For the deployments, the most relevant chapters within this book that discuss them are indicated

BES type		
Reactor unit	*Stand alone BES in industrial or domestic context*	Chapter 19
Retrofit	*BES integrated within existing engineered facility*	Chapter 19
Sediment BES	*BES operated at sediment–water interface*	Chapter 16
Soil MFC	*BES inserted into soil matrix*	Chapter 15
Rhizosphere MFC	*BES operated in root environment*	Chapter 15
Whole cell bio-sensor	*BES using whole microorganisms for sensing purposes*	Chapter 17

The prime target of MFCs has, thus far, been the energy efficient treatment of wastewater (Habermann and Pommer 1991; Kim *et al.* 2004; Liu *et al.* 2004; Rabaey *et al.* 2005b). A considerable number of studies addressed different industrial wastewaters, aside from domestic wastewater (see also Chapter 18). It appears that in most cases, BESs will be integrated within an existing wastewater line of an industrial or domestic application (see Chapter 19) (Pham *et al.* 2006). Highly exciting is the capability to go beyond the production of electrical power but rather perform highly specific bioremediation processes or produce useful products. These processes very often rely on cathodic biocatalysts, the research on which has only emerged very recently (see Chapters 14 and 15). Also the removal of inorganic compounds such as sulfur species is highly feasible in BESs (Chapter 12). In case the products at the cathode are of an organic nature, such as L-glutamic acid (*Hongo and Iwahara* 1979a,b) and propionate (Emde and Schink 1990), BESs can be fully integrated in the emerging biorefinery industry. The scope of this is elaborated in Chapter 22. Table 1.4 indicates a range of already published product outcomes (or proposed outcomes), or bioremediation targets, using BESs.

Table 1.4. Overview of products and bioremediation targets of BESs

Products (reference)	Bioremediation (reference)
Electrical power (Berk and Canfield 1964)	Wastewater treatment (Habermann and Pommer 1991)
Hydrogen (Liu *et al.* 2005)	Uranium removal (Gregory and Lovley 2005)
Formate (Reda *et al.* 2008)	Organochlorine removal (Aulenta *et al.* 2008)
Methane (Clauwaert *et al.* 2008)	Perchlorate reduction (Thrash *et al.* 2007)
Succinate (Park and Zeikus 1999)	Denitrification (Clauwaert *et al.* 2007)
Ethanol (Steinbusch *et al.* 2008)	Leachate treatment (You *et al.* 2006)
Sulfur (Rabaey *et al.* 2006)	Decreased sediment build up (De Schamphelaire *et al.* 2007)
Propionic acid (Emde and Schink 1990)	Methane abatement (De Schamphelaire *et al.* 2008)
Glutamic acid (Hongo and Iwahara 1979a)	Sulfide removal (Rabaey *et al.* 2006)
	Acid mine drainage (Cheng *et al.* 2007)

1.5 ACKNOWLEDGEMENTS

Korneel Rabaey is supported through the ARC Discovery program (DP0879245) and the ARC Linkage program (LP0989159).

REFERENCES

Aelterman, P., Rabaey, K., De Schamphelaire, L., Clauwaert, P., Boon, N. and Verstraete, W. (2008) *Bioenergy*. Wall, J.D., Harwood, C.S. and Demain, A.L. (eds).

Aulenta, F., Canosa, A., Majone, M., Panero, S., Reale, P. and Rossetti, S. (2008) Trichloroethene dechlorination and H_2 evolution are alternative biological pathways of electric charge utilization by a dechlorinating culture in a bioelectrochemical system. *Environ. Sci. Technol.* **42**(16), 6185–6190.

Berk, R.S. and Canfield, J.H. (1964) Bioelectrochemical energy conversion. *Applied Microbiology* **12**, 10–12.

Chang, I.S., Moon, H., Bretschger, O., Jang, J.K., Park, H.I., Nealson, K.H. and Kim, B.H. (2006) Electrochemically active bacteria (EAB) and mediator-less microbial fuel cells. *J. Microbiol. Biotechnol.* **16**(2), 163–177.

Cheng, S., Dempsey, B.A. and Logan, B.E. (2007) Electricity generation from synthetic acid-mine drainage (AMD) water using fuel cell technologies. *Environ. Sci. Technol.* **41**(23), 8149–8153.

Clauwaert, P., Rabaey, K., Aelterman, P., De Schamphelaire, L., The Pham, H., Boeckx, P., Boon, N. and Verstraete, W. (2007) Biological denitrification driven by microbial fuel cells. *Environ. Sci. Technol.* **41**(9), 3354–3360.

Clauwaert, P., Toledo, R., Van der Ha, D., Crab, R., Verstraete, W., Hu, H., Udert, K.M. and Rabaey, K. (2008) Combining biocatalyzed electrolysis with anaerobic digestion. *Water Sci. Technol.* **57**(4), 575–579.

Cohen, B. (1931) The bacterial culture as an electrical half-cell. *J. Bacteriol.* **21**, 18.

De Schamphelaire, L., Rabaey, K., Boon, N., Verstraete, W. and Boeckx, P. (2007) Minireview: the potential of enhanced manganese redox cycling for sediment oxidation. *Geomicrobiol. J.* **24**(7–8), 547–558.

De Schamphelaire, L., Van den Bossche, L., Dang, H.S., Höfte, M., Boon, N., Rabaey, K. and Verstraete, W. (2008) Microbial fuel cells generating electricity from rhizodeposits of rice plants. *Environ. Sci. Technol.* **42**(8), 3053–3058.

Dulon, S., Parot, S., Delia, M.-L. and Bergel, A. (2007) Electroactive biofilms: new means for electrochemistry. *J. Appl. Electrochem.* **37**(1), 173–179.

Emde, R. and Schink, B. (1990) Enhanced Propionate Formation by Propionibacterium-Freudenreichii Subsp Freudenreichii in a 3-Electrode Amperometric Culture System. *Appl. Environ. Microbiol.* **56**(9), 2771–2776.

Freguia, S., Rabaey, K., Yuan, Z.G. and Keller, J. (2008) Syntrophic processes drive the conversion of glucose in microbial fuel cell anodes. *Environ. Sci. Technol.* **42**(21), 7937–7943.

Gregory, K.B., Bond, D.R. and Lovley, D.R. (2004) Graphite electrodes as electron donors for anaerobic respiration. *Environ. Microbiol.* **6**(6), 596–604.

Gregory, K.B. and Lovley, D.R. (2005) Remediation and recovery of uranium from contaminated subsurface environments with electrodes. *Environ. Sci. Technol.* **39**(22), 8943–8947.

Growe, W. (1839) Energiedirektumwandlung. *Phil. Mag. III* **14**, 129.

Habermann, W. and Pommer, E.-H. (1991) Biological fuel cells with sulphide storage capacity. *Appl. Microbiol. Biotechnol.* **35**, 128–133.

Heijnen, J.J. (1999) Bioprocess technology: fermentation, biocatalysis and bioseparation. Flickinger, M.C. and Drew, S.W. (eds). John Wiley & Sons, Inc., New York, pp. 267–291.

Hongo, M. and Iwahara, M. (1979a) Electrochemical studies on fermentation. 1. Application of electro-energizing method to L-glutamic acid fermentation. *Ag. Biol. Chem.* **43**(10), 2075–2081.

Hongo, M. and Iwahara, M. (1979b) Electrochemical studies on fermentation 2. Determination of electro-energizing conditions for L-glutamic acid fermentation. *Ag. Biol. Chem.* **43**(10), 2083–2086.

Ishii, S., Hotta, Y. and Watanabe, K. (2008a) Methanogenesis versus electrogenesis: morphological and phylogenetic comparisons of microbial communities. *Biosci. Biotechnol. Biochem.* **72**(2), 286–294.

Ishii, S., Watanabe, K., Yabuki, S., Logan, B.E. and Sekiguchi, Y. (2008b) Comparison of electrode reduction activities of geobacter sulfurreducens and an enriched consortium in an air-cathode microbial fuel cell. *Appl. Environ. Microbiol.* **74**(23), 7348–7355.

Katz, E., Shipway, A.N. and Willner, I. (2003) *Handbook of Fuel Cells – Fundamentals, Technology and Applications*. Vielstich, W., Gasteiger, H.A. and Lamm, A. (eds). John Wiley & Sons, New York, pp. 355–381.

Kim, B.H., Park, H.S., Kim, H.J., Kim, G.T., Chang, I.S., Lee, J. and Phung, N.T. (2004) Enrichment of microbial community generating electricity using a fuel-cell-type electrochemical cell. *Appl. Microbiol. Biotechnol.* **63**(6), 672–681.

Larminie, J. and Dicks, A. (2000) Fuel cell systems explained, John Wiley & Sons, Chichester.

Liu, H., Grot, S. and Logan, B.E. (2005) Electrochemically assisted microbial production of hydrogen from acetate. *Environ. Sci. Technol.* **39**(11), 4317–4320.

Liu, H., Ramnarayanan, R. and Logan, B.E. (2004) Production of electricity during wastewater treatment using a single chamber microbial fuel cell. *Environ. Sci. Technol.* **38**(7), 2281–2285.

Logan, B., Aelterman, P., Hamelers, B., Rozendal, R., Schroder, U., Keller, J., Freguia, S., Verstraete, W. and Rabaey, K. (2006) Microbial fuel cells: methodology and technology. *Environ. Sci. Technol.* **40**(17), 5181–5192.

Logan, B.E. (2008) Microbial fuel cells. John Wiley & Sons, Hoboken, NJ.

Logan, B.E., Call, D., Cheng, S., Hamelers, H.V.M., Sleutels, T.H.J.A., Jeremiasse, A.W. and Rozendal, R.A. (2008) Microbial electrolysis cells for high yield hydrogen gas production from organic matter. *Environ. Sci. Technol.* **42**(23), 8630–8640.

Lovley, D.R. (2006) Bug juice: harvesting electricity with microorganisms. *Nature Rev. Microbiol.* **4**, 497–508.

Lovley, D.R. (2008) The microbe electric: conversion of organic matter to electricity. *Curr. Opinion Biotechnol.* **19**(6), 564–571.

Nevin, K.P., Richter, H., Covalla, S.F., Johnson, J.P., Woodard, T.L., Orloff, A.L., Jia, H., Zhang, M. and Lovley, D.R. (2008) Power output and columbic efficiencies from biofilms of Geobacter sulfurreducens comparable to mixed community microbial fuel cells. *Environ. Microbiol.* **10**(10), 2505–2514.

Park, D.H. and Zeikus, J.G. (1999) Utilization of electrically reduced neutral red by Actinobacillus succinogenes: Physiological function of neutral red in membrane-driven fumarate reduction and energy conservation. *J. Bacteriol.* **181**(8), 2403–2410.

Park, D.H. and Zeikus, J.G. (2000) Electricity generation in microbial fuel cells using neutral red as an electronophore. *Appl. Environ. Microbiol.* **66**(4), 1292–1297.

Park, H.S., Kim, B.H., Kim, H.S., Kim, H.J., Kim, G.T., Kim, M., Chang, I.S., Park, Y.K. and Chang, H.I. (2001) A novel electrochemically active and Fe(III)-reducing bacterium phylogenetically related to Clostridium butyricum isolated from a microbial fuel cell. *Anaerobe* **7**(6), 297–306.

Pham, T.H., Boon, N., Aelterman, P., Clauwaert, P., De Schamphelaire, L., Vanhaecke, L., De Maeyer, K., Hofte, M., Verstraete, W. and Rabaey, K. (2008) Metabolites produced by *Pseudomonas* sp. enable a Gram-positive bacterium to achieve extracellular electron transfer. *Appl. Microbiol. Biotechnol.* **77**(5), 1119–1129.

Pham, T.H., Rabaey, K., Aelterman, P., Clauwaert, P., De Schamphelaire, L., Boon, N. and Verstraete, W. (2006) Microbial fuel cells in relation to conventional anaerobic digestion technology. *Engng. Life Sci.* **6**(3), 285–292.

Potter, M.C. (1911) Electrical effects accompanying the decomposition of organic compounds. *Proc. R. Soc. London Ser. B, Containing Papers of a Biological Character* **84**(571), 260–276.

Rabaey, K., Boon, N., Siciliano, S.D., Verhaege, M. and Verstraete, W. (2004) Biofuel cells select for microbial consortia that self-mediate electron transfer. *Appl. Environ. Microbiol.* **70**(9), 5373–5382.

Rabaey, K., Lissens, G. and Verstraete, W. (2005a) *Biofuels for Fuel Cells: Biomass Fermentation Towards Usage in Fuel Cells.* Lens, P.N., Westermann, P., Haberbauer, M. and Moreno, A. (eds). IWA Publishing, London.

Rabaey, K., Ossieur, W., Verhaege, M. and Verstraete, W. (2005b) Continuous microbial fuel cells convert carbohydrates to electricity. *Water Sci. Technol.* **52**(1–2), 515–523.

Rabaey, K., Read, S.T., Clauwaert, P., Freguia, S., Bond, P.L., Blackall, L.L. and Keller, J. (2008) Cathodic oxygen reduction catalyzed by bacteria in microbial fuel cells. *Isme Journal* **2**(5), 519–527.

Rabaey, K., Rodriguez, J., Blackall, L.L., Keller, J., Gross, P., Batstone, D., Verstraete, W. and Nealson, K.H. (2007) Microbial ecology meets electrochemistry: electricity-driven and driving communities. *ISME J* **1**(1), 9–18.

Rabaey, K., Van de Sompel, K., Maignien, L., Boon, N., Aelterman, P., Clauwaert, P., De Schamphelaire, L., Pham, T.H., Vermeulen, J., Verhaege, M., Lens, P.N. and Verstraete, W. (2006) Microbial fuel cells for sulfide removal. *Env. Sci. Technol.* **40**(17), 5218–5224.

Reda, T., Plugge, C.M., Abram, N.J. and Hirst, J. (2008) Reversible interconversion of carbon dioxide and formate by an electroactive enzyme. *Proc. Natl. Acad. Sci. USA* **105**(31), 10654–10658.

Richter, H., Lanthier, M., Nevin, K.P. and Lovley, D.R. (2007) Lack of electricity production by Pelobacter carbinolicus indicates that the capacity for Fe(III) oxide reduction does not necessarily confer electron transfer ability to fuel cell anodes. *Appl. Environ. Microbiol.* **73**(16), 5347–5353.

Roller, S.D., Bennetto, H.P., Delaney, G.M., Mason, J.R., Stirling, J.L. and Thurston, C.F. (1984) Electron-transfer coupling in microbial fuel-cells 1. Comparison of redox-mediator reduction rates and respiratory rates of bacteria. *J. Chem. Technol. Biotechnol. B- Biotechnol.* **34**(1), 3–12.

Rozendal, R.A., Hamelers, H.V.M., Euverink, G.J.W., Metz, S.J. and Buisman, C.J.N. (2006) Principle and perspectives of hydrogen production through biocatalyzed electrolysis. *Int. J. Hydrogen Energy* **31**(12), 1632–1640.

Tender, L. (2007) *Harvesting Energy form Marine Sediments to Persistently Power Coastal Environmental Sensor Networks*. UMass, Boston.

Thrash, J.C. and Coates, J.D. (2008) Review: direct and indirect electrical stimulation of microbial metabolism. *Environ. Sci. Technol.* **42**(11), 3921–3931.

Thrash, J.C., VanTrump, J.I., Weber, K.A., Miller, E., Achenbach, L.A. and Coates, J.D. (2007) Electrochemical stimulation of microbial perchlorate reduction. *Environ. Sci. Technol.* **41**(5), 1740–1746.

Torres, C.I., Kato Marcus, A. and Rittmann, B.E. (2007) Kinetics of consumption of fermentation products by anode-respiring bacteria. *Appl. Microbiol. Biotechnol.* **77**(3), 689–697.

van Hees, W. (1965) A bacterial methane fuel cell. *J. Electrochem. Soc.* **112**(3), 258–262.

Wrighton, K.C., Agbo, P., Warnecke, F., Weber, K.A., Brodie, E.L., DeSantis, T.Z., Hugenholtz, P., Andersen, G.L. and Coates, J.D. (2008) A novel ecological role of the Firmicutes identified in thermophilic microbial fuel cells. *ISME J* **2**(11), 1146–1156.

Xing, D.F., Zuo, Y., Cheng, S.A., Regan, J.M. and Logan, B.E. (2008) Electricity generation by Rhodopseudomonas palustris DX-1. *Environ. Sci. Technol.* **42**(11), 4146–4151.

You, S.J., Zhao, Q.L., Jiang, J.Q., Zhang, J.N. and Zhao, S.Q. (2006) Sustainable approach for leachate treatment: Electricity generation in microbial fuel cell. *J. Environ. Sci. Health Part A – Toxic/Hazardous Subst. Environ. Engng.* **41**(12), 2721–2734.

2

Microbial energy production from biomass

Caroline M. Plugge, Jules B. van Lier, Alfons J.M. Stams and David Jetson

2.1 BIOMASS: SOLAR ENERGY STORED IN ORGANIC MATERIAL

Even though energy saving programs will be installed in the near future, the global demand for energy is rapidly increasing. Indeed, energy demand is projected to grow by more than 50% by the year 2025 (Ragauskas *et al.* 2006). Biomass is increasingly recognized as a 'green' energy source. It contains solar energy captured through photosynthesis, performed by trees, plants and other photosynthetic organisms. Through photosynthesis, carbon dioxide is reduced and converted into biomass. When this 'green' energy is collected from biomass, carbon dioxide is again released. Biomass represents a renewable energy source

 Bioelectrochemical Systems: From Extracellular Electron Transfer to Biotechnological Application. Edited by Korneel Rabaey, Largus Angenent, Uwe Schröder and Jürg Keller. ISBN: 9781843392330. Published by IWA Publishing, London, UK.

since it is based on the natural 'short-chain' carbon cycle and can be used and produced in an ongoing renewable fashion. Fossil fuel, on the other hand, required millions of years to be converted from its original biomass form into its fossilized gaseous and liquid forms, i.e. natural gas and crude oil. The use of these fossil fuels liberates the stored carbon as carbon dioxide and thereby contributes to the accumulation of greenhouse gases (GHG) in today's atmosphere.

When using bio-energy on a carbon neutral basis, specifically the amount of biomass that is used for energy purposes equals the newly grown amount, then there should be no net increase in carbon dioxide emissions to the atmosphere. However, production of energy derived from biomass may consume fossil fuels, which are required for example by fertilizer production, planting, harvesting, and possible other agrotechnological treatments and waste (water) processing. The net energy benefit and reduction of GHG emissions with biomass as energy source is site and process specific, and should be evaluated case by case (Pabon 2009). Furthermore, bio-energy conversion technologies play a crucial role in maximizing the energy output from any raw feedstock. With specific technologies, e.g. those aiming at the production of liquid fuels, the energy yield may not be substantial, once the energy required to produce and process the biomass is considered (Ponton 2009).

The 1970s oil crisis gave a major impulse to the search for alternative energy sources. As a result, bio-ethanol production was developed on a large scale in Brazil. Also anaerobic digestion of wastes and wastewaters received major attention as cost-effective treatment systems coupled to bio-energy generation in those countries that implemented restrictive environmental laws, e.g. The Netherlands. Nowadays, the worldwide increasing energy needs plus the ongoing demand for CO_2 neutral fuels represent a renewed strong driving force for the production of energy derived from biological resources. Furthermore, biomass is abundantly available in most parts of the world. This contrasts with traditional fossil fuels, available in only a few, and in some cases, politically unstable parts of the world.

To reduce the CO_2 emissions, the most sensible solution to meet the increased energy demand is the utilization of renewable resources instead of fossil resources. Furthermore, it is expected that most sources of fossil fuels will be depleted by the next century. The foreseen depletion of fossil fuel reserves, together with the growing environmental concerns related with GHG emissions, provide strong incentives to further develop and implement renewable energy production. From the several options in development, such as the utilization of solar, wind, or geothermal energy, the utilization of biomass seems closest at hand. Indeed, according to 2006 statistics, biomass represented 66% of the renewable energy capability in the European Union (Figure 2.1).

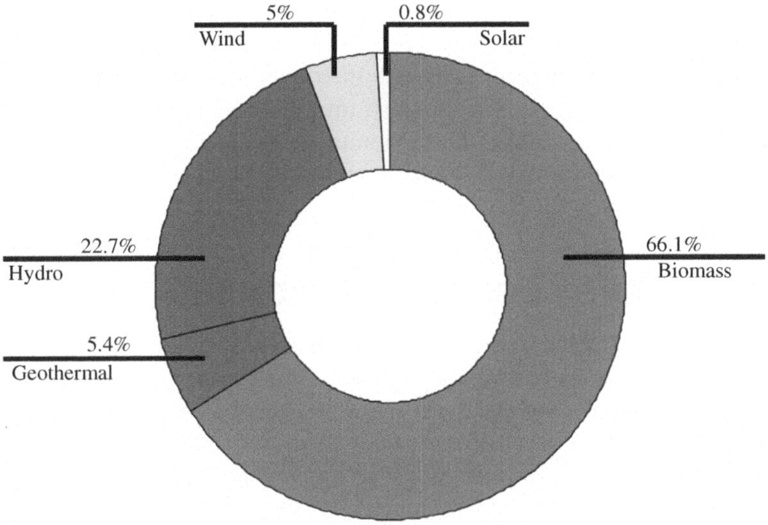

Figure 2.1. Renewable energy sources in European Union (adapted from www.energy.eu)

The ongoing discussion on competitive land use – mainly with respect to first generation biofuels – has raised concerns about the benefits of employing large scale biomass systems for fuel production. Particularly when food crops are used, or when high-quality arable land is used to produce energy crops, the overall benefits are highly questionable. Second generation biofuels are produced from cellulose from dedicated energy crops (e.g. perennial grasses), forestry materials, crop remainders, and organic waste streams. As such, the produced second generation bio-energy is not competitive with food supply chains.

Negative value waste streams are indeed regarded as the most optimum substrate for bio-energy production. Moreover, advances in the conversion processes will improve the sustainability of biofuels, through higher efficiencies and reduced environmental impacts. So far, the relatively low price of fossil fuels still hampers the large-scale application of the bio-based alternatives. However, even after the collapse of energy prices in 2008, which immediately followed the extreme high-peak fuel prices in the same year, a general increase in energy prices for the coming decades is expected. The willingness to pay high prices for energy in the future is a great uncertainty. Regarding the use of biomass, it seems more realistic to extract all components from biomass, including food crops and energy precursors, in order to produce a multi-functional 'basket' of products each with their own, maybe marginal, return on investment. Such an approach would lead towards the creation of biorefineries, facilities that integrate biomass conversion processes and equipment to produce different goods such as food, fuels, materials and chemicals (Octave and Thomas, 2009).

2.2 THE ENERGY CONTENT OF BIOMASS

The conversion of light energy to chemical energy is the key process in photosynthesis. Biomass is assimilated from carbon dioxide, water, nitrogen and a number of mineral compounds, using solar energy for its formation. The composition of biomass varies from species to species and can be generalized by the formula $C_nH_aO_bN_c$. The potential energy content depends on the average oxidation state of the carbon in the biomass, i.e. the number of electrons that can be released per weight of substrate when the compound is fully oxidized. Thus, to assess the potential energy one needs to determine the amount of electrons that can be recovered from the substrate, as well as the energy associated with them.

An adequate approach to estimate the maximum possible electron transfer from any particular compound is the measurement of the chemical oxygen demand (COD). Per atom C, the number of electrons made free in the complete oxidation of $C_nH_aO_bN_c$ amounts to: $4+(a-2b-3c)/n$, in which 4 is the maximum number of electrons per fully reduced carbon atom, subsequently corrected by the presence of oxidized hydrogen and reduced oxygen and nitrogen with a valence of 2 and 3, respectively (van Lier *et al.* 2008). It can be easily calculated that complete oxidation of oil and lipid-rich biomass may generate much more energy per weight of dry matter than the oxidation of sugar rich biomass. On average, for organic matter, the energy content is about 13.5 MJ/kg biomass COD.

The net energy yield of a conversion process, however, depends on the type of oxidative processes applied. Technologies for the production of energy and fuels from biomass can be classified as thermal (Table 2.1) or fermentative. In general, process selection to release the biomass-enclosed energy largely depends on the nature of the organic matter and the type of energy carrier which is aimed at. For instance, the structural formula of wood and wet herbal biomass is not very different, so theoretically the biomass-enclosed energy content is rather similar. However, when for example applying incineration, the net energy generation is obviously determined by the water content of the biomass as well.

Table 2.1. Thermal biomass conversion processes

Process	Dry matter content	Expected efficiency	P (bar)	T (°C)	Products
Hydrothermal upgrading (HTU)	≈ 15%	70%	180	300	Bio-oil/ Bio-crude, biodiesel, gas, water
Pyrolysis	≈ 90%	n.a.**	n.a.	400	Pyrolysis oil, gas+coal
Incineration*	≈ 35–40%	n.a.	n.a.	n.a.	CO_2, H_2O
(Supercritical) Gasification	(≈ 15%) >30%	n.a.	300	450–700	Syngas: CO, CH_4, H_2, N_2

* applied at 5 kW–2 GW plant size
** not available

Fermentation and anaerobic respiration processes produce well defined compounds such as ethanol, methane, acetone/butanol/ethanol (ABE) and hydrogen which can be used as energy carriers, feedstock for chemical industries, or may be utilized as such. These products can be characterized by the change in Gibbs free energy of their formation reactions from glucose (Table 2.2). Biological production of liquid and gaseous energy carriers takes place at moderate conditions of pressure and temperature, and does not generate toxic compounds.

Table 2.2. Gibbs free energy values ($\Delta G'^{0}$) for product formation from the various fermentation processes from glucose. All calculations are based on stoichiometrical reactions per mol glucose using ($\Delta G'^{0}$ values of formation for all components as given by Thauer *et al.* (1977)

Fermentation process	$\Delta G'^{0}$ (kJ/mol glucose)
Ethanol	−225.9
Methane	−393.0
ABE*	−209.3
H_2	+3.2

*Stoichiometry for ABE is glucose:butanol:acetone:ethanol: acetate:H_2 = 1:0.5:0.3:0.1:0.1:2.6

In addition, valuable by-products are often produced, of which the significance has not yet been studied in detail. It is easy to understand that after anaerobic fermentation of lignocellulose, lignin plus the proteinaceous microbial biomass will be found among the by-products. Besides these major by-products, other valuable smaller fractions, such as amino acids and vitamins, may be produced. The recovery of these products may add to the total economic feasibility of the overall biological process. Biological fermentation processes offer the possibility of favoring a specific end product, such as a liquid or gaseous energy carrier, or a specific side product of interest, such as polyalkanoates, depending on the process conditions applied. The mentioned features in fact form the quintessence of the application of fermentation for a balanced conversion of biomass.

2.3 BIO-ALCOHOL PRODUCTION FROM BIOMASS

Biologically produced alcohols, most commonly ethanol, and less commonly propanol and butanol, are produced by microorganisms. At present, the fermentation of sugars to ethanol is the best established process for conversion of biomass to energy. Ethanol production from sugar cane started in Brazil and the United States in the early 1970s. The Brazilian production capacity in 2005 was 19,200 million litres, representing close to a 50% share within the Brazilian fuel market (van Haandel 2005). In Brazil, most alcohol distilleries are 'autonomous', i.e. they use sugar cane juice directly for fermentation. There are also 'attached' distilleries that use spent molasses from sugar mills to produce alcohol. Alcohol production from sugar cane is a significant economic activity in Brazil. The planted area of cane fields for alcohol production alone is over 3 million hectares. The employed workforce for the agricultural, industrial and commercial activities is about 1 million people (van Haandel and van Lier 2006). In many rural areas the production of sugar and alcohol is the main source of employment. It is likely that alcohol production will even further increase in the near future as the use of alcohol as an additive or mixture to gasoline is implemented or planned to be implemented in industrialized countries such as the USA, Germany and Japan. As sugar cane can be produced in large areas of the country, the alcohol production potential is much larger than the actual production rate. Competitive land use and preservation of bio-diversity are not yet setting the Brazilian agenda for bio-energy matters. In fact, gasoline prices of the passed years have made alcohol a competitive fuel in Brazil: no subsidies are given and the price is set by the market, normally hovering at around 70% of the gasoline price, which is its equivalence point in terms of energy efficiency. In Brazil, on average, 13 tons of sugarcane generates 800 kg of liquid ethanol (van Haandel 2005). The United States is another major producer of bioethanol. The key feedstock here is maize. During 2008 about 34,000 million litres were produced, surpassing Brazil's production capacity (RFA, 2009).

Many reports and reviews have been published on the microbial production of ethanol, focussing on the diversity of bacteria, yeasts, and fungi used for the production of ethanol (Lin and Tanaka 2006). Production of ethanol from sugar derived from starch and sucrose has been commercially dominated by the yeast *Saccharomyces cerevisiae*. *Saccharomyces cerevisiae* can produce ethanol at concentrations as high as 18% of the fermentation broth. *Saccharomyces cerevisiae* is also generally recognized as safe (GRAS) as a food additive for human consumption and is therefore ideal for producing alcoholic beverages and for leavening bread. This yeast can grow both on simple sugars, such as glucose, and on the disaccharide sucrose. It must be noted, however, that sugars derived from biomass contain a mixture of hexoses (primarily glucose) and pentoses (primarily xylose) and most wild-type strains of *S. cerevisiae* do not metabolize xylose (Gray *et al.* 2006).

Saccharomyces cerevisiae metabolizes glucose by the Embden–Meyerhof (EM) pathway. Beside this pathway, the Entner–Doudoroff (ED) pathway is another means of glucose catabolism in prokaryotes, such as *Zymomonas*. The high ethanol yield and productivity observed for *Zymomonas* sp. are a consequence of its distinctive physiology. The ED pathway yields only half as much ATP per mole of glucose as the EM pathway, as it uses 6-phosphogluconate dehydrase and 2-keto-3-deoxyglucosephophate aldolase to form pyruvate from glucose. As a consequence, *Zymomonas* produces less biomass than yeast, and more carbon is funnelled to fermentation products (Lin and Tanaka 2006). Like yeast, *Z. mobilis* is GRAS and has straightforward nutritional needs. It is so well suited for ethanol production that it was advocated as superior to *S. cerevisiae* in the 1970s and 1980s. Despite its advantages as an ethanol producer, *Z. mobilis* is – likewise to *S. cerevisiae* – not well suited for all of the biomass resource conversions because it ferments only C_6 sugars such as glucose, fructose and sucrose and not C_5 sugars. Also, when *Z. mobilis* is grown on synthetic media containing glucose, the specific rates of sugar uptake and ethanol production are much higher than when fructose or sucrose are used as substrate. This is due to a less efficient metabolization of fructose and sucrose. Overall, *S. cerevisiae* is still preferred by the industry because of the yeast robustness.

2.4 ANAEROBIC METHANOGENIC DIGESTION: WASTE STABILIZATION PLUS RENEWABLE ENERGY SOURCE

2.4.1 Process performance

During anaerobic degradation, most of the chemical energy present in organic compounds is conserved as methane, if no other electron acceptors such as SO_4^{2-} and NO_3^- are present. The Gibbs free energy that is available from this fermentation (reaction 1 for glucose) is used by the microorganisms to grow and is released as heat. From the comparison of the Gibbs free energy values between reactions 2 and 3, it is clear that close to 85% of the intrinsic energy contained in the glucose is channeled to methane.

The methane produced has a high calorific value, and can be easily harvested when produced under controlled conditions. As such, methane is not released in the atmosphere as GHG, but exploitable as bio-fuel.

$$(1)\ \ \text{Glucose} \rightarrow 3CO_2 + 3CH_4 \qquad \Delta G'^0 = -393\,\text{kJ/reaction}$$

$$(2)\ \ 3CH_4 + 6O_2 \rightarrow 3CO_2 + 6H_2O \qquad \Delta G'^0 = -2428\,\text{kJ/reaction}$$

$$(3)\ \ \text{Glucose} + 6O_2 \rightarrow 6CO_2 + 6H_2O \qquad \Delta G'^0 = -2821\,\text{kJ/mol}$$

In methanogenic environments, the stoichiometry of CO_2 and CH_4 formation is dependent on the oxidation state of the organic compound. The treatment of more reduced organic waste compounds such as lipids will result in a higher CH_4 yield than the treatment of more oxidized compounds, such as polysaccharides. The stoichiometry of methane formation can be predicted from the structural formula of such compounds:

$$C_nH_aO_bN_c + (n - a/4 - b/2)H_2O \rightarrow (n/2 + a/8 - b/4)CH_4 + (n/2 - a/8 + b/4)CO_2 + cNH_3$$

When digesting mixed substrates or wastes with unknown organic compounds, the ratio between the COD and total organic carbon (TOC) predicts the CH_4 content of the final products according to: CH_4 (%) = 18.75×COD/TOC (van Lier *et al.* 2008). Obviously, the exact methane content depends on reactor alkalinity, substrate biodegradability and presence of alternative electron acceptors.

During degradation of organic matter in environments in which the availability of inorganic electron acceptors is limiting, organic material serves both as electron donor and electron acceptor, ultimately resulting in the production of a mixture of completely oxidized as well as completely reduced carbon compounds, i.e. CO_2 and CH_4, a mixture known as biogas. Although Alessandro Volta in 1776 already noted the relationship of biogas ('Aria infiamabile') formation to water-saturated decaying plants, it took almost a century before it was demonstrated by Béchamp – a student of Pasteur – in 1886 and Poppoff in 1873 that biogas formation from decomposing plant material was of microbial origin (Barker 1956). Later, this process was termed anaerobic digestion (Boone 1991). From the end of the last century onwards, AD is applied in man-made environments for both energy generation as well as a cost-effective method for waste stabilization (Lettinga 1996; Lettinga *et al.* 1999; van Lier 2008). The positive energy balance on the one hand and the increasing world-wide energy demand on the other hand, resulted in a growing interest in anaerobic digestion as an environmental friendly waste-treatment–energy-recovery system. In addition to solid waste digestion, anaerobic digestion is more and more recognized as a mature alternative for the treatment of slurries, liquid wastes and wastewater.

Biogas formation represents one of the most versatile types of bioenergy, and can be produced from wastes and organically polluted wastewaters. The produced bio-methane can be used in multiple applications such as vehicle fuel, heat and electricity production. When applying anaerobic digestion to a waste, a 'dual energy benefit' is at hand: little or no energy is required for the stabilization, and an energy-rich end-product is produced, i.e. biogas. Figure 2.2 shows the biogas cycle through which wastes are stabilized, and renewable energy is produced. Recently, an inventory was made of the AD potentials for generating 'green' energy in Europe (Tilche and Galatola 2008). This study identified that the potential production of bio-methane in the European Union is in the order of 3.9×10^9 m^3 per year. The latter assessment only takes into consideration the biogas from the anaerobic digestion of available animal waste, landfills and sewage sludge. From landfills and sewage sludge it is possible to efficiently harvest the bio-methane with some simple modifications. In energy terms, the sum of bio-methane from these sources in Europe corresponds to about 380 PJ/y (or about 9 Mtonnes of oil equivalents). If vehicle use is considered, the latter amount may substitute 9.07×10^9 kg of gasoline, corresponding to the driving of 1.31×10^{11} km with light vehicles each year. Considering an average yearly distance per EU citizen-car of 9757 km, this amount of fuel corresponds to the average needs of 13.4 million citizens. The overall amount of bio-methane would cover about 48% of the 2010 EU target for biofuels (5.75% of all transport fuels). With respect to the more recent 2020 10% EU target for biofuels, the calculated bio-methane would cover almost 1/3 of it (29%) (Tilche and Galatola 2008).

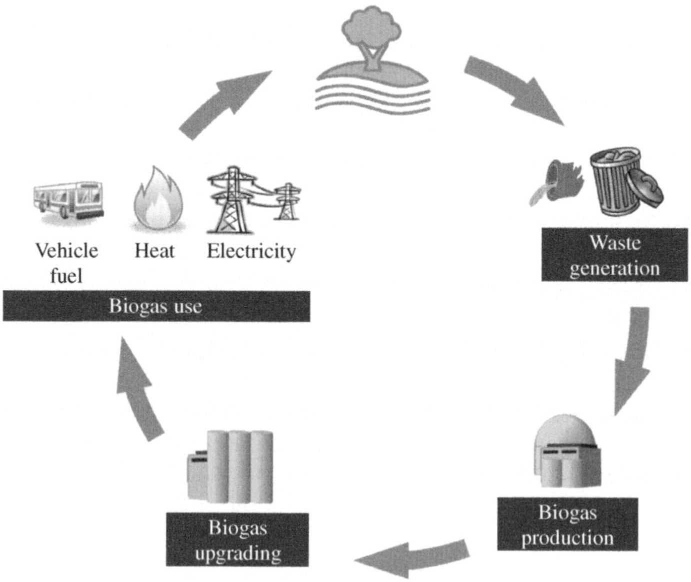

Figure 2.2. The biogas-carbon cycle

2.4.2 The microbiology of methanogenesis

Methanogenic Archaea are thus far the only prokaryotic group found to produce methane (Whitman *et al.* 2006). These microorganisms possess a unique biochemistry, which enables them to derive metabolic energy from the methanogenic pathway (Whitman *et al.* 2006). Compared to most other microbes, methanogens have a very restricted metabolism. The most important energy yielding reactions of methanogens are:

$$4H_2 + HCO_3^- + H^+ \rightarrow CH_4 + 3H_2O \qquad \Delta G'^0 = -135.6\,\text{kJ/reaction}$$

$$4\,\text{Formate}^- + H_2O + H^+ \rightarrow CH_4 + 3HCO_3^- \qquad \Delta G'^0 = -130.4\,\text{kJ/reaction}$$

$$4\,\text{Methanol} \rightarrow 3CH_4 + HCO_3^- + H_2O + H^+ \qquad \Delta G'^0 = -314.8\,\text{kJ/reaction}$$

$$\text{Acetate}^- + H_2O \rightarrow CH_4 + HCO_3^- \qquad \Delta G'^0 = -31.0\,\text{kJ/reaction}$$

Most of the described species of methanogens are relatively specialized. *Methanobrevibacter* sp. is only able to use H_2 as an electron donor and CO_2 as an acceptor for growth, whereas *Methanosaeta* sp. only uses acetate as energy substrate. *Methanosarcina* species are more versatile; they can use $H_2 + CO_2$, acetate, methanol, methylated amines and pyruvate for growth. Ethanol ($+ CO_2$) and isopropanol ($+ CO_2$) can only be utilized by a few methanogenic strains (Whitman *et al.* 2006).

2.4.3 The importance of extracellular electron transfer in AD

As a consequence of the limited range of substrates used by methanogens, the anaerobic breakdown of complex organic matter is carried out by closely cooperating communities of different physiological types of anaerobic microorganisms. Biopolymers are hydrolyzed and fermented. The products formed are funnelled to precursors of methanogenesis (Figure 2.3). Polysaccharides yield monomeric and oligomeric C6 and C5 sugars, while proteins are converted to amino acids and small peptides. Lipids are degraded to glycerol and long-chain fatty acids. Subsequently, fermentative bacteria degrade the hydrolysis products, easily degradable compounds like sugars, amino acids, purines, pyrimidines and glycerol, to a variety of fatty acids, carbon dioxide, formate and hydrogen. Then, acetogenic bacteria degrade higher fatty acids to acetate, carbon dioxide and hydrogen, and formate. These compounds are then the substrates for methanogens (Schink and Stams 2006). All these processes, fermentation, acetogenesis and methanogenesis, will take place at the same time, but often, because of the different growth rates and activities of the microorganisms present, the different processes are partially uncoupled, which results in the accumulation of organic acids (such as propionate and butyrate).

For thermodynamic reasons these communities of microorganisms can only metabolize these accumulated organic acids when the hydrogen concentration is kept low by the methanogens. These energetic constraints were first recognized by Bryant and colleagues (1967) when they discovered that what they originally believed to be methanogenic pure culture *Methanobacillus omelianskii* was in fact a co-culture of an ethanol-degrading bacterium and a methanogen. Propionate and butyrate degradation under methanogenic conditions will be discussed in more detail, to illustrate this phenomenon further.

Propionate and butyrate are important products of sugar fermentation, whereas propionate and branch-chain fatty acids are formed during amino acid degradation. Degradation of these fatty acids is a very sensitive process, which is easily disturbed by changes in the environmental conditions. The degradation is carried out by proton-reducing acetogenic bacteria. The Gibb's free energy change under standard conditions is highly positive (reactions below). Therefore, methanogens are needed to consume the produced hydrogen, through extracellular electron transfer. This means that these proton-reducing acetogens are obligatory dependent on methanogens, resulting in obligate syntrophic conversions.

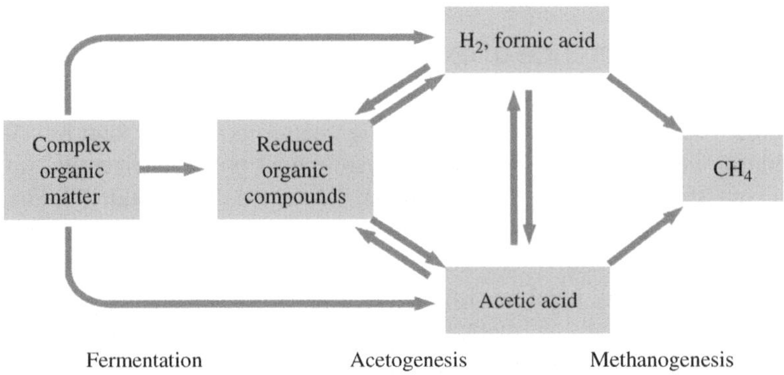

Figure 2.3. Carbon flow during the anaerobic methanogenic conversion of organic matter

$$\text{Propionate}^- + 3H_2O \rightarrow \text{Acetate}^- + HCO_3^- + H^+ + 3H_2 \qquad \Delta G'^0 = +76.1\ \text{kJ/reaction}$$

$$\text{Butyrate}^- + 2H_2O \rightarrow 2\text{Acetate}^- + H^+ + 2H_2 \qquad \Delta G'^0 = +48.1\ \text{kJ/reaction}$$

$$H_2 + HCO_3^- + H^+ \rightarrow CH_4 + 3H_2O \qquad \Delta G'^0 = -135.6\ \text{kJ/reaction}$$

$$\text{Acetate} + H_2O \rightarrow CH_4 + HCO_3^- \qquad \Delta G'^0 = -31\ \text{kJ/reaction}$$

It can be calculated that fatty acid oxidation can only proceed within very narrow ranges of the hydrogen partial pressure, typically between 10^{-3} and 10^{-5} atmosphere (1 to 100 Pa) (Schink 1997). At such low gradient, high conversion rates are only possible if hydrogen-producing acetogens and hydrogen-consuming methanogens are in close vicinity of each other. Since anaerobic biomass tends to aggregate, the average inter-bacterial distance is typically in the nanometre to micrometre-range. Disruption of the structure results in a severe drop in the specific methanogenic activity (Schmidt and Ahring 1996).

Overall, methanogenesis is a dynamic process in the sense that methanogens strongly influence the metabolism of fermentative and acetogenic bacteria by means of this extracellular electron transfer (Schink and Stams 2006; Stams *et al.* 2006). However, most importantly, decomposition is always directed to CH_4, CO_2, NH_3 and minor amounts of H_2S, provided that all functional groups of microorganisms are present.

2.4.4 Application of anaerobic digestion

2.4.4.1 Anaerobic Digestion (AD) for solid waste

Anaerobic digestion for the treatment of sewage sludge, manure and other kinds of organic waste has already been applied for several decades. In fact, the first applications of anaerobic sludge stabilization originate from the end of the 19th century (McCarty 2001). Farm-scale digesters were developed in countries like India and China for biogas generation from livestock manure. The produced energy is used for cooking and lighting on the farms. At present, the technology is strongly supported by governmental aid programs to, for example, combat deforestation in ecologically sensitive regions. For instance, farmer families in the Himalayan Mountains depend on firewood for cooking, which consequently leads to severe soil erosion and floodings in down-stream countries. AD is nowadays used by thousands of these families in Nepal to meet their daily energy requirement.

The production of 'green energy' is also of interest in Western countries, where the growing concern related to GHG emissions and depletion of fossil fuel resources lead further development of the AD technology. A classic example is the conventional landfill were municipal solid waste (MSW) is stored and slowly mineralized under anaerobic conditions. In fact, a landfill can be considered as a huge anaerobic digester, producing valuable and recoverable methane. Different full-scale landfill-biogas projects are realized, producing 100 Nm^3 of CH_4 per ton of municipal solid waste (Themelis and Ulloa 2007).

Several European governments like those from Germany, Austria and Denmark are subsidising centralized biogas plants for the treatment of manure, industrial wastes, MSW, and energy crops. In Denmark, farmers who supply waste to the centralized digester, receive a stabilized residue that consists of valuable material for fertilization and soil conditioning purposes. Whereas aerobic composting requires an energy input of 20–35 kWh_e per ton of MSW, AD produces 75–150 kWh_e per ton, equivalent to 100–200 m^3 biogas per ton.

The energy content of the biogas is 20–25 MJ/m^3 with a CH_4 content of 55–70% (by volume) with CO_2 as the remainder. In addition to CH_4 and CO_2, small fractions of other gases, such as N_2, NH_3, H_2, H_2S, O_2, and trace volatile organic carbons (VOCs) are found, but usually not exceeding 1–2% (per volume). This reveals an important environmental advantage of controlled stabilization of organic wastes under anaerobic conditions: a controlled production of the biofuel CH_4 and reduced emissions of toxic and odorous gases to the atmosphere. By introducing source separation at the household, the energy content of the digestible fraction of MSW has gradually increased. As a result, compact technologies have been developed, dedicated to energy recovery and production of anaerobic compost. AD for MSW treatment is performed in several types of reactor systems, which can roughly be divided into continuous flow systems and batch-type systems (International Energy Agency 1996; De Baere 2001).

2.4.4.2 AD for wastewater treatment

Energy production can be considered as the driving force behind the development of AD technologies for solid waste. However, for industrial and domestic wastewaters, environmental concerns, governmental legislation, and avoidance of the use of fossil energy for treatment were the key drivers for developing AD. The oil crises in the early seventies and the positive energy balance of AD processes resulted in a strong interest in adapting the AD technology for wastewater treatment.

Although anaerobic waste(water) treatment is developed for environmental engineering purposes, it may significantly contribute to the energy demand of an industry. For instance, an alcohol distillery can derive its entire industrial energy demand from its own waste(water), whereas the organic pollution load to the environment is diminished by about 75–80% (the ultimate biodegradability of these types of wastewater). In some countries, such as India, methane production from wastewater is governmentally supported as an alternative renewable energy source. The involved authorities regard the concomitant reduction of the environmental pollution as an additional advantage. The applied loading potentials, potential energy output and concomitant CO_2 emission reductions related to modern anaerobic reactor systems are presented in Table 2.3. Taking the wastewater characteristics into account and the anaerobic reactor data from Table 2.3, it can be easily calculated that typical food and beverage plants may generate 0.5–1.5 MW from their wastewater.

Since only a little energy can be obtained during anaerobic conversion of the substrates, the growth rate of the methanogenic biomass is rather low. The reactions below show that the Gibbs free energy changes of anaerobic glucose oxidation are lower than that of aerobic glucose oxidation.

Table 2.3. Loading potentials, potential energy output as well as potential CO_2 emission reduction of modern high-rate anaerobic reactor systems. In-between values can be obtained by linear interpolation. Calculations are based on 80% CH_4-COD recovery relative to influent COD load and 40% electric energy recovery using a modern combined heat and power (CHP) generator (after van Lier *et al.* 2008)

Loading capacity (kg COD/m^3 reactor/d)	10–35
Energy output (MJ/m^3 reactor/d)	55–390
Electric power output (kW/m^3 reactor)	0.25–1.7
CO_2 emission reduction (ton CO_2/m^3 reactor.y) (based on coal driven power plant)	1.9–13

The $\Delta G'^{0}$ of glucose oxidation coupled to methane production is only −404 kJ, which is sufficient for the net synthesis of 5–6 ATP. It is known that the oxidation of 1 molecule of glucose in the presence of O_2 leads to the formation of 38 molecules of ATP.

$$\text{Glucose} + 6O_2 \rightarrow 6HCO_3^- + 6H^+ \qquad \Delta G^{0'} = -2844\,\text{kJ}$$

$$\text{Glucose} + 3NO_3^- + 3H_2O \rightarrow 6HCO_3^- + 3NH_4^+ \qquad \Delta G^{0'} = -1796\,\text{kJ}$$

$$\text{Glucose} + 3SO_4^{2-} + 3H^+ \rightarrow 6HCO_3^- + 3HS^- \qquad \Delta G^{0'} = -453\,\text{kJ}$$

$$\text{Glucose} + 12S + 12H_2O \rightarrow 6HCO_3^- + 12HS^- + 18H^+ \qquad \Delta G^{0'} = -333\,\text{kJ}$$

$$\text{Glucose} + 3H_2O \rightarrow 3HCO_3^- + 3H^+ + 3CH_4 \qquad \Delta G^{0'} = -404\,\text{kJ}$$

As a consequence of the low energy yield, the net bacterial growth yield of anaerobic Bacteria and Archaea is extremely low. Thus, to sufficiently stabilize the organic wastewater constituents, large-scale reactor systems would be needed to enable bacterial growth for COD conversion if no sludge retention were applied. The conventional AD reactor systems such as the Completely Stirred Tank Reactor (CSTR), which are generally used for the semi-solid wastes, are therefore economically not feasible for the anaerobic treatment of low suspended solids containing wastewaters coming from industries and municipalities. In order to increase the amount of viable sludge inside the reactor, i.e. to uncouple the biomass or solids retention time (SRT) from the hydraulic retention time (HRT), several types of 'high-rate' reactors were developed for the treatment of organically polluted wastewaters. A first improvement to the CSTR system was the addition of an external clarifier, creating an anaerobic contact process, the homologue of the well-known aerobic activated sludge process. For the above-named reasons, a breakthrough in the development of anaerobic 'high-rate' reactors occurred during the early seventies when energy prices increased to a very high level. Various full- scale reactor systems were implemented for the treatment of wastewaters coming from the agro-food industry, e.g. sugar-refineries, beer-breweries, potato-processing factories and slaughterhouses. Such wastewaters are characterized by a relatively high organic matter content.

Since anabolic reactions hardly contribute to the overall anaerobic conversion process, the biomass concentration in the reactor became a key parameter for reactor development. In fact, 80–95% of the anaerobic process can be ascribed to catabolic conversions. The latter indicates that a high degree of sludge retention will automatically lead to a high loading capacity of the bioreactor (see Table 3), and thus to an economically attractive system. Indeed, the 'high-rate' reactors, such as the Upflow Anaerobic Sludge Blanket (UASB) reactor, the Fluidized Bed (FB) reactor, the Anaerobic Filter (AF), are characterized by a high biomass content, in which the biomass is immobilized into dense (granular) aggregates or biofilms. Reviews of these systems have been presented by various researchers (e.g. Speece 1996; van Lier 2008). The system with the widest application is undoubtedly the UASB. At present, more than 2500 full-scale anaerobic high-rate reactors are in operation of which nearly 80% are granular sludge-bed based. Modern sludge-bed reactor systems such as the Expanded Granular Sludge Bed (EGSB) reactor and the Internal Circulation (IC) reactor, demonstrated the potential of AD under more extreme conditions. Nowadays, toxic chemical wastewaters, as well as wastewaters with temperatures below 15°C and over 45°C can be treated using AD technology (Frankin *et al.* 1994; Lettinga *et al.* 1999; van Lier *et al.* 1997). Modern design full-scale loading rates reach up to 40 kg COD per m^3 reactor per day.

2.4.4.3 Overall benefits and constraints of anaerobic digestion

The advantages of anaerobic waste(water) treatment are obvious. Organic waste is dealt with in a sustainable and controlled manner, yielding the energy-rich biofuel methane, which can be used for various purposes. Anaerobic wastewater treatment has the following striking advantages over conventional activated sludge systems (van Lier 2008):

1. Reduction of excess sludge production up to 90%.
2. Up to 90% reduction in space requirement when using expanded sludge bed systems.
3. High applicable COD loading rates reaching 20–35 kg COD per m^{-3} reactor volume per day^{-1}, meaning smaller reactor volumes.
4. No use of fossil fuels for treatment whereas conventional aerobic activated sludge systems require an energy input for aeration close to 1 kWh/kg COD removed.
5. Production of about 13.5 MJ CH_4 energy/kg COD removed, giving 1.5 kWh-electric (assuming 40% electric conversion efficiency).
6. Rapid start-up in about 1 week, when using granular anaerobic sludge as seed material.
7. No or very little use of chemicals.
8. Plain technology with high treatment efficiencies.
9. Anaerobic sludge can be stored unfed, with only little loss of activity. Reactors can be then operated during agricultural campaigns only, e.g. 4 months per year in the sugar industry.
10. Excess sludge is very compact and is easily dewatered.
11. Granular excess sludge has a market value as seed material for new reactors.
12. High rate systems facilitate water recycling in factories.

Constraints of the technology are that anaerobic treatment is associated with sulphide and odour formation and that nitrification is not possible. An aerobic post- treatment is nowadays considered an essential part for achieving full treatment. Efficient methods to deal with sulphide and odour problems have been developed recently. Sulphide can be converted in a micro-aerobic process to elementary sulphur, and the recently observed anaerobic ammonium oxidation (i.e. anammox) can be implemented in combination with existing anaerobic treatment technologies (Janssen *et al.* 2009; Op den Camp *et al.* 2006). An anaerobic process is easily disturbed and methanogenic archaea are more susceptible for toxic compounds. However, once immobilized in biofilms or granules, this susceptibility of methanogens is largely diminished.

Technology, based on the use of microbial fuel cells (MFC's) and microbial electrolysis cells (MEC's) as bioelectrochemical systems (BES), is a promising concept for bioconversion processes towards electricity and hydrogen. When compared with anaerobic digestion, this technology has some specific advantages, such as application in the treatment of low concentration substrates at temperatures below 20°C, where anaerobic digestion generally is unsuccessful. This provides some specific application niches of the microbial fuel cell technology where it does not compete but rather complements anaerobic digestion technology. Moreover, the use of BES may have added value in the production of hydrogen and organic chemicals.

2.5 BIO-HYDROGEN PRODUCTION FROM BIOMASS

Biological hydrogen production has many common elements with anaerobic methanogenic digestion. It is important to note that biological hydrogen formation by mixed anaerobic communities in fact can be considered as an uncoupling of hydrogen formation from methane production. Hydrogen gas has potential as a bio-fuel, as during its combustion no CO_2 – a GHG – is produced. However, hydrogen is only a true 'green' fuel if it is produced from renewable sources like wind, sunlight, geothermal energy or biomass. A wide variety of microorganisms is able to form hydrogen in both light-dependent and light-independent processes, such as dark fermentations. Hydrogen is formed in the fermentation of complex biomass, but for large-scale production, thus far only hydrogen formation from polysaccharides is feasible (Kengen *et al.* 2009).

Considerable effort has been dedicated to optimizing operational environmental conditions that maximize hydrogen production. To prevent consumption of hydrogen by, for example, fermenting bacteria, homoacetogens and methanogens bioreactors are sparged with nitrogen to release the hydrogen. To date, optimizing the production of bio-hydrogen has focused only on a relatively small fraction of the total amount of hydrogen that could be released from biomass and waste (Angenent *et al.* 2004).

Methods to improve fermentative hydrogen production have been reviewed by Nath and Das (2004). Improvements clearly can be made by the proper choice of the fermentative bacterium, by applying genetic modification to redirect biochemical pathways in such a way that fewer side products are formed and by a proper process design and process operation (gas sparging, creating a large biofilm surface, etc.). These measures optimized hydrogen formation to maximally only four molecules of hydrogen per molecule of glucose and may result in an increase of the hydrogen production rate (reaction 2 below).

Carbohydrates are the main substrates for fermentative bacteria that produce hydrogen. Cellulose and hemicellulose are the most abundant polysaccharides available in nature, and therefore, glucose and xylose are the predominant monomeric sugars used for hydrogen formation (Kapdan and Kargi 2006).

Starch and sucrose may also be available in large amounts as these are used as storage material in various plants. Fermentation of glucose and xylose results in hydrogen formation, but in addition, depending on the microorganism, also acetate, butyrate, lactate, ethanol and some other organic compounds (e.g. formate, butanediol, succinate) are produced. Under standard conditions complete oxidation of glucose to CO_2 and H_2 is thermodynamically not possible. A main breakthrough would be achieved if acetate could be efficiently converted to carbon dioxide and hydrogen.

$$\text{Glucose} + 12\text{H}_2\text{O} \rightarrow 6\text{HCO}_3^- + 6\text{H}^+ + 12\text{H}_2 \qquad \Delta G'^0 = +3.2\,\text{kJ/reaction}$$

$$\text{Glucose} + 2\text{H}_2\text{O} \rightarrow 2\text{Acetate}^- + 2\text{HCO}_3^- + 4\text{H}_2 \qquad \Delta G'^0 = -206\,\text{kJ/reaction}$$

Another way to introduce energy in a dark fermentation is via Bioelectrochemical Systems (BES) (Logan *et al.* 2006). The concept of electricity-mediated electrolysis of organic compounds was recently introduced by Liu *et al.* (2005) and Rozendal and coworkers (2006). They showed that in a fuel cell, acetate can yield hydrogen and carbon dioxide and that this process of hydrogen formation is much more cost-effective than electricity mediated electrolysis of water to form hydrogen and oxygen. In a rough calculation they showed that the equivalent of one hydrogen is needed to produce the electricity for the complete electrolysis of acetate. This means that a net yield of 10 molecules of hydrogen per molecule of glucose can be obtained. But, important for practical applications, hydrogen formation is not restricted to (poly)saccharides as substrates, rather all kinds of organic waste components can serve as substrates.

2.6 FUTURE PERSPECTIVES

Anaerobic methanogenic treatment of organic waste and wastewater contributes to reducing our dependency on fossil fuels, providing an efficient waste treatment with low energy requirement; meanwhile bio-energy is produced in the form of methane. Bio-methane formation is nowadays a relevant form of bio-energy, and it is likely to become an important source of renewable energy in the future. Anaerobic treatment is a relatively cheap and simple technology that can be implemented at any scale and at any place (Lettinga 1996; McCarty 2001).

For bio-ethanol and bio-alcohol production, sustainable methods for production, collection and distribution need to be ensured, to avoid environmental problems. Cellulosic biomass from agricultural residues represents a very promising source of feedstock for the production of bio-ethanol and bio-alcohols. Moreover, other renewable chemicals (e.g. phenol based chemicals) and a range of commercial products can be produced from the waste streams and greatly expand current production.

To make bio-hydrogen a realistic addition to the world wide renewable energy sources, the challenge is to increase the H_2/glucose ratio beyond the current maximum of four. To achieve this, future research and technology development should focus on metabolic engineering to decrease the amount of acetate produced or more specifically on acetate oxidation by biocatalyzed electrolysis.

Developing and proving the feasibility of integrated biological treatment systems based on anaerobic and (micro-) aerobic biological treatment methods, complemented with specific physical-chemical methods may provide abundant clean energy. As such, decentralized solutions are becoming increasingly attractive. A recent example of bio-catalyzed electricity production in a photosynthetic algal microbial fuel cell (PAMFC) showed that electricity production can be attractive because these combined systems generate useful energy in a renewable, sustainable, and efficient manner (Strik *et al.* 2008). Another example of integrated approach is the generation of bio-electricity from residual COD in the effluent of an anaerobic digestion.

Overall, the scarcity of the natural resources will be an important driver for the selection of sustainable environmental technologies and energy recovery systems at any place in the near future. A large-scale introduction of biofuels can bring together interests of environment and agriculture and as such it can be an important step towards a sustainable society for the future.

REFERENCES

Angenent, L.T., Karim, K., Al-Dahhan, M.H., Wrenn, B.A. and Domíguez-Espinosa, R. (2004) Production of bioenergy and biochemicals from industrial and agricultural wastewater. *Trends in Biotechnol.* **22**, 477–485.

Barker, H.A. (1956) *Bacterial Fermentations*. Wiley, New York.

Boone, D.R. (1991) Ecology of methanogenesis. In *Microbial Production and Consumption of Greenhouse Gases: Methane, Nitrogen Oxides and Halomethanes*. ASM Washington, pp. 57–70.

Bryant, M.P., Wolin, E.A., Wolin, M.J., and Wolfe, R.S. (1967) *Methanobacillus omelianskii*, a symbiotic association of two species of bacteria. *Arch Microbiol.* **59**, 20–31.

De Baere, L. (2001) Full-scale experience with digestion facilities treating 25.000 and 50.000 ton of biowaste per year. *Proc. 9th World Cong. Anaerobic Digestion*. Antwerp-Bel, pp. 65–68.

Gray, K.A., Zhao, L. and Emptage, M. (2006) Bioethanol. *Curr. Opion. Chem. Biol.* **10**, 141–146.

Frankin, R.J., Koevoets, W.A.A. and Versprille, A.I. (1994) Application of the Biobed[R] system for formaldehyde containing dimethyl-therephtalate (= DMT) waste water. In *Poster Paper Pre-prints of the Seventh International Symposium on Anaerobic Digestion*. Cape Town, January 23–27, 1994, South Africa, pp. 244–247.

International Energy Agency (IEA) (1996) Municipal Solid Waste, IEA Bioenergy, Energy Recovery from Solid Waste Task, Anaerobic Digestion Activity; Resource Development Associates, 240 Ninth Street NE, Washington, DC.

Janssen, A.J.H., Lens, P.N.L., Stams, A.J.M., Plugge, C.M., Sorokin, D.Y., Muyzer, G., Dijkman, H., Van Zessen, E., Luimes, P. and Buisman, C.J.N. (2009) Application of bacteria involved in the biological sulfur cycle for paper mill effluent purification. *Sc. Tot. Environ.* **407**, 1333–1343.

Kapdan, I.K. and Kargi, F. (2006) Bio-hydrogen production from waste materials. *Enz. Microbiol. Technol.* **38**, 569–582.

Kengen, S.W.M., Goorissen, H.P., Verhaart M, van Niel, E.W.J., Claassen, P.A.M. and Stams, A.J.M. (2009) Biological hydrogen production by anaerobic microorganisms. In *Biofuels*. Wiley, Chichester, UK, pp. 197–222.

Lettinga, G. (1996) Sustainable Integrated Biological Wastewater Treatment. *Wat. Sci. Technol.* **33**(3), 85–98.

Lettinga, G., Rebac, S., Parshina, S., Nozhevnikova, A., van Lier, J.B. and Stams, A.J.M. (1999) High-rate anaerobic treatment of wastewater at low temperatures. *Appl. Env. Microbiol.* **65**, 1696–1702.

Lin, Y. and Tanaka, S. (2006) Ethanol fermentation from biomass resources: current state and prospects. *Appl. Microbiol. Biotechnol.* **69**, 627–642.

Liu, H., Grot, S. and Logan, B.E. (2005) Electrochemically assisted microbial production of hydrogen from acetate. *Env. Sci. Technol.* **39**, 4317–4320.

Logan, B.E., Hamelers, B., Rozendal, R., Schröder, U., Keller, J., Freguia, S., Aelterman, P., Verstraete, W. and Rabaey, K. (2006) Microbial fuel cells: methodology and technology *Env. Sci. Technol.* **40**, 5181–5192.

McCarty P.L. (2001) The development of anaerobic treatment and its future. *Wat. Sci. Technol.* **44**, 149–156.

Nath, K. and Das, D. (2004) Improvement of fermentative hydrogen production: various approaches. *Appl. Microbiol. Biotechnol.* **65**, 520–529.

Octave, S. and Thomas, D. (2009) Biorefinery: toward an industrial metabolism, *Biochimie* **91**(6), 659–664.

Op den Camp, H.J.M., Kartal, B., Guven, D., van Niftrik, L.A.M.P., Haaijer, S.C.M., van der Star, W.R.L., van de Pas-Schoonen, K.T., Cabezas, A., Ying, Z., Schmid, M.C., Kuypers, M.M.M., van de Vossenberg, J., Harhangi, H.R., Picioreanu, C., van Loosdrecht, M.C.M., Kuenen, J.G., Strous, M. and Jetten, M.S.M. (2006) Global impact and application of the anaerobic ammonium-oxidizing (anammox) bacteria. *Biochem. Soc. Trans.* **34**, 174–178.

Pabon, C.P. (2009) Anaerobic digestion in sustainable biomass chains. PhD thesis Wageningen University, Subdepartment of Environmental Technology, pp. 236.

Ponton, J.W. (2009) Biofuels: thermodynamic sense and nonsense. *J. Cleaner Prod.* **17**, 896–899.

Ragauskas, A.J., Williams, C.K., Davison, B.H., Britovsek, G., Cairney, J., Eckert, C.A., Frederick Jr., W.J., Hallett, J.P., Leak, D.J., Liotta, C.L., Mielenz, J.R., Murphy, R.,

Templer, R. and Tschaplinski, T. (2006) The path forward for biofuels and biomaterials. *Science* **311**, 484–489.

Rozendal, R.A., Hamelers, H.V.M., Euverink, G.J.W., Metz, S.J. and Buisman, C. (2006) Principle and perspectives of hydrogen production through biocatalyzed electrolysis. *Int. J. Hydrogen Energy*, **31**, 1632–1640.

RFA (Renewable Fuels Association) (2009) Ethanol industry statistics. Washington, DC. (www.ethanolrfa.org).

Schink, B. (1997) Energetics of syntrophic cooperations in methanogenic degradation. *Microbiol. Mol. Biol. Rev.* **61**, 262–280.

Schink, B. and Stams, A.J.M. (2006) Syntrophism among prokaryotes. In *The Prokaryotes: an Evolving Electronic Resource for the Microbiological Community*. Springer-Verlag, New York, 3rd edn., Vol. 2, pp. 309–335.

Schmidt, J. and Ahring, B. (1996) Granular sludge formation in upflow anaerobic sludge blanquet (UASB) reactors. *Biotechnol. Bioeng.* **49**, 229–246.

Speece, R.E. (1996) *Anaerobic Biotechnology for Industrial Wastewaters*. Archae Press, Nashville Tennessee.

Stams, A.J.M., de Bok, F.A.M., Plugge, C.M., van Eekert, M.H.A., Dolfing, J. and Schraa, G. (2006) Exocellular electron transfer in anaerobic microbial communities. *Environ. Microbiol.* **8**, 371–382.

Strik, D.B.T.B., Terlouw, H., Hamelers, H.V.M. and Buisman, C. (2008) Renewable sustainable biocatalyzed electricity production in a photosynthetic algal microbial fuel cell (PAMFC). *Appl. Microbiol. Biotechnol.* **81**, 659–668.

Thauer, R.K., Jungermann, K. and Decker, K. (1977) Energy conservation in chemotrophic anaerobic bacteria. *Bacteriol. Rev.* **41**, 100–180.

Themelis, N.J. and Ulloa, P.A. (2007) Methane generation in landfills. *Renew. Energy* **32**(7), 1243–1257.

Tilche, A. and Galatola, M. (2008) The potential of bio-methane as bio-fuel/bio-energy for reducing greenhouse gas emissions: a qualitative assessment for Europe in a life cycle perspective. *Water Sci. Technol.* **57**(11), 1683–1692.

van Haandel, A.C. (2005) Integrated energy production and reduction of the environmental impact at alcohol distillery plants. *Wat. Sci. Technol.* **52**, 49–57.

van Haandel, A.C. and van Lier, J.B. (2006) Vinasse treatment for energy production and environmental protection at alcohol distillery plants. In *Proc. of DIVER 2006, Int. Conf. on Production and Uses of Ethanol*. June 19–23, 2006, Havana, Cuba.

van Lier, J.B., Rebac, S. and Lettinga, G. (1997) High-rate anaerobic wastewater treatment under psychrophilic and thermophilic conditions. *Wat. Sci. Technol.* **35**(10), 199–206.

van Lier, J.B. (2008) High-rate anaerobic wastewater treatment: diversifying from end-of the pipe treatment to resource oriented conversion techniques. *Wat. Sci. Technol.* **57**(8), 1137–1148.

van Lier, J.B., Mahmoud, N. and Zeeman, G. (2008) Anaerobic wastewater treatment. In *Biological Wastewater Treatment, Principles, Modelling and Design*. IWA Publishing, London, UK, pp. 415–456.

Whitman, W., Bowen, T. and Boone, D.R. (2006) The methanogenic bacteria. In *The Prokaryotes*. 3rd edn., Vol. 3, pp. 165–207.

3

Enzymatic fuel cells and their complementarities relative to BES/MFC

Zana Zulic and Shelley D. Minteer

3.1 INTRODUCTION

Biofuel cells convert energy obtained from biochemical reactions into electrical energy by the use of different types of biological catalysts (living organisms, organelles, or enzymes)(Arechederra and Minteer 2008; Cooney *et al.* 2008; Moseley 2001). Biofuel cells are advantageous, because they do not require a pure fuel and will operate at room temperature and at a neutral pH (Moseley 2001). In addition, living organisms and/or their enzymes have the ability to catalyze a wide range of reactions, enabling a biofuel cell to use a wider range of substrates as fuels. The three main types of biofuel cells reported in literature since the 1960s include microbial fuel cells, organelle-based fuel cells and enzyme-based fuel

 Bioelectrochemical Systems: From Extracellular Electron Transfer to Biotechnological Application. Edited by Korneel Rabaey, Largus Angenent, Uwe Schröder and Jürg Keller. ISBN: 9781843392330. Published by IWA Publishing, London, UK.

cells (Arechederra and Minteer 2008; Cooney *et al.* 2008; Karube *et al.* 1981). As shown in Figure 3.1, the only true difference between each type of fuel cell is the type of biocatalyst used at the anode of the fuel cell. However, the properties of the biocatalysts affect the properties of the biofuel cells.

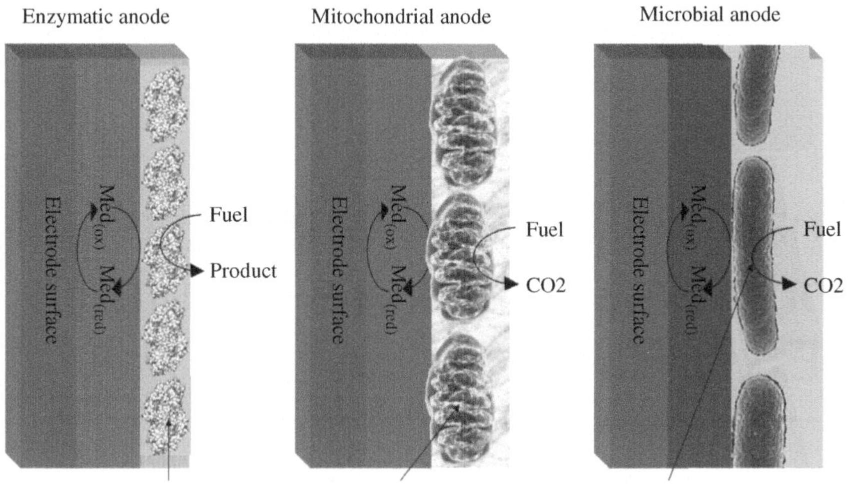

Figure 3.1. Comparison of the anodes of each of the three different types of biofuel cells: Enzymatic, Organelle/Mitochondrial, and Microbial

As discussed in other chapters of this book, bioelectrochemical systems such as microbial fuel cells utilize whole living microorganisms that act as micro-reactors and allow multiple substrates to be oxidized deeply or completely (Bullen *et al.* 2006). Some of the most common microbes applied to a microbial fuel cell include yeast cells as well as various strains of bacterial cells (Dewan *et al.* 2008; Matsunaga and Suzuki 1980; Suzuki *et al.* 1979). Some of the compounds employed as fuels include glucose, acetate, and components of wastewater (Habermann and Pommer 1991; Mohan *et al.* 2008a; Mohan *et al.* 2008b).

Many microorganisms are able to completely oxidize biofuels to carbon dioxide and water due to the metabolic processes such as the citric acid cycle and glycolysis that they can undergo. The exact metabolic process that a cell undergoes depends on the substrate that is being oxidized (Sokic-Lazic and Minteer 2008). The metabolic systems of the cell contain all the necessary enzyme pathways or cascades to undergo these metabolic processes, with no need for additional electrocatalyst addition to the microbial fuel cell system.

Even though the microbial fuel cell may contain all its biocatalysts in their natural environment, two main issues still remain with microbial fuel cells: the slow mass transfer across the cell wall of the microorganism (Karube *et al.* 1981; Sokic-Lazic and Minteer 2008) and lower volumetric catalytic activity (Karube *et al.* 1981; Sokic-Lazic and Minteer 2008) as shown by the data in Tables 3.1 and 3.2 for microbial biofuel cells and enzymatic biofuel cells, respectively.

Table 3.1. Power density of various microbial fuel cells in literature

System (oxidation/reduction)	Organism(s)	Power density ($\mu W\ cm^{-2}$)	Reference
MET Acetate/O_2	Suspended *E. coli*, anaerobic	1.14	(Park and Zeikus 2000)
MET and DET Glucose/ferricyanide	Mixed culture from anaerobic methanogenic sludge	431	(Rabaey *et al.* 2004)
DET Lactate/O_2	Suspended Shewanella putrefaciens and *E. coli*, anaerobic	0.02	(Kim *et al.* 2002)
MET Glucose/O_2	Suspended *E. coli* or sewage sludge, anaerobic	84.5	(Park and Zeikus 2003)
Glucose	Sludge from waste water, anaerobic	0.25	(Park *et al.* 2007)
N/A	N/A	240	(Logan *et al.* 2007)
MET Glucose/potassium ferricyanide	*Klebsiella pneumoniae*	0.697	(Vijay and Meyyappan 2008)
MET Acetate/O_2	*Geobacter sulfurreducens*	188	(Nevin *et al.* 2008)

In contrast, enzymatic biofuel cells have employed many of the same fuels (carbohydrates, lactate, etc.), but have used isolated and purified oxidoreductase enzymes to catalyze the oxidation of the fuels. The most common enzyme employed in the enzyme-based biofuel cells is glucose oxidase, which oxidizes

Table 3.2. Power density of various enzymatic fuel cells in literature

System (oxidation/reduction)	Biocatalyst(s)	Power density ($\mu W\ cm^{-2}$)	Reference
MET Glucose/cumene peroxide	GOx/MP-11	520	(Katz *et al.* 1999a)
MET EtOH or MeOH	ADH, ADH + AldDH, or ADH + formaldehyde dehydrogenase + FDH	1160–2120	(Akers *et al.* 2005)
MET CH_3OH to H_2CO to HCOOH to CO_2 reducing NAD^+ NADH reducing BV_2^+, BV^+/oxygen	ADH, AldDH, FDH and diaphorase	680	(Palmore *et al.* 1998)
MET Glucose (or plant sugars)/O_2	GOx/Box	440	(Mano *et al.* 2003a)
MET Ethanol	ADH, AldDH, IDH, KDH, SDH, MDH	1010	(Sokic-Lazic and Minteer 2008)
MET Ethanol/O_2	ADH, AldDH, bilirubin oxidase with membrane	830	(Topcagic and Minteer 2006)
DET Lactose	CDH	32	(Tasca *et al.* 2008)
DET Fructose/O_2	Laccase, D-Fructose dehydrogenase	850	(Kamitaka *et al.* 2007)
MET Glucose/O_2	Glucose dehydrogenase, bilirubin oxidase	1450	(Sakai *et al.* 2009)
MET Glucose/O_2	Glucose oxidase, laccase, bilirubin oxidase	250	(Atanassov *et al.* 2004)
MET Ethanol/O_2	ADH, AldDH, Bilirubin oxidase without membrane	390	(Topcagic and Minteer 2006)

glucose to gluconolactone, although many dehydrogenase enzymes are also used. Although these enzymes typically provide higher current densities (improved mass transport and higher volumetric catalytic activity), they provide lower efficiencies, because they only partially oxidize the substrate/fuel and they typically have much shorter active lifetimes, because the enzymes cannot be regenerated in the electrochemical cell. For instance, if a microbial fuel cell is capable of completely oxidizing glucose, it will harness all 24 electrons, but an enzymatic biofuel cell employing glucose oxidase will only harness 2 of the 24 electrons and therefore will have lower efficiency. In the third type of biofuel cell, the mitochondria of the living cell are isolated and employed to catalyze the oxidation of the biofuel (i.e. pyruvate). Mitochondria contain the Kreb's cycle metabolic pathway and are able to completely oxidize pyruvate to carbon dioxide, but they have a subcellular membrane and, similar to complete cells, have lower volumetric catalytic activity, so they have lower current density. Typically, the current density will be between an enzymatic fuel cell and a microbial fuel cell (Arechederra and Minteer 2008). Table 3.3 contains a comparison of the characteristic of enzymatic, microbial, and organelle-based biofuel cells.

Table 3.3. General characteristics of enzymatic, microbial, and organelle-based biofuel cells

	Enzymatic fuel cell	Microbial fuel cell	Organelle fuel cell
Power density	High (mW/cm£)	Low (μW/cm£)	Average (between enzymatic and microbial fuel cell performance)
Efficiency	Low (single step 2e oxidation)	High (complete oxidation or deep oxidation)	High (complete oxidation)
Selectivity	High (enzyme specificity)	Low (many microbes consume wide range of substrates/fuels)	Average (capable of oxidizing substrates of mitochondria)
Lifetime	Average (10 days–1 year)	Long (3–5 years)	Short (days)
Scalability	Miniature (typically 0.01 to 10 cm^2)	Scalable to any size	Miniature (~1–5 cm^2)

3.2 SIMILARITIES BETWEEN TYPES OF MICROBIAL AND ENZYMATIC BIOFUEL CELLS

Several early microbial fuel cells were designed and some of them have analogous counterparts in enzymatic fuels cells. The main types of microbial fuel cells are: (1) bioreactor-style with traditional fuel cell, (2) microbe in anode solution, (3) microbe and mediator in anode solution, (4) immobilized microbe with mediator in solution, (5) immobilized microbe with immobilized mediator, (6) microbe in solution with immobilized mediator, and (7) direct electron transfer-based microbial anode.

3.2.1 Bioreactor design

The first type of microbial fuel cell involves a microbial bioreactor producing fuel separated from the anodic side of the electrochemical cell (Katz *et al.* 2003). In this case, the biocatalytic part of the device (the bioreactor) is not directly placed inside the electrochemical cell.

The electrochemical part operates under conditions independent of those of the biological part of the device, allowing them to operate separately and at different pH and temperature conditions, since fuel cells typically operate better at higher temperatures and more extreme pHs than microbes could survive.

An example of a device that applies this approach adds a bioreactor to produce hydrogen gas from biomass to the front end of a H_2/O_2 fuel cell (Katz *et al.* 2003). However, this kind of application of the micro-organisms to the fuel cell can be classified as a bioprocessing operation and the system altogether is not considered a biofuel cell, because the fuel cell is operating in the absence of the biocatalyst and the electrocatalyst is still a precious metal. Instead, the biofuel produced by the microorganism in the bioreactor is fed into the fuel cell (Bullen *et al.* 2006). Similar systems have also been developed with enzymatic bioreactors (Harmand *et al.* 2004; Zhang and Mielenz 2007), but they are more rare in the literature than other types of enzymatic fuel cells. This is mainly due to the fact that enzyme-based bioprocessing is typically used for simplistic, single step chemical reactions (i.e. industrial use of glucose isomerase) and fuel production from biomass typically involves a complicated series of reactions.

3.2.2 *In-situ* bioreactor style

In a second type of microbial fuel cell, the biocatalyst (microbe) is suspended directly in the anodic compartment of a fuel cell, allowing the fuel cell to utilize products formed from the microbial oxidation of the fuel. In this type of biofuel cell, the operational temperature and pH conditions of the electrochemical cell/ battery have to be controlled to ensure that they are consistent with microbe sustainability, making them very different from those in a conventional fuel cell (i.e. typically 20–37°C and near neutral pH) (Katz *et al.* 2003). A traditional precious metal-based H_2/O_2 fuel cell can operate under these conditions as well, but it results in much lower fuel cell performance (lower current and power densities) (Katz *et al.* 2003). There don't seem to be any analogous enzymatic fuel cells systems in the literature for this type of microbial fuel cell. The main reason for the fact that there is no analogous enzymatic fuel cell is that proteins/enzymes easily passivate the precious metal surfaces of traditional fuel cells, which results in a loss of fuel cell performance.

3.2.3 Catalyst in anolyte solution

A third type of microbial fuel cell involves adding an additional redox species to the microbe suspension in the anode. This additional redox species helps transfer the electron between the microbe and the electrode/conductor. These redox species are called mediators. In this case, the electrons are not transferred to the natural electron acceptor in the microbial system, but instead flow to the artificial acceptor and then the artificial acceptor actually transfers the electron(s) to the current collector/electrode surface. This changes the biological process that occurs within the microorganism. In order to eliminate competition between the natural electron acceptor and the artificial electron relay, it is suggested to completely remove the microorganism's natural electron acceptor (Katz *et al.* 2003). This type of mediated electron transfer can be identified as diffusive mediated electron transfer (MET) since the mediator can diffuse across the cell membrane in both directions (Bullen *et al.* 2006). In this type of microbial fuel cell, the microbes operate under anaerobic conditions to allow electron transfer to the mediator and then to the anode (Katz *et al.* 2003). A similar system is seen with enzymatic bioelectrodes whereas small molecule, redox active mediators in solution can act as artificial electron relays between the biocatalyst (the enzyme) and the electrode (Blaedel and Jenkins 1975; *Kim et al.* 2006). There is a wealth of examples of this type of enzymatic biofuel cell in both the early enzymatic biofuel cell literature and the recent literature. Mediators commonly employed are quinines (Islam *et al.* 2004; Tamaki *et al.* 2007), ferrocene, nickelocene, and other similar organometallic complexes (Forrow and Walters 2004; Liu *et al.* 2007; Muchindu and Ngila 2007), and organic dyes (Brunetti *et al.* 2000; Kulys *et al.* 1994). These mediators are employed for both bioelectrodes used in enzymatic biofuel cells and biosensors (i.e. blood glucose/sugar sensors). Mediators are typically chosen for their ability to interact with the enzyme/coenzyme, to have reversible electrochemistry at the current collector/electrode surface, and to be stable in solution for multiple turnovers and a wide range of pH. However, it is important to note that as of today, there has been no ideal mediator found for enzymatic biofuel cells.

3.2.4 Immobilized catalyst and/or mediator

An alternative design of microbial fuel cells that was later designed included the immobilization of the microorganism and/or a redox mediator onto electrodes. (Davis and Higson 2007). Allen and Bennetto utilized a system where *Proteus vulgaris* was immobilized on an electrode to generate currents from carbohydrates (Kim *et al.* 2000b). This strategy allows for faster responses to fuel/substrate additions and improved efficiencies. Instead of immobilizing the microorganism, the mediator can be immobilized. When comparing immobilized mediators to mediators suspended in solution, it was found that the currents increased in microbial systems. When Park and Zeikus immobilized neutral red on an anode and acetate was used as the fuel and *E. coli* as the microorganism, the power output doubled (Davis and Higson 2007). This type of mediated electron transfer is considered as non-diffusive MET, because the mediator does not penetrate the cell membrane.

A similar system is seen with enzymatic bioelectrodes where immobilized mediators (redox polymers, conducting polymers, or redox hydrogels) are used to transfer the electron from the enzyme to the electrode (Nishikiori *et al.* 2008; Sato *et al.* 2005). However, in enzymatic systems, currents typically decrease with immobilized mediators versus mediators in solution, because electron transport through the redox polymer or hydrogel is frequently slower than in solution. However, those engineering problems have been addressed by some electrochemists for some systems and are being examined for all enzymatic systems. Also, researchers in the enzymatic biofuel cell community have explored immobilizing the enzyme at the electrode surface instead of having purified enzyme in the anode solution and they have found that both the efficiency and the lifetime of the enzymatic biofuel cell is increased via immobilization of the enzyme (Akers *et al.* 2005; Kim *et al.* 2003; Mano *et al.* 2003b). The majority of enzymatic biofuel cells today employ a mediator and immobilize both the mediator and the enzyme at the electrode surface (Barton *et al.* 2004; Cooney *et al.* 2008; *Kim et al.* 2006; Moehlenbrock and Minteer 2008).

3.2.5 Direct electron transfer catalysts

A mediator-less microbial fuel cell utilizes electrochemically active bacteria capable of transferring electrons to the electrode directly (Bullen *et al.* 2006). This is commonly called direct electron transfer (DET) and has been observed for some oxidoreductase enzymes as well (i.e. glucose oxidase, PQQ-dependent alcohol dehydrogenase, laccase, etc.) (Ivnitski *et al.* 2006; Treu and Minteer 2005). Although researchers have effectively demonstrated efficient direct electron transfer at laccase biocathodes, direct electron transfer anodic systems have been more difficult and although they appear in the literature, they typically have significantly lower power/current density due to low electron transfer efficiency. The future of enzymes in biofuel cells will be focused on either forming more stable mediator systems or solving the low electron transfer efficiency of DET of these oxidoreductase enzymes (Cooney *et al.* 2008), because these two problems have been plaguing commercialization of enzymatic biofuel cells.

3.3 CATALYST SOURCES FOR MET AND DET SYSTEMS

A wide variety of microorganisms have been applied to MET system type microbial fuel cells to generate electrical power. Park and Zeikus suspended a culture of anaerobic *Actinobacillus succinogenes* at a graphite felt anode where glucose was used as the fuel and thionin as the mediator. The current density generated from this system was 8.5 μA/cm^2 along with a power density of 5.3 μW/cm^2 (Park and Zeikus 2000). Another group studied *Proteus vulgaris* at an RVC anode which utilized sugar as the fuel and thionin as a diffusing mediator (Kim *et al.* 2000a). This system generated a current density of 13.2 μA/cm^2 and a power density of 5.26 μW/cm^2. Other microorganisms applied to a MET-based microbial fuel cell system include *Desulphovibrio desulphuricans, Erwina dissolven, E. coli, Gluconobacter oxydans, Klebsiella pneumoniae, Lactobacillus plantarum, Proteus mirabilis, Pseudomonas aeruginosa, Streptococcus lactis, Shewanella putrefaciens,* and *Desulphovibrio vulgaris*. (Bullen *et al.* 2006; Du *et al.* 2007). Many of these microorganism sources have been used to isolate enzymes for enzymatic fuel cells. For instance, most of the DET-based dehydrogenase enzymes are isolated from *Gluconobacter oxydans* (i.e. PQQ-dependent alcohol dehydrogenase, aldehyde dehydrogenase, and lactate dehydrogenase) (Treu *et al.* 2008; Treu and Minteer 2008). The commercial sources of alcohol oxidation enzymes are isolated from yeast (*Saccharomyces cerevisiae),* lactic dehydrogenase is from *Lactobacillus,* glucose dehydrogenase is from *Pseudomonas*. However, the only reason for using microbial sources of enzymes has been the ease of production and purification and not properties inherent in the microbe. However, future research in the field of enzymatic biofuel cells will likely focus on enzyme sourcing from thermophiles and extremophiles to increase the stability of the enzymes and the resulting enzymatic biofuel cell.

It is important to note that some of these microorganisms may have the ability for direct electron transfer (DET) between the microorganism and the electrode/current collector, allowing for faster electron transfer (Bullen *et al.* 2006). Schröeder *et al.* employed anaerobic *E. coli* and glucose was used as the fuel at a variety of anodes. The current density generated from this DET system (immobilized mediator on anode) was 2600 $\mu A/cm^2$ and the power density 1200 $\mu W/cm^2$ (Schröeder *et al.* 2003). Another group, Kim *et al.* utilized *Shewanella putrefaciens* as the microorganism and lactate as the fuel. This system resulted in the generation of a current density of 0.71 $\mu A/cm^2$ and power density of 0.02 $\mu W/cm^2$ (Kim *et al.* 2002). The microbes *Rhodoferax ferrireducens, Geobacter metallireducins, Geobacter sulfurreducens, and Aeromonas hydrophila* have also been applied to DET-based microbial fuel cells in the past (Bullen *et al.* 2006; Du *et al.* 2007). Although these microbes do oxidative metabolism of fuel/ substrate and enzymatic biofuel cell researchers are interested in enzymatic DET, researchers are not using microbes capable of DET for enzyme isolation and purification, because the DET mechanism for a microbe is different than for an enzyme. DET for a microbe means that the cell wall or other microbial conduits (nanowires, etc.) can transfer electrons between the microbe and the electrode, whereas enzymatic DET means that the enzyme can transfer (usually through multiple metal centers in the enzyme) the electron from the enzyme active site to the electrode. This difference in process means that researchers are interested in a particular protein structure that is advantageous for DET.

3.4 COMPARISON OF PROPERTIES OF MICROBIAL AND ENZYMATIC FUEL CELLS

The application of whole microorganisms to a fuel cell has its advantages as well as its disadvantages as compared to utilizing isolated enzymes. One of the advantages of the use of whole microorganisms is the elimination of the need to isolate and purify enzymes, which is typically costly and time intensive and sometimes results in enzyme instability. In addition, the microorganisms are able to work efficiently, since the enzymes within the intact and viable microorganisms are able to function in their natural environment with all natural metabolic pathways intact and natural cofactor regeneration (Katz *et al.* 2003). Microbial fuel cells also allow more thorough oxidation of the biofuels and are less sensitive to poisoning and loss of activity (Davis and Higson 2007). Finally, research has indicated a lifetime of the microbial fuel cell up to five years (Kissinger and Heineman 1996). On the contrary, microbial fuel cells suffer from low volumetric catalytic activity of the whole organism and low power densities due to slow mass transport of the fuel across the cell wall (Kissinger and Heineman 1996). In addition, in some cases it can be difficult to use the electrons generated by the cell. In this case, the problem may be solved by using non-toxic mediators that will aid in the transfer of electrons across the cell membrane of the microorganisms and an active method for transporting the mediator across cell membranes (Davis and Higson 2007). Therefore, researchers are trying to develop methods for improving the efficiency of the enzyme-based biofuel cells, so it is more similar to the efficiency of the microbial fuel cell, while maintaining the high volumetric catalytic activity and high power/current densities. The major research focuses in enzymatic fuel cells have been toward designing enzyme immobilization strategies that increase stability and enzyme lifetime, while addressing low efficiency by attempting to incorporate cascades or metabolic pathways of enzyme to ensure deep or complete oxidation of biofuels.

3.5 ENZYMES EMPLOYED IN ENZYMATIC BIOFUEL CELLS

A fuel cell is considered to be an enzymatic biofuel cell if it relies on isolated enzymes to catalyze either the oxidation of the fuel at the anode and/or the reduction of oxygen or hydrogen peroxide at the cathode. Some of the fuels employed in enzyme-based biofuel cells are ethanol, glycerol, and glucose (Akers *et al.* 2005; Yahiro *et al.* 1964). The enzymes employed in biofuel cells are typically oxidoreductase enzymes that catalyze redox reactions. However, sometimes other classifications of enzymes (i.e. invertase or kinases) are employed to transform the fuel or fuel byproduct into a product that can be oxidized by an oxidoreductase enzyme (Sokic-Lazic and Minteer 2008). It is important to note that enzymes are attractive catalysts, because they are renewable, relatively inexpensive, biodegradeable, and they are simple enough to retain their original structure at the end of the reaction (Molinari *et al.* 1995; Silverman 2000). The purified form of the enzymes used in enzyme-based biofuel cells need to be extracted and purified from some natural biological source, which most frequently is a microorganism, such as bacteria or yeast cells. Oxidoreductase enzymes are classified by the coenzyme or cofactor that the enzyme depends on for the opposing the redox half reaction. Some of the oxidoreductase enzymes employed in biofuel cells are FAD^+-dependent or $NAD(P)^+$-dependent enzymes. For instance, for alcohol dehydrogenase, the oxidation half reaction of alcohol to aldehyde only occurs when a reduction half reaction can occur and that is the reduction of NAD to NADH. Similarly, if an enzyme is FAD-dependent then the oxidation reaction can only occur when FAD is reduced to $FADH_2$. These enzymes are typically mediated electron transfer

(MET) enzymatic systems, although FAD-dependent enzymes have been shown to be capable of direct electron transfer (DET); however inefficiently. Typically, the mediator will oxidize the NAD(P)H or $FADH_2$ to regenerate the cofactor (NAD(P) or FAD). The reduced form of the mediator will then be oxidize at the electrode surface. These systems are typically limited by the reversible electrochemistry of the mediator. Researchers have recently discovered direct electron transfer (DET) enzymatic systems such as PQQ-dependent enzymes (Akers *et al.* 2005; Pizzariello *et al.* 2002; Yuhash*i et al.* 2005) and those enzymes are being explored, because currently, enzyme lifetime in the biofuel cell is longer than mediator lifetime in the biofuel cell.

Enzyme-based biofuel cells utilizing various MET-based enzymes have been reported in the literature. The most widely studied biofuel cell is the FAD^+-dependent glucose oxidase based biofuel cell that converts glucose to gluconolactone and peroxide at the anode (Katz *et al.* 1999a; Katz *et al.* 1999b; Willner *et al.* 1998a; Willner *et al.* 1998b). This specific enzyme can be purified from *Penicillium notatum* (Gorniak and Kaczkowski 1974). In addition to FAD^+-dependent glucose oxidase, NAD^+-dependent alcohol dehydrogenase is also a MET-system based enzyme that utilizes methanol as the substrate at the anode and produces formaldehyde. NAD^+-ADH has been isolated from a specific yeast strain called *Saccharomycosis lipolytica* (Barth and Kuenkel 1979).

Many of the more recent reported biofuel cells utilize a DET system that allows for elimination of the mediator and more efficient electron transfer. A recent study applied PQQ-dependent alcohol dehydrogenase at the bioanode to convert ethanol to acetaldehyde (Treu and Minteer 2005) or to oxidize glycerol (Arechederra *et al.* 2007). PQQ-dependent ADH has been isolated from various gram-negative acetic acid bacteria such as *Acetobacter*, *Gluconobacter*, *Pseudomonas* and *Commamonas* (Austin 1967; Moore *et al.* 2004; Price and Stevens 1989; Richards 1977; Silverman 2000; Stenesh 1998; Storm and Koshland 1970). PQQ-dependent glucose dehydrogenase is another DET based enzyme whose substrate is glucose that is converted to gluconolactone. It can be found within *Acinetobacter calcoaceticus* (Oubrie *et al.* 1999). These systems show promise in enzymatic biofuel cells, but the entire metabolic pathways for oxidation of biofuels are not present in PQQ-dependent enzymes, so these enzymes are useful for partial oxidation-based biofuel cells, but not as useful for complete oxidation of biofuels (i.e. ethanol, glucose, pyruvate).

Enzymes are specific for fuel utilization or oxygen reduction allowing the elimination of the polymer electrolyte membrane, which is the most expensive fabrication material, along with creating increased resistance. Biofuel cells without a polymer electrolyte membrane separator are typically called: membrane-less or compartment-less (Kim *et al.* 2003; Mano *et al.* 2003a;

Topcagic and Minteer 2006). They have become increasingly popular in the enzymatic biofuel cell literature as the problems with crossover and traditional polymer electrolyte membranes have plagued the commercialization of direct methanol fuel cells and other low temperature fuel cells.

Enzymes are sensitive to heat and pH, and exposure to high temperatures or harsh pH will lead to denaturation by hydrolyzing the interchain linkage of the protein (Price and Stevens 1989). Due to the enzyme's instability issues, most enzyme-based biofuel cell lifetime's only extend from hours to a few days. Hence leading to the creation of short lifetime biofuel cells (Calabrese-Barton *et al.* 2004; *Mano et al.* 2003a). Like a whole microorganism, an enzyme can be either suspended in solution or immobilized on an electrode in the biofuel cell (Davis and Higson 2007).

The early enzyme-based biofuel cells contained enzymes in solution that led to a biofuel cell lifetime of a few hours due to the enzyme's instability in solution, which is one of the most common issues associated with enzymatic biofuel cells. In later work, immobilized enzymes on an electrode allowed greater stability and lifetime of the enzyme as well as biofuel cell lifetime as high as a few months (Bullen *et al.* 2006; Calabrese-Barton *et al.* 2004). In addition, some enzyme-based biofuel cells utilize mediators for electron transfer between the enzyme and the electrode, such as organic dyes (Palmore and Whitesides 1994). Utilizing mediators in the systems introduces an additional step in the reaction chains, but they also allow for greater efficiency of the biofuel cell (Kim *et al.* 2006).

Biofuel cells designed with immobilized NAD^+-dependent enzymes on the surface of the electrode employ redox mediators in the system in order to reduce the high overpotential that the NADH oxidation causes (Blaedel and Jenkins 1975). The NAD^+ has a short lifetime in solution and a limited life time in a semipermeable membrane (Blaedel and Jenkins 1975). In order to eliminate the need for a redox mediator and to employ a coenzyme that is stable for at least as long as the enzyme is stable, the later work applied a PQQ-dependent enzyme that led to direct electron transfer to the electrode. Since the PQQ-dependent enzymes are membrane-bound enzymes, they require a hydrophobic environment to be stable proteins, which was provided by a hydrophobically modified Nafion membrane, which is a hydrophobic, micellar polymer. The PQQ cofactor has a lower overpotential and is attached to the enzyme; therefore it will lead to increased lifetime and activity and a less complex biofuel cell. Treu *et al.* immobilized PQQ-dependent ADH onto the anode utilizing a micellar polymer to encapsulate the enzyme and protect and constrain it from harsh chemical environments, which allowed the alcohol biofuel cell to have a lifetime of greater than 1 year and a power density of 3.61 mW/cm^2 for a 10 mmol/l ethanol solution in a static system (Treu and Minteer 2005). However, there are currently no enzymatic biofuel cells in the literature that have shown the 5-year lifetimes that are frequently observed for microbial fuel cells. This is because the microbe regenerates enzyme; whereas the immobilization membrane provides a protective environment for the enzyme, but cannot regenerate or repair the enzyme.

3.6 DEEP AND/OR COMPLETE OXIDATION OF FUEL

A microbial fuel cell contains all of its enzymes for complex oxidation processes of a variety of substrates in one main compartment – the microbial cell. On the contrary, most enzyme based biofuel cells were designed employing a single enzyme to partially oxidize a specific fuel (Sokic-Lazic and Minteer 2008). In order to allow an enzyme-based biofuel cell to completely oxidize a biofuel, it is required to immobilize multi-enzyme systems instead of one single enzyme (Sokic-Lazic and Minteer 2008). One main issue associated with this approach is developing a successful immobilization of the whole multi-enzyme system that will lead to more efficient enzyme-based biofuel cells.

The first multi-enzyme fuel cell that oxidized methanol to carbon dioxide and water by employing alcohol dehydrogenase (ADH), aldehyde dehydrogenase (AldDH), formate dehydrogenase (FDH), and diaphorase was developed by Palmore *et al.* (Palmore *et al.* 1998). This system generated a power density of 0.68 mW/cm^2 and an open circuit potential of 0.8 V. In another study, Sokic-Lazic and Minteer developed a citric acid cycle biomimic on a carbon electrode where all the enzymes and cofactors of the citric acid cycle were incorporated to completely oxidize ethanol to carbon dioxide (Sokic-Lazic and Minteer 2008). The results indicated an increase in power density with addition of each additional dehydrogenase, resulting in a total 8.71-fold increase of power density by the six-dehydrogenase system as compared to a single-enzyme system (Sokic-Lazic and Minteer 2008). Therefore, although most enzymatic fuel cells have low efficiency due to single-step oxidation, improved efficiency can be observed from multi-enzyme systems; however, much future work is needed to further develop and understand these multi-enzyme systems.

3.7 CONCLUSIONS

Overall, enzymatic biofuel cells are complementary to microbial fuel cells. They use many of the same components and have many similarities, but the main differences lie in the fact that microbial fuel cells have higher efficiency due to deep or complete oxidation of fuel/substrate and they have very long active lifetimes, but they typically have lower power densities due to low volumetric catalytic activity and mass transport across cellular membranes. Enzymatic biofuel cells typically have higher current/power densities, but they are typically plagued with short lifetimes and low efficiency, although the efficiency can be improved by multi-enzyme cascades for deep oxidation and the lifetimes can be improved by the implementation of direct electron transfer enzyme systems and good enzyme immobilization technology.

3.8 ACKNOWLEDGEMENTS

The authors would like to thank Air Force Office of Scientific Research for funding and Michael Moehlenbrock for figure preparation.

REFERENCES

Akers, N.L., Moore, C.M. and Minteer, S.D. (2005) Development of alcohol/O2 biofuel cells using salt-extracted tetrabutylammonium bromide/Nafion membranes to immobilize dehydrogenase enzymes. *Electrochim. Acta* **50**(12), 2521–2525.

Arechederra, R.L. and Minteer, S.D. (2008) Organelle-based biofuel cells: immobilized mitochondria on carbon paper electrodes. *Electrochimica Acta* **53**(23), 6698–6703.

Arechederra, R.L., Treu, B.L. and Minteer, S.D. (2007) Development of glycerol/O-2 biofuel cell. *J. Power Sources* **173**(1), 156–161.

Atanassov, P., Colon, F. and Rajendran, V. (2004) Glucose–air enzymatic bio-fuel cell. In: Proceedings of the Colloid Chemistry Division, American Chemical Society, Washington DC, p. 207.

Austin, L.G. (1967) Fuel Cells: A Review of Goverment-Sponsored Research, 1950–1964. Office of Technology Utilization, National Aeronautics and Space Administration: Washington DC.

Barth, G. and Kuenkel, W. (1979) Alcohol dehydrogenase (ADH) in yeasts. II. NAD^{+}- and $NADP^{+}$-dependent alcohol dehydrogenases in *Saccharomycopsis lipolytica*. *Z.t Allgemeine Mikrobiol.* **19**(6), 381–390.

Barton, S.C., Gallaway, J. and Atanassov, P. (2004) Enzymatic biofuel cells for Implantable and microscale devices. *Chemical Reviews* **104**(10), 4867–4886.

Blaedel, W.J. and Jenkins, R.A. (1975) Study of the electrochemical oxidation of reduced nicotinamide adenine dinucleotide. *Anal. Chem.* **47**(8), 1337–1343.

Brunetti, B., Ugo, P., Moretto, L.P. and Martin, C.R. (2000) Electrochemistry of phenothiazine and methyl viologen biosensor electron transfer mediators at nanoelectrode ensembles. *J. Electroanal. Chem.* **491**(1–2), 166–174.

Bullen, R.A., Arnot, T.C., Lakeman, J.B. and Walsh, F.C. (2006) Biofuel cells and their development. *Biosensors Bioelectron.* **21**, 2015–2045.

Calabrese-Barton, S., Gallaway, J. and Atanassov, P. (2004) Enzymatic biofuel cells for implantable and microscale devices. *Chem. Rev.* **104**(10), 4867–4886.

Cooney, M.J., Svoboda, V., Lau, C., Martin, G.P. and Minteer, S.D. (2008) Enzyme catalysed biofuel cells. *Energy Environ. Sci.* **1**, 320–337.

Davis, F. and Higson, S.P.J. (2007) Biofuel cells–recent advances and applications. *Biosensors Bioelectron.* **22**, 1224–1235.

Dewan, A., Beyenal, H. and Lewandowski, Z. (2008) Scaling up microbial fuel cells. *Environ. Sci. Technol.* **42**(20), 7643–7648.

Du, Z., Li, H. and Gu, T. (2007) A state of the art review on microbial fuel cells: a promising technology for wastewater treatment and bioenergy. *Biotechnol. Adv.* **25**, 464–482.

Forrow, N.J. and Walters, S.J. (2004) Transition metal half-sandwich complexes as redox mediators to glucose oxidase. *Biosensors Bioelectron.* **19**(7), 763–770.

Gorniak, H. and Kaczkowski, J. (1974) Isolation and characterization of glucose oxidase from mycelium and nutrient of Penicillium notatum cultures. *Bull. Acad. Polonaise Sci., Series Sci. Biol.* **22**(6), 351–355.

Habermann, W. and Pommer, E.H. (1991) Biological fuel cells with sulfide storage capacity. *Appl. Microbiol. Biotechnol.* **35**(1), 128–133.

Harmand, J., Rapaport, A. and Drame, A.K. (2004) Optimal design of two interconnected enzymatic bioreactors. *J. Process Contr.* **14**(7), 785–794.

Islam, M.K., Ohashi, T., Yasukawa, T., Shiku, H., Nishizawa, M., Kosuge, J., Fukasaku, N. and Matsue, T. (2004) Use of a quinone as a mediator at anode in a glucose/O2 biofuel cell. *Chem. Sensors* **20**, 744–745.

Ivnitski, D., Branch, B., Atanassov, P. and Apblett, C. (2006) Glucose oxidase anode for biofuel cell based on direct electron transfer. *Electrochem. Comm.* **8**(8), 1204–1210.

Kamitaka, Y., Tsujimura, S., Setoyama, N., Kajino, T. and Kano, K. (2007) Fructose/dioxygen biofuel cell based on direct electron transfer-type bioelectrocatalysis. *Phys. Chem. Chem. Phys.* **9**(15), 1793–1801.

Karube, I.S., Matsunaga, S.T. and Kuriyama, S. (1981) Biochemical energy conversion by immobilized whole cells. *Ann. NY Acad. Sci.* **91**.

Katz, E., Filanovsky, B. and Willner, I. (1999a) A biofuel cell based on two immiscible solvents and glucose oxidase and microperoxidase-11 monolayer-functionalized electrodes. *New J. Chem.* **23**(5), 481–487.

Katz, E., Shipway, A.N. and Willner, I. (2003) Biochemical fuel cells. In *Handbook of Fuel Cells – Fundamentals, Technology and Applications*. Vielstich, W., Gasteiger, H.A. and Lamm, A. (eds). John Wiley & Sons, Ltd, Jerusalem, pp. 1–27.

Katz, E., Willner, I. and Kotlyar, A.B. (1999b) A non-compartmentalized glucose O_2 biofuel cell by bioengineered electrode surfaces. *J. Electroanal. Chem.* **479**(1), 64–68.

Kim, H.-H., Mano, N., Zhang, Y. and Heller, A. (2003) A miniature membrane-less biofuel cell operating under physiological conditions at 0.5 V. *J. Electrochem. Soc.* **150**(2), A209–A213.

Kim, H.J., Park, H.S., Hyun, M.S., Chang, I.S., Kim, M. and Kim, B.H. (2002) A mediator-less microbial fuel cell using a metal reducing bacterium, Shewanella putrefaciens. *Enzyme Microbial Technol.* **30**(2), 145–152.

Kim, J., Jia, H. and Wang, P. (2006) Challenges in biocatalysis for enzyme-based biofuel cells. *Biotechnol. Adv.* **24**(3), 296–308.

Kim, N., Choi, Y., Jung, S. and Kim, S. (2000a) Development of microbial fuel cells using Proteus Vulgaris. *Bull. Korean Chem. Soc.* **21**(1), 44–48.

Kim, N., Choi, Y., Jung, S. and Kim, S. (2000b) Effect of initial carbon sources on the performance of microbial fuel cells containing *Proteus vulgaris*. *Biotechnol. Bioeng.* **70**(1), 109–114.

Kissinger, P.T. and Heineman, W.R. (1996) *Laboratory Techniques in Electroanalytical Chemistry*, Marcel Dekker, Inc.

Kulys, J., Hansen, H.E., Buch-Rasmussen, T., Wang, J. and Ozsoz, M. (1994) Glucose biosensor based on the incorporation of Meldola Blue and glucose oxidase within carbon paste. *Analyt. Chim. Acta* **288**(3), 193–196.

Liu, S., Yin, Y. and Cai, C. (2007) Immobilization and characterization of glucose oxidase on single-walled carbon nanotubes and its application to sensing glucose. *Chinese J. Chem.* **25**(4), 439–447.

Logan, B.E., Cheng, S. and Watson, V. (2007) Progress and challenges in scale up of electrogenic reactors such as microbial fuel cells. Boston, MA, pp. FUEL-129.

Mano, N., Mao, F. and Heller, A. (2003a) Characteristics of a miniature compartment-less glucose-O2 biofuel cell and its operation in a living plant. *J. Am. Chem. Soc.* **125**(21), 6588–6594.

Mano, N., Mao, F., Shin, W., Chen, T. and Heller, A. (2003b) A miniature biofuel cell operating at 0.78 V. *Chem. Comm. (Cambridge, UK)* **4**, 518–519.

Matsunaga, T.K. and Suzuki, S. (1980) Some observations on immobilized hydrogen-producing bacteria: behavior of hydrogen in gel membranes. *Biotechnol. Bioeng.* **22**.

Moehlenbrock, M.J. and Minteer, S.D. (2008) Extended lifetime biofuel cells. *Chem. Soc. Rev.* **37**, 1188–1196.

Mohan, S.V., Mohanakrishna, G., Reddy, B.P., Saravanan, R. and Sarma, P.N. (2008a) Bioelectricity generation from chemical wastewater treatment in mediatorless (anode) microbial fuel cell (MFC) using selectively enriched hydrogen producing mixed culture under acidophilic microenvironment. *Biochem. Engng. J.* **39**(1), 121–130.

Mohan, S.V., Mohanakrishna, G., Srikanth, S. and Sarma, P.N. (2008b) Harnessing of bioelectricity in microbial fuel cell (MFC) employing aerated cathode through anaerobic treatment of chemical wastewater using selectively enriched hydrogen producing mixed consortia. *Fuel* **87**(12), 2667–2676.

Molinari, F., Villa, R., Manzoni, M. and Aragozzini, F. (1995) Aldehyde production by alcohol oxidation with Gluconobacter oxydans. *Appl. Microbiol. Biotechnol.* **43**, 989–994.

Moore, C.M., Akers, N.L., Hill, A.D., Johnson, Z.C. and Minteer, S.D. (2004) Immobilizing oxidoreductase enzymes in surfactant and symmetrical ammonium treated nafion *Polym. Prepr.* **45**(1), 15–16.

Moseley, P.T. (2001) Fuel cell systems explained, edited by James Larminie, Andrew Dicks. *J. Power Sources* **93**(1–2), 285.

Muchindu, M. and Ngila, J.C. (2007) Laccase enzyme electrode with ferrocene-monocarboxylic acid mediator for determination of phenol. *Asian J. Chem.* **19**(3), 2070–2082.

Nevin, K.P., Richter, H., Covalla, S.F., Johnson, J.P., Woodard, T.L., Orloff, A.L., Jia, H., Zhang, M. and Lovley, D.R. (2008) Power output and columbic efficiencies from biofilms of Geobacter sulfurreducens comparable to mixed community microbial fuel cells. *Environ. Microbiol.* **10**(10), 2505–2514.

Nishikiori, H.K., Kano, K. and Tsujimura, K. (2008) Biochemical fuel cells and their electrodes immobilizing enough amount of enzymes and electron mediators on surface. *Jpn. Kokai Tokkyo Koho*.

Oubrie, A., Rozeboom, H.J., Kalk, K.H., Olsthoorn, A.J.J., Duine, J.A. and Dijkstra, B.W. (1999) Structure and mechanism of soluble quinoprotein glucose dehydrogenase. *EMBO J.* **18**(19), 5187–5194.

Palmore, G.T.R. and Whitesides, G.M. (1994) Microbial and enzymic biofuel cells. *ACS Symp. Ser.* **566**, 271–290.

Palmore, G.T.R., Bertschy, H., Bergens, S.H. and Whitesides, G.M. (1998) A methanol/dioxygen biofuel cell that uses NAD^+-dependent dehydrogenases as catalysts: application of an electro-enzymic method to regenerate nicotinamide adenine dinucleotide at low overpotentials. *J. Electroanal. Chem.* **443**(1), 155–161.

Park, D.H. and Zeikus, J.G. (2000) Electricity generation in microbial fuel cells using neutral red as an electronophore. *Appl. Environ. Microbiol.* **66**(4), 1292–1297.

Park, D.H. and Zeikus, J.G. (2003) Improved fuel cell and electrode designs for producing electricity from microbial degradation. *Biotechnol. Bioengng.* **81**(3), 348–355.

Park, H.I., Mushtaq, U., Perello, D., Lee, I., Cho, S.K. and Yun, M. (2007) Effective and low-cost platinum electrodes for microbial fuel cells deposited by electron beam evaporation. *Energy Fuels* **21**(5), 2984–2990.

Pizzariello, A., Stred'ansky, M. and Miertus, S. (2002) A glucose/hydrogen peroxide biofuel cell that uses oxidase and peroxidase as catalysts by composite bulk-modified bioelectrodes based on a solid binding matrix. *Bioelectrochemistry* **56**(1–2), 99–105.

Price, N.C. and Stevens, L. (1989) *Fund. Enzymol.* Oxford University Press, New York.

Rabaey, K., Boon, N., Siciliano, S.D., Verhaege, M. and Verstraete, W. (2004) Biofuel cells select for microbial consortia that self-mediate electron transfer. *Appl. Environ. Microbiol.* **70**(9), 5373–5382.

Richards, F.M. (1977) Areas, volumes, packing and protein structure. *Ann. Rev. Biophys. Bioeng.* **6**, 151.

Sakai, H., Nakagawa, T., Tokita, Y., Hatazawa, T., Ikeda, T., Tsujimurac, S. and Kano, K. (2009) A high-power glucose/oxygen biofuel cell operating under quiescent conditions. *R. Soc. Chem.* **2**, 133–138.

Sato, F., Togo, M., Islam, M.K., Matsue, T., Kosuge, J., Fukasaku, N., Kurosawa, S. and Nishizawa, M. (2005) Enzyme-based glucose fuel cell using Vitamin K3-immobilized polymer as an electron mediator. *Electrochem. Comm.* **7**(7), 643–647.

Schröeder, U., Niessen, J. and Scholz, F. (2003) A generation of microbial fuel cells with current outputs boosted by more than one order of magnitude. *Angew. Chemie, Int. Edn.* **42**(25), 2880–2883.

Silverman, R.B. (ed.) (2000) *The Organic Chemistry of Enzyme Catalyzed Reactions.* Academic Press, San Diego.

Sokic-Lazic, D. and Minteer, S.D. (2008) Citric acid cycle biomimic on a carbon electrode. *Biosens. Bioelectron.* **24**, 945–950.

Stenesh, J. (1998) *Biochemistry: Foundations of Biochemistry/Biomolecules.* Plenum Publishing, New York.

Storm, D.R. and Koshland, D.E.J. (1970) Source for the special catalytic power of enzymes: orbital steering. *Proc. Natl. Acad. Sci. USA* **66**, 445.

Suzuki, S.K., Matsunaga, I.T., Suzuki, S., Karube, I. and Matsunaga, T. (1979) Application of a biochemical fuel cell to wastewaters. *Biotechnol. Energy Prod. Conserv. Biotechnol. Bioeng. Symp.* **8**(501).

Tamaki, T., Ito, T. and Yamaguchi, T. (2007) Immobilization of hydroquinone through a spacer to polymer grafted on carbon black for a high-surface-area biofuel cell electrode. *J. Phys. Chem. B* **111**(34), 10312–10319.

Tasca, F., Gorton, L., Harreither, W., Haltrich, D., Ludwig, R. and Noll, G. (2008) Direct electron transfer at cellobiose dehydrogenase modified anodes for biofuel cells *J. Phys. Chem. C* **112**(26), 9956–9961.

Topcagic, S. and Minteer, S.D. (2006) Development of a membraneless ethanol/oxygen biofuel cell. *Electrochim. Acta* **51**(11), 2168–2172.

Treu, B.L. and Minteer, S.D. (2005) Improving the lifetime, simplicity, and power of an ethanol biofuel cell by employing ammonium treated nafion membranes to immobilize PQQ-dependent alcohol dehydrogenase. *Polymeric Mater.: Sci. Eng.* **92**, 192–193.

Treu, B.L. and Minteer, D.S. (2008) Isolation and purification of PQQ-dependent lactate dehydrogenase from Gluconobacter and use for direct electron transfer at carbon and gold electrodes. *Bioelectrochemistry* **74**, 73–77.

Treu, B.L., Arechederra, R.L. and Minteer, S.D. (2008) Bioelectrocatalysis of ethanol via PQQ-dependent dehydrogenases utilizing carbon nanomaterial supports. *J. Nanosci. Nanotechnol.* (in press).

Vijay, M. and Meyyappan, R.M. (2008) Enhanced power generation in microbial based biological fuel cell using methylene blue as an electronophore. *Chem. Engng World* **43**(4), 58–61.

Willner, I., Arad, G. and Katz, E. (1998a) A biofuel cell based on pyrroloquinoline quinone and microperoxidase-11 monolayer-functionalized electrodes. *Bioelectrochem. Bioenerg.* **44**(2), 209–214.

Willner, I., Katz, E., Patolsky, F. and Buckmann, A.F. (1998b) Biofuel cell based on glucose oxidase and microperoxidase-11 monolayer-functionalized electrodes. *J. Chem. Soc., Perkin Trans. 2: Phys. Org. Chem.* **8**, 1817–1822.

Yahiro, A.T., Lee, S.M. and Kimble, D.O. (1964) Bioelectrochemistry. I. Enzyme utilizing biofuel cell studies *Biochim. Biophys. Acta* **88**(2), 375–383.

Yuhashi, N., Tomiyama, M., Okuda, J., Igarashi, S., Ikebukuro, K. and Sode, K. (2005) Development of a novel glucose enzyme fuel cell system employing protein engineered PQQ glucose dehydrogenase. *Biosensors Bioelectron.* **20**(10), 2145–2150.

Zhang, Y.H.P. and Mielenz, J. (2007) *Bio-hydrogen Production by an Artificial Enzymatic Pathway*. USPTO (ed.), United States.

4

Shuttling via soluble compounds

Enrico Marsili and Xiaoming Zhang

4.1 INTRODUCTION

In electroactive bacteria (EAB), the redox strategy which employs membrane-associated compounds (direct electron transfer, see Chapter 4) to transfer electrons to the electrode in Microbial Fuel Cells (MFC) and other Bioelectrochemical Systems (BES) is usually considered more energetically efficient with respect to endogenous soluble redox shuttle compounds (mediated electron transfer).

However, an increasing number of reports have proved that mediated electron transfer to the electrode is widely spread among pure culture and mixed consortia. Particularly, it has been demonstrated that:

- several strains are capable of producing and excreting soluble redox shuttles when exposed to electrodes maintained at oxidative potentials and in the absence of soluble electron acceptors,

 Bioelectrochemical Systems: From Extracellular Electron Transfer to Biotechnological Application. Edited by Korneel Rabaey, Largus Angenent, Uwe Schröder and Jürg Keller. ISBN: 9781843392330. Published by IWA Publishing, London, UK.

- accumulation of such redox shuttles increases several-fold the electron transfer rate to the electrode,
- soluble redox shuttle compounds are non-specific and those produced by a strain can be used by other strains in a mixed consortium to facilitate electron transfer to the electrode.

Is there any correlation between bacterial genus and the preferred electron transfer strategy? Most membrane-bound electron transfer agents were found in proteobacteria. On the other hand, endogenous redox shuttle compound mediators seem to be a more general approach to extracellular electron transfer, and have been observed across all bacteria.

In addition to the two strategies outlined above, microorganisms can also use redox mediators already available in the environment. It has been shown that many bacteria can use soil components (e.g. humic acids or organic matter) to increase the extracellular electron transfer rate to electrodes. Also, cysteine available in soils functions as a redox mediator between dissimilatory iron reducing bacteria and Fe(III). Finally, inorganic species such as S^0/HS^- redox couples may serve as soluble redox shuttles under certain environmental conditions, as has been observed in bioremediation studies (see also Chapter 14).

Modeling of extracellular electron transfer in EAB biofilms is still in its early stages. Detailed and sophisticated biofilms models are available, but little is known about the diffusion-reaction behaviour of soluble redox shuttles. It has been recently proposed that the electrical charge of the redox shuttles, both in their reduced and oxidized form, strongly affects the diffusion of redox shuttles in biofilms and in solution, and ultimately the energetic cost of the redox strategy based on soluble redox shuttles.

In this chapter, we will present to the reader the state-of-the-art in mediated electron transfer in BES. After a fundamental description of redox shuttles (4.2) and a review of early experiments in mediated electron transfer (4.3), we will review the exogenous (4.4) and the endogenous redox shuttles (4.4) found in EAB biofilms, according to their chemical properties. We will then provide insight about the best experimental practices which enable us to identify soluble redox shuttles in BES (4.6). After a discussion of the energetic implications of the mediated extracellular electron transfer strategy (4.7), we will briefly mention the relevance of soluble redox shuttles to innovative BES, such as high current density MFC, biocatalysts and biosensors (4.8).

4.2 REDOX SHUTTLES

In a typical inorganic electrochemical reaction, the half-cell reactants react reversibly and rapidly with the electrode, so that the electrode potential reflects the activity of the electron in equilibrium with the couple (Bard and Faulkner 2001). When dealing with biological molecules (e.g. redox enzymes), either in solution or adsorbed to the electrode, this assumption is not realistic, because the redox centres are often buried in the enzyme. In most cases, the distance between redox centres and the electrode results in a high overpotential which drastically reduces the electron transfer rate at the electrode (Armstrong 2002). The use of soluble redox shuttles (Lojou *et al.* 2005) enables us to reduce such overpotential and permit: (a) determination of redox processes in enzymes that cannot interface directly with the electrode; (b) accelerate electron transfer rate at the electrode. In short, a redox mediator catalyzes the equilibration between the electrode and the enzyme redox center(s). A typical redox mediator is a low molecular weight water-soluble redox couple of relatively high reactivity. The high reactivity translates into a low overpotential both for receiving electrons from the electron donor and for donating electrons to the final electron acceptor. An ideal redox mediator enables us to harvest most of the potential energy available in a given electron transfer process. By choosing or synthesizing the appropriate mediator, it is possible to transfer electrons between most redox-active biomolecules and the electrode.

This simple concept has been largely applied to unveil redox processes in enzymes which are not able to directly interface with the electrode. Furthermore, mediated electron transfer between redox proteins and the electrode is the working principle of enzyme-based biofuel cells, which have been successfully used to power small electronic devices (Sakai *et al.* 2009). We refer the reader to Chapter 3 for further details on this topic.

A similar conceptual framework can be applied to mediated electron transfer between viable bacteria and extracellular substrates, such as metal surfaces and electrodes. Extracellular electron transfer can occur via small molecules that act either to chelate insoluble metals and deliver them to intracellular metal oxidoreductases or by themselves, acting as electron shuttles between bacteria and extracellular substrates (Gralnick and Newman 2007; Kappler and Straub 2005). From the standpoint of BES, we are particularly interested in electron transfer between the bacteria and the electrodes. Depending on the electrode potential, the electrode can be either the terminal electron acceptor or the primary electron donor. In the first case (electrode as terminal electron acceptor), several compounds, exogenous or endogenous have been identified as electron shuttles. It has been observed that some of these compounds deliver electrons to the same membrane-associated compounds which are relevant to direct electron transfer and this blurs the distinction between direct and mediated pathways (Lies *et al.* 2005). Little research has been published about mediated electron transfer between bacteria and electrodes when the electrode is the primary electron donor (Aulenta *et al.* 2007).

Ideally, without side reactions, a redox mediator can perform many cycles and transfer many electrons between a viable cell and the electrode in its lifetime. As a consequence, small concentrations (micromolar or submicromolar) of such agents have a considerable effect on the turnover of the terminal oxidant in a given environment (Hernandez *et al.* 2004; Marsili *et al.* 2008a). Furthermore, redox mediators lower the overpotential with respect to the electrode (Rabaey *et al.* 2004) and the electrochemical losses.

In MFCs aimed at electricity production, the costs and the environmental burdens associated with the addition of organic exogenous redox mediators are very high (Chaudhuri and Lovley 2003; Du *et al.* 2007). For this reason, recent research has focused on the identification of environmental and endogenous redox shuttles. These mediators are of the utmost importance in long shelf-life BES such as MFCs which need to be deployed in remote environments (Donovan *et al.* 2008).

4.3 EARLY EXPERIMENTS

Since the early years of bioelectrochemistry, exogenous redox mediators have been used to investigate redox protein behaviour (Frew and Hill 1988). The first generation of MFC 'rediscovered' in the early '80s was based on fermentative bacteria (non-electroactive) and employed exogenous redox mediators such as thionine and resorufin (Bennetto *et al.* 1983) and ferric chelate (Tanaka *et al.* 1983). Almost a decade ago, a number of reports described anodic pure strain (Chaudhuri and Lovley 2003; Kim *et al.* 2002; Park *et al.* 2001) and mixed consortia (Gil *et al.* 2002; Rabaey *et al.* 2003) capable of coupling organic substrate oxidation with mediator-less electron transfer to electrodes. At this time, the mechanism of electron transfer in these MFCs was not well understood and research focused on MFC application as sensors or for renewable electricity production. It is important to point out that MFCs are somewhat cumbersome devices, and they do not permit us to maintain consistent environmental conditions throughout the duration of an experiment. For example, it may require several days before the anodic community grows and the anode potential becomes constant. Furthermore, oxygen diffusion from the cathode compartment through the cationic membrane may alter the growth pattern and metabolism of microorganisms at the anode. It is now understood that environmental conditions affect endogenous redox mediators production (Bond and Lovley 2005). The use of potentiostat-controlled electrochemical cells enables *in situ* electrochemical analysis of microorganisms and their secondary metabolites, among which are endogenous redox mediators (see Section 4.7).

In the following years, the development of miniature MFCs (Iafelice *et al.* 2004; Ringeisen *et al.* 2006) and fast and precise potentiostat-based methods (Fricke *et al.* 2008; Marsili *et al.* 2008b), enabled more precise controlling of growth conditions and permitted multiple experiments in a short time.

4.4 EXOGENOUS REDOX MEDIATORS

4.4.1 Artificial mediators

Artificial mediators are easy to dose and their redox potential may be adjusted over a wide range, by careful design of the molecule. In turn, this enables increasing potential difference between anode and cathode and, consequently, power production. Currently, artificial mediators are employed only in specific cases, for example when attempting to degrade a complex substrate with non-electroactive microorganisms. Several mediators, such as resazurin, enable conversion of cellulose into electrical energy at the anode of an MFC inoculated with the fermentative strain *Clostridium cellulolyticum*. The artificial mediator reduces the electron transfer overpotential and rescues electrons across the bacterial membrane (Sund *et al.* 2007). Addition of the humic acid analogue anthraquinone 2-6-disulfonate (AQDS) enables sustained generation of electricity in gram-positive microorganisms (Milliken and May 2007). Artificial redox mediators are also needed in yeast MFC, since the electron transfer chain is located at the mitochondria within the cytoplasm (Schaetzle *et al.* 2008) and common yeasts such as *Saccharomices cerevisiae* are not known to produce endogenous redox mediators (Gunawardena *et al.* 2008). The use of redox mediators in cathodic reaction (transfer of electrons from the electrode) has been investigated only recently, since it is more difficult maintaining viable cells under these conditions. For example, it was shown that an electrochemical cell with a solid electrode polarized at −500 mV vs. Standard Hydrogen Electrode (SHE), in combination with the artificial redox mediator methyl viologen, can efficiently transfer electrons to microorganisms which respire chlorinated solvents (Aulenta *et al.* 2007).

Besides their application in BES, artificial mediators with a different degree of hydrophobicity/lipophilicity may be used to investigate the electron transfer chain at different subcellular locations in eukaryotic cells. For example, a combination of hydrophilic and lipophilic redox mediators enabled us to observe selectively redox processes that happen inside and outside the bacterial membrane of human breast epithelial cells (Liu *et al.* 2000).

Devices such as those described here need continuous addition of artificial redox mediators. In order to avoid dispersion of mediators in the environment (and the associated costs), artificial redox mediators can be immobilized on the electrode, which obviously minimizes release in the environment (Adachi *et al.* 2008; Lowy *et al.* 2006). However, such a solution is not optimal for thick biofilms and processes with a high concentration of suspended biomass, since the immobilized mediators will be able to shuttle electrons (either in cathodic or anodic reactions) only to the first layer(s) of the biofilm.

As with any other redox mediator, the main advantage of artificial redox mediators is their non-specificity. A single mediator can be used by several bacterial species in a mixed consortium to transfer electrons to the electrode, providing that the actual potential of the redox mediator couple is between the oxidation potential of the organic substrate and the electrode potential. This lack of specificity enables us to extract electrons from nearly every gram-negative bacterium and makes artificial redox mediators the best choice in laboratory BES, where the goal is to extract information from the *in vivo* redox process. Artificial mediators permit us also to study gram-positive bacteria and other organisms with thick membranes or cell walls, which are not suitable for direct electron transfer to the electrode. Artificial mediators still hold great promise for preparing novel electrode material, which can enhance the electron transfer rate at the electrode by orders of magnitude (Adachi *et al.* 2008).

4.4.2 Natural redox mediators in the subsurface environment

BES deployed in subsurface environments, such as sediment MFC and benthic unattended generators (BUG) (Tender *et al.* 2002) are usually operating at low concentration of organic carbon and are surrounded by complex organic matter, humic acids (HA), and a host of S-containing compounds. All these species may play a role in extracellular electron transport between bacteria and the electrode.

HAs are heterogeneous, soluble, high-molecular-weight organic compounds. They derive from enzymatic depolymerization and oxidation of plant biopolymers (Wershaw 1999) and are ubiquitous in terrestrial and aquatic environments. HAs are resistant to microbial degradation and therefore not generally involved in microbial metabolism, especially in anoxic habitats (McKnight and Aiken 1998). Dissimilatory metal reducing bacteria can use HAs as extracellular electron acceptors (Lovley *et al.* 1996) and also as redox mediators in the extracellular respiration of Fe(III) (Lovley *et al.* 1998). In fact, HAs contain quinone structures, which may be reduced to a corresponding hydroquinone. Hydroquinone can then transfer electrons to the electrode, and also to other bacteria (Stams *et al.* 2006; Fredrickson *et al.* 2000). Since unfractionated HAs can serve both as electron acceptors and electrode mediators, it is still debated whether they contribute to Fe(III) reduction in low organic carbon contents environments such as subsurface sediments (O'Loughlin 2008). A thorough microbial ecology investigation of BES deployed in marine and freshwater environments has shown that humic acids and organic matter are the substrate for consortia of microorganisms, which degrade these complex materials into simple organic substrates. These substrates, in turn, are oxidized by *Geobacteraceae* (in freshwater sediments) and *Desulfuromonas* sp. (in marine sediments), which transfer electrons to the anodes through membrane-associated compounds (Lovley 2006).

Cysteine is an essential amino acid commonly encountered in soils, and a widely used reducing agent in growth media. Cysteine has been shown to mediate interspecies electron transfer in cocultures of *Geobacter sulfurreducens* and *Wolinella succinogenes* (Kaden *et al.* 2002). Furthermore, addition of cysteine to a pure *G. sulfurreducens* culture increased 8 to 11 times the reduction rate of extracellular, poorly crystalline Fe(III). In contrast to common complexing agents (e.g. NTA, EDTA), which form stable Fe(II) complexes, cysteine increases microbial reduction of iron by acting as an electron shuttle (Doong and Schink 2002).

The S^0/H_2S couple with a standard redox potential of –220 mV vs. SHE represents a suitable alternative to cystine/cysteine as redox mediator. Straub and Schink have shown that the sulfur-reducing bacterium *Sulfurospirillum deleyianum* coupled formate oxidation to CO_2 with Fe(III) reduction to Fe(II) in the presence of small thiosulfate concentration. The oxidized sulfur species resulting from electron transfer to Fe(III) have not been identified, but they are likely to be sulfur or thiosulfate (Straub and Schink 2002). The role of the redox couple S^0/S^{2-} (sulphur/sulphide) as a redox mediator system has been assessed in the reductive dechlorination of trichloroacetic acid (TCA) to dichloroacetic acid (DCA) (De Wever *et al.* 2000), but no evidence has been found in BES. In addition, other redox couples such as CO_2/formate, $2H^+/H_2$, NO_3^-/NO_2^-, MnO_2/Mn^{2+} and Fe^{3+}/Fe^{2+} are also found in anaerobic respiration (Schroder 2007).

4.5 ENDOGENOUS REDOX MEDIATORS

Instead of (or in addition to) membrane-associated redox compounds, many electroactive bacteria can reduce electrodes through endogenous soluble redox mediators, which transfer electrons back and forth across bacterial membrane between cells and the electrode surface (Gralnick and Newman 2007).

The discovery rate of new endogenous redox mediators in electroactive microorganisms is very fast. This is due to a renewed research interest in BES in general and MFC in particular, but also to the establishment of microbial culture methods and electrochemical characterization procedures which enable reproducible and rapid experiments with electroactive bacteria (Schroder 2007; Marsili *et al.* 2008b). Until recently, the production and secretion of soluble redox mediator were determined through insoluble metal reduction assays. The equivalence between electrode reduction and insoluble metal oxide reduction has not been fully demonstrated. In a recent study it was demonstrated that *Pelobacter carbinolicus*, although able to reduce insoluble metal oxide, is not able to reduce electrodes (Richter *et al.* 2007). *Pelobacter* reduces metals through direct electron transfer. In this case, the interaction between the membrane-bound electron transfer agent and the metal or the electrode may be affected by the surface characteristic of the solid oxidant. In case of mediated electron transfer, it is more probable that the soluble redox mediator does not 'see' the difference between a solid metal oxide and an electrode.

In numerous studies, the secretion and the accumulation of a soluble redox mediator is the only explanation of results. However, only a few endogenous redox mediators have been identified with certainty. In fact, identification of redox mediators is a very complex task. They are usually found in low concentration and in very complex biological mixtures such as those characteristic of mixed microbial consortia or domestic and animal wastewater. Plus, even if the accumulation of a redox active species is observed, a great deal of work is necessary to prove that the redox species under investigation is produced in order to increase electron transfer rate at the electrode. Finally, electrochemistry alone can only provide hints about the chemical skeleton of the putative redox mediator. Complete and undisputed identification of a redox mediator requires use of spectrophotometry, chromatography and other separation/analysis techniques. In the following section, we will group the endogenous redox shuttles by their chemical characteristics. For those redox shuttles whose identification has not yet been performed, we will discuss the preliminary results.

Since we define the mediators as species which are not consumed during the electron transfer process (Rabaey and Verstraete 2005b; Schröder 2007), we will not include in the following discussion the metabolic intermediates (i.e. formate) which facilitate electron transfer in microbial consortia (see for example de Bok *et al.* 2004).

4.5.1 Known microbially produced redox mediators

4.5.1.1 Phenazines

Although not the first soluble redox mediators to be identified, these redox-active species, produced by the γ-proteobacterium *Pseudomonas* sp., have been extensively investigated, mainly because of their anti-fungal properties. Besides this function, natural antibiotics from soil microorganisms may help in intercellular signalling and redox homeostasis in order to ensure the persistence of microorganisms in the environment under high cell density and low nutrient conditions (Price-Whelan *et al.* 2006). The root isolate *Pseudomonas chlororaphis* PCL1391 was able to reduce poorly crystalline iron and manganese oxides, by producing phenazine-1-carboxamide (PCN). Each PCN molecule underwent numerous electron transfer cycles in its lifetime (Hernandez *et al.* 2004). PCN-mutants of *P. chlororaphis* lost their iron-reduction capability, which was restored by PCN addition. Multiple phenazines produced by *P. aeruginosa*, among which pyocyanin was characterized, promote electrode reduction and electrical power production in a two-chamber MFC inoculated with a mixed consortium from an anaerobic, methanogenic sludge (Rabaey *et al.* 2004). In following research, both PCN and pyocyanin enable extracellular electron transfer to the electrode of MFC inoculated with *Pseudomonas aeruginosa* KRP1 (Rabaey *et al.* 2005). Interestingly, pyocyanin was used as a soluble redox shuttle by several other bacteria in a mixed consortium MFC (Figure 4.1.). The non-specificity of redox mediators is extremely interesting from a fundamental and applied standpoint. As for the first, a possible explanation is that phenazines (and maybe other redox mediators) rescue electrons from the electron transfer chain at a similar subcellular location for each species. From an applied standpoint, it is worth noting that endogenous redox mediators enable non-mediator-producing strains, and even Gram-positive bacteria (Pham *et al.* 2008) to transfer electron extracellularly.

4.5.1.2 Flavins

Given the extensive knowledge on *Shewanella* sp. physiology and genetics and its relevance to biogeochemistry, this organism has been long investigated to characterize its extracellular electron transfer mode. First, it was proposed that unidentified excreted quinones mediate extracellular electron transfer (Newman and Kolter 2000). However, only recently, flavins have been identified as the main endogenous redox shuttles in *Shewanella* sp. First, Von Canstein *et al.* showed that multiple *Shewanella* strains use FMN and riboflavin to mediate reduction of insoluble Fe(III) oxides (Von Canstein *et al.* 2008). Marsili *et al.*

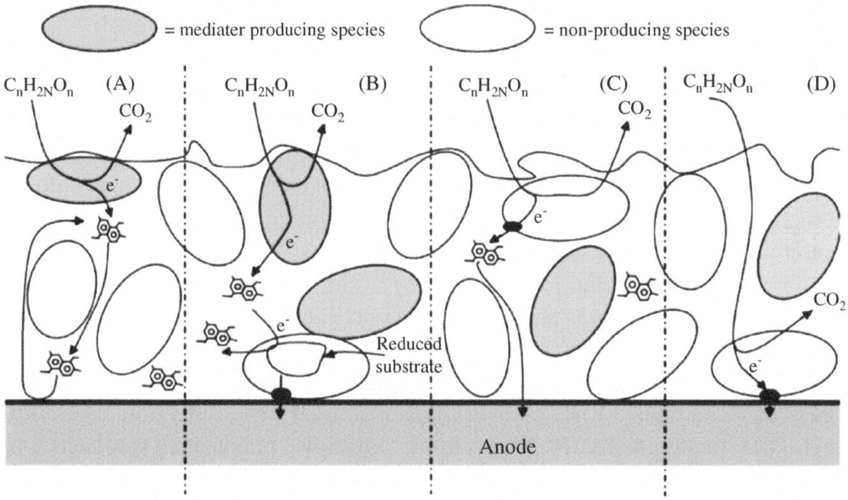

Figure 4.1. Possible working mechanism of self-mediated electron transfer in a mixed species biofilm. Grey bacteria are capable of producing soluble mediators; white bacteria are not. Black membrane circles (B) represent membrane-bound compounds capable of transferring electrons from the bacteria to the extracellular environment. (A) Redox mediator producing bacteria use mediators to reach the anode. (B) A bacterium uses a reduced redox mediator as electron donor (hypothesized) and the anode as electron acceptor. (C) A bacterium uses a redox mediator produced by another bacterium to reach the anode. (D) A bacterium transfers electrons to the anode through membrane bound mediators (reproduced from Rabaey *et al.* 2005)

determined clearly how batch cultures of *Shewanella oneidensis* MR-1 and MR-4 accumulate flavins at concentrations of 250–500 nM and use them as a soluble redox mediators for extracellular electron transfer to electrodes maintained at oxidative potential (Marsili *et al.* 2008a).

4.5.1.3 Quinones

The importance of the microbially excreted quinone–hydroquinone redox couples in the extracellular reduction of naphthalenesulfonates by *Pseudomonas* sp. BN6 was discussed by Nortemann *et al.* (1994). Also the role of 2-amino-3-carboxy-1,4-naphthoquinone (ACNQ) in the extracellular reduction was investigated (Hernandez and Newman 2001). Other naphthoquinones were isolated from *Sphingomonas xenophaga* BN6 during anaerobic degradation of naphthalene-2-sulfonate and identified by HPLC/MS as 4-amino-1,2-naphthoquinone and

4-ethanolamino-1,2-naphthoquinone (Keck *et al.* 2002). In another study, the acidobacterium *Geothrix fermentans* was shown to produce a mediator which increases extracellular iron reduction. Thin layer chromatography results suggested that the mediator was a water-soluble quinone (Nevin and Lovley 2002). In follow-up research (Bond and Lovley 2005), the putative mediators increased the electron transfer rate to solid electrodes, as revealed by the higher operative potential of a conventional MFC. The mediator has not yet been identified. This was the first case of an endogenous mediator produced in bacteria outside Proteobacteria.

4.5.1.4 Cytochromes and soluble enzymes

Although few authors suggest that soluble enzymes and extracellular cytochromes produced by *Geobacter sulfurreducens* may serve as redox mediators in Fe(III) reduction and interspecies electron transfer, this result is debatable, given the high energetic cost of this strategy (Luu and Ramsay 2003).

4.5.1.5 Melanin

Shewanella algae BrY can couple H_2 oxidation to Fe(III) reduction. *S. algae* BrY produces extracellular melanin, which serves as redox mediator in Fe(III) reduction (Turick *et al.* 2002). While the evidence for the production of extracellular melanin is convincing, it is questionable to classify melanin as a redox mediator, because of the highly complex chemical structure. In a following study (Turick *et al.* 2003) it was observed that microbially produced melanin associated with the bacterial membrane. Removal of membrane-associated melanin reduced 10-fold the Fe(III) reduction rate, while the addition of free melanine restored only 20% of the Fe(III) reduction capability.

4.5.1.6 Other mediators

In an old study, an MFC inoculated with a mixed culture of *Lactobacillus plantarum*, *Streptococcus lactis,* and *Erwinia dissolvens* was found to produce ferric chelate compounds which mediated electron transfer to the anode (Vega and Fernandez 1987). In a very recent paper, Wang *et al.* found that the actinobacterium *Propionibacterium freudenreichii* produced a redox mediator, 1,4-dihydroxy-2-naphthoic acid, only in the presence of an electrode as electron acceptor at an appropriate potential. Furthermore, when the electrode was maintained at an oxidative potential (0.4 vs. SHE) and glucose fermentation was performed, the molar ratio of the final product shifted from 2:3 to 1:1 acetate to propionate (Wang *et al.* 2008).

Even *Escherichia coli*, which were until then considered not-electroactive, excrete a redox mediator (Wang *et al.* 2007). *E. coli* produced such mediator when grown under anaerobic conditions, in a continuous, two compartment MFC fed with glucose. The presence of a redox mediator was confirmed through cyclic voltammetry. Further attempts were made to identify the putative redox mediator and it was shown through GC-MS that the mediators were three polar species: ethyl 2-aminoacetimidate, methyl-4-ethoxybutanoate, and 3-isobutyl-hexahydropyrrolo [1,2-a]pyrazine-1,4-dione (Zhang *et al.* 2008).

4.5.2 Unidentified endogenous mediators

As researchers focus their attention on soluble redox mediators, they have been found outside the known 'electroactive bacteria', such as Proteobacteria. In a recent study, an enriched phototrophic consortium was found to produce one or more redox mediators. Indole was determined through fluorescence spectroscopy as the most abundant electron transfer group (Cao *et al.* 2008). Yet another unidentified redox mediator that promoted extracellular iron reduction was discovered in the hyperthermophilic archaeon *Pyrobaculum aerophilum* (Feinberg and Holden 2006).

4.6 METHODS FOR IDENTIFICATION OF SOLUBLE REDOX SHUTTLES

The identification and quantification of soluble redox shuttles in anodic electroactive bacteria is a challenging task, because of (a) the low concentration of redox shuttles in growing cultures, (b) the chemical complexity of the medium, especially when wastewater-grown mixed consortia are investigated and (c) the existence of multiple pathways for extracellular electron transfer within the mixed consortium or even in the same strain which is under investigation. In fact, several bacteria, such as *Geobacter* and *Shewanella* may use more than one 'link' to interface with the electrode and it is possible that certain growth conditions trigger metabolic pathways not yet well understood.

Because of the above described problems, only a few redox shuttles have been identified with certainty. In most other cases, an unidentified redox shuttle is the only explanation of the experimental observations. Identification of endogenous redox shuttles in extracellular electron transfer requires establishment of a reproducible methodology, ideally valid for pure species and mixed consortia. Following published results, we mention here key procedures which may be applied where the presence of a soluble redox mediator is hypothesized.

4.6.1 Potentiostat-controlled electrochemical cells

Electrochemical cells permit control of working electrode potential and show higher reproducibility and accuracy compared to MFC (see also Chapter 8).

4.6.2 Environmental conditions

Working electrodes should be set at oxidative values and higher than the limiting potential that is expected for the electron transfer process (e.g. 0.2–0.3 V vs. SHE). Higher potentials may result in bacterial membrane stress and even irreversible oxidation of cell components.

4.6.3 Batch experiments

Endogenous redox mediator concentrations are usually in the micromolar to submicromolar concentration range. In order to accumulate sufficient redox mediator for the detection, long batch experiments (up to 7–14 days) should be preferred (Bond and Lovley 2005; Marsili *et al.* 2008b).

4.6.4 Media formulation

Numerous growth media used for cultivation of environmental microorganisms contain electroactive ingredients, such as resazurin (Caccavo *et al.* 1995), vitamins, reducing agents, and high concentrations of metals, which may act as redox mediators. Elimination (if possible) of these components will reduce the occurrence of false positives in redox mediator investigation. Also, minimal media are preferred over concentrated media since they will provide a lower base signal in electrochemical analysis.

4.6.5 Electrochemical methods

Voltammetry, particularly pulse voltammetry with time enables us to determine accumulation of endogenous redox shuttles, either adsorbed at the electrode or in the solution. Scan rate analysis enables calculation of thermodynamic and kinetic parameters of the dominant electron transfer process. Electrochemical Impedance Spectroscopy permits us also to follow the biofilm growth and its impedance in the presence and in absence of the endogenous redox mediator.

4.6.6 Medium change

This experiment can be easily performed in potentiostat-controlled electrochemical cells poised at oxidative potential. Once the extracellular electron transfer rate (as determined by chronoamperometry) has reached a stationary value and the electroactive biofilm (as determined for example by electrical impedance spectroscopy) has reached its maximum thickness, the following procedure permits us to assess whether extracellular electron transfer is due to soluble redox mediator(s) or to membrane-associated compounds. The growth medium in the electrochemical cell is removed and substituted with fresh medium. In the presence of an endogenous soluble redox shuttle, this will result in a substantial drop in electron transfer rate. Upon reinsertion of the sterile-filtered (cell-free) medium, the anodic (oxidation) current should be almost completely recovered.

4.6.7 Chemical structure of the mediator

Electrochemistry gives only a hint about the chemical skeleton of the mediator. Addition of the putative redox mediator should result in increased extracellular electron transfer at the same potential. However, the definitive identification needs to be performed by spectroscopic methods and/or by HPLC/MS.

From a microbiological standpoint, the final confirmation of the importance of a certain molecule as mediator is the creation of a mutant and its complementation, following the fundamental molecular microbiology principles.

4.7 RELEVANCE OF SOLUBLE REDOX MEDIATORS SHUTTLE TO MICROBIAL METABOLISM

As described earlier, soluble redox shuttle compounds are non-specific to the strain that produces them and, within a bacterial consortium, they can mediate extracellular electron transfer between other bacterial strains and the insoluble electron acceptor (oxidized metal or electrode). Therefore, the presence of endogenous soluble redox mediator is a selective advantage for all the microbial community. It is also interesting to note that many bacteria, conventionally included among the 'non-electroactive bacteria', are able to produce and excrete redox shuttle compounds at certain growth phases (e.g. stationary) and under challenging conditions (lack of soluble electron acceptors). These observations suggest that the metal/electrode reducing pathway based on soluble redox shuttle compounds is a very general energy-conserving strategy, and it is likely that many other microorganisms have devised analogous electron transfer chains. From an energetic standpoint, endogenous soluble shuttling compound mediators seem to be an expensive strategy for electron disposal. In order to interface with the oxidized metal particles or with the electrode, the cells need to synthesize or overproduce a molecule which will be washed out after a short time. Therefore, one expects that most bacteria use membrane-bound electron transfer agents. Instead, such a strategy for electron disposal has been found mostly in proteobacteria. On the other side, numerous endogenous redox shuttling compounds have been recently discovered and they are spread across different bacterial groups. This observation brought us to reconsider early assumptions about the bioenergetics of redox shuttling. Careful calculations have shown that the energy required to produce a redox shuttle compound can be a small part of the total energetic balance (Marsili *et al.* 2008a).

It is also worth noting that electroactive bacteria lacking soluble electron acceptors overproduce redox active species with the right electrochemical potential (i.e. between the organic substrate and the oxidized metal/electrode potentials) already available in the cells (Hernandez *et al.* 2004; Marsili *et al.* 2008a). Also, this observation suggests that extracellular electron transfer through soluble redox shuttle is not as specific an energy-conserving strategy, as the adaptation of membrane-bound redox shuttles.

Electroactive bacteria deposit layers of redox shuttles at the electrode. While this hinders the electrode surface and decreases extracellular electron transfer rate, it represents an efficient way of reducing washing out of the redox mediator. Furthermore, soluble redox mediators are not electroneutral and their diffusion through the medium and the biofilm water channels in the vicinity of the electrode maintained at oxidative potential is affected by their charge (Rabaey *et al.* 2007). For this reason, washing out will be slower than for other neutral species.

Production of soluble redox shuttles enables distant cells to interface with the metal/electrode. This increases the concentration of active biomass and in turn, permits higher bacterial growth than in the case of direct electron transfer. This characteristic is a competitive advantage when organic substrate is not limiting. In Fe(III)-rich oligotrophic environments such as subsurface sediments, it is more likely that direct electron transfer is the common energy-conserving strategy. Under such conditions, bacteria will form thin and sparse biofilms (Cordas *et al.* 2008). In the presence of abundant substrate, we are more likely to have thick biofilm and planktonic biomass with species which produce soluble redox shuttles.

In mixed consortia, soluble redox shuttles enable interspecies electron transfer, which is never observed when direct electron transfer species are selected through use of minimal media.

In view of the above considerations, it appears that production of endogenous redox mediators is a suitable strategy for the development of the electroactive community. We believe that further research will show even more the extent to which mediated electron transfer is adopted in bacteria. Particularly, the development of high-throughput devices (Iafelice *et al.* 2004) will enable characterization of still unexplored environments and will bring the discovery of novel soluble redox shuttling compounds.

4.8 SOLUBLE REDOX SHUTTLES IN BIOELECTROCHEMICAL DEVICES

4.8.1 Microbial fuel cells

MFCs have been thoroughly described in other chapters of this book. Here, we will briefly mention how soluble redox mediators affect electricity production in MFCs. MFC anodes inoculated with bacteria which produce soluble redox shuttling compounds do not need artificial redox mediators and have a lower environmental impact. This is crucial to application of MFCs in powering of environmental sensors and permits design of low power, self-sustained facilities, such as those employed in space and military missions. Since bacterial cells do not need to be in direct contact with the electrode, the electroactive biomass concentration is higher compared to MFCs based on direct electron transfer. Furthermore, the presence of one or more redox shuttle-producing strain permits other non-electroactive strains to dispose electrons in the electrode, by so increasing the power output. In order to obtain the maximum power output, a continuous MFC based on mediated electron transfer needs to be fed with concentrated substrates (e.g. animal waste) and operated at long residence time, so that the endogenous redox mediator can accumulate in the anodic compartment until it becomes non-limiting. Thanks to the non-specificity of the redox shuttle, such MFC can potentially convert into electrical energy a wider variety of substrates compared to MFCs based on direct electron transfer.

4.8.1.1 Biosensors

Among others, enzymes, antibodies, nucleic acids, tissues, viable microorganisms and polysaccharides have been tested as sensing elements for sensors based on mediated electron transfer (Mohanty and Kougianos 2006). Amperometric biosensors based on redox enzymes (e.g. catalase, glucose oxidase) have been widely researched and applied in recent decades. The high concentration of electroactive biomass in devices based on endogenous soluble redox shuttles would increase sensitivity and enable fast detection of low concentration of substrates and metabolic inhibitors (e.g. xenobiotics, pollutants, pesticides) for both on-line and off-line biosensors. Although selectivity and reproducibility problems need to be addressed, this technology holds great promise for the rapid detection of pollutants in wastewater (see Chapter 17) and monitoring of bioremediation processes (Tront *et al.* 2008).

4.8.1.2 Electrodes modified with redox mediators

Bacterial strains which produce soluble redox mediators can be used to coat high-surface electrodes with mono- or multiple layers of redox mediator. Such functionalized, low-cost electrodes may find applications as redox biocatalysts for laboratory and commercial use, or may simply enable investigation of transmembrane processes in non-electroactive bacteria. Since redox mediators adsorb to the electrode with weak interaction and can be easily desorbed under normal experimental conditions, further research is needed to devise stable electrode modification protocols, for example through intermediate self-assembled monolayers.

REFERENCES

Adachi, M., Shimomura, T., Komatsu, M., Yakuwa, H. and Miya, A. (2008) A novel mediator–polymer-modified anode for microbial fuel cells. *Chem. Commun.* 2055–2057.

Armstrong, F.A. (2002) Insights from protein film voltammetry into mechanisms of complex biological electron-transfer reactions. *J. Chem. Soc., Dalton Trans.* 661–671.

Aulenta, F., Catervi, A., Majone, M., Panero, S., Reale, P. and Rossetti, S. (2007) Electron transfer from a solid-state electrode assisted by methyl viologen sustains efficient microbial reductive dechlorination of TCE. *Environ. Sci. Technol.* **41**(7), 2554–2559.

Bennetto, H.P., Stirling, J.L., Tanaka, K. and Vega, C.A. (1983) Anodic reactions in microbial fuel cells. *Biotech. Bioeng.* **25**(2), 559–568.

Bond, D.R. and Lovley, D.R. (2005) Evidence for involvement of an electron shuttle in electricity generation by *Geothrix fermentans. Appl. Environ. Microbiol.* **71**, 2186–2189.

Cao, X., Huang, X., Boon, N., Liang, P. and Fan, M.Z. (2008) Electricity generation by an enriched phototrophic consortium in a microbial fuel cell. *Electrochem. Comm.* **10**, 1392–1396.

Chaudhuri, S.K. and Lovley, D.R. (2003) Electricity generation by direct oxidation of glucose in mediatorless microbial fuel cells. *Nature Biotech.* **21**(10), 1229–1232.

Cordas, C.M., Tiago Guerra, L., Xavier, C. and Moura, J.J.G. (2008) Electroactive biofilms of sulphate reducing bacteria. *Electrochim. Acta* **54**, 29–34.

De Bok, F.A.M., Plugge, C.M. and Stams, A.J.M. (2004) Interspecies electron transfer in methanogenic propionate degrading consortia. *Wat. Res.* **38**, 1368–1375.

De Wever, H., Cole, J.R., Fettig, M.R., Hogan, D.A., and Tiedje, J.M. (2000) Reductive dechlorination of trichloroacetic acid by *Trichlorobacter thiogenes* gen. nov., sp. nov. *Appl. Environ. Microbiol.* **66**, 2297–2301.

Donovan, C., Dewan, A., Heo, D. and Beyenal, H (2008) Batteryless, wireless sensor powered by a sediment microbial fuel cell. *Environ. Sci. Technol.* **42**(22), 8591–8596.

Doong, R.A. and Schink, B. (2002) Cysteine-mediated reductive dissolution of poorly crystalline iron(III) oxides by Geobacter sulfurreducens. *Environ. Sci. Technol.* **36**, 2939–2945.

Du, Z., Li, H. and Gu, T. (2007) A state-of-the-art review on microbial fuel cells: A promising technology for wastewater treatment and bioenergy. *Biotech. Adv.* **25**, 464–482.

Feinberg, L.F. and Holden, J.F. (2006) Characterization of dissimilatory Fe(III) versus NO3-reduction in the hyperthermophilic archaeon *Pyrobaculum aerophilum. J. Bacteriol.* **188**, 525–531.

Fredrickson, J.K., Zachara, J.M., Kennedy, D.W., Duff, M.C., Gorby, Y.A., Li, S.M.W. and Krupka, K.M. (2000) Reduction of U(VI) in goethite (a-FeOOH) suspensions by a dissimilatory metal-reducing bacterium. *Geochim. Cosmochim. Acta* **64**(18), 3085–3098.

Frew, J.E. and Hill, H.A.O (1988) Direct and indirect electron transfer between electrodes and redox proteins. *Eur. J. Biochem.* **172**, 261–269.

Fricke, K., Harnisch, F. and Schröder, U. (2008) On the use of cyclic voltammetry for the study of anodic electron transfer in microbial fuel cells. *Energy Environ. Sci.* **1**, 144–147.

Gil, G.C., Chang, I.S., Kim, B.H, Kim, M., Jang, J.K., Park, H.S. and Joo Kim, H.J. (2003) Operational parameters affecting the performance of a mediator-less microbial fuel cell. *Biosens. Bioelec.* **18**, 327–334.

Gralnick, J.A. and Newman, D.K. (2007) Extracellular respiration. *Mol. Microb.* **65**(1), 1–11.

Gunawardena, A.G., Sandun, F. and Filip, T. (2008) Performance of a yeast-mediated biological fuel cell. *Int. J. Mol. Sci.* **9**, 1893–1907.

Hernandez, M.E. and Newman, D.K. (2001) Extracellular electron transfer. *Cell. Mol. Life Sci.* **58**, 1562–1571.

Hernandez, M.E., Kappler, A. and Newman, D.K. (2004) Phenazines and other redox-active antibiotics promote microbial mineral reduction. *Appl. Environ. Microbiol.* **70**, 921–928.

Iafelice, B., Ferrarini, V. and Guerrieri, R. (2004) A parallel electronic sensor for phenotypic screening in energy production. *IEEE Proceedings* **1**, 68–71.

Kaden, J., Galushko, A.S. and Schink, B. (2002) Cysteine-mediated electron transfer in syntrophic acetate oxidation by cocultures of *Geobacter sulfurreducens* and Wolinella succinogenes. *Arch Microbiol.* **178**, 53–58.

Kappler, A. and Straub, K.L. (2005) Geomicrobiological cycling of iron. *Rev. Mineral. Geochem.* **59**, 85–108.

Keck, A., Rau, J., Reemtsma, T., Mattes, R., Stolz, A. and Klein, J. (2002) Identification of quinoide redox mediators that are formed during the degradation of naphthalene-2-sulfonate by *Sphingomonas xenophaga* BN6. *Appl. Environ. Microbiol* **68**(9), 4341–4349.

Kim, H.J., Park, H.S., Hyun, M.S., Chang, I.S., Kim, M. and Kim, B.H. (2002) A mediator-less microbial fuel cell using a metal reducing bacterium, *Shewanella putrefaciens*. *Enz. Microb. Tech.* **30**, 145–152.

Lies, D.P., Hernandez, M.E., Kappler, A., Mielke, R.E., Gralnick, J.A. and Newman, D.K. (2005) *Shewanella oneidensis* MR-1 uses overlapping pathways for iron reduction at a distance and by direct contact under conditions relevant for biofilms. *Appl. Environ. Microbiol.* **71**, 4414–4426.

Liu, B., Rotenberg, S.A. and Mirkin, M.V. (2000) Scanning electrochemical microscopy of living cells: Different redox activities of nonmetastatic and metastatic human breast cells. *Proc. Natl. Acad. Sci.* **97**(18), 9855–9860.

Lojou, E., Giudici-Orticoni, M.T. and Bianco, P. (2005) Direct electrochemistry and enzymatic activity of bacterial polyhemic cytochrome c3 incorporated in clay films. *J. Electroanal. Chem.* **579**, 199–213.

Lovley, D.R. (2006) Microbial fuel cells: novel microbial physiologies and engineering approaches. *Curr. Opin. Biotech.* **17**, 327–332.

Lovley, D.R., Coates, J.D., Blunt-Harris, E.L., Phillip, E.J.P. and Woodward, J.C. (1996) Humic substances as electron acceptors for microbial respiration. *Nature* **442**, 445–448.

Lovley, D.R., Fraga, J.L., Blunt-Harris, E.L., Hayes, L.A., Phillips, E.J.P. and Coates, J.D. (1998) Humic substances as a mediator for microbially catalyzed metal reduction. *Acta Hydrochim. Hydrobiol.* **26**(3), 152–157.

Lowy, D.L., Tender, L.M., Zeikus, J.G., Park, D.H. and Lovley, D.R. (2006) Harvesting energy from the marine sediment–water interface II: Kinetic activity of anode materials. *Biosens. Bioelec.* **21**, 2058–2063.

Luu, Y.S. and Ramsay, J.A. (2003) Review: microbial mechanisms of accessing insoluble Fe(III) as an energy source. *World J. Microbiol. Biotech.* **19**, 215–225.

Marsili, E., Baron, D.B., Shikhare, I.D., Coursolle, D., Gralnick, J.A. and Bond, D.R. (2008a) Shewanella secretes flavins that mediate extracellular electron transfer. *Proc. Natl. Acad. Sci.* **105**, 3968–3973.

Marsili, E., Rollefson, J.B., Baron, D.B., Hozalski, R.M. and Bond, D.R. (2008b) Microbial biofilm voltammetry: direct electrochemical characterization of catalytic electrode-attached biofilms. *Appl. Environ. Microbiol.* **74**(23), 7329–7337.

McKnight, D.M. and Aiken, G.R. (1998) Sources and ages of aquatic humus. In *Aquatic Humic Substances: Ecology and Biogeochemistry*. Hessen, D.O. and Tranvik, L.J. (eds). Berlin: Springer, pp. 9–39.

Milliken, C.E. and May, H.D. (2007) Sustained generation of electricity by the spore-forming, Gram-positive, Desulfitobacterium hafniense strain DCB2. *Appl. Microbiol. Biotechnol.* **73**, 1180–1189.

Mohanty, S.P. and Kougianos, E. (2006) Biosensors: A tutorial review. *IEEE Potentials* **25**(2), 35–40.

Nevin, K.P. and Lovley, D.R. (2002) Mechanisms for accessing insoluble Fe(III) oxide during dissimilatory Fe(III) reduction by *Geothrix fermentans*. *Appl. Environ. Microbiol.* **68**(5), 2294–2299.

Newman, D.K. and Kolter, R. (2000) A role for excreted quinones in extracellular electron transfer. *Nature* **405**, 94–98.

Nortemann, B., Kuhm, A.E., Knackmuss, H.J. and Stolz, A. (1994) Conversion of substituted naphthalenesulfonates by *Pseudomonas* sp. BN6. *Arch. Microbiol.* **161**, 320–327.

O'Loughlin, E.J. (2008) Effects of electron transfer mediators on the bioreduction of lepidocrocite (γ-FeOOH) by *Shewanella putrefaciens* CN32. *Environ. Sci. Technol.* **42**(18), 6876–6882.

Park, H.S., Kim, B.H., Kim, H.S., Kim, H.J., Kim, G.T., Kim, M., Chang, I.S., Park, Y.K. and Chang, H.I. (2001) A novel electrochemicallyactive and Fe(III)-reducing bacterium phylogenetically related to *Clostridium butyricum* isolated from a microbial fuel cell. *Anaerobe* **7**, 297–306.

Pham, T.H., Boon, N., Aelterman, P., Clauwaert, P., Schamphelaire, L.D., Vanhaecke, L., De Maeyer, K., Höfte, M., Verstraete, W. and Rabaey, K. (2008) Metabolites produced by *Pseudomonas* sp. enable a Gram-positive bacterium to achieve extracellular electron transfer. *Applied Microbiology and Biotechnology* **77**(5), 1119–1129.

Price-Whelan, A. Dietrich, L.E.P. and Newman, D.K. (2006) Rethinking 'secondary' metabolism: physiological roles for phenazine antibiotics. *Nature Chem. Biol.* **2**(2), 71–78.

Rabaey, K. and Verstraete, W. (2005b) Microbial fuel cells: novel biotechnology for energy generation. *Trends Biotech.* **23**, 291–298.

Rabaey, K., Lissens, G., Siciliano, S.D. and Verstraete, W. (2003) A microbial fuel cell capable of converting glucose to electricity at high rate and efficiency. *Biotech. Lett.* **25**, 1531–1535.

Rabaey, K., Boon, N., Siciliano, S.D., Verhaege, M. and Verstraete, W. (2004) Biofuel cells select for microbial consortia that self-mediate electron transfer. *Appl. Environ. Microbiol.* **70**, 5373–5382.

Rabaey, K., Boon, N., Hofte, M. and Verstraete, W. (2005). Microbial phenazine production enhances electron transfer in biofuel cells. *Environ. Sci. Technol.* **39**(9), 3401–3408.

Rabaey, K., Rodrıguez, J., Blackall, L.L., Keller, J., Gross, P., Batstone, D., Verstraete, W. and Nealson, K.H. (2007) Microbial ecology meets electrochemistry: electricity-driven and driving communities. *ISME J.* **1**, 9–18.

Richter, H., Lanthier, M., Nevin, K.P. and Lovley, D.R. (2007) Lack of electricity production by *Pelobacter carbinolicus* indicates that the capacity for Fe(III) oxide reduction does not necessarily confer electron transfer ability to fuel cell anode. *Appl. Environ. Microbiol.* **73**(16), 5347–5353.

Ringeisen, B.R., Henderson, E., Wu, P.K., Pietron, J., Ray, R., Little, B., Biffinger, J.C. and Jones-Meehan, J.M. (2006) High power density from a miniature microbial fuel cell using. *Shewanella oneidensis* DSP10. *Environ. Sci. Technol.* **40**(8), 2629–2634.

Sakai, H., Nakagawa, T., Tokita, Y., Hatazawa, T., Ikeda, T., Tsujimura, T. and Kano, K. (2009) A high-power glucose/oxygen biofuel cell operating under quiescent conditions. *Energy Environ. Sci.* **2**, 133–138.

Schaetzle, O., Barriere, F. and Baronian, K. (2008) Bacteria and yeast as catalysts in microbial fuel cells: electron transfer from micro-organisms to electrodes for green electricity. *Energy Environ. Sci.* **1**, 607–620.

Schröder, U. (2007) Anodic electron transfer mechanisms in microbial fuel cells and their energy efficiency. *Phys. Chem. Chem. Phys.* **9**, 2619–2629.

Stams, A.J.M., de Bok, F.A.M., Plugge, C.M. van Eekert, M.H.A., Dolfing, J. and Schraa, G. (2006) Exocellular electron transfer in anaerobic microbial communities. *Environ. Microbiol.* **8**(3), 371–382.

Straub, K.L. and Schink, B. (2002) Ferrihydrite-dependent growth of *Sulfurospirillum deleyianum* through electron transfer via sulfur cycling. *Appl. Environ. Microbiol.* **70**(10), 5744–5749.

Sund, C.J., McMasters, S., Crittenden S.R., Harrell, L.E. and Sumner, J.J. (2007) Effect of electron mediators on current generation and fermentation in a microbial fuel cell. *Appl. Microbiol. Biotechnol.* **76**, 561–568.

Tanaka, K., Vega, C.A. and Tamamushi, R. (1983) Thionine and ferric chelate compounds as coupled mediators in microbial fuel cells. *Bioelectrochem. Bioenerget.* **11**(4–6), 289–297.

Tront, J.M., Fortner, J.D., Plotze, M., Hughes, J.B. and Puzrin, A.M. (2008) Microbial fuel cell technology for measurement of microbial respiration of lactate as an example of bioremediation amendment. *Biotechnol. Lett.* **30**(8), 1385–1390.

Turick, C.E., Tisa, L.S. and Caccavo, F. (2002) Melanin production and use as a soluble electron shuttle for Fe(III) oxide reduction and as a terminal electron acceptor by *Shewanella algae* BrY. *Appl. Environ. Microbiol.* **68**, 2436–2444.

Turick, C.E., Caccavo, F. and Tisa, L.S. (2003) Electron transfer from *Shewanella algae* BrY to hydrous ferric oxide is mediated by cell-associated melanin. *FEMS Microbiol. Lett.* **220**, 99–104.

Vega, C.A. and Fernandez, I. (1987) Mediating effect of ferric chelate compounds in microbial fuel cells with *Lactobacillus plantarum, Streptococcus lactis, and Erwinia dissolvens*. *Bioelectrochem. Bioenerg.* **17**, 217–222.

Von Canstein, H., Ogawa, J., Shimizu, S. and Lloyd, J.R. (2008) Secretion of flavins by *Shewanella* species and their role in extracellular electron transfer. *Appl. Environ. Microbiol.* **74**(3), 615–623.

Wang, Y.F., Tsujimura, S., Cheng, S.S. and Kano, K. (2007) Self-excreted mediator from *Escherichia coli* K-12 for electron transfer to carbon electrodes. *Appl. Microbiol. Biotech.* **76**, 1439–1446.

Wang, Y.F., Masuda, M., Tsujimura, S. and Kano, K. (2008) Electrochemical regulation of the end-product profile in Propionibacterium freudenreichii ET-3 with an endogenous mediator. *Biotech. Bioeng.* **101**, 579–586.

Wershaw, R.L. (1999) Molecular aggregation of humic substances. *Soil Sci.* **164**(11), 803–813.

Zhang, T., Cui, C., Chen, S., Yang, H. and Shen, P. (2008) The direct electrocatalysis of *Escherichia coli* through electroactivated excretion in microbial fuel cell. *Electrochem. Comm.* **10**, 293–297.

5

A survey of direct electron transfer from microbes to electronically active surfaces

Orianna Bretschger, Yuri A. Gorby and Kenneth H. Nealson

5.1 INTRODUCTION

Energy flow from electron donors (fuels) to electron acceptors (oxidants) powers the bioenergetics of all living organisms. Internal electron transfer occurs when the terminal electron acceptor is soluble and can permeate the cellular membrane. However, extracellular electron transfer is required when solid phase compounds, such as iron and manganese oxides, or large molecular weight compounds such as humic acids, serve as terminal electron acceptors. The biological reduction of solid-phase metal-oxides, known as dissimilatory metal

 Bioelectrochemical Systems: From Extracellular Electron Transfer to Biotechnological Application. Edited by Korneel Rabaey, Largus Angenent, Uwe Schröder and Jürg Keller. ISBN: 9781843392330. Published by IWA Publishing, London, UK.

reduction, was independently reported in two landmark papers that introduced the model microorganisms, *Shewanella oneidensis* MR-1 (Myers and Nealson 1988a) and *Geobacter metallireducens* GS-15 (Lovley and Phillips 1988). Now, two decades later, the study of how bacteria perform extracellular electron transfer (EET) is still under way and has been expanded to multiple organisms, environmental conditions, molecular interactions, geological and electrochemical signals. What started with a novel description of anaerobic bacterial respiration has now morphed into an interdisciplinary field that connects physics, microbiology, chemistry, geology and engineering.

Several EET strategies have emerged during decades of research. These include: (1) production or use of small soluble molecules that 'shuttle' electrons outside of the cell membrane to terminal electron acceptors (Lovley *et al.* 1996; Nevin and Lovley 2002; Hernandez and Newman 2001; Newman and Kolter 2000; Marsili *et al.* 2008; Bond and Lovley 2005; von Canstein *et al.* 2008); (2) exogenous or bacterially-produced organic ligands that liberate soluble organic-metal compounds from insoluble metal-oxides for use as terminal electron acceptors (DiChristina *et al.* 2005); (3) direct interaction of cellular membranes with electronically active surfaces (Arnold *et al.* 1988; Myers and Nealson 1990; Nealson and Saffarini 1994; Lovley 1991); or (4) construction of electronically conductive extracellular appendages referred to as bacterial 'nanowires' (Reguera *et al.* 2005; Gorby *et al.* 2006). Each proposed strategy has been thoroughly debated and yet a clear understanding of the components and mechanisms involved in extracellular electron transfer has not been established. Given the complexity of microbial ecosystems and the environments in which these organisms function, it is not unreasonable to assume that several EET mechanisms may be employed by any given organism. While elucidating this process is a complex task, understanding the factors that induce and control EET processes is needed to reveal how they shape natural ecosystems and can be practically employed for optimizing bioelectrochemical systems and other biologically catalyzed applications.

An elegant description of EET by soluble compounds has been provided in Chapter 4. Hence, this chapter will address EET mechanisms that require physical connections between bacterial constructs (outer-membranes and exocellular appendages) and electronically active surfaces.

5.2 EXTRACELLULAR ELECTRON TRANSFER – MICROBIAL CONNECTIONS

While it is commonly accepted that EET occurs within complex communities and across diverse phyla, the in-depth study of EET mechanisms has revolved around two model organisms: *Shewanella oneidensis* MR-1 (Venkateswaran *et al.* 1999) and *Geobacter sulfurreducens* (Caccavo *et al.* 1994). The genomes of both microorganisms were among the first environmental isolates to be sequenced (Heidelberg *et al.* 2002; Methe *et al.* 2003), making them genetically tractable for the study of EET.

5.2.1 Localized sites for membrane associated EET

The process of EET by direct interaction of membrane-bound electroactive compounds with electronically active surfaces has primarily been studied in dissimilatory metal reducing bacteria (DMRB). It has been demonstrated that redox active proteins, such as *c*-type cytochromes and iron-sulfur proteins, are localized to the outer-membrane and can act as direct conduits for electron flow to solid-phase electron acceptors. These proteins, their regulation, and their assemblies in functional electron transport chains have been primarily studied in *Shewanella* and *Geobacter* strains.

5.2.1.1 Shewanella *cytochromes*

The first explorations into *Shewanella's* genetic framework associated with EET to metal-oxides were carried out by Myers, Myers and Nealson (Myers and Nealson 1988b; Myers and Nealson 1990; Myers and Myers 1992; Myers and Myers 1997; Myers and Myers 2000). These studies suggested that EET was facilitated by a specific *c*-type cytochrome (CymA) that moved electrons from the quinone pool to other decaheme *c*-type cytochromes (MtrC and OmcA). Another protein, MtrB, was implicated as part of the electron transport system in *Shewanella* (Myers and Myers 2001; Myers and Myers 2002). These early studies paved the way for more in-depth hypotheses relative to EET in *Shewanella.* When the full *Shewanella* genome was assembled and annotated, forty-two genes encoding *c*-type cytochromes were identified, fourteen of which contained four or more hemes (Mowat and Chapman 2005). This revelation furthered the hypothesis that heme-centered proteins play a critical role in *Shewanella* EET processes, and that several electron transport pathways may exist via cytochrome interactions throughout the inner-membrane, periplasm and outer-membrane (Meyer *et al.* 2004).

The respiratory possibilities associated with this plethora of *c*-type cytochromes were examined using global mRNA patterns in *S. oneidensis* MR-1 cells exposed to a variety of metal and non-metal terminal electron acceptors (Beliaev *et al.* 2005). Differential gene expression was observed in response to available soluble electron acceptors, e.g. TMAO, DMSO and thiosulfate. However, differential expression of genes encoding decaheme cytochromes under different metal-oxide reducing conditions was inconclusive. This result was attributed to the 'low-specificity and the opportunistic nature of the metal-reducing electron transport pathways' (Beliaev *et al.* 2005).

Researchers formulated a working hypothesis that the same set of genes was involved with EET to all solid-phase electron acceptors, including electrodes in bioelectrochemical systems (BES) such as microbial fuel cells (MFCs). This hypothesis was tested using several in-frame single deletion mutants of *S. oneidensis* MR-1 that were deficient in their ability to produce one of the forty-two hypothesized *c*-type cytochromes (Meyer *et al.* 2004). In addition to these cytochrome mutants, a few mutants deficient in regulatory proteins, protein secretion and methylation were evaluated for their ability to reduce iron(III)- and Mn(IV)-oxides, and produce current in a MFC (Bretschger *et al.* 2007). These data suggested that while *S. oneidensis* MR-1 may feature functional redundancy relative to EET, the roles that each cytochrome plays are not uniform across solid-phase electron acceptors. For example, among the forty single deletion mutants that were evaluated, only six mutations displayed similar functional trends in MFC current production, and iron(III)- and Mn(IV)-oxide reduction. Work performed by Richter *et al.*, provided further evidence that EET is not uniformly applied to solid-phase electron acceptors by demonstrating that *Pelobacter carbinolicus* was unable to generate electricity in MFCs but could successfully reduce solid-phase iron(III)-oxide (Richter *et al.* 2007).

While these studies yielded some insights related to the functional role(s) of specific *c*-type cytochromes, they provided little insight into the nature of the electron transport components involved. A functional multi-protein electron transport chain would require an understanding about protein–protein interactions as well as cytochrome specificity.

Protein–protein interactions among cytochromes have been explored by several researchers, yielding hypotheses of how these proteins transfer electrons to the outer-membrane. The crystallization and characterization of a tetraheme *c*-type cytochrome encoded by *S. oneidensis* MR-1 was conducted by Leys *et al.* (2002) and yielded a perspective on electron transfer through small *c*-type cytochromes. They surmised that protein structure affects the efficiency of electron transfer among these multi-heme proteins. The authors proposed that 'the common "heme core"' in multi-heme proteins is a consequence of the limited number of ways bis-histidine-coordinated heme groups can be packed within direct contact, rather than of a evolutionary pressure to maximize certain electronic effects between porphyrin planes'. Furthermore, the authors suggested that this protein had 'a promiscuous "electron-harvesting"' role in contrast to a predetermined electron path along a "heme wire" with specific points for entry and delivery and the associated need for specific interaction sites with electron donor and acceptor, respectively' (Leys *et al.* 2002). To our knowledge, this was the only report relating electron efficiency to small multi-heme cytochrome packing.

Subsequent research appeared to focus on the more classical electron path interpretation of EET, using the decaheme proteins OmcA and MtrC as a model. These two proteins were hypothesized to form a complex within the electron transport chain and the authors (Shi *et al.* 2006) demonstrated accelerated reduction rates of chelated iron(III) in the presence of this complex relative to each individual protein. To date, MtrA (the multiheme cytochrome present in the periplasm), MtrB (the transmembrane protein involved in incorporating MtrC and OmcA to the outer-membrane), and MtrC and OmcA (decaheme cytochromes found outside of the cell membrane) have all been implicated as a complex that is needed to transfer electrons from the periplasmic region to the cell exterior. Given the proposed membrane location of this complex, it was hypothesised that it acts 'only as a grounding wire and cannot act as part of the proton motive energy-conserving system of the cell which is confined to the inner membrane' (Ross *et al.* 2007). The molecular orientation of these four proteins, and the other thirty-eight *c*-type cytochromes encoded by *Shewanella*, is still under consideration, however new insights about the redox mechanisms of multi-heme proteins are being elucidated.

Wigginton *et al.* 2007 conducted a series of experiments using scanning tunnelling microscopy (STM) and tunnelling spectroscopy to characterize specific electron transfer properties of MtrC and OmcA. The authors found that each protein had a unique ability to 'mediate tunnelling current despite having very similar molecular and biochemical properties' (Wigginton *et al.* 2007). This was the first study to demonstrate that STM and tunnelling spectroscopy could be employed as effective techniques for analyzing the redox processes of environmentally relevant proteins. These techniques have since been applied to studying EET reactions of a different kind, which will be discussed in the following section.

5.2.1.2 Geobacter *cytochromes*

The journey to elucidate EET mechanisms in *Geobacter sulfurreducens* was a parallel effort that followed a path similar to the *Shewanella* research. The assembled and annotated genome of *G. sulfurreducens* revealed the potential for over one hundred *c*-type cytochromes and the implication that many terminal electron acceptors could be utilized for respiration by this organism. The initial work suggesting that *G. sulfurreducens* incorporated multi-heme cytochromes in the outer-membrane during anaerobic respiration was performed by Seeliger *et al.* (1998). These authors found a specific 9.6-kDa, tri-heme, *c*-type cytochrome in the membrane fraction, periplasmic region, and growth media when fumarate acted as the electron acceptor. The report indicated that the purified cytochrome was able to reduce solid-phase and soluble iron(III) compounds. It was also reported that this cytochrome could as an electron 'shuttle' between *G. sulfurreducens* and *Wolinella succinogenes* during co-culture experiments (Seeliger *et al.* 1998). Subsequent work by Lloyd *et al.* (1999) refuted this work and suggested that the 9.6-kDa periplasmic cytochrome did not act as a shuttle to reduce solid-phase iron(III) compounds, but rather a 41-kDa membrane-bound protein was responsible for the observed reduction reaction (Lloyd *et al.* 1999).

Later work with *G. sulfurreducens* identified the putative decaheme *c*-type cytochrome, OmcB, as a possible electron carrier or outer-membrane terminal reductase for iron(III)-oxides, although its primary function remained ambiguous (Leang *et al.* 2003). In 2007, evidence was reported that OmcB and a multi-copper protein (OmpB) were both outer-membrane surface proteins that functioned as terminal reducatases for iron(III)-oxides. The data indicated that OmcB was partially exposed, and OmpB was evenly distributed, along the cell surface (Qian *et al.* 2007); however, clearly defined functionality of these proteins and how they cooperated within an electron transport chain was not elucidated by this study.

Mehta *et al.* (2005) also suggested that multi-heme *c*-type cytochromes play a critical role in mediating the reduction of extracellular metal-oxides by *G. sulfurreducens*. In this report, two *c*-type cytochromes, OmcE and OmcS were identified and predicted to be outer membrane proteins with four and six heme groups, respectively. Deletion of either identified gene decreased the ability of *G. sulfurreducens* to reduce iron(III)-oxides, but the capacity was recovered when genes were expressed *in trans*. The *omcT* gene, located immediately downstream of *omcS*, was also found to encode a hexaheme *c*-type cytochrome that was 62% identical to OmcS. These data suggested that the *omcS* gene was either transcribed by itself or co-transcribed with *omcT*, while *omcT* was only co-transcribed with *omcS*. Deletion of *omcT* affected the expression of *omcS* and lowered the rates of metal-oxide reduction by *G. sulfurreducens*. Deletion of *omcE*, *omcS* or *omcT*, however, did not have any effect on the ability of *G. sulfurreducens* to reduce the soluble chelated iron(III) compound, ferric citrate. These results offered another pathway for outer membrane reduction of insoluble metal-oxides by *G. sulfurreducens* via the OmcS, OmcE and OmcT *c*-type cytochromes (Mehta *et al.* 2005; Shi *et al.* 2007).

Further work exploring the possible electron transport chain in *G. sulfurreducens* was reported by Kim and Lovley (2008) where the relationships between identified *c*-type cytochromes in the inner membrane (MacA), the periplasm (PpcA) and the outer-membrane (OmcB, OmcS and OmcE) were evaluated (Kim and Lovley 2008). Previous reports (Butler *et al.* 2004; Lloyd *et al.* 2003; Leang *et al.* 2003) hypothesized that MacA, PpcA, and OmcB were required for the reduction of soluble, chelated iron(III) and solid-phase iron(III)-oxides; and that OmcS and OmcE were only necessary for solid-phase iron(III)-oxide reduction. The basis for these proposed functions was derived from evaluating the iron(III) reduction abilities of deletion mutants relative to each hypothesized cytochrome protein. However, research done by Kim (2005, 2006) and co-workers showed that gene deletions relative to specific *c*-type cytochromes in *G. sulfurreducens* could lead to incorrect transcription, inhibited synthesis, or post-translational modification of other cytochromes. Subsequently, Kim and Lovley (2008) reported that MacA was not directly involved in electron transfer but that OmcB remained an important component in the reduction of iron(III) by *G. sulfurreducens*.

While it is clear that *c*-type cytochromes play an essential role in EET to different solid-phase electron acceptors, there appears to be much functional redundancy within this family of proteins; and the pathway(s) for electron transport from the inner-membrane to the outer-membrane remain undefined in both model organisms *Geobacter* and *Shewanella*. However, the most recently proposed EET pathways for both organisms are illustrated in Figure 7.1.

Electron transfer via *c*-type cytochromes is not the only mechanism that has been proposed for EET by *Shewanella* or *Geobacter*. Both *G. sulfurreducens*, *S. oneidensis* MR-1, and other microorganisms produce electrically conductive 'nanowires', which are hypothesized to facilitate electron transfer from cells to solid phase electron acceptors.

5.2.2 Bacterial nanowires

The recent discovery that numerous microorganisms ranging from DMRB to oxygenic photosynthetic cyanobacteria produce electrically conductive appendages called bacterial nanowires is rapidly reshaping our fundamental understanding of extracellular electron transfer in microbial communities (Gorby *et al.* 2006; Reguera *et al.* 2005). Only a limited number of laboratories presently conduct research on bacterial nanowires. Within these groups, most researchers agree that the knowledge about composition and conductive properties of these extracellular structures is limited, but that the physiological, environmental and industrial implications are profound.

5.2.2.1 Geobacter *nanowires*

The first report of bacterial nanowires was relative to the bacterium *Geobacter sulfurreducens* (Reguera *et al.* 2005) and its type IV pili, or 'geopili'. Conductivity of thin filaments with diameters of about 8 nm and measuring tens of microns in length was confirmed using conductive atomic force microscopy. The authors suggested that *G. sulfurreducens* uses type IV pili to transfer the electrons directly from the outer membrane to extracellular metal-oxides. It was postulated that these pili receive electrons from the proteins associated with the inner membrane, periplasm and/or outer membrane because type IV pili are anchored to the inner membrane and extend across the periplasm and outer membrane into the extracellular space. Therefore, multiheme cytochromes, such as OmcE and OmcS, might transfer electrons to pili (Reguera *et al.* 2005). However, no biochemical or biophysical data are available to substantiate this speculation (Shi *et al.* 2007).

Research into the function of *G. sulfurreducens* nanowires has suggested that the type IV pili are necessary for reducing solid-phase metal-oxides (Reguera *et al.* 2005). Research has also demonstrated that nanowires are necessary for maximum power production in MFC systems, but are not absolutely required for producing minor amounts of current (Reguera *et al.* 2006). Reguera *et al.* (2006) presented data that showed a pilin-deficient *G. sulfurreducens* mutant was still able to generate electricity in MFC systems, but that current production, and observed biomass at the MFC anode surface, were much lower than that of wild type. When the pilin gene (*pilA*) was expressed *in trans* in the pilin-deficient mutant, current production increased and more biomass was generated, via cell stacking, at the anode. Although these data are clearly relevant to BES technology, the authors noted that the observed stacking function of *G. sulfurreducens* cells had 'no known counterpart in iron(III)-oxide reduction in sedimentary environments, where it is not advantageous for cells to permanently attach to metal oxides' (Reguera *et al.* 2006). Geopili may also play a structural role in biofilms, as suggested by data relating degenerate biofilm formation on non-conductive substrates with deletion of the *pilA* gene (Reguera *et al.* 2007). Further research is under way to develop mechanistic and functional descriptions of *Geobacter* nanowires. This information will be particularly useful for comparing the fundamental properties of nanowires produced by organisms of different genera and metabolic guilds.

5.2.2.2 Shewanella *nanowires*

Nanowires from *Shewanella oneidensis* strain MR-1 were first observed from chemostat cultures operated under conditions of electron acceptor limitation (Gorby *et al.* 2006). Conductivity of *Shewanella* nanowires was initially confirmed using scanning tunneling microscopy (STM). These observations revealed thin filaments measuring about 8 nm in diameter and tens of microns in length. Filaments typically appeared in bundles, or 'nanocables', that assumed a branched morphology.

At first glance, *Shewanella* nanowires were morphologically distinct from *Geobacter* nanowires. However, bundling of nanowires is likely a result of cultivation conditions, as both organisms demonstrate nanocable formation when cultivated under conditions of electron acceptor limitation. Figure 5.1 contains SEM images of *Shewanella oneidensis* MR-1 (Figure 5.1A) and *Geobacter sulfurreducens* PCA (Figure 5.1B). Branched morphologies of nanocables with diameters ranging from 50 to 150 nm are displayed by both organisms. Hence, morphology should not be used as a reliable criterion for drawing distinctions between nanowires from these metal reducing bacteria. However, apparent compositional differences appear to exist between *Geobacter* and *Shewanella* nanowires. Many microscopists would suggest that these structures are nothing more than dehydration artifacts. But, as published early on, nanowires can be observed under fully hydrated conditions using non-specific fluorescent protein stains (Gorby *et al.* 2006).

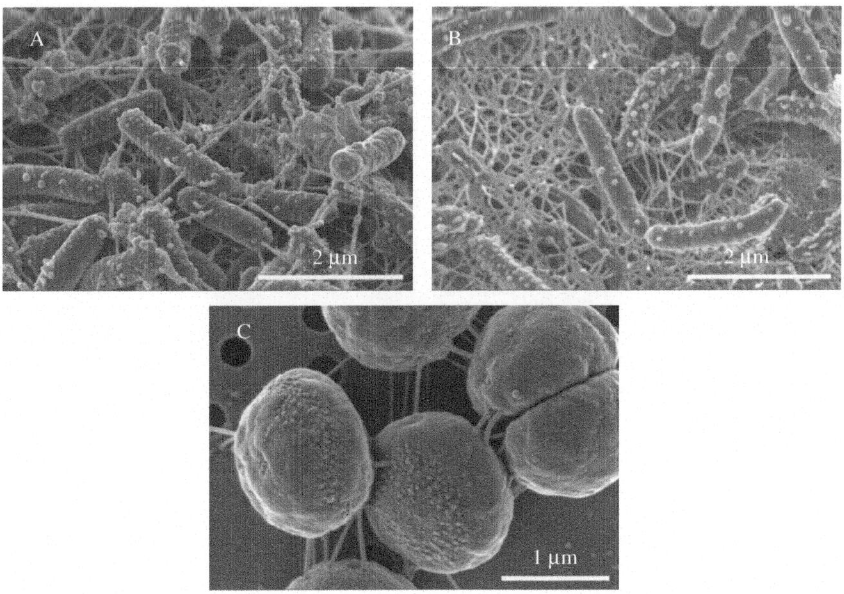

Figure 5.1. SEM micrographs of: (A) *Shewanella oneidensis* MR-1 nanowires; (B) *Geobacter sulfurreducens* nanowire; and (C) Synechocystis PCC 6803

5.2.2.3 Nanowires produced by other microorganisms

Electrically conductive nanowires are not restricted to metal reducing bacteria and have been confirmed in the oxygenic cyanobacterium *Synechocystis* PCC 6803 (Figure 5.1C) and the thermophylic syntroph *Pelotomaculum thermopropionicum* (Gorby *et al.* 2006). Structures that are morphologically similar to nanowires produced by *Shewanella* and the other laboratory cultures have also been observed in biofilms and cell aggregates from environmental samples and different pure cultures. However, the conductivity of these structures has not been verified so we apply the term 'putative nanowires' to these observed extracellular appendages. For example, putative nanowires were observed in *Desulfovibrio vulgaris* cultivated under sulfate limited conditions (Figure 5.2A). These structures span between cells and the surfaces of biogenic minerals, presumably iron sulfides. Figure 5.2B is an example of a sulfur-oxidizing streamer sampled from the opening of a hydrothermal vent at the bottom of Yellowstone Lake. Putative nanowires extend from cells into a vast integrated network that included sulfur rich minerals. Conductivity measurements and other characterization techniques are planned for these samples and a number of others in which 'putative nanowires' have been observed.

Figure 5.2. SEM micrographs of "putative" microbial nanowires: (A) *Desulfovibrio vulgaris* samples prepared by critical point drying (Dwayne Elias, University of Missouri, Columbia); and (B) Sulfur oxidizing microbial community prepared and observed using cryo-SEM approaches (Rich Macur, University of Montana)

5.2.3 Nanowire characterization

It is reasonable to state that our knowledge of bacterial nanowires is exceedingly limited. Many questions remain about the composition, mechanisms, and role(s) of bacterial nanowires. For example, what are they made of? How do they compare across genera? By what mechanisms do electrons flow? Are nanowires ubiquitous in natural environments? Do they participate in interspecies charge transfer reactions? Are they involved in cell signaling? Answers to these questions can only be provided by applying proven characterization techniques to study nanowires produced by a broad range of microorganisms. However some preliminary data related to these questions have been collected for the model organisms *Shewanella* and *Geobacter*.

5.2.3.1 Composition

Shewanella nanowires are believed to be partially composed of *c*-type cytochromes that assemble along the length of pilin structural proteins. Gorby *et al.* (2006) demonstrated that the absence of outer-membrane cytochromes, MtrC and OmcA, resulted in pili-structures that were not conductive via STM analysis. Furthermore, extracellular appendages were non-conductive in the absence of a gene encoding pseudopilin of the Type 2 secretion pathway (*gspG*), which is hypothesized to be involved with transporting a variety of preassembled proteins (including MtrC and OmcA) to the cell exterior. Deletion of genes encoding a pre-pilin peptidase (*pilD*) produced a mutant that did not make extracellular appendages. These data support the following conclusions about *Shewanella* nanowires: (1) peptidase gene(s) are important for synthesis or assembly of structural protein nanowire components, (2) a Type 2 secretion pathway transports decaheme, *c*-type cytochromes to the cell exterior and possibly along the length of extracellular structures; and (3) nanowire conductivity is a result of an assemblage of redox-active *c*-type cytochromes that coat the structural pili components. More research is needed to validate these conclusions.

Reguera *et al.* (2005) and Lovley (2008) hypothesized that geopilins are composed of unique proteins that do not incorporate metalloproteins, but still enable unimpeded electron flow. Presently, there are no reported biophysical models that describe the unique conductive properties of geopili proteins (Shi *et al.* 2007), but genetic studies of the PilA protein are ongoing (Lovley 2008).

5.2.3.2 Regulation

The environmental conditions that result in nanowire production seem to be more clearly defined through experimentation than nanowire composition. Several respiratory microbes have been analyzed in our laboratory, and so far, each one has formed nanowires under conditions of electron acceptor limitation. For example, soluble electron acceptors including oxygen, fumarate, nitrate, and chelated forms of iron(III) must be limited in terms of molar concentration to induce nanowire formation. Solid phase electron acceptors such as iron(III)- and

Mn(IV)-oxides, or electrodes, must be limited by surface area availability to induce nanowire formation.

5.2.3.3 Conductivity

Conductivity of bacterial nanowires has only been reported perpendicular to the long axis and confirmation of conductivity down the length of a nanowire is still required to elucidate electron transport mechanisms (Gorby *et al.* 2008). Reports have suggested that measurements taken along the length of geopili structures were acquired by laying the filaments across 'interdigitated electrodes'. An interdigitated electrode resembles two metallic combs that fit together by interlocking the teeth; however in the case of the electrode, the teeth do not physically touch and are therefore electrically isolated. After bacterial filaments were laid across the electrode surfaces, AFM techniques were employed and a potential was applied to one 'digit', or tooth, that was in contact with the bacterial filament. The current flow through the bacterial filament was monitored at a neighboring (electronically disconnected) 'digit' that was contacting the same filament. According to Lovley (2008), this method has proven to be technically challenging and the results are difficult to interpret.

Conductivity of nanowires produced by *Shewanella oneidensis* MR-1 was first confirmed using scanning tunneling microscopy (STM) (Gorby *et al.* 2006). A more recent publication provided additional conductive AFM data and partially described the conductive properties through a single *Shewanella* nanowire bundle (El-Naggar *et al.* 2008). In this study the authors utilized AFM techniques to provide quantitative measurements of conductivity at discrete locations along different *Shewanella* nanowires. These data suggested that *Shewanella* nanowires display a 'non-linear electrical transport behavior, where the voltage dependence of the conductance reveals peaks indicating discrete energy levels with higher electronic density of states' (El-Naggar *et al.* 2008). Further analysis suggested that the observed features within the density of states reflected the electronic structure of the compounds associated with bacterial nanowires; and the authors proposed that electron transport proceeds by using delocalized energy states across the nanowire, i.e. electron hopping. El-Naggar *et al.* (2008) also addressed the notion that contact resistance between the nanowires and substrate electrodes may contribute to differences in the overall resistances from point to point; however contact resistance does not impede the reproducible observations of those density–of-states features that are reflective of electronic structure.

Conductivity along the *length* of a bacterial nanowire has yet to be reported. Advanced nanofabrication and nanoelectrical techniques provide hope for characterizing conductivity along the length of bacterial nanowires. New techniques such as the recently described two-color, time-resolved, surface-enhanced resonance Raman spectroelectrochemical approach (Kranich *et al.* 2008) could also be applied to investigate electron transfer and protein dynamics. Developing a mechanistic understanding of electron transfer within and along these macromolecular structures will hopefully advance our fundamental understanding of energy transfer within diverse microbial communities.

5.2.3.4 Function

Although bacterial nanowires appear to have conductive properties, the primary function for these appendages is still not clear. For example, *Geobacter* nanowires may act as terminal reductases for metal-oxides or structural components in biofilms depending on what the terminal electron acceptor may be (Reguera *et al.* 2005; Reguera *et al.* 2007). Yet, alternatives to these hypotheses also exist, and it is possible that bacterial nanowires could: (1) function as part of an advanced cell-signaling system; (2) could facilitate intercellular or interspecies electron transfer; or (3) may be related to cellular bioenergetics.

5.2.3.5 Prevalence

The occurrence of nanowires throughout the environment has been hypothesized and may indeed be prevalent in many different ecosystems including pathogenic biofilms, extreme environments and at different redox interfaces; however confirmation of this conjecture must still be provided.

5.3 ECOLOGICAL SIGNIFICANCE OF EXTRACELLULAR ELECTRON TRANSFER

Extracellular electron transfer is an important aspect of energy transfer and transformation in microbial systems, especially those where solid phase substrates, such as iron or manganese minerals or electrodes, or soluble humic acids serve as electron acceptors. In the absence of exogenous or endogenous electron shuttles, direct cell contact or electrically conductive appendages may be necessary for electron transfer to occur. Recent data showing that DMRB produce bacterial nanowires, and that these wires may be a strategy for electron transfer to redox reactive minerals or electrode surfaces, are reasonably accepted within the scientific community. Research has shown that electron movement to solid-phase electron acceptors is significantly impacted by removing genes that encode for conductive geopili in *Geobacter* or genes that encode for electron transport proteins (cytochromes) associated with nanowires in *Shewanella* (Bretschger *et al.* 2007; Reguera *et al.* 2006). However, residual function can still be observed in some mutant strains and these data suggest that the residual metal reduction and/or electrogenic activity relied on one or mechanisms including: (1) dissolved electron shuttles; and/or (2) direct cell-to-mineral or cell-to-electrode contact. In any case, the ecological significance of EET is profound.

Extracellular electron transfer is almost certainly important for organisms other than DMRB, although scientific evidence supporting this claim is somewhat scarce. Furthermore, the observation that nanowires have been detected in organisms ranging from oxygenic cyanobacteria to thermophilic methanogenic co-cultures supports the hypothesis that these structures may serve roles of ecological significance. For example, electrochemical activity in sediment batteries and wastewater microbial fuel cells has been observed as a function of mixed-community biofilms attached to electrodes (Bond *et al.* 2002; Holmes *et al.* 2004; Reimers *et al.* 2001; Tender *et al.* 2002). Genomic analyses of these mixed communities typically revealed significant populations of organisms other than metal reducers (Aelterman *et al.* 2008); and imagery provided evidence for extracellular appendages, i.e. putative nanowires, occurring throughout these communities.

Scanning electron micrographs (SEM) of samples that have been carefully prepared almost invariably reveal branched extracellular features that morphologically resemble nanowires produced by DMRB. These observations support a reasonably sound hypothesis that bacteria participate in intercellular electron transfer reactions that may, in part, be facilitated by direct cell contact, nanowires, conductive matrices or electron shuttles.

Notable are methanogenic syntrophic consortia, such as those composed of *Pelotomaculum thermopropionicus* and *Methanothermobacter thermoautotrophicus* (Imachi *et al.* 2007; Imachi *et al.* 2002; Ishii *et al.* 2005; Kosaka *et al.* 2006), which participate in a process known as interspecies hydrogen transfer. Thin filaments were reported to physically bind these two syntrophs and allow for efficient transfer of hydrogen (Ishii *et al.* 2005). Subsequent investigations revealed that these structures are in fact electrically conductive nanowires (Gorby *et al.* 2006). Although these results do not conclusively demonstrate that interspecies electron transfer occurs, it does support an exciting and experimentally testable hypothesis.

Beyond what is known from defined laboratory cultures, the possibility that conductive nanowires may be present, and perhaps even prevalent, in diverse microbial communities is scientifically provocative and experimentally testable.

Researchers in many fields, from maxillofacial surgeons to biogeophycisists, are beginning to suspect that the extracellular structures they have observed in countless micrographs may be electrically conductive nanowires and not simply artefacts of sample preparation.

For example, subsurface sediments and redox-reactive minerals typically exhibit rough topographies, complete with pore spaces and mineralogical features that are too small to allow for direct contact by cell surfaces. Conductive nanowires, which have been reported to grow 10–100 microns in length, could facilitate electron transfer to minerals located at distances or in locations that are otherwise inaccessible by direct cell contact. These electron transfer processes could involve the reduction of solid-phase electron acceptors and/or oxidation of insoluble electron donors such as pyrite at circumneutral conditions.

As a second example, biofilms that colonize the oral cavity and promote a variety of pathogenic conditions produce structures that could be considered putative nanowires (Sedghizadeh *et al.* 2008). Having electrically conductive filaments within thick biofilms could facilitate access to electron acceptors such as oxygen, that would be present at the surface but unavailable at depth. Charge transfer through these putative nanowires could promote bone dissolution by removing electrons to oxygen while the corresponding protons remain near the bone surface forming corrosive acids.

These hypotheses must be substantiated and much more work remains for developing a compositional and mechanistic understanding of EET throughout diverse environments. But we are encouraged by the data and ideas originating from all researchers studying EET processes, the availability of technologies to experimentally address such compelling ideas, and look forward to a greater understanding about microbially driven EET processes.

REFERENCES

Aelterman, P., Rabaey, K., De Schamphelaire, L., Clauwaert, P., Boon, N. and Verstraete, W. (2008) Microbial fuel cells as an engineered ecosystem. In *Bioenergy*. Wall, J.D., Harwood, C.S. and Demain, A.L. (eds). ASM Press.

Arnold, R.G., DeChristina, T.J. and Hoffman, M.R. (1988) Reductive dissolution of Fe(III) oxides by *Pseudomonas* sp. 200. *Biotechnol. Bioengng.* **32**, 1081–1096.

Beliaev, A.S., Klingeman, D.M., Klappenbach, J.A., Wu, L., Romine, M.F., Tiedje, J.M., Nealson, K.H., Fredrickson, J.K. and Zhou, J. (2005) Global transcriptome analysis of *Shewanella oneidensis* MR-1 exposed to different terminal electron acceptors. *J. Bacteriol.* **187**(20), 7138–7145.

Bond, D.R., Holmes, D.E., Tender, L.M. and Lovley, D.R. (2002) Electrode-reducing microorganisms that harvest energy from marine sediments. *Science* **295**(5554), 483–485.

Bond, D.R. and Lovley, D.R. (2005) Evidence for involvement of an electron shuttle in electricity generation by *Geothrix fermentans*. *Appl. Environ. Microbiol.* **71**(4), 2186–2189.

Bretschger, O., Obraztsova, A., Sturm, C.A., Chang, I.S., Gorby, Yuri A., Reed, S.B., Culley, D.E., Reardon, C.L., Barua, S., Romine, M.F., Zhou, J., Beliaev, A.S., Bouhenni, R., Saffarini, D., Mansfeld, F., Kim, B.-H., Fredrickson, J.K. and Nealson, K.H. (2007) An exploration of current production and metal oxide reduction by *Shewanella oneidensis* MR-1 wild type and mutants. *Appl. Environ. Microbiol.* **70**(21), 7003–7012.

Butler, J.E., Kaufmann, F., Coppi, M.V., Nunez, C. and Lovley, D.R. (2004) MacA a diheme *c*-type cytochrorne involved in Fe(III) reduction by *Geobacter sulfurreducens*. *J. Bacteriol.* **186**(12), 4042–4045.

Caccavo, F., Lonergan, J., Lovley, D.R., Davis, M., Stoltz, J.F. and McInerney, J. (1994) *Geobacter sulfurreducens* sp. nov., a hydrogen- and acetate-oxidizing dissimilatory metal-reducing microorganism. *Appl. Environ. Microbiol.* **60**, 3752–3759.

DiChristina, T.J., Fredrickson, J.K. and Zachara, J.M. (2005) Enzymology of electron transport: energy generation with geochemical consequences. *Rev. Mineral. Geochem.* **59**(1), 27–52.

El-Naggar, M.Y., Gorby, Y.A., Xia, W. and Nealson, K.H. (2008) The molecular density of states in bacterial nanowires. *Biophys. J.* **95**(1), L10–12.

Gorby, Y., Mclean, J., Korenevsky, A., Rosso, K., El-Naggar, M.Y. and Beveridge, T.J. (2008) Redox-reactive membrane vesicles produced by *Shewanella*. *Geobiology* **6**(3), 232–241.

Gorby, Y.A., Yanina, S., McLean, J.S., Rosso, K.M., Moyles, D., Dohnalkova, A., Beveridge, T.J., Chang, I.S., Kim, B.H., Kim, K.S., Culley, D.E., Reed, S.B., Romine, M.F., Saffarini, D.A., Hill, E.A., Shi, L., Elias, D.A., Kennedy, D.W., Pinchuk, G., Watanabe, K., Ishii, S.I., Logan, B., Nealson, K.H. and Fredrickson, J.K. (2006) Electrically conductive bacterial nanowires produced by *Shewanella oneidensis* Strain MR-1 and Other Microorganisms. *Proc. Natl. Acad. Sci. US* **103**(30), 11358–11363.

Heidelberg, J.F., Paulsen, I.T., Nelson, K.E., Gaidos, E.J., Nelson, W.C., Read, T.D., Eisen, J.A., Seshadri, R., Ward, N., Methe, B., Clayton, R.A., Meyer, T., Tsapin, A., Scott, J., Beanan, M., Brinkac, L., Daugherty, S., DeBoy, R.T., Dodson, R.J., Durkin, A.S., Haft, D.H., Kolonay, J.F., Madupu, R., Peterson, J.D., Umayam, L.A., White, O., Wolf, A.M., Vamathevan, J., Weidman, J., Impraim, M., Lee, K., Berry, K., Lee, C., Mueller, J., Khouri, H., Gill, J., Utterback, T.R., McDonald, L.A., Feldblyum, T.V., Smith, H.O., Venter, J.C., Nealson, K.H. and Fraser, C.M. (2002) Genome sequence of the dissimilatory metal ion-reducing bacterium *Shewanella oneidensis*. *Nature Biotechnol.* **20**(11), 1118–1123.

Hernandez, M.E. and Newman, D.K. (2001) Extracellular electron transfer. *Cell. Molec. Life Sci.* (*CMLS*) **58**(11), 1562.

Holmes, D.E., Bond, D.R., O'Neil, R.A., Reimers, C.E., Tender, L.R. and Lovley, D.R. (2004) Microbial communities associated with electrodes harvesting electricity from a variety of aquatic sediments. *Microb. Ecol.* **48**(2), 178–190.

Imachi, H., Sakai, S., Ohashi, A., Harada, H., Hanada, S., Kamagata, Y. and Sekiguchi, Y. (2007) *Pelotomaculum propionicicum* sp. nov., an anaerobic, mesophilic, obligately syntrophic, propionate-oxidizing bacterium. *Int. J. Syst. Evol. Microbiol.* **57**(Pt 7), 1487–1492.

Imachi, H., Sekiguchi, Y., Kamagata, Y., Hanada, S., Ohashi, A. and Harada, H. (2002) *Pelotomaculum thermopropionicum* gen. nov., sp. nov., an anaerobic, thermophilic, syntrophic propionate-oxidizing bacterium. *Int. J. Syst. Evol. Microbiol.* **52**(Pt 5), 1729–35.

Ishii, S., Kosaka, T., Hori, K., Hotta, Y. and Watanabe, K. (2005) Coaggregation facilitates interspecies hydrogen transfer between *Pelotomaculum thermopropionicum* and *Methanothermobacter thermautotrophicus*. *Appl. Environ. Microbiol.* **71**(12), 7838–7845.

Kim, B.-C., Leang, C., Ding, Y.-H.R., Glaven, R.H., Coppi, M.V. and Lovley, D.R. (2005) OmcF, a putative c-type monoheme outer membrane cytochrome required for the expression of other outer membrane cytochromes in *Geobacter sulfurreducens*. *J. Bacteriol.* **187**(13), 4505–4513.

Kim, B.-C. and Lovley, D.R. (2008) Investigation of direct vs. indirect involvement of the *c*-type cytochrome MacA in Fe(III) reduction by *Geobacter sulfurreducens*. *FEMS Microbiol. Lett.* **286**(1), 39–44.

Kim, B.-C., Qian, X., Leang, C., Coppi, M.V. and Lovley, D.R. (2006) Two putative *c*-type multiheme cytochromes required for the expression of OmcB, an outer membrane protein essential for optimal Fe(III) reduction in *Geobacter sulfurreducens*. *J. Bacteriol.* **188**(8), 3138–3142.

Kosaka, T., Uchiyama, T., Ishii, S., Enoki, M., Imachi, H., Kamagata, Y., Ohashi, A., Harada, H., Ikenaga, H. and Watanabe, K. (2006) Reconstruction and regulation of the central catabolic pathway in the thermophilic propionate-oxidizing syntroph *Pelotomaculum thermopropionicum*. *J. Bacteriol.* **188**(1), 202–210.

Kranich, A., Ly, H.K., Hildebrandt, P. and Murgida, D.H. (2008) Direct observation of the gating step in protein electron transfer: electric-field-controlled protein dynamics. *J. Am. Chem. Soc.* **130**(30), 9844–9848.

Leang, C., Coppi, M.V. and Lovley, D.R. (2003) OmcB, a c-type polyheme cytochrome, involved in Fe(III) reduction in *Geobacter sulfurreducens*. *J. Bacteriol.* **185**(7), 2096–2103.

Leys, D., Meyer, T.E., Tsapin, A.S., Nealson, K.H., Cusanovich, M.A. and Van Beeumen, J.J. (2002) Crystal structures at atomic resolution reveal the novel concept of "electron-harvesting" as a role for the small tetraheme cytochrome c. *J. Biol. Chem.* **277**(38), 35703–35711.

Lloyd, J.R., Blunt-Harris, E.L. and Lovley, D.R. (1999) The periplasmic 9.6-kilodalton c-type cytochrome of *Geobacter sulfurreducens* is not an electron shuttle to Fe(III). *J. Bacteriol.* **181**(24), 7647–7649.

Lloyd, J.R., Leang, C., Myerson, A.L.H., Coppi, M.V., Cuifo, S., Methe, B., Sandler, S.J. and Lovley, D.R. (2003) Biochemical and genetic characterization of PpcA, a periplasmic *c*-type cytochrome in *Geobacter sulfurreducens*. *J. Biochem.* **369**, 153–161.

Lovley, D.R. (1991) Dissimilatory Fe(III) and Mn(IV) reduction. *Microbiol. Molec. Biol. Rev.* **55**(2), 259–287.

Lovley, D.R. (2008) Extracellular electron transfer: Wires, capacitors, iron lungs, and more. *Geobiology* **6**, 225–231.

Lovley, D.R., Coates, J.D., Blunt-Harris, E.L., Phillips, E.J.P. and Woodward, J.C. (1996) Humic substances as electron acceptors for microbial respiration. *Nature* **382**(6590), 445–448.

Lovley, D.R. and Phillips, E.J.P. (1988) Novel mode of microbial energy metabolism: Organic carbon oxidation coupled to dissimilatory reduction of iron or manganese. *Appl. Environ. Microbiol.* **54**(6), 1472–1480.

Marsili, E., Baron, D.B., Shikhare, I.D., Coursolle, D., Gralnick, J.A. and Bond, D.R. (2008) Shewanella secretes flavins that mediate extracellular electron transfer. *Proc. Natl Acad. Sci. US* **105**(10), 3968–3973.

Mehta, T., Coppi, M.V., Childers, S.E. and Lovley, D.R. (2005) Outer membrane c-type cytochromes required for Fe(III) and Mn(IV) oxide reduction in *Geobacter sulfurreducens*. *Appl. Environ. Microbiol.* **71**(12), 8634–8641.

Methe, B.A., Nelson, K.E., Eisen, J.A., Paulsen, I.T., Nelson, W., Heidelberg, J.F., Wu, D., Wu, M., Ward, N., Beanan, M.J., Dodson, R.J., Madupu, R., Brinkac, L.M., Daugherty, S.C., DeBoy, R.T., Durkin, A.S., Gwinn, M., Kolonay, J.F., Sullivan, S.A., Haft, D.H., Selengut, J., Davidsen, T.M., Zafar, N., White, O., Tran, B., Romero, C., Forberger, H.A., Weidman, J., Khouri, H., Feldblyum, T.V., Utterback, T.R., Van Aken, S.E., Lovley, D.R. and Fraser, C.M. (2003) Genome of *Geobacter sulfurreducens*: Metal reduction in subsurface environments. *Science* **302**(5652), 1967–1969.

Meyer, T.E., Tsapin, A.I., Vandenberghe, I., De Smet, L., Frishman, D., Nealson, K.H., Cusanovich, M.A. and Van Beeumen, J.J. (2004) Identification of 42 possible cytochrome *c* genes in the *Shewanella oneidensis* genome and characterization of six soluble cytochromes. *OMICs-a J. Integr. Biol.* **8**(1), 57–77.

Mowat, C.G. and Chapman, S.K. (2005) Multi-heme cytochromes – new structures, new chemistry. *Dalton Trans.* 3381–3389.

Myers, C.R. and Myers, J.M. (1992) Localization of cytochromes to the outer membrane of anaerobically grown *Shewanella putrefaciens* MR-1. *J. Bacteriol.* **174**(11), 3429–3438.

Myers, C.R. and Myers, J.M. (1997) Cloning and sequence of cymA, a gene encoding a tetraheme cytochrome *c* required for reduction of Iron(III), fumarate, and nitrate by *Shewanella putrefaciens* MR-1. *J. Bacteriol.* **179**(4), 1143–1152.

Myers, C.R. and Myers, J.M. (2002) MtrB is required for proper incorporation of the cytochromes OmcA and OmcB into the outer membrane of *Shewanella putrefaciens* MR-1. *Appl. Environ. Microbiol.* **68**(11), 5585–5594.

Myers, C.R. and Nealson, K.H. (1988a) Bacterial manganese reduction and growth with manganese oxide as the sole electron acceptor. *Science* **240**, 1319–1321.

Myers, C.R. and Nealson, K.H. (1988b) Microbial reduction of manganese oxides: interactions with Iron and Sulfur. *Geochim. Cosmochim. Acta* **52**(11), 2727–2732.

Myers, C.R. and Nealson, K.H. (1990) Respiration-linked proton translocation coupled to anaerobic reduction of manganese(IV) and Iron(III) in *Shewanella putrefaciens* MR-1. *J. Bacteriol.* **172**(11), 6232–6238.

Myers, J.M. and Myers, C.R. (2000) Role of the tetraheme cytochrome CymA in anaerobic electron transport in cells of *Shewanella putrefaciens* MR-1 with normal levels of menaquinone. *J. Bacteriol.* **182**(1), 67–75.

Myers, J.M. and Myers, C.R. (2001) Role for outer membrane cytochromes OmcA and OmcB of *Shewanella putrefaciens* MR-1 in reduction of manganese dioxide. *Appl. Environ. Microbiol.* **67**(1), 260–269.

Nealson, K.H. and Saffarini, D. (1994) Iron and manganese in anaerobic respiration: environmental significance, physiology, and regulation. *Ann. Rev. Microbiol.* **48**(1), 311–343.

Nevin, K.P. and Lovley, D.R. (2002) Mechanisms for Fe(III) oxide reduction in sedimentary environments. *Geomicrobiol. J.* **19**, 141–159.

Newman, D.K. and Kolter, R. (2000) A role for excreted quinones in extracellular electron transfer. *Nature* **405**(6782), 94–97.

Qian, X., Reguera, G., Mester, T. and Lovley, D.R. (2007) Evidence that OmcB and OmpB of *Geobacter sulfurreducens* are outer membrane surface proteins. *FEMS Microbiology Letters* **277**(1), 21–27.

Reguera, G., McCarthy, K.D., Mehta, T., Nicoll, J.S., Tuominen, M.T. and Lovley, D.R. (2005) Extracellular electron transfer via microbial nanowires. *Nature* **435**(7045), 1098.

Reguera, G., Nevin, K.P., Nicoll, J.S., Covalla, S.F., Woodard, T.L. and Lovley, D.R. (2006) Biofilm and nanowire production leads to increased current in *Geobacter sulfurreducens* fuel cells. *Appl. Environ. Microbiol.* **72**(11), 7345–7348.

Reguera, G., Pollina, R.B., Nicoll, J.S. and Lovley, D.R. (2007) Possible nonconductive role of *Geobacter sulfurreducens* pilus nanowires in biofilm formation. *J. Bacteriol.* **189**(5), 2125–2127.

Reimers, C.E., Tender, L.M., Fertig, S. and Wang, W. (2001) Harvesting energy from the marine sediment water interface. *Environ. Sci. Technol.* **35**(1), 192–195.

Richter, H., Lanthier, M., Nevin, K.P. and Lovley, D.R. (2007) Lack of electricity production by *Pelobacter carbinolicus* indicates that the capacity for Fe(III) oxide reduction does not necessarily confer electron transfer ability to fuel cell anodes. *Appl. Environ. Microbiol.* **73**(16), 5347–5353.

Ross, D.E., Ruebush, S.S., Brantley, S.L., Hartshorne, R.S., Clarke, T.A., Richardson, D.J. and Tien, M. (2007) Characterization of protein-protein interactions involved in iron reduction by *Shewanella oneidensis* MR-1. *Appl. Environ. Microbiol.* **73**(18), 5797–5808.

Sedghizadeh, P.P., Kumar, S.K.S., Gorur, A., Mastin, C. and Boros, A. (2008) Toxic epidermal necrolysis with a rare long-term oral complication requiring surgical intervention. *Oral Surg., Oral Med., Oral Path., Oral Radiol. Endodont.* **105**(4), e29–e33.

Seeliger, S., Cord-Ruwisch, R. and Schink, B. (1998) A periplasmic and extracellular *c*-type cytochrome of *Geobacter sulfurreducens* acts as a ferric iron reductase and as an electron carrier to other acceptors or to partner bacteria. *J. Bacteriol.* **180**(14), 3686–3691.

Shi, L., Chen, B., Wang, Z., Elias, D.A., Mayer, M.U., Gorby, Y.A., Ni, S., Lower, B.H., Kennedy, D.W., Wunschel, D.S., Mottaz, H.M., Marshall, M.J., Hill, E.A., Beliaev, A.S., Zachara, J.M., Fredrickson, J.K. and Squier, T.C. (2006) Isolation of a high-affinity functional protein complex between OmcA and MtrC: Two outer membrane decaheme c-type cytochromes of *Shewanella oneidensis* MR-1. *J. Bacteriol.* **188**(13), 4705–4714.

Shi, L., Thomas, S.C., Zachara, J.M. and Fredrickson, J.K. (2007) Respiration of metal (hydr)oxides by *Shewanella* and *Geobacter*: a key role for multihaem *c*-type cytochromes. *Molec. Microbiol.* **65**(1), 12–20.

Tender, L.M., Reimers, C.E., Stecher, H.A., Holmes, D.E., Bond, D.R., Lowy, D.A., Pilobello, K., Fertig, S.J. and Lovley, D.R. (2002) Harnessing microbially generated power on the seafloor. *Nature Biotechnol.* **20**(8), 821–825.

Venkateswaran, K., Moser, D.P., Dollhopf, M.E., Lies, D.P., Saffarini, D.A., MacGregor, B.J., Ringelberg, D.B., White, D.C., Nishijima, M., Sano, H., Burghardt, J., Stackebrandt, E. and Nealson, K.H. (1999) Polyphasic taxonomy of the genus *Shewanella* and description of *Shewanella oneidensis* sp. nov. *Int. J. System. Bacteriol.* **49**, 705–724.

von Canstein, H., Ogawa, J., Shimizu, S. and Lloyd, J.R. (2008) Secretion of flavins by *Shewanella* species and their role in extracellular electron transfer. *Appl. Environ. Microbiol.* **74**(3), 615–623.

Wigginton, N.S., Rosso, K.M. and Jr, M.F. (2007) Mechanisms of electron transfer in two decaheme cytochromes from a metal-reducing bacterium. *J. Phys. Chem. B* **111**(44), 12857–12864.

6

Genetically modified microorganisms for bioelectrochemical systems

Miriam Rosenbaum and Largus T. Angenent

6.1 INTRODUCTION

Biotechnological utilization of microbial processes normally starts with the discovery, investigation, and understanding of naturally occurring microbial reactions. Biomolecular tools can help us, not only to gain better understanding of protein reactions; they also enable us to influence reaction properties. For efficient practical applications, modern biological techniques allow us to tailor these microbial reactions to optimize the desired functionality.

We are still in the infancy of our understanding of microbial reactions in microbial fuel cells (MFCs), or more generally in bioelectrochemical systems (BES). One major pathway of electron transfer to a MFC anode is the direct

 Bioelectrochemical Systems: From Extracellular Electron Transfer to Biotechnological Application. Edited by Korneel Rabaey, Largus Angenent, Uwe Schröder and Jürg Keller. ISBN: 9781843392330. Published by IWA Publishing, London, UK.

extracellular electron transfer by means of cell membrane bound redox-proteins. Thus far, *Shewanella oneidensis* and *Geobacter sulfurreducens* have been extensively studied for extracellular respiration with an anode. Both are known for their ability to respire with solid extracellular electron acceptors, such as mineral oxides of Fe, Mn, and U, and are model representatives of the group of dissimilatory metal-reducing bacteria (DMRB) (Nealson *et al.* 2002; Weber *et al.* 2006). Common to those organisms is a chain of redox-proteins (mostly cytochrome *c*-type) to conduct electrons across the cell envelope. The true pathway of the electrons and the importance of specific proteins is still under investigation, but researchers have identified strong suspects that may be involved in this respiration process. However, a genomic screen for both species shows a vast number of coded *c*-type cytochromes, which hinder standard biomolecular knock-out studies to reveal the function of certain proteins. The possibility of expressing the most promising assembly of *c*-type cytochromes in a heterologous host (*Escherichia coli*) will be discussed with regard to both scientific (investigative) and technological (improvement) interest.

Biomolecular techniques help us clarify the essential electron transfer reaction steps; the first approaches are under way to engineer and design microorganisms that can utilize complex fuels and convert them into electricity in much more efficient ways. This chapter will discuss recent work on the investigation and improvement of these extracellular respiration reactions using genetic tools and give an outlook of what may be possible in the near future.

6.2 EXTRACELLULAR RESPIRATION IN *SHEWANELLA ONEIDENSIS* AND *GEOBACTER SULFURREDUCENS*

Shewanella or *Geobacter* spp. respiration with extracellular electron acceptors requires the transport of catabolic electrons across the cell envelope. Proposed electron transfer pathways for both organisms are shown in Figure 6.1. Common to extracellular respiration and aerobic or other anaerobic respiration pathways is reoxidation of NAD(P)H by cell-membrane bound dehydrogenases. Protons are pumped over the cell membrane to generate a proton motive force for subsequent ATP generation, while electrons are fed into a quinone/menaquinone pool located in the cell membrane. For both species, the subsequent steps of the electron pathway are not yet completely proven, and several possible pathways are discussed in the literature. As a first step in this chapter, we will reflect on the most probable reaction steps.

In *S. oneidensis*, cell membrane bound CymA, which is a tetraheme cytochrome, takes up electrons from the quinone/menaquinone pool to reduce

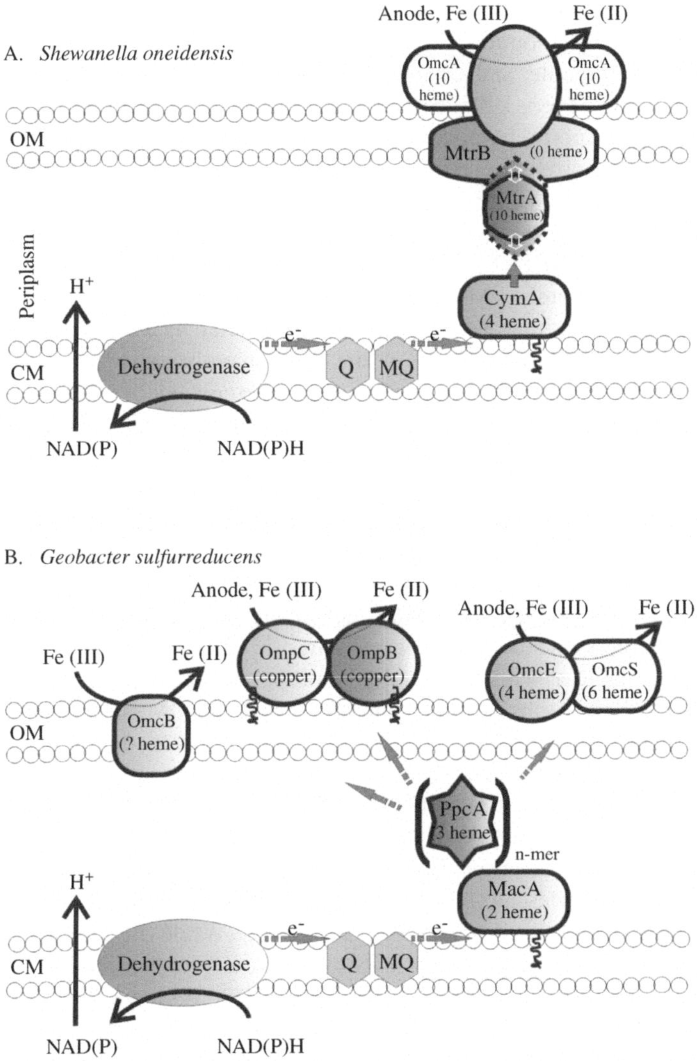

Figure 6.1. Proposed electron transfer steps for extracellular respiration with mineral oxides or MFC anodes in *Shewanella oneidensis* MR-1 (A) and *Geobacter sulfurreducens* (B). OM – outer membrane, CM – cell (inner) membrane, Q and MQ – quinone/menaquinone pool; all other proteins are discussed in the main text

periplasmic reductases (Myers and Myers 2000) (Figure 6.1A). It is involved in a broad range of respiration reactions, for example with Fe(III), Mn(IV), nitrate, nitrite, fumarate, and DMSO (Gescher *et al.* 2008; Murphy and Saltikov 2007; Myers and Myers 1997; Schwalb *et al.* 2002, 2003). From this point on, the electron transport chain for extracellular respiration is uncoupled from proton transport across the inner membrane. CymA reduces the periplasmic decaheme protein MtrA. MtrA can act as a terminal reductase to reduce soluble Fe(III) (e.g. Fe(III)-citrate), and it most likely functions as a linking protein in the periplasm to shuttle electrons between CymA at the inner membrane and outer-membrane proteins MtrB and MtrC (Pitts *et al.* 2003). Very recent studies have shown that MtrA is weakly associated with MtrB, and thus anchored to the periplasmic side of the outer membrane (Ross *et al.* 2007). Interaction and complex formation studies also revealed a strong complex being formed between the noncytochrome *c* protein MtrB and decaheme cytochrome MtrC (Ross *et al.* 2007). MtrC (also called OmcB), on the other hand, has been found to form a complex with an OmcA dimer (Shi *et al.* 2006). MtrC and OmcA have been proven to be terminal reductases that are crucial for the reduction of insoluble iron (hydr)oxides and electron transfer to MFC electrodes (Beliaev *et al.* 2001; Bretschger *et al.* 2007; Hartshorne *et al.* 2007; Myers and Myers 2001; 2003). MtrB is crucial for the correct localization of MtrC and OmcA in the outer membrane (Myers and Myers 2002). This information, together with the complex formation studies, suggests that non-cytochrome MtrB forms a sleeve around MtrA and MtrC and enables the electron transfer (or tunnelling) from periplasmic MtrA to the outer membrane MtrC, while MtrC either directly reduces external electron acceptors or passes the electrons on to OmcA (Figure 6.1A).

In *Geobacter sulfurreducens*, MacA, an inner membrane anchored diheme cytochrome, receives electrons from the quinone/menaquinone pool (Butler *et al.* 2004; Kim and Lovley 2008) (Figure 6.1B). MacA reduces the periplasmic triheme cytochrome PpcA (Lloyd *et al.* 2003). While in *S. oneidensis* the electron transfer is most probably uncoupled from proton transfer beyond the reactions in the inner membrane, Pessanha *et al.* (2006) have shown that PpcA has the potential to contribute to the cells' proton motive force, and thus energy (ATP) generation by transferring electrons coupled to trans-membrane proton transport. It was proposed that PpcA polymerizes in a chain-like conformation to conduct electrons as a periplasmic 'nanowire' from the inner membrane to outer membrane reductases (Pokkuluri *et al.* 2004). Several different terminal reductases are crucial for either metal oxide reduction only, or electron transfer to metal oxides and anode electrodes. Thus, the cytochromes OmcE and OmcS with four and six heme groups, respectively, which are located on the outside of the outer membrane of *G. sulfurreducens,* seem to be essential for respiring when using a MFC anode as the terminal electron acceptor (Holmes *et al.* 2006; Mehta *et al.* 2005). Both, also are necessary for efficient reduction of Fe(III) (hydr)oxides. On the other hand, a deletion mutation of outer membrane cytochrome OmcB only affects the reduction of extracellular Fe(III) and not the electron transfer to an anode (Holmes *et al.* 2006; Leang *et al.* 2003). Another pair of reductases, which are even less understood but apparently important, are multi-copper proteins OmpC and OmpB (both being crucial for Fe(III) reduction and electron transfer to anodes) (Holmes *et al.* 2008). Thus, the specific protein functions can be very different and leave much room for investigation.

6.3 SCIENTIFIC MOTIVATION FOR HETEROLOGOUS GENE EXPRESSION

The genomes of *S. oneidensis* MR-1 and *G. sulfurreducens* PCA have been completely sequenced and are available to the public (see reference to Comprehensive Microbial Resources). A genome wide search for both organisms predicts a total of 42 and 111 membrane-associated *c*-type cytochromes for *S. oneidensis* and *G. sulfurreducens,* respectively (Methe *et al.* 2003; Meyer *et al.* 2004). The high numbers of proteins involved in potential anaerobic respiration processes, make functional studies within these organisms very complicated. The enzymatic pathways presented in Section 6.2 are not completely proven. One reason being that deletion mutations often do not lead to an overall loss of function in the mutant because other cytochromes take over and allow the reaction to happen. It is also becoming more and more obvious that results we gain from iron or manganese oxide reduction experiments are not necessarily similar for electron transfer to anodes. For example, Bretschger *et al.* (2007) showed that *S. oneidensis* knock-out mutations of the outer membrane reductases OmcA/MtrC only repress iron oxide reduction and electron transfer rates to the anode, but not manganese oxide reduction rates; while knock-outs of periplasmic proteins MtrA and MtrB limit the ability to reduce all three types of electron acceptors. Obviously, some proteins participate in a broad range of enzymatic pathways, while others are very specific for certain reactions.

Knock-out studies also demonstrate that *c*-type cytochromes, which are crucial for Fe(III) reduction or electron transfer to anodes (Figure 6.1), often require the expression of other cytochromes that are not directly involved in the respiration reaction to allow full functionality. For example, *G. sulfurreducens* *c*-type cytochromes OmcG and OmcH are hypothesized to post-translationally affect OmcB. As mentioned earlier, OmcB has been proven to be essential for Fe(III) reduction in *G. sulfurreducens*, and therefore knocking out OmcG and OmcH will reduce reduction rates indirectly (Kim *et al.* 2006). In addition, the small outer membrane cytochrome OmcF from *G. sulfurreducens* influences expression levels of essential cytochromes for Fe(III) reduction: in an OmcF-deficient strain no transcripts of OmcB could be detected, while OmcS was upregulated (Kim *et al.* 2005). Recently, a possibly similar function of expression control of OmcB, and thus indirect involvement in the reduction reaction, has also been described for MacA, which is a protein that had been considered a key-player in the electron transfer chain of *G. sulfurreducens* (Kim and Lovley 2008).

A number of nonspecific proteins are essential to ensure proper post-translational folding and membrane localization of the terminal reductases. First, the respiratory proteins require transport over the inner membrane of the bacteria by the evolutionary conserved Sec pathway (most proteins designated for export leave the cytoplasm via this translocation system). Second, the proteins mature with the help of cytochrome *c* maturation complexes, which assist the correct attachment of heme molecules to the proteins within the periplasmic space (Feissner *et al.* 2006). Third, if necessary, they are then transported over the outer membrane, which is most probably facilitated by the Type II secretion system (Shi *et al.* 2008).

Overall, due to the large number of *c*-type cytochromes and the special processing requirements of the proteins, electron transfer reactions to extracellular electron acceptors in *S. oneidensis* and *G. sulfurreducens* are very complex, which makes functional studies of individual enzymes difficult.

One way to study the minimal enzymatic requirement for respiration with extracellular electron acceptors is the heterologous expression of candidate genes in a host strain without the machinery to reduce external electron acceptors. Genetic engineering star organism *E. coli* is perfectly suitable for this task, because the wild-type strain has no intrinsic ability to reduce external electron acceptors and does not posses outer membrane *c*-type cytochromes. More importantly, a broad range of biomolecular tools are already available for manipulating and screening *E. coli* cell lines. The genes of potential enzyme candidates can be transferred into *E. coli* and studied for expression, folding, localization, and *in-vitro* functionality. Via step by step addition of the most promising proteins into the heterologous host, we may be able to build the minimum chain of proteins that will allow *E. coli* to respire with extracellular terminal electron acceptors. Once this minimal required protein chain is established in one *E. coli* working cell line, the protein expression can be tuned and optimized to maximize product yield (current production). The gene assembly can then also be transferred into a broad variety of available *E. coli* cell lines that have been optimized for certain purposes, such as specific substrate degradation or product formation to broaden the application range of these new highly productive strains.

Although our toolkit for genetic engineering in *E. coli* is tremendous, our existing techniques to delegate post-translational modifications are still very scarce, making such a research effort difficult. Membrane-associated or secreted proteins often require extensive post-translational maturation to become functional. It is a scientific challenge to express one membrane protein in a heterologous host while ensuring correct maturation and localization. Expressing a series of proteins, which all need to be functionalized in very specific spaces within the cell envelop and, in addition, to allow functional interaction between the individual proteins, is an even greater task.

6.4 METHODS AND CHALLENGES FOR HETEROLOGOUS GENE EXPRESSION IN *E. COLI*

As mentioned in the last section, the scientific motivation for heterologous expression of *c*-type cytochromes is considerable. Following from this scientific interest, some first attempts to express cytochromes from both organisms – *S. oneidensis* and *G. sulfurreducens* – in *E. coli* have been published. The strategies and successes of the different approaches will be reflected in this section.

One general requirement for the functional expression of *c*-type cytochromes is the successful attachment of heme with the help of a cytochrome *c* biogenesis system in the periplasm. In the microbial world this maturation process is realized by either System I (α- and γ-proteobacteria and archea) or System II (gram-positive bacteria, ε- and β-proteobacteria and cyanobacteria) (Kranz *et al.* 1998; Thony-Meyer 1997). *E. coli*'s own cytochrome *c* maturation complex System I (coded by the gene cluster *ccmABDCEFGH*) is not expressed under aerobic conditions and, under normal expression rates, most probably would also not have the capacity to allow the maturation of several multi-heme cytochromes (up to 12 heme molecules per cytochrome). Successful over-expression of System I and System II in *E. coli* has been shown before (Arslan *et al.* 1998; Feissner *et al.* 2006) and co-expression with the target cytochrome *c* is regarded as an important requirement for the heterologous expression of multi-heme cytochromes (Herbaud *et al.* 2000).

To the best of our knowledge, only individual recombinant cytochrome *c* expressions in *E. coli* have been published, indicating how challenging and complex the functional expression of cytochrome *c* assemblies is.

For *S. oneidensis* the expression of the following cytochromes has been studied:

MtrA – was the first *Shewanella* cytochrome *c* to be heterologously expressed in *E. coli* (Pitts *et al.* 2003). The protein was correctly targeted to the periplasm and matured with System I cytochrome *c* assembly complex, which was co-expressed from the pEC86 vector (Arslan *et al.* 1998). Analysis with magnetic circular dichroism gave evidence that all ten heme molecules were covalently attached. *In-vivo* functional assays showed that MtrA was able to receive metabolic electrons from the inner membrane and reduce the soluble electron acceptor Fe(III)-NTA in the periplasm or donate electrons to other host cell oxidoreductases (e.g. the nitrite reductase NafA).

CymA – was cloned and expressed in *E. coli* by Gescher *et al.* (2008) using a pBAD202 vector and likewise co-expressed with the pEC86 plasmid containing the *E. coli* System I cytochrome *c* maturation genes. The protein was functionally expressed in a Δ*napCDEF* mutant as periplasmic NapC could perform similarly to CymA. The CymA-expressing strain was able to grow on non-fermentable glycerol and with Fe(III)-nitrilotriacetic acid (Fe(III)-NTA) as soluble terminal electron acceptor, showing that CymA itself is not only a electron-mediating enzyme, but also has terminal reductase activity. However, *in-vivo* respiration could only be shown for Fe(III)-NTA. Other soluble electron acceptors that can be reduced by *S. oneidensis,* such as AQDS, could not be used.

OmcA – was co-expressed with pEC86 in two *E. coli* cell lines: a K12 derivative and a B-type strain (Donald *et al.* 2008). Both strains produced comparable amounts of OmcA, which was located in inner and outer membrane fractions, but not in the soluble fraction (cytoplasm and periplasm). In both strains, the heterologously produced OmcA is functional for soluble Fe(III) reduction, however, only the B-type strain BL21 was able to reduce insoluble amorphous iron (hydr)oxide. These findings indicate that only in the B-type strain is OmcA exposed to the outside of the cell, while in the K12 cell line it is stuck on the inner side of the outer membrane. This is consistent with previous reports on a cryptic type II protein secretion pathway across the outer membrane in *E. coli* K12 strains (Francetic and Pugsley 1996). Deletion mutations within the type II secretion pathway of the B-type strain resulted in a similar phenotype to the K12 strain, confirming that the type II secretion pathway is a crucial factor for the successful expression of outer membrane cytochromes. The successful secretion of recombinant proteins with the host secretion system has been found to be rather rare as the substrate recognition of the secretion complex is very specific

(Filloux 2004). Thus, successful trans-membrane transport in the B-type strain without the requirement for the synthesis of a fusion protein to simplify recognition (as described by d'Enfert and Pugsley 1987), is a crucial piece of information for the functional expression of other potentially important outer membrane cytochromes.

Besides these individual protein expression approaches, Londer *et al.* (2008), recently published six new expression vectors that allow high-throughput parallel expression of cytochrome *c*. The potential of the new vectors was shown with the exemplary expression of 30 hypothetical *S. oneidensis c*-type cytochromes with four or less heme molecules (predicted via genome sequence analysis). The authors reported a high success rate with 26 of the 30 cytochromes being expressed. However, their cloning effort only incorporated the soluble part of the proteins to allow for best purification. The correct localization in the membrane was not the goal of this study.

For *G. sulfurreducens* the expression of the following c-type cytochromes has been studied:

PpcA – was first cloned and heterologously expressed in 2002 by Londer *et al.* (2002). This early work resulted in a fully matured (co-expression of System I genes) and *in-vitro* functional enzyme. Since then, PpcA has been extensively studied in *E. coli* and *G. sulfurreducens*. The structure of PpcA has been resolved by X-ray diffraction (Pokkuluri *et al.* 2004) and, recently, it has been studied via nuclear magnetic resonance (Morgado *et al.* 2008). Both works suggest that PpcA linearly polymerizes, with the three heme molecules of each monomer forming a chain within the polymer resulting in an electron-transferring 'nanowire' in the periplasm. Recently, a new expression system was developed, which allows isotopic labelling of the cytochrome and promises further elucidation of structural functions (Fernandes *et al.* 2008). Since PpcA is one of the hypothesized key-proteins in the centre of the electron transport chain of *G. sulfurreducens* (see Figure 6.1), the knowledge of the *in-vivo* function is of high interest.

Further, **two dodecaheme cytochromes (GSU1996 and GSU0592)** – representing a new class of cytochromes – have been expressed in and purified from *E. coli* (Londer *et al.* 2006). Processing was performed by System I expressed from pEC86. Mass spectrometry analysis indicates that all 12 hemes were attached. These cytochromes' true function, to date, is unknown, but sequence analyses predict that their structure could be a one-dimensional chain of four homologous domains, each containing three heme molecules. Thus, these dodecaheme cytochromes could be the 'one-protein' version of the predicted polymerized PpcA chain and form a biological 'nanowire' through the cell membrane.

From the expression of these individual cytochromes in *E. coli* we can see that: (I) the co-expression of a cytochrome *c* biogenesis system is an important and successful way to produce multi-heme proteins in a heterologous host; (II) inner membrane or periplasmic cytochromes can be expressed and localized correctly within the heterologous host organism; and (III) B-type *E. coli* strains open a promising new alternative to the construction of fusion proteins for the proper localization of outer membrane proteins via the type II secretion pathway. The big remaining challenge is the application of these strategies to express multiple cytochromes in the same host cell and to allow their functional interaction.

One further issue for the recombinant expression of multiple cytochrome *c* proteins and their accessories that has not been addressed yet, but is well known for the expression of other large enzymatic complexes, are crowding conditions within the cell. Diffusion processes within the cell are three to fourfold slower for small molecules and even slower for large proteins (Conrado *et al.* 2008). Thus, we have to consider if expressed proteins experience a 'traffic jam' in the cytoplasm and how we can overcome this challenge.

6.5 BIOTECHNOLOGICAL APPLICATIONS – DESIGNING THE 'SUPER BUG'

6.5.1 The 'super bug' for BES applications

In section 6.4, we broadly discussed scientific motivations for the transfer of the enzymatic machinery for extracellular respiration into *E. coli*. Most of the previous work has been performed because of scientific objectives. However, with growing interest in *S. oneidensis* and *G. sulfurreducens* as biocatalysts in BES, the incentive for recombinant expression of the transmembrane cytochrome *c* assembly has changed. This section will discuss different approaches to use biomolecular engineering to improve BES performance. Hereby, we will not only discuss the heterologous expression of the proteins required for extracellular respiration, but also other possible metabolic changes that will help to design highly efficient and versatile biocatalysts for BES applications.

Beginning with cytochrome *c* proteins: for biotechnological applications it is desirable to respire extracellularly in an easy-to-grow, easy-to-maintain, and easy-to-manipulate host microbe. With the successful over-expression of the minimum required set of cytochrome *c* proteins, we wish to maximize direct electron transfer to a BES anode. While extracellular respiration of *S. oneidensis* is mainly based on the oxidation of lactate (oxidation to acetate and CO_2 gives 4 electrons) and that of *G. sulfurreducens* is based on acetate oxidation (delivering a theoretical maximum of 8 electrons), engineered *E. coli* could presumably completely oxidize glucose to CO_2 and generate 24 electrons. Thus, more complex and efficient 'fuels' could be directly converted into electricity and the improved fuel conversion efficiency would result in better BES performance. The biocatalyst potentially could be designed based on the fuel stream for the BES to further improve efficiency. In addition, if biocatalysts are specifically designed to convert certain fuels (or chemical species) with a high efficiency, BES could also find new applications as environmental or industrial biosensors. Thus, BES current generation could be used as a qualitative and quantitative measure for one specific chemical substrate that is available to the genetically designed biocatalyst during a sensing event.

It should be noted that although this chapter examined the recombinant expression of proteins involved in extracellular respiration in *E. coli*, other host organisms, with other desirable substrate degradation abilities, may also be considered for this purpose. However, *E. coli* is the model organism that is most suitable for breaking ground in this field, because the biomolecular tools for heterologous expression first need to be developed for the other microbes.

While the protein machinery for extracellular respiration would allow *E. coli* to use an anode as the terminal electron acceptor when in direct cell-to-electrode contact, other microorganisms can transfer electrons even when they are not in direct contact. Several bacteria are known to produce soluble electron shuttle molecules (mediators) to pass metabolic electrons to electron acceptors that are not in close cell proximity. For example, *S. oneidensis,* in addition to its cell wall cytochromes, produces soluble riboflavin molecules that can shuttle electrons to external electron acceptors (Marsili *et al.* 2008; von Canstein *et al.* 2008). In Addition, *Pseudomonas aeruginosa* produces phenazine-type redox shuttles to dispose of electrons (Pham *et al.* 2008; Rabaey *et al.* 2005; Rabaey *et al.* 2004). Thus, it is conceivable to further engineer *E. coli* to produce its own specific mediators and use an electrode at a distance as terminal electron acceptor (recently, it was published that *E. coli* could use certain metabolites as redox-shuttles, however, the proof that the suggested compounds actually mediate electrons to an electrode is still lacking (Zhang *et al.* 2008)).

A completely different objective would be to engineer a host microbe that can accept electrons from an electrode rather than donate. Such a bacterium would use the cathode as electron (energy) donor to perform biochemical reactions. In classical BES this would correspond to the reduction of oxygen or alternative electron acceptors (inclusive contaminants) (He and Angenent 2006; Rabaey and Keller 2008). More recently, in microbial electrolysis cells (MECs) this can be the production of valuable chemicals. Thus far, the production of chemicals with MECs mainly focused on cathodic hydrogen production (Rozendal *et al.* 2006). However, the reaction rate of microbial electrolysis processes at this point is greatly limited by the bioelectrochemical process. This lower rate of chemical production may render bioelectrochemical systems (both MFCs and MECs) economically feasible when high-value products, such as fatty acids or vitamins are generated on the cathode with engineered microbes as biocatalysts. One example for a cathodic electron-accepting microorganism is *G. sulfurreducens* (Gregory *et al.* 2004). Thus, a genetic engineering effort could again start with heterologous cloning of *G. sulfurreducens* genes into other host organisms that produce certain valuable chemicals, followed by engineering towards higher efficiency of the electron accepting process. With this approach the objective of BES would switch from a power generating device to a biochemical synthesis reactor.

6.5.2 The 'super bug' for bioremediation applications

Before *S. oneidensis* and *G. sulfurreducens* were discussed as model organisms for direct electron transfer in MFCs, they were intensively studied because of their potential importance in bioremediation processes. *Shewanella* spp. are of great interest due to their ability to respire with a great variety of soluble and insoluble electron acceptors (Nealson *et al.* 2002). Their capacity to reduce a broad range of often toxic or radioactive heavy metal compounds into less environmentally threatening insoluble states, or to reduce halogenated organics is of high interest at contaminated sites (Hau and Gralnick 2007). However, the natural abundance and activity of *Shewanella* spp. is limited to aquatic habitats, while bioremediation applications are also desirable in soil environments. In addition, the rates of bioremediation are low even in aquatic habitats owing to the relatively low quantity of the biocatalysts.

Genetically modified bacteria could be tailored to exclusively live off the reduction of the contaminating species. These organisms could be applied in high concentrations to contaminated sites of, for example, mines, military land, volcanoes, or flooded areas. In a comparably short time, toxic species could be detoxified or converted into less harmful states (e.g. reduction of U(VI) to insoluble U(IV) or reduction of Cr(VI) to oxide forming Cr(III)). If we could force such bacteria to exclusively live off the contaminating species, the controlled host organism would simply die off once the site is clean.

Besides the scientific challenge, these ideas challenge the legal and ethical boundaries of today's society. However, the application of such clean up biocatalysts in more industrial settings (e.g. in closed systems) is conceivable. Heavy metal or radioactive industrial waste waters could be treated with specialized microorganisms to reduce and precipitate heavy metal salts in a new, effective and inexpensive way.

6.6 CLOSING REMARKS

In this chapter we presented the possibilities and opportunities of genetic engineering tools to the scientific investigation of physiological processes in BES. In addition, we discussed the design of desired microbial biocatalysts for biotechnological applications. We mainly focused on the recombinant expression of the cytochrome *c* apparatuses, which allow *S. oneidensis* and *G. sulfurreducens* to respire with extracellular electron acceptors in *E. coli*. Due to the complexity of multi-protein expression, post-translational modification, and localization, the scientific challenges of this approach are tremendous. Yet, some important breakthroughs with single protein expressions during recent years have delivered us the important biomolecular tools to get closer to this goal. It should be noted that the application of biomolecular tools to study or improve bacteria for BES applications are certainly not limited to the proteins of extracellular electron transfer and to recombinant expression in *E. coli*. Genetic engineering approaches can target any desired metabolic pathway and there are no theoretical limits to overexpression in the target microbe or recombinant expression in any host organism, albeit the necessary genetic toolkits may need to be developed. This very young field of biological engineering leaves a lot of room for dreaming. Scientific reality will reveal what is practically possible and in what time frame we can achieve our goals.

6.7 ACKNOWLEDGEMENTS

The financial support for M.R. was provided through a specific collaborative agreement between L.T.A. and the Fermentation Biotechnology Research Unit, USDA, Agricultural Research Service, Peoria, Illinois. L.T.A. acknowledges support through US NSF CAREER grant #0645021.

REFERENCES

Arslan, E., Schulz, H., Zufferey, R., Kunzler, P. and Thony-Meyer, L. (1998) Overproduction of the *Bradyrhizobium japonicum c*-type cytochrome subunits of the *cbb3* oxidase in *Escherichia coli*. *Biochem. Biophys. Res. Commun.* **251**(3), 744–747.

Beliaev, A.S., Saffarini, D.A., McLaughlin, J.L. and Hunnicutt, D. (2001) MtrC, an outer membrane decaheme *c* cytochrome required for metal reduction in *Shewanella putrefaciens* MR-1. *Molec. Microbiol.* **39**(3), 722–730.

Bretschger, O., Obraztsova, A., Sturm, C.A., Chang, I.S., Gorby, Y.A., Reed, S.B., Culley, D.E., Reardon, C.L., Barua, S., Romine, M.F., Zhou, J., Beliaev, A.S., Bouhenni, R., Saffarini, D., Mansfeld, F., Kim, B.H., Fredrickson, J.K. and Nealson,

K.H. (2007) Current production and metal oxide reduction by *Shewanella oneidensis* MR-1 wild type and mutants. *Appl. Environ. Microbiol.* **73**(21), 7003–7012.

Butler, J.E., Kaufmann, F., Coppi, M.V., Nunez, C. and Lovley, D.R. (2004) MacA, a diheme *c*-type cytochrome involved in Fe(III) reduction by *Geobacter sulfurreducens. J. Bacteriol.* **186**(12), 4042–4045.

Comprehensive Microbial Resource (CMR) at the J. Graig Venter Insitute http://cmr.jcvi.org/tigr-scripts/CMR/CmrHomePage.cgi

Conrado, R.J., Varner, J.D. and DeLisa, M.P. (2008) Engineering the spatial organization of metabolic enzymes: mimicking nature's synergy. *Curr. Opin. Biotechnol.* **19**(5), 492–499.

d'Enfert, C. and Pugsley, A.P. (1987) A gene fusion approach to the study of pullulanase export and secretion in *Escherichia coli. Mol. Microbiol.* **1**(2), 159–168.

Donald, J.W., Hicks, M.G., Richardson, D.J. and Palmer, T. (2008) The *c*-type cytochrome OmcA localizes to the outer membrane upon heterologous expression in *Escherichia coli. J. Bacteriol.* **190**(14), 5127–5131.

Feissner, R.E., Richard-Fogal, C.L., Frawley, E.R., Loughman, J.A., Earley, K.W. and Kranz, R.G. (2006) Recombinant cytochromes *c* biogenesis systems I and II and analysis of haem delivery pathways in *Escherichia coli. Mol. Microbiol.* **60**(3), 563–577.

Fernandes, A.P., Couto, I., Morgado, L., Londer, Y.Y. and Salgueiro, C.A. (2008) Isotopic labeling of *c*-type multiheme cytochromes overexpressed in *E. coli. Protein Expr. Purif.* **59**(1), 182–188.

Filloux, A. (2004) The underlying mechanisms of type II protein secretion. *Biochim. Biophys. Acta* **1694**(1–3), 163–179.

Francetic, O. and Pugsley, A.P. (1996) The cryptic general secretory pathway (gsp) operon of *Escherichia coli* K-12 encodes functional proteins. *J. Bacteriol.* **178**(12), 3544–3549.

Gescher, J.S., Cordova, C.D. and Spormann, A.M. (2008) Dissimilatory iron reduction in *Escherichia coli*: identification of CymA of *Shewanella oneidensis* and NapC of *E. coli* as ferric reductases. *Mol. Microbiol.* **68**(3), 706–719.

Gregory, K.B., Bond, D.R. and Lovley, D.R. (2004) Graphite electrode as electron donors for anaerobic respiration. *Environ. Microbiol.* **6**(6), 596–604.

Hartshorne, R.S., Jepson, B.N., Clarke, T.A., Field, S.J., Fredrickson, J., Zachara, J., Shi, L., Butt, J.N. and Richardson, D.J. (2007) Characterization of *Shewanella oneidensis* MtrC: a cell-surface decaheme cytochrome involved in respiratory electron transport to extracellular electron acceptors. *J. Biol. Inorg. Chem.* **12**(7), 1083–1094.

Hau, H.H. and Gralnick, J.A. (2007) Ecology and biotechnology of the genus Shewanella. *Ann. Rev. Microbiol.* **61**, 237–258.

He, Z. and Angenent, L.T. (2006) Application of bacterial biocathodes in microbial fuel cells. *Electroanalysis* **18**(19–20), 2009–2015.

Herbaud, M.L., Aubert, C., Durand, M.C., Guerlesquin, F., Thony-Meyer, L. and Dolla, A. (2000) *Escherichia coli* is able to produce heterologous tetraheme cytochrome *c(3)* when the *ccm* genes are co-expressed. *Biochim. Biophys. Acta* **1481**(1), 18–24.

Holmes, D.E., Chaudhuri, S.K., Nevin, K.P., Mehta, T., Methe, B.A., Liu, A., Ward, J.E., Woodard, T.L., Webster, J. and Lovley, D.R. (2006) Microarray and genetic analysis of electron transfer to electrodes in *Geobacter sulfurreducens*. *Environ. Microbiol.* **8**(10), 1805–1815.

Holmes, D.E., Mester, T., O'Neil, R.A., Perpetua, L.A., Larrahondo, M.J., Glaven, R., Sharma, M.L., Ward, J.E., Nevin, K.P. and Lovley, D.R. (2008) Genes for two multicopper proteins required for Fe(III) oxide reduction in *Geobacter sulfurreducens* have different expression patterns both in the subsurface and on energy-harvesting electrodes. *Microbiology* **154**, 1422–1435.

Kim, B.C. and Lovley, D.R. (2008) Investigation of direct vs. indirect involvement of the *c*-type cytochrome MacA in Fe(III) reduction by *Geobacter sulfurreducens*. *FEMS Microbiol. Lett.* **286**(1), 39–44.

Kim, B.C., Leang, C., Ding, Y.H., Glaven, R.H., Coppi, M.V. and Lovley, D.R. (2005) OmcF, a putative *c*-type monoheme outer membrane cytochrome required for the expression of other outer membrane cytochromes in *Geobacter sulfurreducens*. *J. Bacteriol.* **187**(13), 4505–4513.

Kim, B.C., Qian, X., Leang, C., Coppi, M.V. and Lovley, D.R. (2006) Two putative *c*-type multiheme cytochromes required for the expression of OmcB, an outer membrane protein essential for optimal Fe(III) reduction in *Geobacter sulfurreducens*. *J. Bacteriol.* **188**(8), 3138–3142.

Kranz, R., Lill, R., Goldman, B., Bonnard, G. and Merchant, S. (1998) Molecular mechanisms of cytochrome *c* biogenesis: three distinct systems. *Molec. Microbiol.* **29**(2), 383–396.

Leang, C., Coppi, M.V. and Lovley, D.R. (2003) OmcB, a *c*-type polyheme cytochrome, involved in Fe(III) reduction in *Geobacter sulfurreducens*. *J. Bacteriol.* **185**(7), 2096–2103.

Lloyd, J.R., Leang, C., Hodges Myerson, A.L., Coppi, M.V., Cuifo, S., Methe, B., Sandler, S.J. and Lovley, D.R. (2003) Biochemical and genetic characterization of PpcA, a periplasmic *c*-type cytochrome in *Geobacter sulfurreducens*. *Biochem. J.* **369**(Pt 1), 153–161.

Londer, Y.Y., Pokkuluri, P.R., Tiede, D.M. and Schiffer, M. (2002) Production and preliminary characterization of a recombinant triheme cytochrome *c(7)* from *Geobacter sulfurreducens* in *Escherichia coli*. *Biochim. Biophys. Acta* **1554**(3), 202–211.

Londer, Y.Y., Pokkuluri, P.R., Orshonsky, V., Orshonsky, L. and Schiffer, M. (2006) Heterologous expression of dodecaheme "nanowire" cytochromes *c* from *Geobacter sulfurreducens*. *Protein Expr. Purif.* **47**(1), 241–248.

Londer, Y.Y., Giuliani, S.E., Peppler, T. and Collart, F.R. (2008) Addressing *Shewanella oneidensis* "cytochromome": the first step towards high-throughput expression of cytochromes *c*. *Protein Expr. Purif.* **62**(1), 128–137.

Marsili, E., Baron, D.B., Shikhare, I.D., Coursolle, D., Gralnick, J.A. and Bond, D.R. (2008) *Shewanella* secretes flavins that mediate extracellular electron transfer. *Proc. Natl. Acad. Sci. USA* **105**(10), 3968–3973.

Mehta, T., Coppi, M.V., Childers, S.E. and Lovley, D.R. (2005) Outer membrane *c*-type cytochromes required for Fe(III) and Mn(IV) oxide reduction in *Geobacter sulfurreducens*. *Appl. Environ. Microbiol.* **71**(12), 8634–8641.

Methe, B.A., Nelson, K.E., Eisen, J.A., Paulsen, I.T., Nelson, W., Heidelberg, J.F., Wu, D., Wu, M., Ward, N., Beanan, M.J., Dodson, R.J., Madupu, R., Brinkac, L.M., Daugherty, S.C., DeBoy, R.T., Durkin, A.S., Gwinn, M., Kolonay, J.F., Sullivan,

S.A., Haft, D.H., Selengut, J., Davidsen, T.M., Zafar, N., White, O., Tran, B., Romero, C., Forberger, H.A., Weidman, J., Khouri, H., Feldblyum, T.V., Utterback, T.R., Van Aken, S.E., Lovley, D.R. and Fraser, C.M. (2003) Genome of *Geobacter sulfurreducens*: metal reduction in subsurface environments. *Science* **302**(5652), 1967–1969.

Meyer, T.E., Tsapin, A.I., Vandenberghe, I., de Smet, L., Frishman, D., Nealson, K.H., Cusanovich, M.A. and van Beeumen, J.J. (2004) Identification of 42 possible cytochrome *c* genes in the *Shewanella oneidensis* genome and characterization of six soluble cytochromes. *Omics* **8**(1), 57–77.

Morgado, L., Bruix, M., Orshonsky, V., Londer, Y.Y., Duke, N.E.C., Yang, X.J., Pokkuluri, P.R., Schiffer, M. and Salgueiro, C.A. (2008) Structural insights into the modulation of the redox properties of two *Geobacter sulfurreducens* homologous triheme cytochromes. *Biochim. Biophys. Acta, Bioenerg.* **1777**(9), 1157–1165.

Murphy, J.N. and Saltikov, C.W. (2007) The *cymA* gene, encoding a tetraheme *c*-type cytochrome, is required for arsenate respiration in *Shewanella* species. *J. Bacteriol.* **189**(6), 2283–2290.

Myers, C.R. and Myers, J.M. (1997) Cloning and sequence of *cymA*, a gene encoding a tetraheme cytochrome *c* required for reduction of iron(III), fumarate, and nitrate by *Shewanella putrefaciens* MR-1. *J. Bacteriol.* **179**(4), 1143–1152.

Myers, C.R. and Myers, J.M. (2002) MtrB is required for proper incorporation of the cytochromes OmcA and OmcB into the outer membrane of *Shewanella putrefaciens* MR-1. *Appl. Environ. Microbiol.* **68**(11), 5585–5594.

Myers, J.M. and Myers, C.R. (2000) Role of the tetraheme cytochrome CymA in anaerobic electron transport in cells of *Shewanella putrefaciens* MR-1 with normal levels of menaquinone. *J. Bacteriol.* **182**(1), 67–75.

Myers, J.M. and Myers, C.R. (2001) Role for outer membrane cytochromes OmcA and OmcB of *Shewanella putrefaciens* MR-1 in reduction of manganese dioxide. *Appl. Environ. Microbiol.* **67**(1), 260–269.

Myers, J.M. and Myers, C.R. (2003) Overlapping role of the outer membrane cytochromes of *Shewanella oneidensis* MR-1 in the reduction of manganese(IV) oxide. *Lett. Appl. Microbiol.* **37**(1), 21–25.

Nealson, K.H., Belz, A. and McKee, B. (2002) Breathing metals as a way of life: geobiology in action. *Antonie van Leeuwenhoek* **81**(1–4), 215–222.

Pessanha, M., Morgado, L., Louro, R.O., Londer, Y.Y., Pokkuluri, P.R., Schiffer, M. and Salgueiro, C.A. (2006) Thermodynamic characterization of triheme cytochrome PpcA from *Geobacter sulfurreducens*: evidence for a role played in e-/H+ energy transduction. *Biochemistry* **45**(46), 13910–13917.

Pham, T.H., Boon, N., De Maeyer, K., Hofte, M., Rabaey, K. and Verstraete, W. (2008) Use of *Pseudomonas* species producing phenazine-based metabolites in the anodes of microbial fuel cells to improve electricity generation. *Appl. Microbiol. Biotechnol.* **80**(6), 985–993.

Pitts, K.E., Dobbin, P.S., Reyes-Ramirez, F., Thomson, A.J., Richardson, D.J. and Seward, H.E. (2003) Characterization of the *Shewanella oneidensis* MR-1 decaheme cytochrome MtrA: expression in *Escherichia coli* confers the ability to reduce soluble Fe(III) chelates. *J. Biol. Chem.* **278**(30), 27758–27765.

Pokkuluri, P.R., Londer, Y.Y., Duke, N.E., Long, W.C. and Schiffer, M. (2004) Family of cytochrome *c*7-type proteins from *Geobacter sulfurreducens*: structure of one cytochrome *c*7 at 1.45 A resolution. *Biochemistry* **43**(4), 849–859.

Rabaey, K. and Keller, J. (2008) Microbial fuel cell cathodes: from bottleneck to prime opportunity? *Water Sci. Technol.* **57**(5), 655–659.

Rabaey, K., Boon, N., Hofte, M. and Verstraete, W. (2005) Microbial phenazine production enhances electron transfer in biofuel cells. *Environ. Sci. Technol.* **39**(9), 3401–3408.

Rabaey, K., Boon, N., Siciliano, S.D., Verhaege, M. and Verstraete, W. (2004) Biofuel cells select for microbial consortia that self-mediate electron transfer. *Appl. Environ. Microbiol.* **70**(9), 5373–5382.

Ross, D.E., Ruebush, S.S., Brantley, S.L., Hartshorne, R.S., Clarke, T.A., Richardson, D.J. and Tien, M. (2007) Characterization of protein-protein interactions involved in iron reduction by *Shewanella oneidensis* MR-1. *Appl. Environ. Microbiol.* **73**(18), 5797–5808.

Rozendal, R.A., Hamelers, H.V.M., Euverink, G.J.W., Metz, S.J. and Buisman, C.J.N. (2006) Principle and perspectives of hydrogen production through biocatalyzed electrolysis. *Int. J. Hydrogen Energy* **31**(12), 1632–1640.

Schwalb, C., Chapman, S.K. and Reid, G.A. (2002) The membrane-bound tetraheme *c*-type cytochrome CymA interacts directly with the soluble fumarate reductase in *Shewanella. Biochem. Soc. Trans.* **30**(4), 658–662.

Schwalb, C., Chapman, S.K. and Reid, G.A. (2003) The tetraheme cytochrome CymA is required of anaerobic respiration with dimethyl sulfoxide und nitrite in *Shewanella oneidensis. Biochemistry* **42**(31), 9491–9497.

Shi, L., Chen, B., Wang, Z., Elias, D.A., Mayer, M.U., Gorby, Y.A., Ni, S., Lower, B.H., Kennedy, D.W., Wunschel, D.S., Mottaz, H.M., Marshall, M.J., Hill, E.A., Beliaev, A.S., Zachara, J.M., Fredrickson, J.K. and Squier, T.C. (2006) Isolation of a high-affinity functional protein complex between OmcA and MtrC: Two outer membrane decaheme *c*-type cytochromes of *Shewanella oneidensis* MR-1. *J. Bacteriol.* **188**(13), 4705–4714.

Shi, L., Deng, S., Marshall, M.J., Wang, Z., Kennedy, D.W., Dohnalkova, A.C., Mottaz, H.M., Hill, E.A., Gorby, Y.A., Beliaev, A.S., Richardson, D.J., Zachara, J.M. and Fredrickson, J.K. (2008) Direct involvement of type II secretion system in extracellular translocation of *Shewanella oneidensis* outer membrane cytochromes MtrC and OmcA. *J. Bacteriol.* **190**(15), 5512–5516.

Thony-Meyer, L. (1997) Biogenesis of respiratory cytochromes in bacteria. *Microbiol. Mol. Biol. Rev.* **61**(3), 337–376.

von Canstein, H., Ogawa, J., Shimizu, S. and Lloyd, J.R. (2008) Secretion of flavins by *Shewanella* species and their role in extracellular electron transfer. *Appl. Environ. Microbiol.* **74**(3), 615–623.

Weber, K.A., Achenbach, L.A. and Coates, J.D. (2006) Microorganisms pumping iron: anaerobic microbial iron oxidation and reduction. *Nature* **4**, 752–764.

Zhang, T., Cui, C., Chen, S., Yang, H. and Shen, P. (2008) The direct electrocatalysis of *Escherichia coli* through electroactivated excretion in microbial fuel cell. *Electrochem. Commun.* **10**(2), 293–297.

7

Electrochemical losses

Uwe Schröder and Falk Harnisch

7.1 INTRODUCTION

The most striking feature of fuel cells is their potentially high energy conversion efficiency. Theoretically, in an ideal fuel cell the entire free energy, ΔG, of the fuel oxidation can be converted into electricity. This is the great advantage over heat-engine-based energy conversion processes, in which the theoretical efficiency is restricted by the Carnot theorem, $\varepsilon = (T_h - T_c)/T_h)$, with T_h and T_c being the temperatures of the hot and of the cold reservoir, respectively. This circumstance was emphasized originally in 1894 by Ostwald, who foretold and demanded the replacement of conventional thermo-dynamic machines by galvanic elements for the future of electricity generation (Ostwald 1894).

The above mentioned ideal fuel cell that allows a complete transformation of chemical energy into electricity is characterized by perfect electrodes, by electrochemical reactions that proceed at infinite rates, and by a negligible internal resistance. Of course, practical fuel cells – which comprise conventional as

 Bioelectrochemical Systems: From Extracellular Electron Transfer to Biotechnological Application. Edited by Korneel Rabaey, Largus Angenent, Uwe Schröder and Jürg Keller. ISBN: 9781843392330. Published by IWA Publishing, London, UK.

well as biological fuel cells – are far away from this ideal picture. Their internal resistance can be significant, and the electrochemical reactions proceed with only finite rates. Even under open circuit conditions we do not usually observe the theoretical open circuit potential. As a consequence, the energetic efficiency of a fuel cell is often far below the theoretical value. The optimization of the fuel cell performance requires the knowledge of the individual electrochemical losses in the cell for their elimination – or at least their reduction.

Although bioelectrochemical systems are in principle very similar to conventional fuel cells and share the major types of electrochemical losses, they possess a number of peculiarities that distinguish them from their chemical counterparts. The following chapter will, on a phenomenological basis, summarize and describe the typical, major electrochemical losses and will provide information on particular characteristics of bioelectrochemical systems. The chapter is not intended to provide an exhaustive overview on the respective literature; it is rather aimed to illustrate the principle processes on selected examples. The description of the losses will mainly focus on microbial fuel cells, but most of the phenomena also apply to microbial electrolysers.

7.2 INDIVIDUAL ELECTROCHEMICAL LOSSES

The performance of fuel cells is usually characterized by means of so-called polarization curves, in which the cell voltage (and often also the individual electrode potentials) is plotted as a function of the current density (Larminie and Dicks 2000). Figure 7.1 shows a typical polarization curve and the major segments that are decisive for the fuel cell performance. The regions stand for performance losses caused by limitations of quite different electrochemical nature: Region A represents losses caused by cathode and anode activation (electrode kinetics), region B is governed by ohmic resistances, region C represents mass transport limitation, and the decrease of the cell voltage at open circuit, D, is caused by so-called 'internal curents'.

In the following sections, the origin and the characterisitcs of the individual electrochemical losses will be explained and summarized. The description is mainly based on conventional fuel cell knowledge, with the peculiarities of bioelectrochemical systems being highlighted.

It is worth mentioning that, due to the multidisciplinarity of the world of fuel cells and biofuel cells, scientists from different fields use different terms for one and the same process or phenomenon. Thus, terms like overpotential, overvoltage, voltage drop, polarization, irreversibility, or simply loss, are terms that are being used synonymously for the description of electrochemical fuel cell losses.

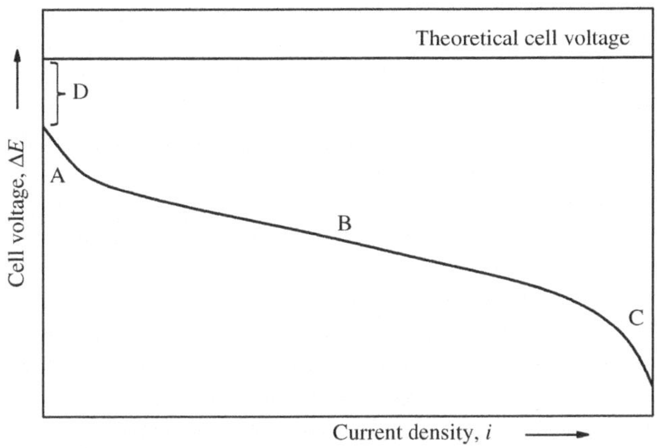

Figure 7.1. Typical fuel cell polarisation curve showing the effects of electrode activation (A), ohmic resistance (B), mass transport (C) and 'internal curents' (D)

7.2.1 Activation polarization

Many electrocatalytic reactions are hampered by comparably slow electrode kinetics (Vetter 1967). This slow kinetics may be caused either by a slow electron transfer step itself, or by any slow process directly linked to the electron transfer step, e.g. educt adsorption or product desorption. An example of such slow electrode kinetics is the electrocatalytic oxygen reduction (see Chapter 13) or the oxidation of methanol (Sammes 2006). The activation polarization thus occurs, and is a measure for, the slowness of an electrochemical reaction. The qualitative and quantitative characterization of activation losses is realized by the analysis of so-called Tafel plots. For a detailed description of the Tafel analysis the reader may be referred to Chapter 8.2.

It has to be emphasized that bioelectrocatalytic reactions are generally quite different to the conventional, heterogeneous electrocatalysis at metal electrodes. In the latter case, the actual catalytic oxidation or reduction and the electron transfer step to the electrode proceed at one and the same location, e.g. the surface of a platinum particle (Bockris and Khan 1993). Here, the reaction kinetics can indeed be evaluated by means of the Tafel analysis. The situation is quite different for bioelectrocatalysis by, e.g. living microorganisms. Here, the substrate oxidation proceeds at different redox enzymes than the electron transfer to the fuel cell anode. Both reactions may have rather different rates. What is important to notice is that the activation polarization will 'see' only the final step of the electron transfer – i.e. it will describe the electrochemical reversibility, e.g. the outer membrane cytochrome or mediator oxidation reaction

(Vincent *et al.* 2007). The rate of the biocatalytic substrate oxidation or reduction will not influence the activation polarization, but may have large impact as a concentration polarization.

7.2.1.1 Means to decrease the activation polarization

1. For chemical electrocatalysts – the use of more effective catalysts: this measure aims at the increase of the exchange current density, i_0. It may be achieved by either increasing the turn-over rate of the individual electrocatalytic centre (using more efficient electrocatalysts), or by increasing the number of active centres (e.g. using higher catalyst loads) (for the example of the ORR see e.g. Zhao *et al.* 2006; Cheng 2006). For a BES, this would require either the increase of the electrochemical reversibility of, e.g. the terminal electron transferring outer membrane cytochrome, or the increase of their coverage ('load') at the electrode surface (see Chapter 6).
2. Increase of real surface area: at a constant absolute exchange current, I_0, the increase of the real surface area, A, leads to proportional decrease of the respective current density, i_0, and thus of the electrode polarization. The increase of surface area can be achieved by increasing the electrode roughness, or using high surface area 3D electrode materials (e.g. Rabaey *et al.* 2005 (anode), Freguia *et al.* 2007 (cathode)).
3. For bioelectrochemical systems, the use of higher reaction temperatures, as commonly practiced in e.g., solid oxide fuel cells (Sammes 2006), is of only limit applicability, since most biocatalysts have a limited tolerance towards higher temperatures. However, especially at the oxygen cathode, higher oxygen concentrations (e.g. by using open air cathodes (Sell *et al.* 1989)) or an increased oxygen pressure (Fornero *et al.* 2008) may be used.

7.2.2 Ohmic polarization

Ohmic losses occur due to the resistance of both the flux of electrons through the electrode materials and the flux of ions in electrolyte solution and separator membrane. Although in bench-top bioelectrochemical systems the electrolyte resistance may cause the most obvious ohmic loss, any upscaling will result in an increasing impact of contact resistances and the resistance of the electrode materials themselves (see Chapter 10).

Both, the ion and electron flow resistance obey Ohm's law. Thus, the ohmic polarization is proportional to the current: $\eta_{\text{ohm}} = I\ R$. Here, I represents the current flow, and R is the total cell resistance, comprising electronic, ionic and contact resistances. The effect of a dominating ohmic resistance is illustrated in Figure 7.2. This figure depicts a so-called power curve, as derived from a polarization curve by multiplying the cell voltage by the respective current density values.

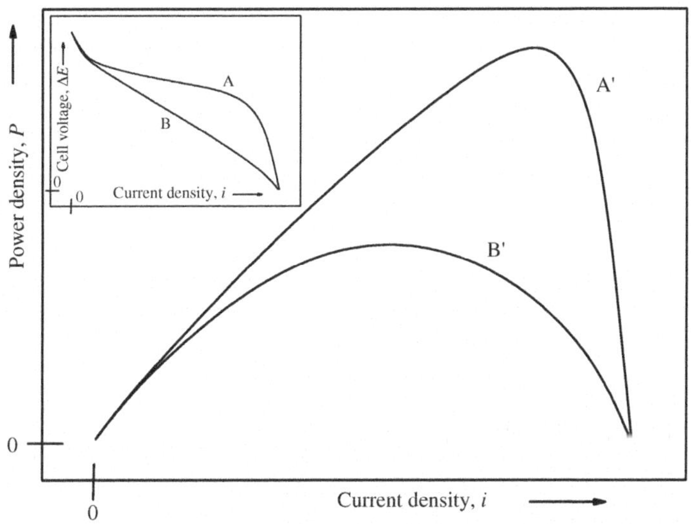

Figure 7.2. Typical power curves as derived from the respective polarisation plots (inset) showing the effect of an increasing ohmic resistance. Curves A, A′: low ohmic resistance, curves B, B′: ohmic resistance domination

A desirable power curve shape is similar to that presented as curve A′, in Figure 7.2. It possesses an asymmetric shape, with a long, linear up-slope, and a steep decrease of the power after the maximum power point (MPP) (Larminie and Dicks 2000). Ideally (at zero internal ohmic resistance) the power density increases with increasing current density until mass transfer limitations lead to a strong decrease of the cell voltage and thus of the resulting power. The higher the internal ohmic resistance – and thus the steeper the slope of the polarization plot (inset of Figure 7.2) – the more the maximum power density decreases and the maximum power point shifts towards lower current densities. The power curve becomes symmetrical, bell shaped. Fuel cells that possess such power performance characteristics are highly inefficient. The maximum power point is reached far before effective substrate consumption takes place; hence the substrate usage efficiency is low.

To illustrate the influence of electrolyte composition and pH, Table 7.1 exemplarily shows specific resistances of different electrolyte solutions and of a Nafion membrane operated in these electrolyte solutions (Harnisch *et al.* 2008). From this table, the following conclusions can be drawn: as is to be expected, the resistance of the electrolyte solution increases from acidic towards neutral pH, and further increases at decreasing electrolyte concentration. Most remarkably, also the ion exchange membrane exhibits a strong dependence on electrolyte pH and concentration – and this dependence is much stronger than that of the respective electrolyte solution: thus, the electrolyte resistance of a pH 7, 10 mmol/l, phosphate buffer solution is 70 times higher than of 100 mmol/l HNO_3; the specific resistance of Nafion immersed in the respective electrolyte solutions, increases by a factor 700.

Table 7.1 Specific resistance of different electrolyte solutions and of a Nafion ion exchange membrane in these solutions (adapted from Harnisch *et al.* 2008)

	$\rho_{membrane}/\Omega cm$	$\rho_{electrolyte}/\Omega cm$
pH 1 (0.1 mol/l HNO_3)	18.0	10
pH 2.5 (0.1 mol/l HNO_3/KNO_3)	360.7	68.5
pH 7 (0.1 mol/l KNO_3)	961.7	95.2
pH 7 (0.1 mol/l KH_2PO_4/K_2HPO_4)	502.7	86.9
pH 7 (0.05 mol/l KH_2PO_4/K_2HPO_4)	573.8	142.5
pH 7 (0.03 mol/l KH_2PO_4/K_2HPO_4)	1781	260.4
pH 7 (0.01 mol/l KH_2PO_4/K_2HPO_4)	12568	701.8

7.2.2.1 Means to decrease the ohmic polarization

There are several means to decrease the ohmic losses in a bioelectrochemical system. The use of electrolyte solutions of a higher ionic strength and thus with a higher ionic conductivity may be appealing – and in some studies this approach has been used to produce outstanding MFC performances, but considering just the expenses for the use of highly concentrated electrolyte salts such an approach is rather questionable.

Further, the decrease of anode-to-cathode spacing, the use of optimised ratios of electrode-to membrane sizes, the avoidance of contact resistances and the use of current collectors are important means to decrease the internal resistance. These measures are discussed in Chapter 10.

7.2.3 Concentration Polarization (Mass transfer and reaction polarization)

The concentration polarization is often simply illustrated as a mass transfer limitation, referring to the limitation of fuel and oxidant supply to the respective electrode surface. Yet, the situation, especially in bioelectrochemical devices, is much more complex. The following processes can be assumed to be significantly contributing to the concentration polarization. Their share and contribution may, however, differ significantly between different bioelectrochemical systems. The list may not be exhaustive, and for a given system further concentration limitations may apply. For clarity we discuss the effects mainly for biofilm based anodes and aqueous chemical oxygen reduction cathodes.

(1) *Rate of biocatalytic substrate conversion (biocatalytic turnover rate)*
Plainly speaking, the biocatalytic activity of an electroactive (anodic) microbial biofilm supplies the electrode with an electrochemically accessible reduction equivalent (Schröder 2007). This may be in the form of a reduced soluble bacterial redox mediator, a reduced outer membrane cytochrome or a reduced fermentation product. The biocatalytic rate, i.e. the rate with which the biocatalyst transforms the substrate into the reduction equivalent, is a major limitation (in the form of a reaction polarization) in the performance of bioelectrochemical devices.

(2) *Rate of substrate and oxidant transport*
Especially in microbial biofilms the diffusion of a substrate through the extracellular polymer substance (EPS) matrix is – due to lower diffusion coefficients – low compared to aqueous solutions. Thus, the substrate supply of the biocatalyst within the biofilm may become rate limiting.

At the cathode, and especially when aqueous oxygen cathodes are used, the limited oxygen solubility in water (~6 mg l^{-1} (Lide 2004)) often causes strong mass transfer limitations (e.g. Zhao *et al.* 2006 and Cheng *et al.* 2006). This effect may further increase when an electroactive microbial biofilm is used as oxygen reduction catalyst (see also Chapter 13).

(3) *Rate of 'electron transport'*
Like the substrate, the electrons – in the form of a given reduction equivalent – have to be transferred from any cell in the bacterial biofilm to the electrode. Depending on the electron transfer mechanism (mediator diffusion, electron hopping between bacteria or along a microbial nanowire, etc.) the rate of electron transfer may be assumed to differ considerably; however, a transport limitation, especially in thicker electroactive biofilms, has to be expected.

(4) *Rate of proton transport*
Nearly all of the relevant anodic and cathodic electrochemical reactions involve the formation or the consumption of protons. At pH 7, a 'normal' pH for bioelectrochemical devices, any proton formation or consumption leads to a strong relative pH change at/near the electrode surface (Vetter 1967). Following the Nernst equation for the respective reaction, such a pH change will significantly affect the electrode performance. This effect is especially pronounced at low concentrations of the used buffer solutions, which at low concentrations have only a limited capacity to compensate occurring pH gradients. As an example, it has been demonstrated for the cathodic oxygen reduction reaction that with a decreasing buffer capacity of the cathodic electrolyte solution the cathode performance is significantly reduced. This is exemplarily shown in Figure 7.3 for the oxygen reduction at iron–phthalocyanine in air saturated pH 7 phosphate buffer solution (Zhao *et al.* 2006). As one can easily imagine, the effect is even more pronounced, when non-buffering electrolyte solutions are used.

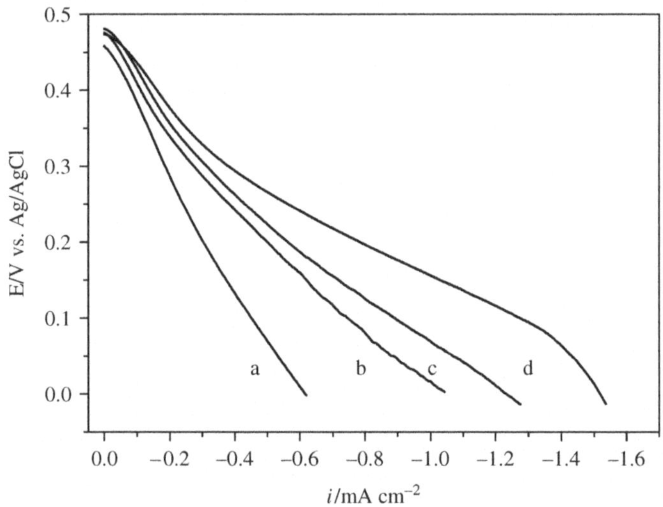

Figure 7.3. Galvanodynamic polarisation plots of the electrocatalytic oxygen reduction at FePc modified graphite foil in (a) 0.01 mol/l, (b) 0.05 mol/l, (c) 0.1 mol/l, and (d) 0.5 mol/l oxygen saturated pH 3.3 phosphate electrolyte solutions (NaH_2PO_4-H_3PO_4). (Adapted from Zhao *et al.* 2006)

Analogously, any anodic substrate oxidation is accompanied by an equivalent liberation of protons. Similar to the substrate diffusion into the biofilm, the reverse proton transport out of the biofilm may be considered as a limitation (Torres *et al.* 2008).

7.2.3.1 Means to decrease the concentration polarization

As to be derived from the sections above that illustrate the complexity of sources, there is no straightforward means to decrease the concentration polarisation in BES. If applicable, better (or more intensive) reactor mixing enhances the mass transfer within the system; however, the necessity of additional energy input has to be taken into consideration.

On the electrolyte side, increasing the buffer concentration (and thus buffer capacity) and lowering the electrolyte pH have been demonstrated to improve the BES performance. Yet, the first approach requires an increasing input of the buffer related chemicals, the second approach will require the use of acid-tolerant microorganisms.

To increase mass transfer within a biofilm, an increasing biofilm conductivity and density is required, which may be achieved, e.g., by a recently proposed high shear biofilm acclimatisation procedure (Pham *et al.* 2008).

7.2.4 Reactant crossover – 'internal currents'

In most bioelectrochemical devices, as in the great majority of low-temperature fuel cells, an ion exchange membrane is used to prevent the mixing of anolyte and catholyte, whilst allowing unhindered charge balancing ion transfer between anode and cathode compartments. Although designed for a selective ion transfer, any membrane possesses a certain unselective permeability for the fuel (substrate) and for oxygen. The occurring fluxes lead to parasitic processes (often referred to as internal currents) in the respective compartments that can result in a considerable lowering of the fuel cell performance (Sammes 2006). This may be illustrated for the example of a platinum based oxygen cathode (Figure 7.4). When oxidizable matter diffuses to the cathode, it can be directly oxidized at the Pt particle by providing electrons for the oxygen reduction reaction at the same catalyst moiety. This results in a continuous electron flow and thus perpetual substrate/oxygen turnover even under open circuit conditions. As depicted in Figure 7.4A, one may interpret this reaction as an internal short circuit reaction. It not only lowers the coulombic efficiency of the device – as Figure 7.4B illustrates, the formation of a so-called mixed potential (which lies between the redox potential of the oxidant and that of the respective substrate) may also considerably lower the redox potential of the cathode.

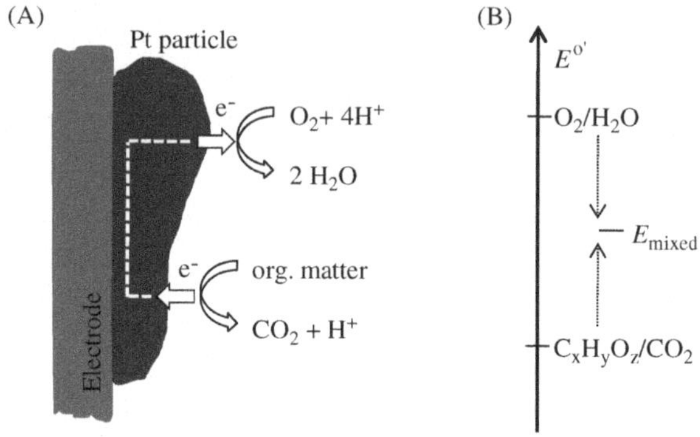

Figure 7.4. Schematic illustration of the principle of internal currents by means of a platinum based oxygen cathode

An analogous situation can be found in the biotic anode compartment: oxygen crossover into the anode compartment leads to aerobic (and thus parasitic) substrate oxidation that lowers the coulombic efficiency and raises the anode potential.

Figure 7.1 illustrates the effect of the internal currents on the polarization curve, they lower the achieved open circuit potential from the theoretical predictable value, represented by region D. It is clear that in the case of membrane-less bioelectrochemical systems, due to the increased cross-over effects, the parasitic internal currents are much more pronounced than in membrane containing systems.

In conventional polymer electrolyte membrane fuel cells, in which generally laminated electrode-membrane assemblies (MEAs) are used, an additional parasitic direct electron transfer across the proton exchange membrane can be observed, which is also referred to as internal current (Sammes 2006). In bioelectrochemical systems, such a process is probably of less importance, since, at least, anode laminated electrode membrane assemblies are not common.

7.2.4.1 Means to decrease internal current losses

From the viewpoint of internal current loss prevention, membrane-less fuel cell operation should be avoided – although membrane-less systems may on the other hand considerably lower the internal ohmic resistance and help avoid pH splitting (see Section 7.2.5).

An important measure to reduce the effects of internal current losses is the use of more selective electrocatalysts that have less cross-selectivity. This would mainly apply for, e.g. oxygen cathodes, where platinum could be replaced by selective enzymes like laccases (Schaetzle 2009).

7.2.5 The pH splitting between anode and cathode

A major problem in bioelectrochemical systems is the occurrence of pH gradients that build up between the anode (acidification) and the cathode compartment (alkalization) of the device upon its operation (e.g. Rozendal *et al.* 2006; Harnisch *et al.* 2008). This process is caused by a strong involvement of electrolyte metal ions in the charge balancing ion transfer between anode and cathode – in a 0.1 mol/l, pH 7 electrolyte solution, the excess of metal ions to protons is 10^6! Since, however, the anodic substrate oxidation is proton liberating, and cathodic processes, like the oxygen reduction and the hydrogen evolution, are proton consuming, only a sole proton transfer would maintain pH neutral conditions in both chambers.

Assuming a one proton, one electron oxidation, a pH decrease of one unit would lead to a shift of the anode potential of +59 mV (assuming Nernstian behavior). The cathodic oxygen reduction on the other hand is proton consuming, here the increase of the pH leads to a negative potential shift (−59 mV/pH unit). Especially in the case of MFC air cathodes, pH values of up to 13 have been reported (Rozendal *et al.* 2007), which can be attributed to the low volume of permeated electrolyte at the cathode side. Thus, the pH splitting between anode and cathode leads to a considerable deminution of the cell voltage of the bioelectrochemical device. The phenomenon is physically inherent for all types of ion exchange membranes (Harnisch *et al.* 2008 and Harnisch *et al.* 2009).

7.2.5.1 Means to prevent the pH splitting

Some strategies to overcome the pH splitting have recently been proposed. They include, e.g. the use of membrane-less cell designs, as well as an effluent loop concept (Freguia *et al.* 2008). The latter approach is based on the use of the effluent of the anode chamber as the influent for the cathode, in order to counterbalance the proton consumption during the oxygen reduction reaction at the cathode by the acidified anode solution.

7.3 METHODS

7.3.1 Experimental strategies for the recording of polarization plots

For an overall characterization of the performance of a bioelectrochemical system, polarization experiments (as described above) are a valuable tool. They provide a first insight into basic limitations of the respective electrochemical device. For the recording of a polarization plot the respective bioelectrochemical device can either be connected to a variable external resistor in combination with a multimeter or can be connected to a potentiostat (Larminie and Dicks 2000). Both approaches are equivalent. Usually, the fuel cells are, starting from open circuit conditions, operated at gradually decreasing external resistance, and the respective cell voltage values are recorded. Although the current flow can be recorded separately, it is usually sufficient to calculate the current from the cell voltage and the respective resistance, following $I = \Delta E/R$. In order to yield additional information on the individual (cathode and anode) electrode potentials it is advisable to place a reference electrode (e.g. an Ag/AgCl electrode) in the vicinity of the respective electrode, against which these potentials can be measured.

A most crucial point in the measuring of polarization plots is to avoid transient (time-dependent) conditions and to assure steady-state values. Whereas at high ohmic resistances – close to open circuit – this is usually not a problem, it becomes highly important at low external resistances and larger current densities. This may be briefly illustrated by means of an electroactive microbial biofilm anode: a microbial biofilm, looked at as a whole, is able to oxidize substrates at a certain rate and to transfer the electrons to the anode. At open circuit, reduction equivalents are accumulated in the biofilm. When the curcuit is closed, an electron transfer takes place that is based on both the direct bioelectrocatalytic activity and the accumulated charge. At high resistance values, the current flow may be low enough to maintain the state of charge accumulation – the observed current is constant. Yet, the lower the external resistance and the higher the resulting current density, the more the accumulated charge is consumed, and the performance decreases until a steady state is achieved that is governed by the bioelectrocatalytic activity of the system. Only when this steady state condition is reached, are physically significant data on the bioelectrocatalytic performance yielded.

It is difficult and probably not advisable to recommend an exact time that has to be waited for reaching steady state conditions. This time span very much depends on the respective electrochemical system. There is, however, the possibility of verifying the polarization data. Thus, after recording the polarization curve from open circuit towards short curcuit, a reverse scan may be recorded. Ideally, the resulting polarization curve should follow the trace of the forward scan. Any remarkable deviation (usually in the negative direction) will indicate an insufficient sustainability of the system's performance.

7.3.1.1 Current interrupt technique

For a detailed qualitative and quantitative analysis of the individual fuel cell losses, different sophisticated tools and techniques, such as Tafel analysis and electrochemical impedance spectroscopy, but also cyclic voltammetry, are available. These techniques are described in detail in Chapters 8a–c. An alternative technique that yields accurate quantitative results (Lee *et al.* 1998) and that can be performed using low cost equipment is the current interrupt technique (Figure 7.5). It allows us to distinguish the ohmic resistance of the electrochemical cell from the activation resistance.

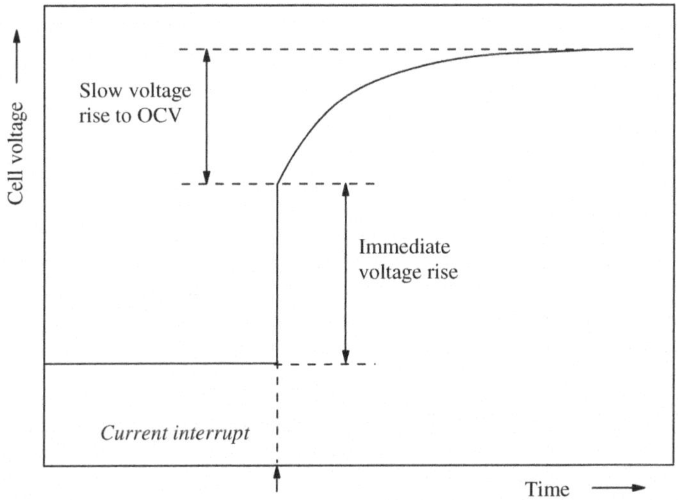

Figure 7.5. Schematic illustration of a current interrup experiment

In order to perform a current interrupt experiment, the microbial fuel cell is operated at a current at which the concentration polarization is negligible and a reasonably high signal can be achieved. Under these conditions, the polarization is governed by the sum of ohmic and activation losses (region B of Figure 7.1). When the current circuit is suddenly opened ($I = 0$), the ohmic losses will immediately reduce to zero (for $E_{\text{ohmic}} = I\,R$), whereas the double-layer related processes require a certain time span (within millisecond range) to equilibrate. The immediate voltage rise after the current interrupt can thus be directly related to the internal ohmic resistance of the system, whereas the following, slower voltage rise can mainly be attributed to the electrode activation polarization.

7.4 CONCLUSIONS

Many electrochemical losses in bioelectrochemical systems and their origin are just being understood, and many still are to be understood. There is no silver bullet for their elimination. All processes are deeply interlaced, and improving one parameter may result in the deterioration of another. Thus, for example,. operating a microbial fuel cell without an ion exchange membrane may lead to the desired significant decrease of the internal ohmic resistance, yet, at the same time it means accepting considerable internal current based losses. So far, microbial fuel cells and microbial electrolysis cells are far from optimized. Considerable efforts, both, from fundamental research and technological development are required to minimize the electrochemical losses in these bioelectrochemical systems to unfold their real potential.

REFERENCES

Bockris, J.O'M. and Khan, S.U.M. (1993) *Surface Electrochemistry: A Molecular Level Approach.* Springer.

Cheng, S., Liu, H. and Logan, B.E. (2006) Power densities using different cathode catalysts (Pt and CoTMPP) and polymer binders (Nafion and PTFE) in single chamber microbial fuel cells. *Environ. Sci. Technol.* **40**, 364–369.

Fornero, J.J., Rosenbaum, M., Cotta, M.A. and Angenent, L.T. (2008) Microbial fuel cell performance with a pressurized cathode chamber. *Environ. Sci. Technol.* **42**, 8578–8584.

Freguia, S., Rabaey, K., Yuan, Z. and Keller, J. (2007) Non-catalyzed cathodic oxygen reduction at graphite granules in microbial fuel cells. *Electrochim. Acta* **53**, 598–603.

Freguia, S., Rabaey, K., Yuan, Z. and Keller, J. (2008) Sequential anode-cathode configuration improves cathodic oxygen reduction and effluent quality of microbial fuel cells. *Wat. Res.* **42**, 1387–1396.

Harnisch, F., Schröder, U., and Scholz, F. (2008) The suitability of monopolar and bipolar ion exchange membranes as separators for biological fuel cells. *Environ. Sci. Technol.* **42**, 1740–1746.

Harnisch, F., Warmbier, R., Schneider, R. and Schröder, U. (2009) Modeling the ion transfer and polarization of ion exchange membranes in bioelectrochemical systems. *Bioelectrochem.* **75**, 136-141.

Larmini, J. and Dicks, A. (2000) *Fuel Cell Systems Explained*, 2nd edn. Wiley and Sons Ltd.

Lee, C.-G., Nakano, H., Nishina, T., Uchida, I. and Kuroe, S. (1998) Characterization of a 100 cm^2 class molten carbonate fuel cell with current interruption. *J. Electrochem. Soc.* **145**, 2747–2751.

Lide, R.L. (ed.) (2004) *CRC Handbook of Chemistry and Physics*, 85th ed. CRC Press, Boca Raton.

Ostwald, W. (1894) Die wissenschaftliche Elektrochemie der Gegenwart und die technische der Zukunft. *Z. Elektrochem.* **4**, 122–125.

Pham, H.T., Boon, N., Aelterman, P., Clauwaert, P., De Schamphelaire, L., van Oostveldt, P., Verbeken, K., Rabaey, K. and Verstraete, W. (2008) High shear enrichment improves the performance of the anodophilic microbial consortium in a microbial fuel cell. *Microb. Biotechnol.* **1**, 487–496.

Rabaey, K., Clauwaert, P., Aelterman, P. and Verstraete, W. (2005) Tubular microbial fuel cells for effcient electricity generation. *Environ. Sci. Technol.* **39**, 8077–8082,

Rozendal, R.A., Hamelers, H.V.M. and Buisman, C.J.N. (2006) Effects of membrane cation transport on pH and microbial fuel cell performance. *Environ. Sci. Technol.* **40**, 5206–5211,

Rozendal, R.A., Hamelers, H.V.M., Molenkamp, R.J. and Buisman, C.J.N. (2007) Performance of single chamber biocatalyzed electrolysis with different types of ion exchange membranes. *Wat. Res.* **41**, 1984–1994,

Sammes, N. (ed.) (2006) *Fuel Cell Technology Reaching Towards Commercialization.* Springer.

Schaetzle, O., Barrière, F. and Schröder, U. (2009). An improved microbial fuel cell with laccase as the oxygen reduction catalyst. *Energy Environ. Sci.* **2**, 96–99

Schröder, U. (2007) Anodic electron transfer mechanisms in microbial fuel cells and their energy efficiency. *PCCP* **9**, 2619–2629.

Sell, D., Kramer, P. and Kreysa, G. (1989) Use of an oxygen gas diffusion cathode and a three-dimensional packed-bed anode in a bioelectrochemical fuel cell. *Appl. Microbiol. Biotechnol.* **31**, 211–213.

Torres, C.I., Marcus, A.K. and Rittmann, B.E. (2008) Proton transport inside the biofilm limits electrical current generation by anode-respiring bacteria. *Biotechnol. Bioengng.* **100**, 872–881.

Vetter, K.J. (1967) *Electrochemical Kinetics: Theoretical and Experimental Aspects.* Academic Press, New York.

Vincent, K.A., Parkin, A. and Armstrong, F.A. (2007) Investigating and exploiting the electrocatalytic properties of hydrogenases. *Chem. Rev.* 107, 4366–4413.

Zhao, F., Harnisch, F., Schröder, U. Scholz, F., Bogdanoff, P. and Herrmann, I. (2006) Challenges and constraints of using oxygen cathodes in microbial fuel cells. *Environ. Sci. Technol.* **40**, 5193–5199.

8

Electrochemical techniques for the analysis of bioelectrochemical systems

Chapter 8 describes three different approaches for the analysis of BESs, each approach is discussed in depth by experts in the field. The three subchapters that can be found under this umbrella are:

8.1. Cyclic voltammetry for the study of microbial electron transfer at electrodes
8.2. Importance of Tafel plots in the investigation of bioelectrochemical systems
8.3. The use of electrochemical impedance spectroscopy (EIS) for the evaluation of the electrochemical properties of bioelectrochemical systems

8.1

Cyclic voltammetry for the study of microbial electron transfer at electrodes

Edward LaBelle and Daniel R. Bond

8.1.1 INTRODUCTION

The fact that certain bacteria can direct electrons to their outer surface, or transfer electrons from their outer surface to their inner compartments, makes it possible to measure the rate of electron flux from a living organism. When enzymes are interfaced with electrodes, either via their own direct mechanisms or mediators, the relationship between applied potential and current flowing at the working electrode can theoretically reveal rate-limiting processes, interfacial kinetics, and reaction stoichiometries (Armstrong 1999, 2005; Heering *et al.* 1997, 1997; Jeuken *et al.* 2002). The use of voltammetry to extract such unique information from bacteria is an emerging field that draws heavily from

© 2010 IWA Publishing. *Bioelectrochemical Systems: From Extracellular Electron Transfer to Biotechnological Application*. Edited by Korneel Rabaey, Largus Angenent, Uwe Schröder and Jürg Keller. ISBN: 9781843392330. Published by IWA Publishing, London, UK.

pioneering work developed first for cytochromes adsorbed to electrodes (Eddowes and Hill 1977; Yeh and Kuwana 1977), and later to study enzymatic catalysis at electrodes (Hirst *et al.* 1996; Sucheta *et al.* 1993). Readers seeking deeper background for the electrochemical theory and models are directed to the discussions of Armstrong, Heering and Hirst (e.g. Armstrong 2005; Armstrong *et al.* 1997; Butt and Armstrong 2008; Léger *et al.* 2003).

Microbial electron transfer to electrodes combines a series of irreversible enzymatic reactions (such as oxidation of acetate via the TCA cycle to yield NADH and reduced menaquinone), with a pathway of reversible electrochemical reactions that allow electrons to pass to the cell surface (Busalmen *et al.* 2008a; Schröder 2007). Electrons must then relay along a series of cytochromes, proteins, bound redox mediators, and/or soluble mediators, until they reach a reversible interfacial reaction at an electrode surface (within approximately 10 Å). A simplified scheme showing an idealized film of bacteria is shown in Figure 8.1.1.

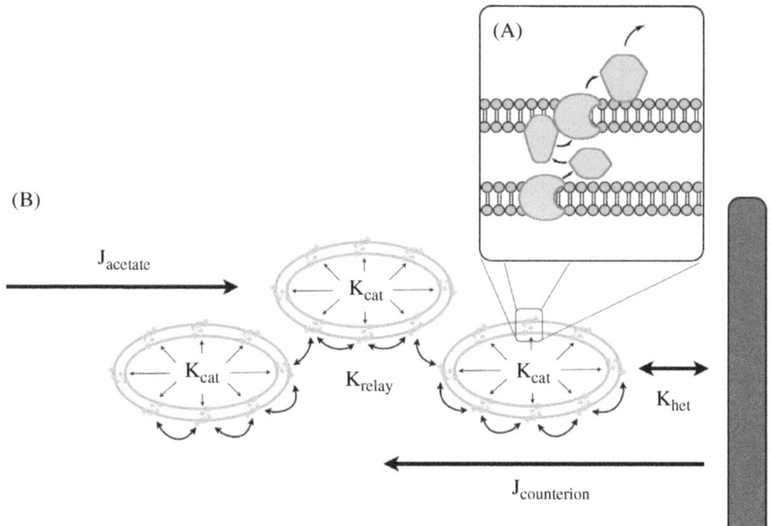

Figure 8.1.1. (A) Cartoon showing possible paths of electron transfer across the inner and outer membranes of an electroactive bacterium. (B) Schematic representation of mass transfer and kinetic limitations in an electrode-reducing biofilm in the absence of any soluble redox-active compounds. The total current is a function of: acetate flux ($J_{acetate}$), irreversible metabolic reactions supplying electrons to the membrane (k_{cat}), electron transfer around and between cells, relaying electrons to the interfacial contact point (k_{relay}), interfacial transfer by proteins actually in contact with the electrode (k_{het}), and counterion flux from the film ($J_{counterion}$)

In the simplest cases of irreversible chemical reactions occurring at an electrode, observed reaction rates are a function of the diffusion coefficient of the donor (how fast compounds can travel to the electrode), the standard rate constant of the reaction (how fast donor reacts when in contact with the electrode), and the electrode potential (which accelerates the rate, if electrons are available to hop from this terminal site). At flat planar electrodes, the rate constant and applied potential of most chemical reactions can be irrelevant, because diffusion quickly becomes rate limiting as the electrode is brought to a favourable potential. Such a combination of factors creates the familiar 'peak' in observed current as electrode potential is swept past the voltage where oxidation of the donor becomes favorable, because the donor is so rapidly depleted near the electrode that the surge in current cannot be sustained.

However, this is not the case with most enzyme-coated and bacteria-coated electrodes (Kano and Ikeda 2000; Marsili *et al.* 2008b). To appreciate this, one can calculate what current density a planar electrode should produce if a donor such as acetate could be rapidly and completely oxidized by a catalyst at the electrode surface. For example, the current of a theoretical rotating disk electrode oxidizing acetate can be calculated according to the Levich equation (Bard and Faulkner 2001), where current only depends upon the concentration of reduced substrate (e.g. 20 mmol/l acetate), rotation rate, the kinematic viscosity of water, the number of electrons per acetate (8), and the diffusion coefficient of acetate (1×10^{-5} cm^2/sec).

One can see from this relationship that current produced is, in fact, not related to any aspect of the catalyst. It is only a function of how fast the electrode is rotated, which influences how fast substrate reaches the electrode. Thus, under conditions analogous to a slowly stirred solution, any electrode coated in a catalyst able to oxidize acetate should easily produce a current of about 50 mA/cm^2. This value is the 'mass transport limited' current, as it is limited only by how fast acetate can arrive at the surface. If bacteria on a surface are not a rate-limiting factor, one would expect similar current densities.

However, if we look at the best rates reported when wild-type strains of dissimilatory Fe(III)-reducing *Geobacter* species have been used to oxidize acetate at well-stirred, well-defined electrodes, values of the order of only ~0.5–1 mA/cm^2 are observed (Bond and Lovley 2003; Ishii *et al.* 2008; Marsili *et al.* 2008b; Min *et al.* 2005; Nevin *et al.* 2008; Reguera *et al.* 2006; Srikanth *et al.* 2008; Torres *et al.* 2007; 2008a; Torres *et al.* 2008b). From this, it is clear that acetate can diffuse to electrodes very quickly in comparison to how fast living cells appear to oxidize fuels and transfer those electrons to electrodes.

This calculation begs the question, if acetate is arriving at the electrode, why aren't bacteria consuming it? By varying the applied potential, and rate of potential change (scan rate), one can dissect this question further. For example, if the final reversible step (k_{het}, in Figure 8.1.1) in which electrons hop to the electrode is sluggish (also described as having a small exchange current), it should respond predictably to a wide window of applied potentials, because potential can increase interfacial electron transfer rates (e.g. help the electrons hop the last gap to the electrode). If the reversible 'k_{relay}' pathway that feeds electrons to this final transfer point is slow, then faster scan rates under appropriate conditions could indicate the time required for electrons to arrive at the surface from more distant locations. Finally, if the irreversible reactions feeding electrons into the relay-and-interfacing pathway (k_{cat}) is insufficient, we would expect a sharp response to applied potential, followed by a flat plateau in the current–potential relationship, as applied potential cannot accelerate an enzyme's rate beyond its V_{max}.

8.1.2 TURNOVER VS. NON-TURNOVER VOLTAMMETRY EXPERIMENTS

The remainder of this section will compare cyclic voltammetry experiments conducted under either *turnover* versus *non-turnover* conditions. Under *turnover* conditions, electron donor is provided to the cells, causing an electron flow that can be varied as a function of electrode potential. The potential is typically changed very slowly (scan rates 1–10 mV/s), so that at each applied potential, all proteins involved in the pathway are oxidized and reduced ('turnover') multiple times. In *non-turnover* conditions, cells are first deprived of all electron donors, then electrons are pushed in and pulled out of cells by changing the electrode potential – inducing, at most, a single oxidation or reduction event by each redox center accessible to the electrode.

8.1.2.1 General considerations

Because diffusion can significantly affect voltammetry, any setup involving electrochemistry and bacteria should strive to create as consistent an environment as possible. For example, in most of the examples described in this chapter, *Geobacter sulfurreducens* cells are allowed to colonize a small electrode of the order of 1 cm^2, which has been polished to expose a uniform surface with features smaller than the bacteria themselves. In very early stages of growth, this produces an array of individual cells on a planar electrode. Analysis during this sub-monolayer phase is closest to the classical model of protein film voltammetry, where each cell behaves as an independent microelectrode surrounded by its own radial diffusion boundaries (Armstrong *et al.* 1993; Picioreanu *et al.* 2007). Once cells begin to crowd, or touch each other, the system enters a regime more akin to a planar electrode, where diffusion may play a larger role. Key decisions to be made in setting up experiments are related to the techniques to be applied. Slow scan rate linear sweep voltammetry is relatively forgiving, as the slow potential steps minimize charging currents (Kim *et al.* 2007; Liu *et al.* 2005). To probe faster scan rates, apply pulse methods, or conduct impedance spectroscopy, one should employ a three electrode arrangement in which all connections are robust, electrodes are in close proximity to each other, and the working electrode surface is uniform (Marsili *et al.* 2008b; Srikanth *et al.* 2008).

As a final comment, one should note the reference used for reporting potentials in the literature. Biochemistry and microbiology are typically discussed in terms of E vs. the Standard Hydrogen Electrode (SHE), for comparison with standard values in biological textbooks and geochemical data. In contrast, it is common in electrochemistry to provide experimental data as collected, with E vs. Ag/AgCl or Hg/HgCl reference. As this review is intended for a biological audience, all values (including those taken from literature) are adjusted to E vs SHE for easy comparison.

8.1.2.2 Voltammetry in the presence of substrates

As mentioned, the first cells attached to the electrode create the only situation similar to an array of individual microelectrodes. Allowing cells to grow as thicker cell layers introduces new factors such as film heterogeneity, long-range electron relay, cell–cell interactions, and diffusion through biofilms. Calculations (based on known rates of electron transfer), as well as biofilm images suggest that the transition from the sub-monolayer to multilayer stage (on polished electrodes) occurs by a current density of approximately 50 $\mu A/cm^2$ for *G. sulfurreducens* (which can easily achieve a current density of 500–1000 $\mu A/cm^2$) (Marsili *et al.* 2008b). The fact that growth of *G. sulfurreducens* continues in an exponential manner near its maximum growth rate, well beyond this monolayer phase strongly suggests that these first additional layers of cells are not causing a significant limitation in the cell–cell relay of electrons to the electrode. To date, no similar information is available for other electrode-reducing bacteria.

A typical catalytic voltammetry experiment can use such a thin film of bacteria. In the simplest case, the level of electron donor is provided in excess, which is easy to verify by increasing the concentration of donor available to cells. In the case of thin films of *Geobacter*, when acetate concentrations are >5 mmol/l, additional acetate has no effect on current density, nor do increased rates of stirring (Marsili *et al.* 2008b; Richter *et al.* 2008). These observations support the idea that the microbial film is not limited by supply of donor to the surface, and factors such as internal oxidation rates or cell–electrode electron transfer rates are dominant controlling influences.

The working electrode is switched to a starting potential (typically held for a brief resting period), then slowly stepped through a series of applied potentials until a maximum is reached, then swept back through that same range. As the goal is to not injure the cells, the potential window should be small enough to limit strong reducing or oxidizing potentials, but wide enough to bracket the oxidation processes of interest. The lowest potential applied is relatively easy to choose, as a current density of essentially zero should be achieved at the point where substrate oxidation is unfavourable (e.g. well below the midpoint potential of the donor). The highest potential applied depends on the response of the bacteria; again the goal is to sweep at least 100 mV above any potential where a change is observed (Bard and Faulkner 2001). Observations with most *Geobacter* strains typically reveal a flat plateau above potentials of +100 mV vs SHE, allowing voltammograms to sweep from as narrow a window as −0.5 V to +0.2 V (Fricke *et al.* 2008; Liu *et al.* 2008). With *Shewanella* sp., a gradual rise in current in the higher potential range, coupled with use of a lower potential donor, requires routine voltammetry to sweep from nearly −0.7 V to +0.4 V (Manohar *et al.* 2008; Marsili *et al.* 2008a). Each investigator should perform preliminary observations to determine an optimal window, demonstrate that repeated sweeps produce similar results, and verify that the process does not harm the cells.

Along with careful choice of the potential window, the scan rate also should be justified. An assumption during slow scan rate voltammetry is that each applied potential achieves a steady-state rate of turnover at that driving force; that it is reflecting physiological information about what the bacteria are capable of, under each imposed condition. Thus, a key definition of catalytic voltammetry is that moderate increases in scan rate do not alter the limiting current achieved at high potentials. Sweeping at 1 mV/s, followed by 2 mV/s, 5 mV/s, etc., and observing a similar response, primarily in the limiting current observed at higher potentials (corrected for the increase in capacitive current) demonstrates that the entire microbial film is being transitioned through a series of potentials slowly enough that diffusion is not affecting steady-state results. It can be desirable to sweep much more rapidly, but in such cases the goal is to 'outrun' catalysis or diffusion, to elucidate turnover or diffusion limitations within the film.

Figure 8.1.2 shows representative data from a growing *G. sulfurreducens* biofilm, where the electrode potential was swept at 1 mV/s. In preliminary experiments, it was determined that sweep rates up to 5 mV/s did not alter the shape or features of the waveform, nor did additional acetate affect results (Marsili *et al.* 2008b). Analysis of these catalytic waves is simplified by plotting the derivative of each sweep as a function of potential, producing a nearly symmetrical peak (or series of peaks) centered at what is often termed the E_{hw} (half-wave potential) or E_{cat} (catalytic center potential) (see Heering *et al.* 1998; Heering *et al.* 1997) for an example of the Butler–Volmer and Marcus theory underlying these and other analyses of catalytic waves).

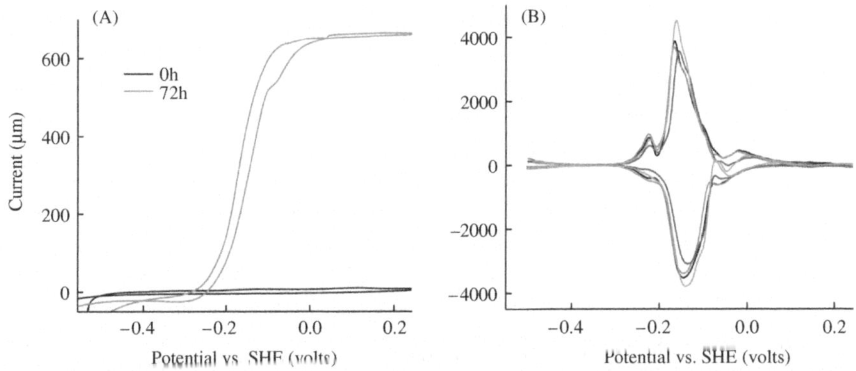

Figure 8.1.2. (A) Cyclic Voltammetry (1 mV/s) of a *Geobacter sulfurreducens* biofilm after 72 h growth at +0.2 V vs. SHE on a 2.5 cm^2 planar graphite electrode. (B) First derivative analysis of four replicate electrodes, revealing midpoint potentials within the catalytic wave

What do these sigmoidal catalytic waves suggest about the physiology of electron transfer to electrodes by *Geobacter?* Starting from the zero current observed at low potentials (below the midpoint potential of acetate), the core reaction does appear irreversible. As potential rises above a threshold (around −0.2 V), the positive anodic current reflects continuous oxidation of acetate, and transfer of electrons to the electrode. This response rises nearly symmetrically around a midpoint of about −0.15 V. In this narrow region (between −0.2 and 0 V) the electrode potential (and likely, the potential of proteins on the exterior of cells) prevents interior reactions from operating at their maximal rate. In other words, between −0.2 and 0 V, the capacity of cells to generate electrons is in excess, but cells are limited by the availability of electron acceptor to take them and pass them to the electrode – the system has an excess of enzyme and is limited by the kinetics at the interface.

However, as the working electrode potential is raised as little as 0.05 V above this midpoint potential, the sharp response in anodic current ceases, and a plateau is reached. This flat region is important, because while interfacial reactions can be accelerated by increases in driving force (as described by Butler-Volmer or Marcus theory), enzymatic reactions have a characteristic V_{max}. The Nernstian response of the anodic current (rising symmetrically around a midpoint with a characteristic width of the derivative peak at half-height) is a hallmark of a system limited by the rate of enzymatic activity, not by interfacial kinetics of electrons hopping from terminal proteins to electrodes (Léger *et al.* 2003). This behavior would be expected from a system where proteins on the outer surface of cells rapidly equilibrate to a potential equal that of the electrode. If it has been previously demonstrated that diffusion to the electrode is not limiting (by varying levels of donor or stirring and finding no effect), then this relationship strongly suggests that enzymatic activity by the film of bacteria is limiting (a conclusion also reached in the discussion of the Levich equation).

As the potential is swept back across the same potential range, a similar voltammogram is observed, as one is essentially repeating the experiment. The only differences may be minor peaks that reflect redox centers accessible to the electrode, but whose redox status is not rate-limiting.

Many voltammograms of bacteria, even of *G. sulfurreducens*, show more complex waveforms than the simple sigmoidal curve. Such results may show more pronounced peaks at the top of the waves, or a large 'surge' in anodic current during the forward scan that is not evident on the reverse scan. While it may be tempting to interpret such peak-like shapes as mediators or additional complexity, these features can be easily created by a mismatch between scan rates and diffusion rates, so that forward voltammograms begin to resemble diffusion-limited peaks at planar electrodes (Busalmen *et al.* 2008b; Yokoyama *et al.* 2002). This kind of behavior would also be expected when thick or porous electrodes are used, as donor concentrations could become depleted during the scan. An example of this kind of behavior was produced in voltammograms of a thin biofilm of *G. sulfurreducens*, simply by exposing it to saturating levels of acetate (producing a sigmoidal curve), vs. diffusion-limiting levels (producing peak-like features) (Marsili *et al.* 2008b).

This again points to the importance of scan rate; if the potential is raised rapidly to make the reaction favourable, and a large reservoir of donor is temporarily available at the electrode, a surge in current will occur. But donor is consumed near the electrode in a Cottrellian fashion (current is proportional to $1/t^{1/2}$), and current decays back to a diffusion-limited rate, creating a 'peak'. Upon the reverse sweep, a more sigmoidal wave is obtained, producing a forward scan with a rapid onset and peak, and a reverse scan without this feature. Importantly, the midpoint potentials of the forward and reverse scans are not identical to those obtained with thin films, because they represent reactions occurring at a variety of local and transient concentrations.

A second 'non-standard' waveshape is that of *Shewanella*, which does not produce a plateau limiting current at higher potential, but continues to rise slowly between 0 and +0.4 V (Marsili *et al.* 2008a). In this case, one hypothesis is that free flavins secreted by cells (which have a midpoint potential of −0.21 V) are responsible for the early Nernstian response, which reaches a limiting current based on the diffusion rate of the soluble mediator. At higher potentials, cytochromes interfaced with the electrode (which have a higher midpoint potential) (Hartshorne *et al.* 2007; Firer-Sherwood *et al.* 2008) begin to contribute to electron transfer. The gradual rise in anodic current suggests that *Shewanella* direct electron transfer is a process with a slow interfacial rate of electron transfer, which requires significant overpotential to overcome.

8.1.2.3 Voltammetry in the absence of substrates

While catalytic voltammetry is obviously related to situations such as continuous electricity production in fuel cells, and produces easily measurable large signals, it represents the interplay between diffusion of donor, irreversible oxidations, and electron transfer. Voltammetry in the absence of electron donor can reveal the potentials of exposed redox proteins, and can often be used to estimate interfacial rate constants or long-range diffusion parameters (Armstrong 1999; 2005; Heering *et al.* 1998; Heering *et al.* 1997; Jeuken *et al.* 2002).

To achieve non-turnover conditions, cells must first be attached to electrodes (typically by growing to the desired density), then starved free of donors by incubating at positive potentials (Fricke *et al.* 2008; Xing *et al.* 2008). This can be problematic, as starving cells may downregulate electron transfer proteins, or detach cells from electrodes. For example, preliminary experiments with *G. sulfurreducens* have shown that starvation for 36 hours (which is needed to completely deplete intracellular stores of thick films) followed by addition of acetate often recovers ~50% of the original limiting current (Marsili *et al.* 2008b). However, once starved, many bacteria respond consistently to repeated scans at variable rates, to provide information about coverage, diffusion of electrons through the film, or interfacial kinetics. In contrast, other strains show a greater propensity to detach or die off, making even simple analysis unreliable.

If a stable, non-catalytic film can be obtained, an initial experiment is to perform a series of slow linear sweeps across the same potential range used for catalytic voltammetry. This should reveal a series of reversible peaks that, after baseline subtraction, are centered at characteristic midpoint potentials for each redox center. While a textbook characteristic of adsorbed redox agents is reversible anodic and cathodic peaks located at identical potentials, it is not uncommon for peaks from adsorbed proteins to be offset by 30–60 mV, possibly due to conformational changes, film heterogeneity, or overpotentials required to reach buried redox centers (Armstrong *et al.* 2000; Hirst and Armstrong 1998; Hirst *et al.* 1998).

To further analyze peak areas or peak heights, it is crucial (and often difficult) to subtract the baseline from non-turnover voltammograms. For an illustration of techniques used to extract the relatively small signals from background capacitance in protein film voltammetry, the reader can look to examples from the laboratories of Armstrong, Elliot, or Léger (Léger *et al.* 2003; Firer-Sherwood *et al.* 2008; Fourmond *et al.* 2009). In most of these approaches, polynomial or cubic spline baselines are fit between user-defined regions (see Figure 8.1.3), in contrast to linear interpolation methods (Fourmond *et al.* 2009). In addition, collection of voltammograms across a range of scan rates using sterile (blank) electrodes can aid in estimation of correct baselines, especially with carbon electrodes.

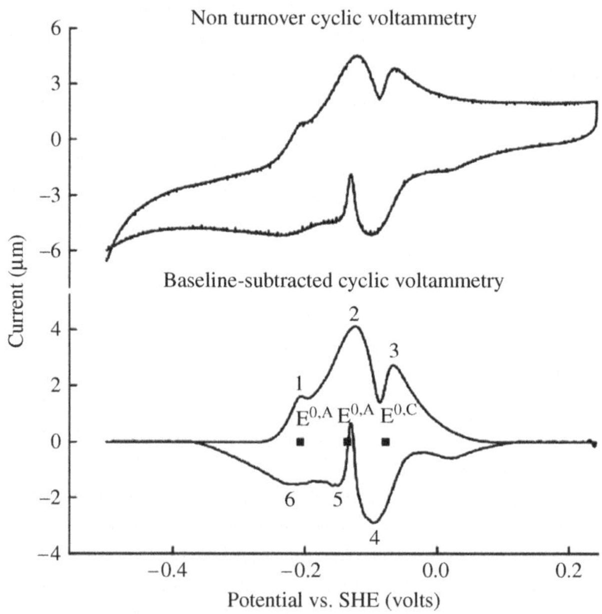

Figure 8.1.3. Raw data from *G. sulfurreducens* grown for 72 h as a biofilm on a 2.5 cm^2 graphite electrode, starved for 36 h in the absence of acetate, and subjected to cyclic voltammetry (1 mV/s) (top). Lower figure shows the baseline-subtracted voltam-mogram (using a cubic spline method in the free DOS program UTILS), and approximate midpoint potentials of putative detected redox centers

The area under each peak indicates surface coverage of redox proteins, (Γ, in mol/cm^2), and can be calculated from the peak area (Q, in Coulombs) via $\Gamma = Q/(nFA)$ (Rusling and Forster 2003). To convert this to mol protein/cm^2, one would need to make assumptions about the average number of redox centers/protein (as many cytochromes postulated to be involved in electron transfer are decaheme or greater). Early voltammetry of both *Geobacter* and *Shewanella* films have produced peaks consistent with a coverage of as much as 10^{-8}–10^{-9} mol electrons/cm^2, which, for an assumption of 10 e^- per protein, suggests 10^{-9}–10^{-10} mol protein are accessible to each cm^2. As a monolayer of 100,000 MW, proteins can pack on a surface with a density of about 10^{-12} mol/cm^2 (Armstrong *et al.* 1997), these observations strongly suggest that redox centers beyond the electrode are rapidly able to accept electrons from the surface-associated members. If confirmed, these unexpectedly high coverage values may reflect relay of electrons to proteins into the periplasm, or even the quinone pool.

At very slow scan rates, all electroactive proteins interfaced with the electrode are assumed to oxidize and reduce, as the potential is slowly increased and decreased. In multicellular films, this requires again that a mechanism exists for electrons to be relayed to more distant proteins from the most closely surface-attached proteins. However, if the potential is changed quickly, more distant redox proteins may not have time to be oxidized. The more formal description of this phenomenon is a transition from thin-film behavior (where the film discharges as a single entity) to semi-infinite diffusion (where only a portion of the film, close to the electrode, discharges). The hallmark of this is a transition from the peak height being proportional to scan rate, to being a linear function of the square root of scan rate (Bard and Faulkner 2001; Fricke *et al.* 2008; Mano *et al.* 2006).

Looking to the future, scan rate analysis may be able to reveal information about interfacial electron transfer, or cell–cell relay properties. As scan rates are increased, the location of anodic and cathodic peaks shift in opposite directions. If the behavior has been confirmed to be thin-film (peak height proportional to scan rate), software or a simple fitting method (described by Laviron 1979) can be used to estimate an average interfacial rate constant of electron transfer for an immobilized redox center (as k_0, with units in s^{-1}). For diffusional redox centers (such as shuttles), or for relay mechanisms showing semi-infinite diffusion, one can also use scan rate analysis based on the method of Nicholson and Shain, but the apparent rate constant (with units in cm/s) cannot yet be derived without further information such as the diffusion coefficient or concentration of redox centers in the film (Nicholson and Shain 1964). Careful analysis of scan rate data, measurements of film discharge rates (chronocoulometry), impedance spectroscopy, and comparison with cells in the non-biofilm (sub-monolayer) stage will likely shed new light on the nature and limitations of cell–cell relay mechanisms, and separate them from cell–electrode electron transfer.

8.1.2.4 Concluding remarks

Considering the complexity of a self-assembling catalyst made of living cells, which performs a series of oxidations and electron transfer reactions that are linked to a series of proteins that naturally interface with electrodes, it is remarkable that techniques such as those described in this and other chapters can be as informative as they are. It should be stressed that, in contrast to the case of most protein film voltammetry, 3-D structures of proteins being studied are not yet known, the number of participating redox centers is an order of magnitude larger, enzymes and cells interact in a network of electron relay mechanisms, while parameters (such as temperature, pH, medium composition) can not be as widely varied to test hypotheses. In all cases, researchers should be cautious in applying theory, techniques, and models to living cells, and strive to create equally defined and controlled experimental conditions to better understand the electron transfer reactions catalyzed by these remarkable organisms.

REFERENCES

Armstrong, F.A. (1999) Electron transfer and coupled processes in protein film voltammetry. *Biochem. Soc. Trans.* **27**(2), 206–210.

Armstrong, F.A. (2005) Recent developments in dynamic electrochemical studies of adsorbed enzymes and their active sites. *Curr. Opinion Chem. Biol.* **9**(2), 110–117.

Armstrong, F.A., Bond, A.M., Buchi, F.N., Hamnett, A., Hill, H.A.O., Lannon, A.M., Lettington, O.C. and Zoski, C.G. (1993) Electrocatalytic reduction of hydrogen peroxide at a stationary pyrolytic-graphite electrode surface in the presence of cytochrome-c peroxidase – a description based on a microelectrode array model for adsorbed enzyme molecules. *Analyst.* **118**(8), 973–978.

Armstrong, F.A., Heering, H.A. and Hirst, J. (1997) Reaction of complex metalloproteins studied by protein-film voltammetry. *Chem. Soc. Rev.* **26**, 169–179.

Armstrong, F.A., Camba, R., Heering, H.A. and Hirst, J., Jeuken, L.J., Jones, A.K., Léger C. and McEvoy, J.P. (2000) Fast voltammetric studies of the kinetics and energetics of coupled electron-transfer reactions in proteins. *Farad. Disc.* **116**, 191–203.

Bard, A.J. and Faulkner, L.R. (2001) *Electrochemical Methods* (2nd edn). Wiley and Sons, Hoboken NJ.

Biffinger, J.C., Pietron, J., Bretschger, O., Nadeau, L.J., Johnson, G.R., Williams, C.C., Nealson, K.H. and Ringeisen, B.R. (2008) The influence of acidity on microbial fuel cells containing *Shewanella oneidensis*. *Biosens. Bioelectron.* **24**(4), 906–911.

Bond, D.R. and Lovley, D.R. (2003) Electricity production by *Geobacter sulfurreducens* attached to electrodes. *Appl. Environ. Microbiol.* **69**(3), 1548–1555.

Busalmen, J.P., Esteve-Nunez, A. and Feliu, J.M. (2008a) Whole cell electrochemistry of electricity-producing microorganisms evidence an adaptation for optimal exocellular electron transport. *Environ. Sci. Technol.* **42**(7), 2445–2450.

Busalmen, J.P., Esteve-Nunez, A., Berna, A. and Feliu, J.M. (2008b) C-type cytochromes wire electricity-producing bacteria to electrodes. *Angew. Chem. Int. Edit.* **47**, 4874–4877.

Butt, J.N. and Armstrong, F.A. (2008) Voltammetry of adsorbed redox enzymes: mechanisms in the potential dimension. *Bioinorganic Electrochemistry.*

Eddowes, M.J. and Hill, H.A.O. (1977) Novel method for the investigation of the electrochemistry of metalloproteins: cytochrome c. *J.C.S. Chem. Comm.* **747**, 771–772

Firer-Sherwood, M., Pulcu, G.S. and Elliott, S.J. (2008) Electrochemical interrogations of the mtr cytochromes from *Shewanella*: Opening a potential window. *J. Biol. Inorg. Chem.* **13**(6), 849–854.

Fourmond, V., Hoke, K.R., and Heering, H.A., Baffert,C., Leroux, F. Bertrand, P. and Léger, C. (2009) SOAS: A free program to analyze electrochemical data and other on-dimensional signals. *Bioelectrochemistry*, in press (epub).

Fricke, K., Harnisch, F. and Schröder, U. (2008) On the use of cyclic voltammetry for the study of anodic electron transfer in microbial fuel cells. *Energy & Environ. Sci.* **1**, 144–147.

Hartshorne, R.S., Jepson, B.N., Clarke, T.A., Field, S.J., Fredrickson, J., Zachara, J., Shi, L., Butt, J.N. and Richardson, D.J. (2007) Characterization of *Shewanella oneidensis* mtrC: A cell-surface decaheme cytochrome involved in respiratory electron transport to extracellular electron acceptors. *J. Biol. Inorg. Chem.* **12**(7), 1083–1094.

Heering, H.A., Weiner, J.H. and Armstrong, F.A. (1997) Direct detection and measurement of electron relays in a multicentered enzyme: Voltammetry of electrode-surface films of *E. coli* fumarate reductase, an iron-sulfur flavoprotein. *J. Am. Chem. Soc.* **119**(48), 11628–11638.

Heering, H.A., Hirst, J. and Armstrong, F.A. (1998) Interpreting the catalytic voltammetry of electroactive enzymes adsorbed on electrodes. *J. Phys. Chem. B* **102**(35), 6889–6902.

Hirst, J. and Armstrong, F.A. (1998) Fast-scan cyclic voltammetry of protein films on pyrolytic graphite edge electrodes: Characteristics of electron exchange. *Anal. Chem.* **70**(23), 5062–5071.

Hirst, J., Sucheta, A., Ackrell, B.A.C. and Armstrong, F.A. (1996) Electrocatalytic voltammetry of succinate dehydrogenase: direct quantification of the catalytic properties of a complex electron-transport enzyme. *J. Am. Chem. Soc.* **118**(21), 5031–5038.

Hirst, J., Duff, J.L.C., Jameson, G.N.L., Kemper, M.A., Burgess, B.K. and Armstrong, F.A. (1998) Kinetics and mechanism of redox-coupled, long-range proton transfer in an iron-sulfur protein. Investigation by fast-scan protein-film voltammetry. *J. Am. Chem. Soc.* **120**(28), 7085–7094.

Ishii, S., Watanabe, K., Yabuki, S., Logan, B.E. and Sekiguchi, Y. (2008) Comparison of electrode reduction activities of *Geobacter sulfurreducens* and an enriched consortium in an air-cathode microbial fuel cell. *Appl. Environ. Microbiol.* **74**(23), 7348–7355.

Jeuken, L.J., Jones, A.K., Chapman, S.K., Cecchini, G. and Armstrong, F.A. (2002) Electron-transfer mechanisms through biological redox chains in multicenter enzymes. *J. Am. Chem. Soc.* **124**(20), 5702–5713.

Kano, K. and Ikeda, T. (2000) Fundamentals and practices of mediated bioelectrocatalysis. *Analytical Sciences* **16**(10), 1013–1021.

Katz, E., Lioubashevski, O. and Willner, I. (2005). Magnetic field effects on bioelectrocatalytic reactions of surface-confined enzyme systems: Enhanced performance of biofuel cells. *J. Am. Chem. Soc.* **127**, 3979–3988.

Kim, J.R., Jung, S.H., Regan, J.M. and Logan, B.E. (2007) Electricity generation and microbial community analysis of alcohol powered microbial fuel cells. *Biores. Technol.* **98**(13), 2568–2577.

Laviron, E. (1979) General expression of the linear potential sweep voltammogram in the case of diffusionless electrochemical systems. *J. Electroanal. Chem.* **101**, 19–28.

Léger, C., Elliott, S.J., Hoke, K.R., Jeuken, L.J.C., Jones, A.K. and Armstrong, F.A. (2003) Enzyme electrokinetics: Using protein film voltammetry to investigate redox enzymes and their mechanisms. *Biochemistry* **42**(29), 8653–8662.

Liu, H., Cheng, S.A. and Logan, B.E. (2005) Production of electricity from acetate or butyrate using a single-chamber microbial fuel cell. *Environ. Sci. Technol.* **39**(2), 658–662.

Liu, Y., Harnisch, F., Fricke, K., Sietmann, R. and Schröder, U. (2008) Improvement of the anodic bioelectrocatalytic activity of mixed culture biofilms by a simple consecutive electrochemical selection procedure. *Biosens. Bioelectron.* **24**(4), 1006–1011.

Mano, N., Soukharev, V. and Heller, A. (2006) A laccase-wiring redox hydrogel for efficient catalysis of O_2 electroreduction. *J. Phys. Chem. B* **110**, 11180–11187.

Manohar, A.K., Bretschger, O., Nealson, K.H. and Mansfeld, F. (2008) The use of electrochemical impedance spectroscopy (EIS) in the evaluation of the electrochemical properties of a microbial fuel cell. *Bioelectrochemistry* **72**(2), 149–154.

Marsili, E., Baron, D.B., Shikhare, I.D., Coursolle, D., Gralnick, J.A. and Bond, D.R. (2008a) *Shewanella* secretes flavins that mediate extracellular electron transfer. *Proc. Natl. Acad. Sci. USA* **105**(10), 3968–3973.

Marsili, E., Rollefson, J.B., Baron, D.B., Hozalski, R.M. and Bond, D.R. (2008b) Microbial biofilm voltammetry: direct electrochemical characterization of catalytic electrode-attached biofilms. *Appl. Environ. Microbiol.* **74**(23), 7329–7337.

Min, B., Cheng, S. and Logan, B.E. (2005) Electricity generation using membrane and salt bridge microbial fuel cells. *Water Res.* **39**(9), 1675–1686.

Nevin, K.P., Richter, H., Covalla, S.F., Johnson, J.P., Woodard, T.L., Orloff, A.L., Jia, H., Zhang, M. and Lovley, D.R. (2008) Power output and columbic efficiencies from biofilms of *Geobacter sulfurreducens* comparable to mixed community microbial fuel cells. *Environ. Microbiol.* **10**(10), 2505–2514.

Nicholson, R.S. and Shain, I. (1964) Theory of stationary electrode polarography. Single scan and cyclic methods applied to reversible, irreversible, and kinetic systems. *Anal. Chem.* **36**(4), 706–723.

Picioreanu, C., Head, I.M., Katuri, K.P., Van Loosdrecht, M.C.M. and Scott, K. (2007) A computational model for biofilm-based microbial fuel cells. *Water Res.* **41**(13), 2921–2940.

Reguera, G., Nevin, K.P., Nicoll, J.S., Covalla, S.F., Woodard, T.L. and Lovley, D.R. (2006) Biofilm and nanowire production leads to increased current in *Geobacter sulfurreducens* fuel cells. *Appl. Environ. Microbiol.* **72**(11), 7345–7348.

Richter, H., Mccarthy, K., Nevin, K.P., Johnson, J.P., Rotello, V.M. and Lovley, D.R. (2008) Electricity generation by *Geobacter sulfurreducens* attached to gold electrodes. *Langmuir* **24**(8), 4376–4379.

Rusling, J.F. and Forster, R.J. (2003). Electrochemical catalysis with redox polymer and polyion-protein films. *J. Colloid Interface Sci.* **262**(1), 1–15.

Schröder, U. (2007) Anodic electron transfer mechanisms in microbial fuel cells and their energy efficiency. *Phys. Chem. Chem. Phys.* **9**(21), 2619–2629.

Srikanth, S., Marsili, E., Flickinger, M.C. and Bond, D.R. (2008) Electrochemical characterization of *Geobacter sulfurreducens* cells immobilized on graphite paper electrodes. *Biotechnol. Bioeng.* **99**(5), 1070–1073.

Sucheta, A., Cammack, R., Weiner, J. and Armstrong, F.A. (1993) Reversible electrochemistry of fumarate reductase immobilized on an electrode surface – direct voltammetric observations of redox centers and their participation in rapid catalytic electron-transport. *Biochemistry* **32**(20), 5455–5465.

Torres, C.I., Kato Marcus, A. and Rittmann, B.E. (2007). Kinetics of consumption of fermentation products by anode-respiring bacteria. *Appl. Microbiol. Biotechnol.* **77**, 689–697.

Torres, C.I., Kato Marcus, A. and Rittmann, B.E. (2008a) Proton transport inside the biofilm limits electrical current generation by anode-respiring bacteria. *Biotechnol. Bioeng.* **100**(5), 872–881.

Torres, C.I., Marcus, A.K., Parameswaran, P. and Rittmann, B.E. (2008b) Kinetic experiments for evaluating the Nernst-Monod model for anode-respiring bacteria (ARB) in a biofilm anode. *Environ. Sci. Technol.* **42**(17), 6593–6597.

Xing, D., Zuo, Y., Cheng, S., Regan, J.M. and Logan, B.E. (2008) Electricity generation by *Rhodopseudomonas palustris* DX-1. *Environ. Sci. Technol.* **42**(11), 4146–4151.

Yeh, P. and Kuwana, T. (1977) Reversible electrode reaction of cytochrome c. *Chem. Lett.* 1145–1148.

Yokoyama, K., Koide, S. and Kayanuma, Y. (2002) Cyclic voltammetric simulation of electrochemically mediated enzyme reaction and elucidation of biosensor behaviors. *Anal. Bioanal. Chem.* **372**, 248–253.

8.2

Importance of Tafel plots in the investigation of bioelectrochemical systems

Daniel A. Lowy

8.2.1 INTRODUCTION

When there is no mass transport limitation of the electroreactive species to the electrode surface, the electrode current is controlled by the interfacial dynamics (Bard and Faulkner 2001). In 1905, Julius Tafel established the relationship between the current density, *i*, produced by an electrode and the overpotential, η, at which the electrode is operated (Eq. 8.2.1):

$$\eta = a + b \log i \tag{8.2.1}$$

where: *a* and *b* are characteristic constants of the electrode system (Kordesch 1981). On its turn, the overpotential, defined by Eq. (8.2.2), is the difference between the potential under load (E) and without a load, i.e., in equilibrium ($E^{0\prime}$):

$$\eta = E - E^{o\prime} \tag{8.2.2}$$

In a semi-logarithmic plot Eq. (8.2.1) yields a straight line, and indicates the extent to which one should change the electrode potential for getting a certain current flow. A plot of the logarithm of current density relative to the overpotential is called the *Tafel plot*, while the resulting straight line is the *Tafel line*.

While Tafel discovered his relationship empirically, later it was derived on a theoretical basis, as a limiting form of the Butler–Volmer equation, Eq. (8.2.3), which predicts the net current flowing through the system

$$i = i_0[e^{-\alpha nF\eta/RT} - e^{(1-\alpha)nF\eta/RT}] \tag{8.2.3}$$

where: i_0 is the exchange current density, α– the electron transfer coefficient, n is the number of electrons involved in the Faradaic charge transfer process, T – absolute temperature, and F Faraday's constant (Kissinger *et al.* 1996). At extreme overpotential values the rate of one of the reactions, and consequently, one term of the Butler–Volmer equation becomes negligible. Thus, when the positive polarization of the electrode exceeds +120 mV/n relative to the equilibrium, only the oxidation reaction will proceed (according to Eq. 8.2.4), the reduction being insignificant. By contrast, at negative polarization values of $\eta < -120$ mV/n, reduction will take place, and no oxidation is allowed. Equation (8.2.4), which applies for anodic polarization, can be linearized as Eq. (8.2.5)

$$i = i_0 e^{(1-\alpha)nF\eta/RT} \tag{8.2.4}$$

$$\ln i = \ln i_0 + (1-\alpha)nF\eta/RT \tag{8.2.5}$$

When converted to 10-base logarithm, and rearranged, Eq. (8.2.5) yields Eq. (8.2.6), which is equivalent to the empirical Tafel relationship (Eq. 8.2.1):

$$\eta = -[2.3RT/(1-\alpha)nF]\log i_0 + [2.3RT/(1-\alpha)nF]\log i \tag{8.2.6}$$

Hence, the physical meaning of the empirical constants a and b in Eq. 1 can be understood from the theoretical treatment (Eq. 8.2.5) as Eqs (8.2.7a) and (8.2.7b) (Bard and Faulkner 2001):

$$a = -[2.3RT/(1-\alpha)nF]\log i_0 \tag{8.2.7a}$$

$$b = 2.3RT/(1-\alpha)nF \tag{8.2.7b}$$

Constant b, termed as the *Tafel slope*, provides information about the reaction mechanism, while constant a mainly reflects the rate constant of the reaction. Rigorously, Eq. (8.2.1) should be written by using the absolute value of the current density, as shown in Eq. (8.2.8):

$$\eta = a \pm b \log |i| \tag{8.2.8}$$

where the plus sign stands for the anodic overpotential, while the minus sign is used for the cathodic overpotential (Bard and Faulkner 2001).

It should be noted here that the Ohmic resistance of the solution may cause deviations from the ideal electrode response described by Eq. (8.2.3). Therefore, in the real case one should amend the overpotential in Eq. (8.2.2) with a term, which accounts for solution resistance, R_s, the density of current flowing through the cell, i, and the exposed surface area of the working electrode, S, as shown in Eq. (8.2.9) (Rocchini 1993, 1996, 1997):

$$\eta = E - E^{o\prime} - R_s i S \tag{8.2.9}$$

Considerations on eliminating the effects of solution resistance by the appropriate positioning of the reference electrode will be made in Section 8.2.2.1.

Use of Tafel plots, as an analytical tool for evaluating fuel cell performance has become popular over the past years; however, most studies are restricted to chemical power sources, rather than being extended to microbial fuel cells. Most likely, authors choose Tafel plots as their electroanalytical technique for uncomplicated electroactive processes, which have a well-defined stoichiometry. For such simple charge transfer reactions one can exploit the linear portion of the Tafel plot for deriving the number of electrons transferred in the process and the exchange current density, i_0 (by extrapolating the linear portion of the plot to $\eta = 0$). The latter reveals the ability of the system to produce current, when moved out from the state of equilibrium.

Steady-state Tafel plots enable assessment of the electrocatalytic activity of various electrodes used in methanol or hydrogen based fuel cells. Hence, Tafel plots provide insight into the reaction mechanism of methanol oxidation on carbon-supported binary and ternary alloys of Pt, Ru, and Sn catalysts, reveal changes in the rate determining step, and when recorded at different temperatures, they allow for comparing the activation energy of different catalysts (Zhu *et al.* 2008). Also, they can identify the composition of catalysts with the greatest activity in methanol oxidation (Eguiluz *et al.* 2008; Innocente and Angelo 2008; Sieben *et al.* 2008; Salazar-Banda *et al.* 2007). Also, this analysis tool can be used for the performance assessment of the anode and cathode during direct methanol fuel cell operating conditions (Kim *et al.* 2006).

Another widespread application of Tafel plots in chemical fuel cells is the monitoring of the reaction kinetics of oxygen reduction. Recent examples of this evergreen topic addressed by means of Tafel plots are the investigation of O_2 reduction in subfreezing polymer electrolyte membrane (PEM) fuel cells (Thompson *et al.* 2007), the study of the oxygen discharge mechanism in concentrated alkaline solution on hydrophobic cathodes (Chirkov and Rostokin 2007), and the long-term durability testing of H_2-air proton exchange membrane fuel cells (Liu and Case 2006).

8.2.2 USE OF TAFEL PLOTS FOR PERFORMANCE EVALUATION OF MICROBIAL FUEL CELLS

Over the past decade, the process of direct electron transfer from anaerobic microorganisms to the anode of a fuel cell has been extensively investigated by means of a wealth of techniques. A relatively small share of this overall effort can be attributed to the utilization of Tafel plots. The less frequent use of this procedure has its explanation in the more complex nature of microbial processes, which often have an ill-defined stoichiometry. This limits the applicability of Tafel plots, and imposes special precautions when utilizing this technique for the evaluation of microbial fuel cells. The unique capability of Tafel plots entirely justifies the adoption of these measures.

Let us review briefly the strengths of Tafel plots in microbial fuel cell applications. By having on their x axes potential values converted to overpotentials, Tafel plots are efficient in comparing the electroactivity of microbial strands that have very different equilibrium potentials. A generic example displayed in Figure 8.2.1 illustrates such a comparison; in Figure 8.2.1(A) Strand A is in equilibrium at +0.35 V vs. Ag(RE), while the formal potential of Strand B is −0.25 V vs. Ag(RE). Upon converting the potentials to volts relative to the open circuit potential of each strand, one obtains a more compact and easy to compare graph (Figure 8.2.1C). As revealed by the same figure, owing to the logarithmic scale of the current axis, Tafel plots allow for scaling in and comparing electrons transfer rates (represented by the current density values). Here, Strand A exhibits 10^3 times faster electron transfer rate than Strand B, as revealed by the magnitudes of current densities: for Strand A it is measured in mA cm^{-2} units, and for Strand B in $\mu A\ cm^{-2}$.

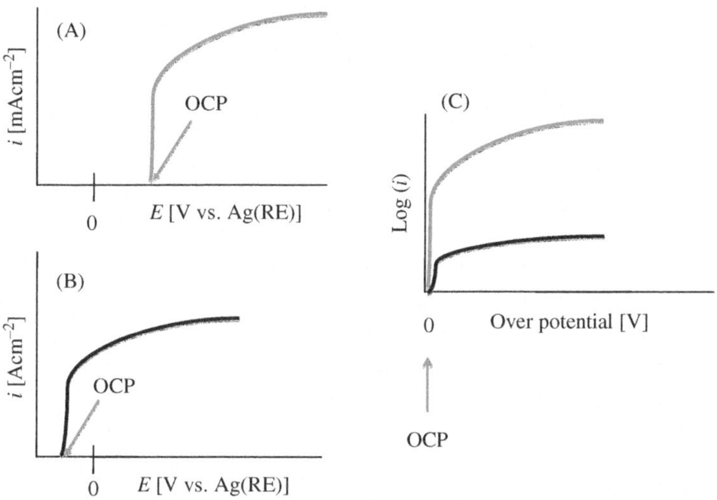

Figure 8.2.1. Comparison of polarization curves for two microbial strands: (A) has current densities of the order of mA cm^{-2}, and is graphed relative to electrode potential (vs. Ag/AgCl, 3 mol/l KCl reference electrode, henceforth abbreviated as Ag(RE); current density of (B) is measured in μA cm^{-2}, and is plotted relative to electrode potential (vs Ag(RE)), while (C) is the combined graph for A and B relative to overpotential

Another very important feature of Tafel plots is that they enable investigation of exclusively one half cell of the fuel cell. Hence, the charge transfer proceeding at the anode can be examined so that there is no need to consider any charge transfer limitation that may arise from the cathode. In other words, Tafel plots reveal the maximum power production of a given fuel cell with one particular anode, as if the cathode were of infinite size. This attribute of Tafel plots will be discussed later in this section.

8.2.2.1 Tafel plots for monitoring the electrocatalytic activity of anode materials toward microbial consortia

A relevant demonstration of how Tafel plots can monitor efficiently slight differences in the electrocatalytic activity of various anodes materials was conducted in benthic microbial fuel cells (BMFCs). Multiple Tafel plots were recorded with modified and plain graphite anodes (connected as the working electrode). A detailed description of the experimental setup was reported (Lowy *et al.* 2006). In our approach, Tafel plots were employed for the relative comparison of the kinetic activity (KA) of various electrodes. Hence, such data was acquired for each modified electrode and compared to similar data obtained for an unmodified electrode of identical size and geometry, in the same system, under identical conditions, and one after the other. Different anodes were positioned at equal distance from the cathode, i.e. from the counter electrode (Lowy and Tender 2006).

It is important to position the Ag(RE) as close to the working electrode (anode) as possible; by this, one can compensate for some of the ohmic potential drop (commonly termed IR drop), discussed in conjunction with Eq. (8.2.9). Proper placement of the reference electrode cannot prevent, however, this IR drop completely, because of the finite physical size of Ag(RE); the so-called uncompensated resistance (R_u) is left behind. Fortunately, most electrochemical workstations in use today allow for IR compensation. One should point out here that for the benthic MFCs (BMFCs) examined by us, the electrolyte solution has been always sufficiently conductive for rendering negligible the potential drop between the anode and the Ag(RE).

By using Tafel plots with all limitations and precautions described above, graphite anodes incorporating microbial oxidants were examined relative to plain graphite. Anodes evaluated include graphite modified by adsorption of anthraquinone-1,6-disulfonic acid (AQDS) or 1,4-naphthoquinone (NQ), a graphite–ceramic composite containing Mn^{2+} and Ni^{2+}, and graphite modified with graphite paste containing microcrystalline magnetite (Fe_3O_4) or, alternatively, Fe_3O_4 and Ni^{2+} (Lowy *et al.* 2006).

In other experiments Tafel plots served for the investigation of a charge transfer catalyst. Examples include an Sb(V) complex, which acts as an electron shuttle in sulfide oxidation, and oxidized graphite anodes, which have a surface enriched in electron transfer mediating quinone and carbonyl groups (Lowy and Tender 2008).

Given that benthic microbial fuel cells are mass transport limited, potential sweep should be extremely slow, typically not exceeding 1 mV s^{-1}; the maximum useful scan rate may be identified by having a reasonable overlay of the current of the Tafel plots obtained by scanning forward and, subsequently, backward. Mild overpotential values are recommended, from $\eta = 0$ mV to $\eta = 100$ (most cases), where $\eta = 0$ corresponds to the open circuit potential (OCP) of the anode vs. the Ag/AgCl, 3 mol/l KCl reference electrode. Non-negligible differences among Tafel plots were noticed for the same type of anode in different cells, presumably due to variability in sediment and cell configuration. At least 8 h were required for re-establishing a stable OCP between successive measurements (Lowy *et al.* 2006).

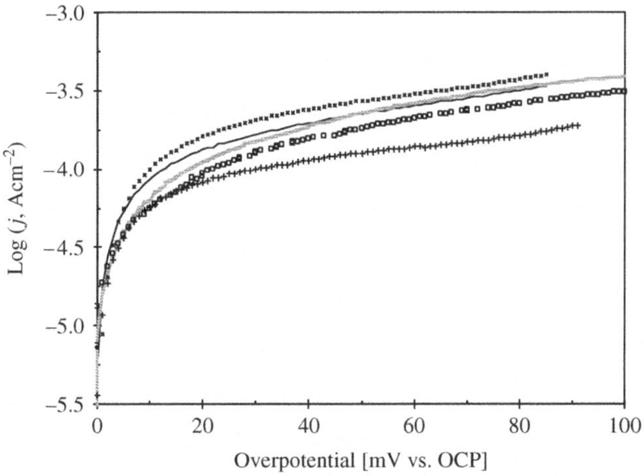

Figure 8.2.2. Tafel plots recorded in a fuel cell equipped with 3 square-shaped anodes equally spaced from the cathode: graphite disk modified with (i) magnetite: ×, (ii) Ni-magnetite: solid black line, (iii) AQDS: solid grey line, (iv) NQ: □, and (v) plain graphite anode: +, used alternatively in conjunction with the same plain graphite disk cathode. Experimental conditions were overpotential range: 0–100 mV (i.e. potential values relative to OCP), scan rate: 1 mV s^{-1}; reference electrode: Ag/AgCl, 3 mol l^{-1} KCl (Lowy *et al.* 2006)

Tafel plots offer a fast and easy way to perform a technique for estimating the kinetic activity of electron transfer shuttles. As sulfide is considered the main inorganic species involved in the electron transfer proceeding at the anode in BMFC (Bond *et al.* 2002; Ryckelynck *et al.* 2005), its fast regeneration is critical for efficient power production. Therefore, the incorporation in the anode material of an efficient sulfide-oxidizing agent can improve significantly power production of MBFCs. Such an effective sulfide oxidizing agent is the antimony (V) complex hexadecylpyridinium bis(chlorinato) Sb(V), reported by Karayannis and co-workers (Prodromidis *et al.*). Again, Tafel plots offer a means for a fast diagnostics of the efficiency of this complex. As revealed by Figure 8.2.3, when using this antimony complex incorporated in the anode the KA increased by a factor of 1.9 relative to plain; such a rise can be termed as significant. Conclusions drawn from Tafel plots were validated by an alternative electrochemical technique, namely by discharge curves recorded in the two-electrode configuration. For carbon paste loaded with antimony complex the maximum power density was *ca.* 4.2 times greater as compared to plain carbon paste (Lowy and Tender 2008).

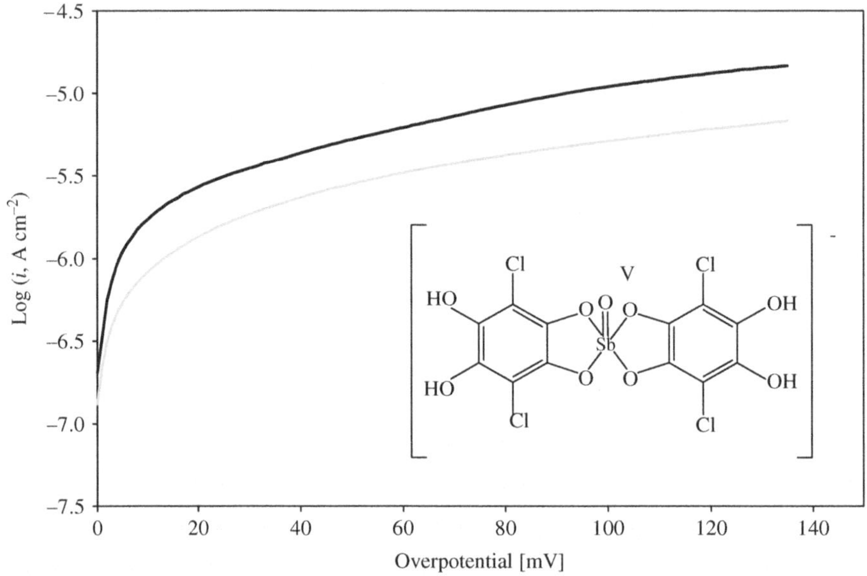

Figure 8.2.3. Tafel plots recorded in a fuel cell equipped with carbon paste anode, which contains Sb(V) complex (black line) relative to a carbon paste electrode with no added antimony complex (grey line). Overpotential range: 0–140 mV (relative to OCP), scan rate: 0.1 mV s^{-1}. Inset shows the structure of the antimony complex

Table 8.1 summarizes results of the laboratory evaluation of eight graphite-based anodes; the electrocatalytic activity (kinetic activity) derived from Tafel plots for each anode was compared to that of plain graphite. Listed kinetic activities are mean values of 5–12 independent Tafel plots recorded in the overpotential range from 0 to 100 mV. In order to determine kinetic activities, current densities were extracted from Tafel plots of modified graphite anodes and plain graphite at 3 overpotential values: $\eta = 60$, 70, and 80 mV, and the average current ratios of modified graphite anodes vs. plain graphite were calculated. By choosing this low overpotential range, the current density values will depend on the charge limited electrochemical process, the effect of mass transfer being negligible. One should emphasize here again that, in order to render Tafel plots a reliable technique, experiments need to be conducted under identical conditions, in a cell equipped with anodes of equal size embedded in the same sediment. Therefore, in several tests we used square-shaped electrodes made of graphite modified with AQDS, graphite with 1,4-NQ, and plain graphite, each electrode having the geometric surface area of 23.2 cm^2. As an additional precaution, the same plain graphite cathode was utilized throughout the experiments, and the anodes were equally spaced from the cathode (Lowy *et al.* 2006; Lowy and Tender 2008).

Table 8.1. Electrocatalytic activity of various chemically modified graphite anodes relative to plain graphite. Relative kinetic activities are based on current density values at three overpotential values, obtained from Tafel plots (Lowy and Tender 2006; Lowy and Tender 2008)

Anode	Open circuit potential *vs.* Ag(RE) [mV]	Kinetic activity
Graphite[a]	−411±16	1.0
NQ-modified graphite	**−411±12**	1.5
{Fe_3O_4+Ni^{2+}} in graphite paste[a]	−472±28	1.7
AQDS-modified graphite	**−403±15**	1.7
Sb(V) complex in carbon paste	−224±20	1.9
Fe_3O_4 in carbon paste[b]	−440±4	2.1
{Mn^{2+}+Ni^{2+}} in graphite-ceramic	−318±25	2.2
Oxidized GC or graphite	**−350±10**	57.8
AQDS-modified oxidized GC or graphite	**−350±10**	217.8

[a]Contains 5 wt.% of microcrystalline Fe_3O_4 + 1 wt.% $NiCl_2.6H_2O$ in ceramic coatings on the surface of plain graphite; Tafel plots recorded in the overpotential range from 0 to 140 mV;
[b]Contains 5 wt.% of microcrystalline Fe_3O_4 in carbon paste; Tafel plots recorded in the overpotential range from 0 to 85 mV.

Upon converting the *x* axis from potential vs. Ag(RE) to overpotential (i.e. potential values relative to the open circuit potential, OCP), the absolute values of OCP are lost (all scans start from a conventional zero value, as shown in Figure 8.2.1C). This is why the second column of Table 8.2.1 lists OCP, data which can assist in deriving the mechanism of microbial oxidation. For example, NQ-modified and AQDS-modified graphite anodes (lines 2 and 4, respectively) are similar to, if not indistinguishable from those of plain graphite anodes (line 1). This suggests the same electron-transfer reactions proceed at these electrodes, and are consistent with use of AQDS as an oxidant of micro-organisms in the family Geobacteracea in solution (Lovley *et al.* 1996) and with XPS analysis of graphite, which indicates that multiple carbon and oxygen containing functional groups (including quinone-like structures) are naturally present on plain graphite surfaces (Wu *et al.* 2002).

Electrochemically oxidized graphite anodes (Regisser *et al.* 1996), shown in line 8 of Table 8.2.1, are rich in quinone-like and carboxylic groups, which facilitate microbial oxidation. There is a strong rationale for utilizing oxidized graphite as the anode material; the surface density of quinone-like and carboxylic functional groups naturally present on the graphite surface can be increased significantly by electrochemical oxidation. For that reason, the performance of oxidized graphite as the anode material in BMFCs was examined via Tafel plots relative to plain graphite.

It was found that microbial consortia colonized oxidized graphite preferentially, and affected the composition of the microbial community, as well (Liu *et al.* 2007). As demonstrated by Tafel plots, oxidized graphite subsequently modified with AQDS exhibits almost 220-fold greater electrocatalytic activity as compared to plain graphite; see line 9 in Table 8.2.1 (Lowy and Tender 2008).

Predictions made based upon Tafel plots scale up in a very reliable manner, hence, they allow for the relevant comparison of anodes of very different size. Examples provided in Table 8.2.1 were obtained with anodes of a projected surface area encompassing the range from 0.018 to 71.3 cm^2. All tabled data are mean values of 5–12 independent experimental runs, the relative kinetic activities being calculated from current density values at three overpotential values, obtained from Tafel plots recorded under identical conditions (Lowy and Tender 2006; Lowy and Tender 2008).

8.2.2.2 Tafel plots for examining charge transfer with microbial pure cultures

Tafel plots can be recorded with pure culture microbial strains, rather than with microbial consortia. Such scans are able to identify the presence of a live microbial community, as compared to the dead microbial community on the same electrode surface (or in the bulk phase). Curve A in Figure 8.2.4 corresponds to two overlaid Tafel plots recorded for a live community of wild type *Geobacter sulfurreducens*, in the presence of acetate, with 11-mercapto-undecanoic (11-MUDA) acid modified gold serving as the working electrode. When compared to curve C, corresponding to dead *Geobacter sulfurreducens*, examined under identical conditions, one measured a 19-fold lower current (as microbes ceased driving the electron transfer process). Tafel plots also evidenced differences between microbial colonization of various electrode surfaces. Plot B in Figure 8.2.4 was recorded on gold modified with a monolayer of 12-dodecanethiol on gold, and showed a different pattern and slightly greater activity of the same microbial strain than with 11-MUDA on gold (curve A).

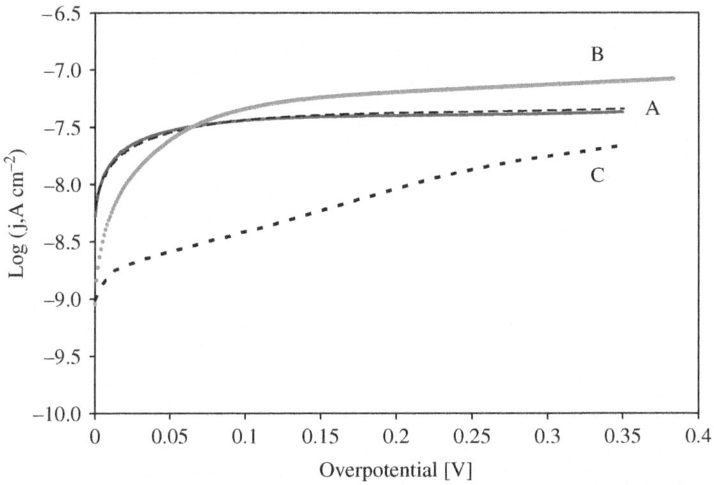

Figure 8.2.4. Tafel plots recorded with wild type *Geobacter sulfurreducens*: A – overlay of two curves acquired with live microbial community, C – plots obtained with dead microbial community on the same surface, and B – curve corresponding to live microbes colonizing a different electrode surface; see details in text (Chaudhuri and Lowy 2004)

When nutrients are being added to the system, the microbial electron transfer is significantly enhanced. Hence, plot A in Figure 8.2.5 was recorded for wild type *Geobacter sulfurreducens* colonizing the electrode surface, at a known initial acetate concentration, while plot B corresponds to the double of the initial acetate concentration. Curve 3 is the control experiment, recorded in the presence of live bacteria and the absence of acetate. Current densities generated in the absence of the nutrient (acetate) are lower by two orders of magnitude (Chaudhuri and Lowy 2004).

8.2.2.3 Estimating the maximum power production from Tafel plots

A series of carefully designed BMFC experiments demonstrated that in essence Tafel plots recorded for the working electrode in a fuel cell are equivalent to the electron transfer proceeding without any counter electrode limitations. In other words, Tafel plots correspond to the ideal scenario, when the working electrode is scanned relative to a counter electrode of infinite size. Displayed in Figure 8.2.6 are two power curves recorded in the 2-electrode configuration, using the same graphite anode plate, imbedded in the sediment, in conjunction with increasing size cathodes, immersed in the overlaying seawater. Particularly suitable for conducting these experiments are the so-called carbon fiber 'bottle brush' electrodes, made of a titanium wire core with carbon fibers oriented in radial direction, having a specific surface of approximately 26 m^2 per 1 m long electrode (Borehole 2006; Hasvold *et al.* 1999).

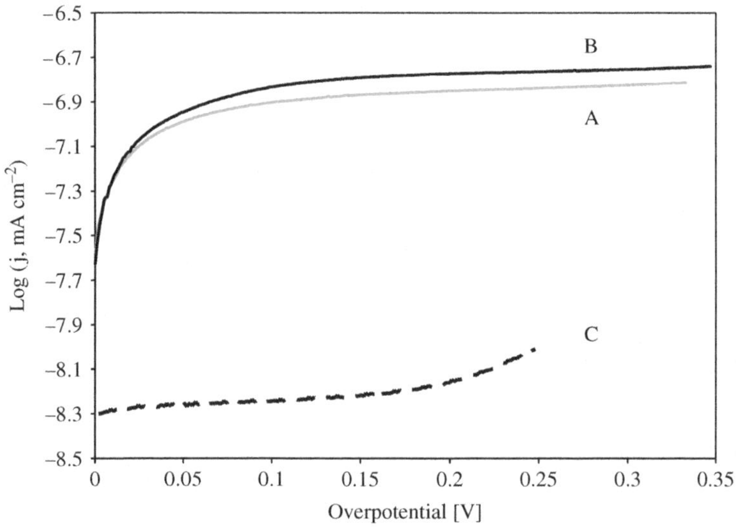

Figure 8.2.5. Tafel plots recorded with pure culture of wild type *Geobacter sulfurreducens* at different nutrient concentrations: A – in the presence of one volume of acetate in the supporting electrolyte; B – with double volume of added acetate; C – no acetate supplied to the electrolyte; see details in text (Chaudhuri and Lowy 2004)

Power curve A in Figure 8.2.6 was recorded with an approximately unity-length cathode (surface area *ca.* 20.0 m^2), while power curve B corresponds to the experiment performed with a double length cathode (surface area *ca.* 46.0 m^2). These two curves are compared to a third power curve, derived from a Tafel plot (curve C, recorded in 3-electrode geometry), obtained as follows: (i) the anode potential values relative to Ag(RE) were adjusted to the cell voltage value, by adding to each the cathode potential measured independently, and then (ii) the power density (cell voltage multiplied with the current density) was plotted relative to the increasing current density (Lowy and Tender 2009).

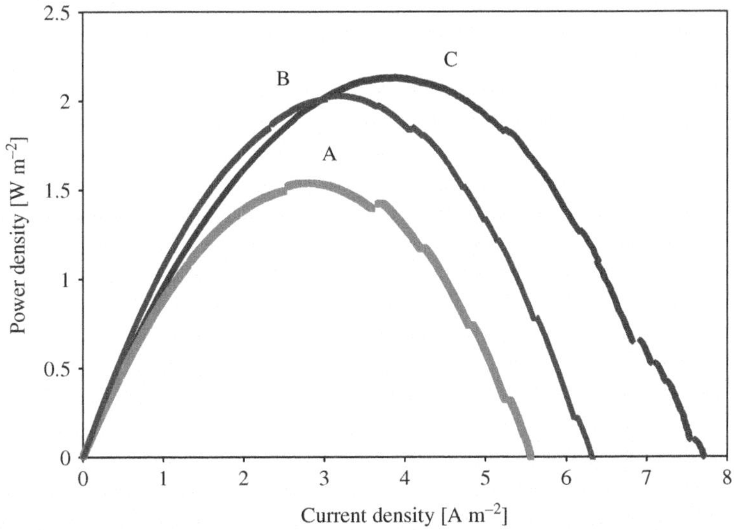

Figure 8.2.6. Power curves with one and two counter electrodes compared to the power curve derived from a Tafel plot; A – power curve from discharge curves, recorded in 2-electrode configuration with single cathode, B – power curve from discharge curves in 2-electrode configuration with double size cathode, C – power curve extracted from a Tafel plot recorded in 3-electrode configuration (Lowy and Tender 2007)

The ability of the Tafel technique of investigating exclusively the half cell of interest (in this case the anode) was sometimes regarded with reservation even by the electrochemistry community. As revealed by Figure 8.2.6, upon increasing the surface area of the counter electrode, the cell performance approaches stepwise the one derived from the Tafel plot. We believe that this experimentally proves the assumption that the use of Tafel plots characterizes the electrode of interest in the fuel cell as if there were no counter electrode limitations. Hence, these results substantiate the usefulness of Tafel plots as a major technique for the investigation of the half cell of interest in any fuel cell, and as an efficient tool for estimating the maximum of power production, when using one particular working electrode in a given system (Lowy and Tender 2007).

REFERENCES

Bard, A.J. and Faulkner, L.R. (2001) *Electrochemical Methods. Fundamentals and Applications*. John Wiley & Sons, Inc., New York, Chichester, Weinheim, Brisbane, Singapore, Toronto, 2nd edn., pp. 92, 100–106.

Bond, D.R., Holmes, D.E., Tender, L.M. and Lovley D.R. (2002) Electrode-reducing microorganisms that harvest energy from marine sediments. *Science* **295**(5554), 483–485.

Borehole Instrument Package (2006) Figures F20 and F21 display structure and side views drawing of the carbon fiber brush. <http://www-odp.tamu.edu/publications/186_IR/chap_03/chap_03.htm> (31 July 31 2006).

Chaudhuri, S.K. and Lowy, D.A. (2004) Unpublished results.

Chirkov, Y.G. and Rostokin V.I. (2007) Hydrophobized oxygen cathode of a fuel cell with a liquid electrolyte: Calculating overall currents and thicknesses. *Russian J. Electrochem.* **43**(2), 146–156.

Eguiluz, K.I.B.; Salazar-Banda, G.R.; Miwa, D.; Machado, S.A.S. and Avaca, L.A. (2008) Effect of the catalyst composition in the Pt-x(Ru-Ir)(1-x)/C system on the electro-oxidation of methanol in acid media. *J. Power Sources* **179**(1), 42–49.

Hasvold, Ø., Johansen, K.H., Mollestad, O., Forseth, S. and Storkersen, N. (1999) The alkaline aluminium hydrogen peroxide power source in the Hugin II unmanned underwater vehicle. *J. Power Sources*, **80**(1–2), 254–260.

Innocente, A.F. and Angelo, A.C.D. (2008) Hydrogen oxidation on ordered intermetallic electrodes covered with CO. *J. Power Sources* **175**(2), 779–783.

Kim, Y.J., Hong, W.H., Woo, S.I. and Lee, H.K. (2006) Analysis of the polarization of a direct methanol fuel cell using a pseudo-reversible hydrogen reference electrode. *J. Power Sources* **159**(1), 491–500.

Kissinger, P.T., Preddy, C.R., Shoup, R.E. and Heineman, W.R. (1996) Fundamental Concepts of Analytical Chemistry. In *Laboratory Techniques in Electroanalytical Chemistry* (eds. Kissinger, P.T. and Heineman, W.R.), Marcel Dekker, New York, Basel, 2nd edn., pp. 37–38.

Kordesch, K. (1981) Electrochemical energy storage. In *Comprehensive Treatise of Electrochemistry*. Vol. 3: *Electrochemical Energy Conversion and Storage* (eds). Bockris, J.O'M., Conway, B.E., Yeager, E. and White, R.E., Plenum Press, New York, London, pp. 139–141.

Liu, D. and Case, S. (2006) Durability study of proton exchange membrane fuel cells under dynamic testing conditions with cyclic current profile. *J. Power Sources* **179**(1), 42–49.

Liu, J.L., Lowy, D.A., Baumann, R.G. and Tender, L.M. (2007) Influence of anode pretreatment on its microbial colonization. *J. Appl. Microbiol.* **102**(1), 177–183.

Lovley, D.R., Coates, J.D., BluntHarris, E.L., Phillips, E.J.P. and Woodward, J.C. (1996) Humic substances as electron acceptors for microbial respiration. *Nature* **382**(6590), 445–448.

Lowy, D.A. (2008) Electrochemical techniques for investigating electron transfer at the electrode surface. *108th General Meeting of the American Society for Microbiology*, Boston, June 4, 2008.

Lowy, D.A. and Tender, L.M. (2007) Importance of cathode size in microbial benthic fuel cells equipped with manganese anodes. *The 234th ACS National Meeting, Division of Fuel Chemistry*, Boston, MA, August 19–23, 2007, FUEL-158.

Lowy, D.A. and Tender, L.M. (2008) Harvesting energy from the marine sediment-water interface. III. Kinetic activity of quinone and antimony-based anode materials. *J. Power Sources* **185**(1), 70–75.

Lowy, D.A. and Tender, L.M. (2009) Role of the cathode size on the efficiency of microbial fuel cells (in preparation).

Lowy, D.A., Tender, L.M., Zeikus, J.G., Park, D.H. and Lovley, D.R. (2006) Harvesting energy from the marine sediment-water interface. II. Kinetic studies on anode materials. *Biosens. Bioelectron.* **21**(11), 2058–2063.

Prodromidis, M.I., Veltsistas, P.G. and Karayannis, M.I. (2000) Electrochemical study of chemically modified and screen-printed graphite electrodes with [(SbO)-O-V(CHL)(2)]Hex. Application for the selective determination of sulphide. *Anal. Chem.* **72**(17), 3995–4002.

Regisser, F., Lavoie, M.-A., Champagne, G.Y. and Bélanger, D.J. (1996) Randomly oriented graphite electrode. 1. Effect of electrochemical pretreatment on the electrochemical behavior and chemical composition of the electrode. *Electroanal. Chem.* **415**(1–2), 47–54.

Rocchini, G. (1993) The influence of the Ohmic drop on the shape of polarization curves. *Corr. Science* **34**(12), 2019–2030.

Rocchini, G. (1996) A theoretical study of the influence of the ohmic drop on polarization curves. *Corr. Science* **38**(4), 655–668.

Rocchini, G. (1997) The computation of the solution resistance from polarization data. *Corr. Science* **39**(10–11), 1861–1872.

Ryckelynck, N., Stecher III, H.A. and Reimers C.E. (2005) Understanding the anodic mechanism of a seafloor fuel cell: interactions between geochemistry and microbial activity. *Biogeochemistry* **76**(1), 113–139.

Salazar-Banda, G.R., Suffredini H.B., Calegaro, M.L., Tanimoto, S.T. and Avaca, L.A. (2006) Sol-gel-modified boron-doped diamond surfaces for methanol and ethanol electro-oxidation in acid medium. *J. Power Sources* **162**(1), 9–20.

Sieben, J.M., Duarte, M.M.E. and Mayer, C.E. (2008) Supported Pt and Pt-Ru catalysts prepared by potentiostatic electrodeposition for methanol electrooxidation. *J. Appl. Electrochem.* **38**(4), 483–490.

Thompson, E.L., Jorne, J. and Gasteiger, H.A. (2007) Oxygen reduction reaction kinetics in subfreezing PEM fuel cells. *J. Electrochem. Electrochem. Soc.* **154**(8), B783–B792.

Wu, Y.P., Jiang, C., Wan, C. and Holze, R. (2002) Modified natural graphite as anode material for lithium ion batteries. *J. Power Sources* **111**(2), 329–334.

Zhu, J., Cheng, F.Y., Tao, Z. and Chen, J. (2008) Electrocatalytic methanol oxidation of $Pt_{0.5}Ru_{0.5-x}Sn_x/C$ (x = 0–0.5). *J. Phys. Chem. C* **112**(16), 6337–6345.

8.3

The use of electrochemical impedance spectroscopy (EIS) for the evaluation of the electrochemical properties of bioelectrochemical systems

Aswin K. Manohar, Zhen He and Florian Mansfeld

8.3.1 INTRODUCTION

Electrochemical Impedance Spectroscopy (EIS) has become one of the most extensively used techniques in the analysis and characterization of electrochemical systems and processes. The development of AC techniques in the analysis of electrochemical systems dates back to the pioneering work by

© 2010 IWA Publishing. *Bioelectrochemical Systems: From Extracellular Electron Transfer to Biotechnological Application*. Edited by Korneel Rabaey, Largus Angenent, Uwe Schröder and Jürg Keller. ISBN: 9781843392330. Published by IWA Publishing, London, UK.

Epelboin and co-workers, who used the 'ac (alternating current) impedance technique' in the evaluation of corrosion and its inhibition (Epelboin *et al.* 1984). Mansfeld and Kendig were the first to use the term 'Electrochemical Impedance Spectroscopy' (EIS) in a paper entitled 'Electrochemical Impedance Spectroscopy of Protective Coatings' (Mansfeld and Kendig 1985). Mansfeld has recently provided a detailed summary of the use of EIS in a large number of cases of corrosion and corrosion inhibition including the interesting phenomenon of corrosion inhibition by bacteria (Mansfeld 2006b). Schiller has presented an introduction to electrochemical instrumentation that includes the instrumentation needed for EIS measurements (Schiller 2006).

In EIS, the properties of an electrode/electrolyte interface are determined as a function of the frequency of a small amplitude AC signal that is applied at a fixed operating point (*E*, *I*) of the system. Since only a small amplitude AC signal, for example 10 mV, is applied, EIS is in principle a non-destructive electrochemical technique (Mansfeld 2006b).

The aim of this chapter is to explain the fundamental principles of EIS and to demonstrate how it can be applied as a very useful tool in determining various electrochemical parameters of microbial fuel cells (MFCs). The two main applications of EIS in MFC studies are the electrochemical characterization of the individual electrodes and the measurement of the internal resistance of the MFC. Various examples that demonstrate the use of this technique in MFC research have been provided.

8.3.2 INSTRUMENTATION AND EXPERIMENTAL APPROACH

Instrumentation: The equipment for the analysis of the dynamic behavior of an electrochemical system consists of a frequency response analyzer (FRA), spectrum analyzer, or network analyzer combined with a digital regulation device to interface the electrochemical cell and a computer system for the collection and analysis of the experimental impedance data. Until a few years ago, measurement of the impedance data involved the difficult task of manually connecting all these instruments and establishing an efficient circuit that could measure the impedance response of a system. In the last few years, great progress has been made in the hardware and software for collection and analysis of impedance data. As a direct consequence of this, potentiostats are available now with all the essential components necessary for performing an EIS experiment incorporated in them. The accompanying software efficiently interfaces the electrochemical system to be measured with a computer connected to the potentiostat (Schiller 2006).

A two-electrode or a three-electrode cell can be used for the measurement of impedance spectra. A general arrangement for the measurement is shown in Figure 8.3.1, where the potentiostat is connected to the frequency response analyzer that facilitates the actual application and measurement of AC signals of varying frequencies. The computer connected to the potentiostat collects and plots the measured data. Modern equipment allows measurement of impedance data from frequencies of about 1 MHz to frequencies less than 1 mHz.

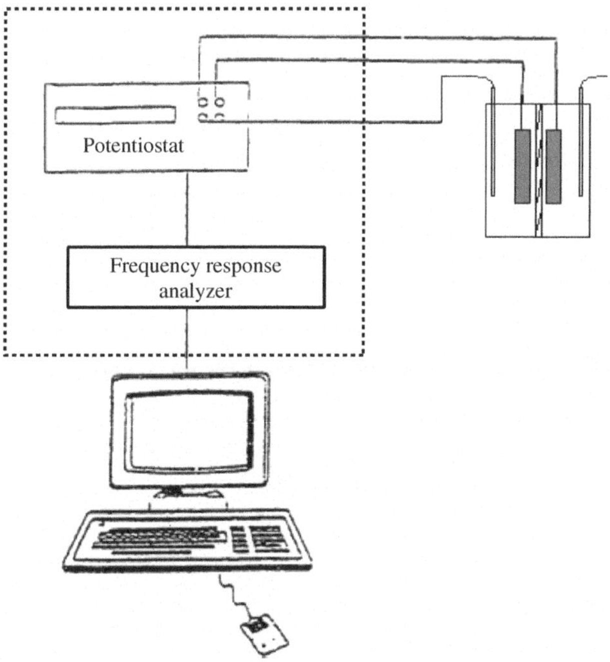

Figure 8.3.1. Schematic sketch of the equipment for electrochemical impedance measurements

Measurement: The measurements of the impedance spectra of the anode and cathode are usually done at their corresponding open-circuit potentials. To measure the impedance spectra of the anode, the working electrode lead of the potentiostat is connected to the anode and the counter electrode to the cathode. The reference electrode lead is connected to a reference electrode placed preferably in the same compartment as the working electrode. When measuring the impedance spectra of the cell, the working electrode lead has to be connected to one electrode, while both the counter and the reference electrode leads are connected to the other electrode.

Choice of experimental variables: Although the measurement of impedance spectra can be performed at very high frequencies, care should be taken to eliminate any phase shift effects caused by the potentiostat and the reference electrode at very high frequencies. These effects might prevent several key features of the impedance spectra from being observed. The high-frequency phase shift effects are due to the resistance and capacitance of the reference electrode used in the impedance measurement and can be eliminated by capacitatively coupling an inert metal wire (Pt, Au) to the reference electrode. This metal wire can act as a pseudo-reference electrode during the impedance measurement at high frequencies (Mansfeld *et al.* 1988). The impedance experiments can be performed either at the open-circuit potential (E_{ocp} or E_{corr}) where no current is flowing through the system and equilibrium is maintained or at any applied potential when a finite current flows and the system has achieved steady state. The potentiostat maintains this operating potential at a constant value during the experiment. The magnitude of the applied AC signal has to be small enough not to perturb the system and maintain the linearity of the system.

8.3.3 DISPLAY AND ANALYSIS OF EIS DATA

The impedance of a system, Z can be represented as $Z = Z' + jZ''$ here Z' and Z'' are the real and imaginary parts of the complex number Z, respectively. The impedance Z is also associated with the phase angle φ which is calculated as:

$$\varphi = \tan^{-1}(-Z''/Z')$$

The experimentally measured impedance values, $Z(j\omega)$ are usually plotted using one of two different methods:

1. Complex plane plots.
2. Bode plots.

In the complex plane plot, the negative values of the imaginary impedance are plotted against the real part of the impedance data.

In the Bode plot, the logarithm of the impedance modulus |Z| and the phase angle are plotted against the logarithm of the frequency f of the applied AC signal. In addition to the complex plane and Bode plot, there are several other methods that can also be employed for the representation of impedance data (Mansfeld 1981; Mansfeld *et al.* 1982).

The Bode plot is favored for the representation of impedance data as it shows the frequency dependence of the impedance and because one can easily determine the number of time constants in the impedance spectrum (Mansfeld 1988). Figure 8.3.2(A) and (B) show respectively the complex plane plot and the Bode plot of the impedance spectrum of the anode in an air–cathode MFC (He *et al.* 2008).

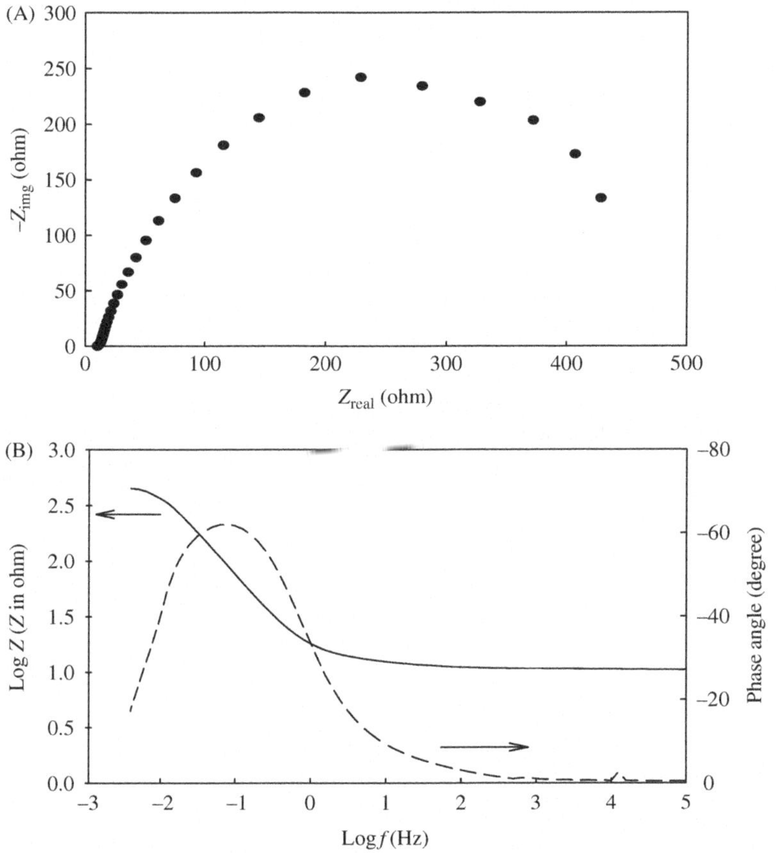

Figure 8.3.2. The impedance of the anode in an air-cathode MFC using acetate solution as electrolyte: (A) the complex plane plot; (B) Bode plot

In order to obtain various electrochemical parameters from the measured impedance spectra, they are fitted to appropriate equivalent circuits (ECs). These ECs contain electrical components like resistors, capacitors and inductors in series and parallel combinations. Each of these components correlates to a particular electrochemical process or property of the system (Jüttner *et al.* 1985; Mansfeld 1981; Mansfeld *et al.* 1982).

A simple charge transfer controlled reaction like the one taking place at the anode whose impedance spectra is shown in Figure 8.3.2, can be represented by the equivalent circuit shown in Figure 8.3.3. R_s is the solution resistance that accounts for all the ohmic components in the system being measured. R_p is the polarization resistance of the electrode and is inversely proportional to the exchange current density (i_o) of the reaction taking place on the electrode surface. C represents the capacitance of the electrode. The impedance of this circuit is given by,

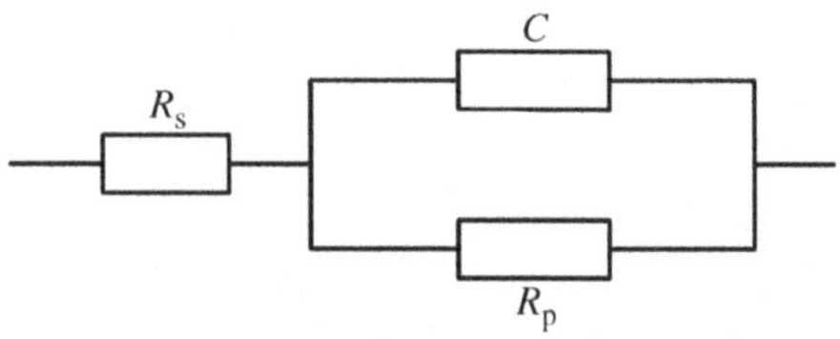

Figure 8.3.3. The equivalent circuit for an anode of a MFC: R_s – solution resistance (ohmic resistance), R_p – polarization resistance and C – Capacitance

$$Z(j\omega) = R_s + \frac{R_p}{1 + j\omega C R_p}$$

where

$$Z' = R_s + \frac{R_p}{1 + (\omega C R_p)^2} \text{ and } Z'' = -\frac{\omega C R_p^2}{1 + (\omega C R_p)^2}$$

Elimination of the frequency ω leads to

$$\left(Z' - R_s - \frac{R_p}{2}\right)^2 + (Z'') = \left(\frac{R_p}{2}\right)^2$$

This represents a semi-circle in the complex plane plot with a diameter equal to R_p.

8.3.4 DETERMINATION OF KEY ELECTROCHEMICAL PARAMETERS FROM IMPEDANCE SPECTRA

For an electrode or any electrochemical system whose equivalent circuit is the same as the one shown in Figure 8.3.3, the various electrochemical parameters can be determined as follows. In the complex plane plot, the high-frequency and the low-frequency intercepts of the real axis correspond to the solution resistance, R_s and sum of the solution resistance and polarization resistance R_s+R_p, respectively.

At very high frequencies, $Z' = R_s$ and $Z'' = 0$

At low frequencies, $Z' = R_s+R_p$ and $Z'' = 0$

The capacitance is calculated using the expression,

$$C = \frac{1}{\omega_m R_p}$$

where $\omega_m = 2\pi f_m$ is the frequency at which the maximum value of Z'' occurs.

R_s and R_p can be obtained from the Bode plot using

$$R_s = \lim_{\omega \to \infty} |Z|$$

$$R_s + R_p = \lim_{\omega \to 0} |Z|$$

The capacitance can be determined from the Bode plot from the impedance value at $\omega = 1$ rad/sec ($f = 0.16$ Hz), since at this frequency $\log|Z| = -\log C$.

In some cases, deviations from ideal capacitive behavior can result in depressed semicircles in the complex plane plot and capacitive regions of slope not equaling -1 in the Bode plot (Mansfeld and Kendig 1983; Mansfeld *et al.* 1985). These spectra can be fitted to ECs with the capacitance replaced by a constant phase element (CPE) which is due to inhomogeneities of the electrode surface (Hsu and Mansfeld 2001; Schmidt *et al.* 1986). A method to convert the measured CPE into a capacitance has been provided Hsu and Mansfeld (2001).

In cases where mass transport becomes important, a Warburg impedance element Z_w can be placed in series with R_p in the equivalent circuit shown in Figure 8.3.3. The impedance of the Warburg element is given by

$$Z_w = \sigma(1 - j)\omega^{0.5}$$

where σ is the Warburg coefficient. The Warburg impedance can be observed as a straight line of slope 1 in the complex plane plot at low frequencies.

Fitting of the experimental impedance data to an appropriate equivalent circuit is usually performed with software developed for impedance data analysis. All impedance data presented in this article have been fitted using the ANALEIS software developed by Shih and Mansfeld (Mansfeld 2006b; Mansfeld *et al.* 1993; Mansfeld *et al.* 1992).

8.3.5 APPLICATIONS OF ELECTROCHEMICAL IMPEDANCE SPECTROSCOPY IN THE STUDY OF MFCS

8.3.5.1 Electrochemical characterization of anode and cathode properties

EIS can be used to determine the electrochemical properties of the individual electrodes in a MFC. The measurement of the impedance spectra of the anode and cathode provides key information that allows analysis of the electrochemical reactions on electrodes and bacterial metabolism (He and Mansfeld 2008).

In MFC systems that have a proton exchange membrane separating the anode and cathode compartments, separate reference electrodes have to be placed in both compartments to facilitate the measurement of the potentials of the individual electrodes without interference from the ohmic resistance of the membrane and the different ohmic resistance of the electrolyte in the two compartments.

Figure 8.3.4 shows the impedance spectra of the anode and cathode of a two-chamber MFC measured at their corresponding open-circuit potentials (E_{an} and E_{ca}). The impedance spectrum of the anode was also measured before the addition of bacteria to the anode compartment. It can be seen that the spectrum of the anode is different in the presence of *Shewanella Oneidensis* MR-1 in the anolyte. In the presence of bacteria, the frequency dependence of the impedance and the phase angle φ at low frequencies demonstrates that the polarization resistance of the anode, R_p^a has decreased considerably. This decrease is accompanied by a sharp decrease of E_{an} from 0.197V to −0.481 V (vs. Ag/AgCl) (Manohar *et al.* 2008a). According to mixed potential theory (Mansfeld 2006a), this decrease of R_p^a and E_{an} in the presence of MR-1 suggests an increase in the rate of the lactate oxidation reaction due to the presence of bacteria at the anode. Comparison of the impedance spectra of the cathode and anode with MR-1 in the anolyte shows that R_p^c is smaller than R_p^a of the anode indicating that the rate of oxygen reduction at the cathode is higher than the rate of oxidation of lactate by bacteria at the anode (Figure 8.3.4 and Table 8.1). The higher capacitance of the cathode is due to the larger active surface area caused by the platinum particles deposited on the graphite. A comparison of the measured and fitted impedance data of the anode with MR-1 in the anolyte is shown in Figure 8.3.5. There is very good agreement between the measured and the fitted impedance data.

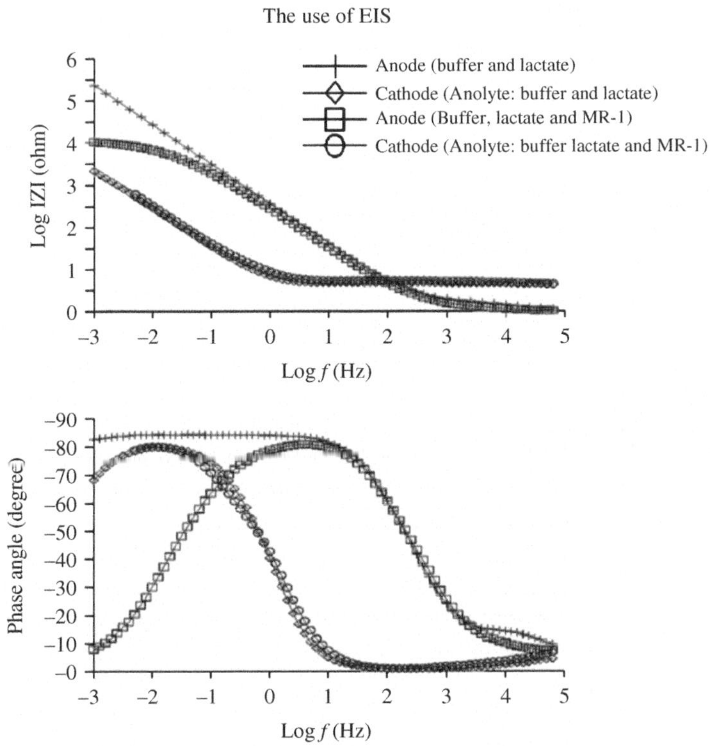

Figure 8.3.4. Impedance spectra of the anode and the cathode of the MFC with and without MR-1 in the anolyte

Table 8.1. Fit parameters for the impedance spectrum of the anode shown in Fig. 8.3.4 with MR-1 in the anolyte

Parameter	Anode
R_p (Ohm)	$1.02*10^4$
C (F)	$9.70*10^{-4}$
R_s (Ohm)	1.1

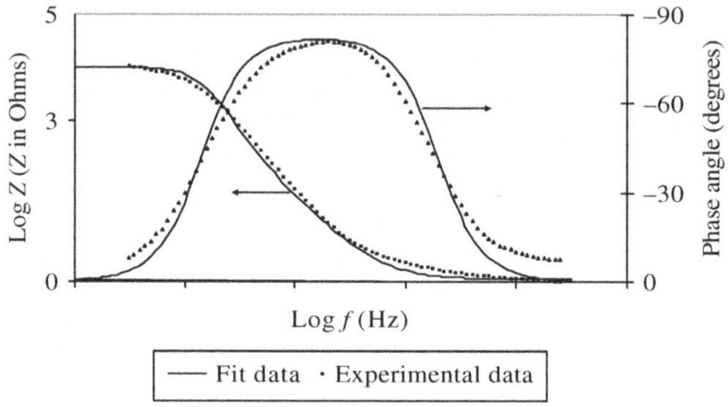

Figure 8.3.5. Comparison of experimental and fitted impedance spectra for the anode with MR-1 in the anolyte (data shown in Fig. 8.3.4)

In a study conducted with *Geobacter sulfurreducens*, the impedance response of a bare graphite paper anode has been compared with the impedance response after the entrapment of bacteria on these electrodes. The study revealed a significant reduction in the polarization resistance of the anode and an increase in the capacitance of the anode in the presence of bacteria. The authors concluded that these observations indicate respectively an increase in the rate of the redox process at the electrode and the presence of a conductive biofilm on the electrode surface (Srikanth *et al.* 2008).

In a recent study, a comparison of the impedance responses of various electrodes such as silver, aluminum, nickel, stainless steel and carbon cloth was conducted (Ouitrakul *et al.* 2007). It was concluded that the metal electrodes had high interfacial impedance and that carbon cloth electrodes and silver electrodes exhibited low R_p. Qiao *et al.* reported that a carbon nanotube/polyaniline composite coated on nickel foam electrodes could reduce the polarization resistance. The EIS results indicated that a higher loading rate of carbon nanotubes further decreased the R_p and increased the reaction rate at the anode as well as electricity production in MFCs (Qiao *et al.* 2007).

It has been suggested that EIS can be used to examine the progress of biofilm formation on an electrode surface. In a recent study, the impedance spectrum of the anode was measured on days 1 and 5 and after 3 weeks. The authors reported that the polarization resistance decreased from 2.61 to 0.48 kΩcm^2 within three weeks, implying that biofilm development improved the kinetics of the electrochemical reactions (Ramasamy *et al.* 2008).

EIS has been used to investigate and understand the effect of experimental conditions on electrochemical reactions taking place at the electrodes in an

MFC. In a rotating cathode MFC, the polarization resistance of the anode increased from 28 to 65 Ω when the cathode electrode was rotating compared with a non-rotating cathode, suggesting that rotation increased the concentration of dissolved oxygen that had a negative effect on the anodic bacterial metabolism (He *et al.* 2007). Meanwhile, the ohmic resistance slightly decreased from 50 to 40 Ω, possibly due to better ion transport by the mixing of water during the rotation.

EIS has also been used to analyze the anode and cathode properties with different electrolyte pH in an air-cathode MFC (He *et al.* 2008). In such a system, the electrolyte pH affects both the anodic and cathodic reactions. It was found that the anodic polarization resistance was the lowest at pH 7, while the cathodic polarization resistance decreased with the increasing electrolyte pH from 5 to 10 (Figure 8.3.6). These results confirmed that the anodic bacterial activity was optimal at a neutral pH and the rate of the electrochemical reaction on the cathode electrode was increased at higher pH (within the test range).

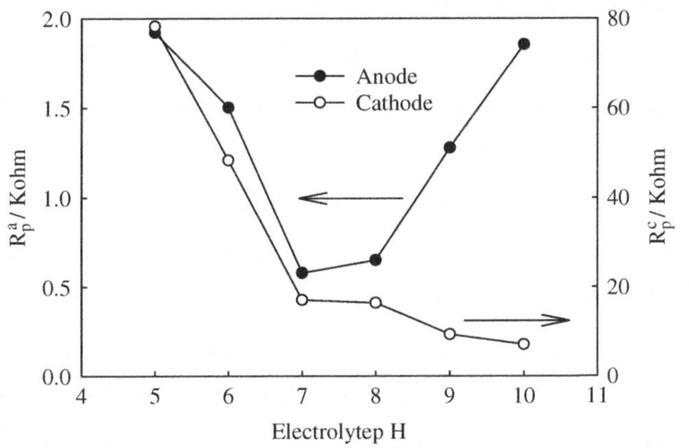

Figure 8.3.6. The polarization resistances of the anode and the cathode of an air-cathode MFC as a function of pH

8.3.5.2 Determination and analysis of the internal resistance R_{in}

The internal resistance of a MFC is defined as

$$R_{in} = R_p^a + R_p^c + R_\Omega = R_p^{cell} \tag{8.3.1}$$

where R_p^a and R_p^c are the polarization resistances of the anode and cathode, respectively and R_Ω contains the ohmic contributions of the electrodes, the anolyte and catholyte, and the membrane. R_p^{cell} is the DC limit of the impedance of the MFC, i.e. the impedance for f –> 0, where f is the frequency of the applied ac signal (Manohar and Mansfeld 2009).

R_{in} is one of the major factors that affect the power output from an MFC. Some of the earliest applications of EIS in MFC research were in determining its internal resistance. EIS provided more precise evaluation of R_{in} than that which could be measured using the current interrupt technique or by calculating the slope of the polarization curve of the cell (Fan *et al.* 2007; He *et al.* 2006; You *et al.* 2007, 2008). A detailed description of the current-interrupt techniques has been provided in Chapter 7. In bioelectrochemical systems like MFCs the microbial activities play an important role and are associated with activation and concentration resistances (He and Mansfeld 2008). Under such conditions, the polarization resistance becomes very important and it is possible that R_{in} cannot be determined accurately using the current interrupt method.

EIS also allows a complete evaluation of the various contributions to R_{in} of an MFC. EIS can also be used to determine the variation of R_{in} with increasing current flow through the cell. This can be accomplished by measuring the impedance spectra of the cell at various applied cell voltages, E_{MFC} (Manohar *et al.* 2008b).

Since R_p of the anode and cathode depend on the current flow in the MFC, it becomes apparent that R_{in} should depend on E_{MFC}. The impedance spectra of the cell at various applied cell voltages can be measured using a two-electrode method.

Manohar *et al.* have used EIS to determine R_{in} of an MFC as a function of cell voltage for two different anolytes (buffer and lactate with and without *Shewanella Oneidensis* MR-1) and two different anode configurations (with and without stainless steel (SS) balls as a packed bed electrode) (Manohar et al., 2008b) (Figure 8.3.7). It was observed that the addition of MR-1 to the anolyte containing buffer and lactate decreased R_{in} significantly (curves 1 and 2). R_{in} also showed a large decrease when SS balls were added to this anolyte due to the much larger total anode area (curve 1 and 3). In the presence of MR-1 and SS balls in the anode compartment R_{in} only decreased at higher cell voltages (curves 2 and 4). The impedance spectra for this case showed that R_p of the anode had become smaller than R_p of the cathode. Any further decrease of R_{in} can therefore only be achieved by decreasing R_p of the cathode. In all four cases a marked decrease of R_{in} with decreasing cell voltage was observed (Figure 8.3.7).

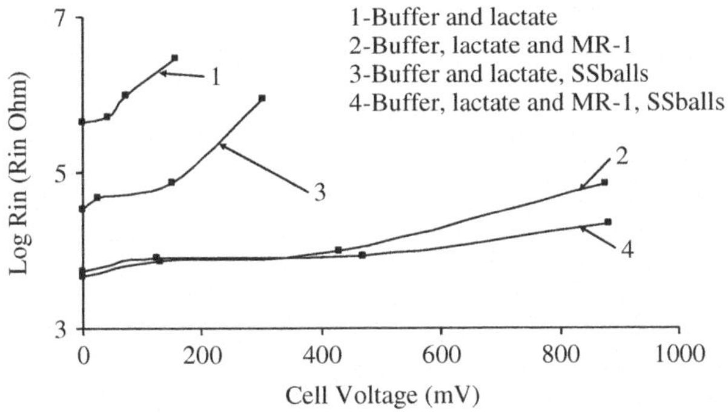

Figure 8.3.7. R_{in} vs cell voltage for MFCs with and without stainless steel balls and MR-1

8.3.6 CONCLUSIONS

EIS is a very powerful technique that along with other electrochemical techniques such as potentiodynamic polarization, cyclic voltammetry and potential sweep can be used to determine several key properties of microbial fuel cells. EIS has been shown to be a very efficient tool in characterizing the electrochemical properties of the electrodes and in determining R_{in} of MFCs. Since EIS is a non-destructive technique, it can be used to monitor changes of an MFC system with time or operating conditions without affecting the properties of the system under investigation.

A complete electrochemical characterization will result in a better understanding of how various MFCs work and identification of various factors that limit the power output of MFCs. EIS could also be employed for characterizing various electrode/electrolyte interfaces, analyzing electrochemical reactions, monitoring biofilm development, examining the effects of different experimental conditions and understanding the bacterial interaction with electrodes.

REFERENCES

Epelboin, I., Gabrielli, C. and Keddam, M. (1984) Non-steady state techniques. In *Comprehensive Treatise of Electrochemistry*. Yeager, E., Bockris, J.O.M., Conway, B.E. and Sarangapani, S. (eds), Plenum Press, New York, p. 62.

Fan, Y., Hu, H. and Liu, H. (2007) Enhanced coulombic efficiency and power density of air-cathode microbial fuel cells with an improved cell configuration. *J. Power Sources* **171**(2), 348–354.

He, Z. and Mansfeld, F. (2008) Exploring the use of electrochemical impedance spectroscopy in microbial fuel cell studies. *Energy Environ. Sci.* **2**(2), 215–219.

He, Z., Huang, Y., Manohar, A.K. and Mansfeld, F. (2008) Effect of electrolyte pH on the rate of the anodic and cathodic reactions in an air-cathode microbial fuel cell. *Bioelectrochemistry* **74**(1), 78–82.

He, Z., Shao, H.B. and Angenent, L.T. (2007) Increased power production from a sediment microbial fuel cell with a rotating cathode. *Biosensors Bioelectron.* **22**(12), 3252–3255.

He, Z., Wagner, N., Minteer, S.D. and Angenent, L.T. (2006) An upflow microbial fuel cell with an interior cathode: assessment of the internal resistance by impedance spectroscopy. *Environ. Sci. Technol.* **40**(17), 5212–5217.

Hsu, C.H. and Mansfeld, F. (2001) Concerning the conversion of the constant phase element parameter Yo into a capacitance. *Corrosion* **57**(9), 747–748.

Jüttner, K., Lorenz, W.J., Paatsch, W., Kendig, M. and Mansfeld, F. (1985) Bedeutung der Dynamischen Systemanalyse für Korrosionsuntersuchungen in Forschung und Praxis. *Mater. Corrosion/Werkst.e Korrosion* **36**(3), 120–130.

Manohar, A.K., Bretschger, O., Nealson, K.H. and Mansfeld, F. (2008a) The polarization behavior of the anode in a microbial fuel cell. *Electrochim. Acta* **53**(9), 3508–3513.

Manohar, A.K., Bretschger, O., Nealson, K.H. and Mansfeld, F. (2008b) The use of electrochemical impedance spectroscopy (EIS) in the evaluation of the electrochemical properties of a microbial fuel cell. *Bioelectrochem.* **72**(2), 149–154.

Manohar, A.K. and Mansfeld, F. (2009) The internal resistance of a microbial fuel cell and its dependence on cell design and operating conditions. *Electrochim. Acta* **54**(6), 1664–1670.

Mansfeld, F. (1981) Recording and analysis of AC impedance data for corrosion studies. I. Background and methods for analysis. *Corrosion* **37**(5), 301–307.

Mansfeld, F. (1988) Concerning the display of impedance data. *Corrosion* **44**(8), 558–559.

Mansfeld, F. (2006a) Classic paper in corrosion science and engineering with a perspective by F. Mansfeld. *Corrosion* **62**(10), 843–855.

Mansfeld, F. (2006b) Electrochemical impedance spectroscopy. In *Analytical Methods in Corrosion Science and Engineering*. Marcus, P. and Mansfeld, F. (eds), CRC Press, pp. 463–505.

Mansfeld, F. and Kendig, M.W. (1983) Determination of the polarization resistance from impedance measurements. *Mater. Corrosion/Werkst. Korrosion* **34**(8), 397–401.

Mansfeld, F. and Kendig, M.W. (1985) Electrochemical impedance spectroscopy of protective coatings. *Mater. Corrosion/Werkst. Korrosion* **36**(11), 473–483.

Mansfeld, F., Kendig, M.W. and Lorenz, W.J. (1985) Corrosion inhibition in neutral, aerated media. *J. Electrochem. Soc.* **132**(2), 290–296.

Mansfeld, F., Kendig, M.W. and Tsai, S. (1982) Recording and analysis of AC impedance data for corrosion studies. II. Experimental approach and results. *Corrosion* **38**(11), 570–580.

Mansfeld, F., Lin, S., Chen, Y.C. and Shih, H. (1988) Minimization of high-frequency phase shifts in impedance measurements. *J. Electrochem. Soc.* **135**(4), 906–907.

Mansfeld, F., Shih, H., Greene, H. and Tsai, C.H. (1993) Analysis of EIS-data for common corrosion processes *ASTM STP* **1188**, 37–53.

Mansfeld, F., Tsai, C.H. and Shih, H. (1992) Software for simulation and analysis of electrochemical impedance spectroscopy (EIS) data. *ASTM STP* **1154**, 186–196.

Ouitrakul, S., Sriyudthsak, M., Charojrochkul, S. and Kakizono, T. (2007) Impedance analysis of bio-fuel cell electrodes. *Biosensors Bioelectron.* **23**(5), 721–727.

Qiao, Y., Li, C.M., Bao, S.J. and Bao, Q.L. (2007) Carbon nanotube/polyaniline composite as anode material for microbial fuel cells. *J. Power Sources* **170**(1), 79–84.

Ramasamy, R.P., Ren, Z.Y., Mench, M.M. and Regan, J.M. (2008) Impact of initial biofilm growth on the anode impedance of microbial fuel cells. *Biotechnol. Bioenging* **101**(1), 101–108.

Schiller, C.-A. (2006) Introduction to electrochemical instrumentation. In *Analytical Methods in Corrosion Science and Engineering*. Marcus, P. and Mansfeld, F. (eds), CRC Press, pp. 361–434.

Schmidt, E., Hitzig, J., Titz, J., Jüttner, K. and Lorenz, W.J. (1986) Inhomogeneous electrodes–a polarization model of the partially blocked reversible metal ion electrode. *Electrochim. Acta* **31**(8), 1041–1050.

Srikanth, S., Marsili, E., Flickinger, M.C. and Bond, D.R. (2008) Electrochemical characterization of Geobacter sulfurreducens cells immobilized on graphite paper electrodes. *Biotechnol. Bioengng.* **99**(5), 1065–1073.

You, S., Zhao, Q., Zhang, J., Jiang, J., Wan, C., Du, M. and Zhao, S. (2007) A graphite-granule membrane-less tubular air-cathode microbial fuel cell for power generation under continuously operational conditions. *J. Power Sources* **173**(1), 172–177.

You, S.J., Zhao, Q.L., Zhang, J., Liu, H., Jiang, J.Q. and Zhao, S.Q. (2008) Increased sustainable electricity generation in up-flow air-cathode microbial fuel cells. *Biosensors Bioelectron.* **23**(7), 1157–1160.

9

Materials for BES

Bruce E. Logan

9.1 INTRODUCTION

Over the past decade there has been a substantial increase in the power densities produced in MFCs. Power densities normalized to the surface area have increased from as little as 0.001 mW/m^2 in reactors using dissolved oxygen at the cathode (which requires substantial energy input for aeration), to nearly 3000 mW/m^2 using passive air cathodes (Logan and Regen 2006; Logan *et al.* 2007). Using ferricyanide as the electrolyte, which allows greater power production as a result, the efficiency of reduction of this chemical at the cathode compared to dissolved oxygen (Oh and Logan 2006); power densities reached levels of 4310 mW/m^2 (Rabaey *et al.* 2004). Power densities have been normalized by the anode or cathode surface area, but usually for reactors with equally-sized electrodes.

It is now clear that for most systems, the performance of the cathode limits maximum power densities more than the anode (Fan *et al.* 2008; Rismani-Yazdi

 Bioelectrochemical Systems: From Extracellular Electron Transfer to Biotechnological Application. Edited by Korneel Rabaey, Largus Angenent, Uwe Schröder and Jürg Keller. ISBN: 9781843392330. Published by IWA Publishing, London, UK.

et al. 2008). To gain increases in power densities, researchers have altered the relative size of the electrodes and, if present, the size of the proton exchange membrane. By reducing the size of the anode relative to the cathode, and normalizing the power density to the projected surface area of the anode, *apparent* maximum power densities can be increased. In traditional H-cell type MFC architecture, it was shown that power increased by 24% when the cathode surface area was increased from 22.5 to 67.5 cm^2 (factor of 3) and that it decreased by 56% when the cathode surface area was reduced to 5.8 cm^2 (Oh *et al.* 2004). In further studies, it was shown that power decreased with cathode size over a change by a factor of 1:19 and that power was much more dependent on cathode size than anode size (Oh and Logan 2006). By increasing the size of the proton exchange membrane from 3.5 to 31 cm^2, power output was increased as well due to a reduction in ohmic resistance. Using air cathodes, power densities increased from 1.04 to 6.9 W/m^2 (relative to the anode size) by reducing the anode to be 1/14th the size of the cathode (Fan *et al.* 2008). In another study, it was found that power output per anode area (power density) increased as the anode was reduced in size to a point where the anode was 1/9th the size of the cathode (Ishii *et al.* 2008). When using ferricyanide, a maximum power density of 4.6 W/m^3 (1.9 W/m^2) was obtained using an anode that was 1/9th the size of the cathode (Nevin *et al.* 2008). However, most of these 'increases' in power production are really only apparent increases as the total power output decreases as anode size is reduced. If the cathode was used instead of the anode to normalize power output in the study by Fan *et al.* (2008), for example, then the power density was only 0.48 W/m^2.

As these power densities have increased, researchers have focused on how to increase power densities per volume of reactor. Originally, systems were not optimized for the volumetric power density. Early studies using two-chambered 'H' type systems, such as the reactor used by Bond and Lovley (2003), produced 0.42 W/m^3 (15 mW/m^2). Volumetric power densities can be optimized by placing the electrodes close to each other, increasing solution conductivity, and reducing the liquid volume (Liu *et al.* 2005a). By putting a cloth separator in between the anode and cathode, and reducing the liquid volume, volumetric power densities as high as 1.55 W/m^3 have been achieved (Fan *et al.* 2007b). However, cloth separators can be degraded by bacteria so additional work is needed to develop efficient and non-biodegradable separator materials. Most MFC studies now focus on maximizing power generation on a volume basis, although it is still useful to report power densities normalized to the surface area of one of the electrodes so that it is possible to compare different systems.

9.1.1 Electrode specific surface areas and material costs

The most important aspects of the MFC reactor design are the surface area of the electrodes and the cost of the materials. Increases in volumetric power densities are mostly related to the specific surface area of the electrodes, which is the total surface area (projected) per volume of reactor. For example, in the early studies by Bond and Lovley (2003) the specific surface area was 27 m^2/m^3. To produce 1.55 W/m^3, Fan *et al.* (2007b) used a reactor that had an anode specific surface area of 280 m^2/m^3. Thus, increasing the packing density of the electrode is a critical feature of MFC designs. For comparison purposes, tricking filters used for wastewater treatment (which require a large air–liquid interface) are commonly built with a specific surface area of 100 m^2/m^3.

The reactor sizes needed for practical applications of MFCs for wastewater treatment will be very large, necessitating the use of extremely low cost materials. As the mass of material used for the electrodes in an MFC increases, so does the cost of the system. For example, if we design a reactor that uses 100 m^2/m^3 of carbon cloth for the anode that costs *ca.* $1,000 per square meter (www.fuelcellstore.com), then our reactor cost for just the anode is $100,000 per cubic meter. This cost simply makes the reactor too expensive to build. Therefore, we must use very low cost materials in MFCs in order to make them practical for large scale applications.

9.2 ELECTRODE MATERIALS FOR MFCS

9.2.1 Anode

Often the same materials are used for the anode and cathodes in MFCs, although the cathodes usually contain catalysts. Common materials in laboratory MFCs are carbon paper, carbon cloth, carbon felt, graphite granules, and reticulated vitreous carbon (RVC) (Bond and Lovley 2003; Chaudhuri and Lovley 2003; He *et al.* 2005; Liu *et al.* 2005b; Liu and Logan 2004; Rabaey *et al.* 2005). The cloth, paper and felt materials can be placed in the reactor at various packing densities, with specific surface areas of ~10–300 m^2/m^3 common (Bond and Lovley 2003; Liu *et al.* 2008; Liu *et al.* 2005a, b; Liu and Logan 2004). The cost of carbon paper and cloth may be due to low production rates. There are reports that prices have decreased to as little as ~\$20/m^2 for materials being produced in bulk in China. Thus, it is possible that reactor costs could substantially decrease using these flat carbon cloth materials.

Graphite granules must touch each other, and have intrinsic low porosities characteristic of a porous medium. Typically, porosities of granular media range from 30 to 50%. The specific surface area of a porous medium made of up spherical particles of diameter d can be calculated as $A_s = 6(1-\theta)/d$ (Logan 1999). Graphite granule anodes (1.5 to 5.5 mm diameter) used in MFCs have had specific surface areas estimated to be between 817 and 2720 m^2/m^3 (Rabaey *et al.* 2005). One of the problems of using granules is the potential for clogging, either from bacteria growing on the electrodes or particles in a wastewater. Maintaining high porosities in an MFC is, therefore, critical for long-term performance.

RVC is a somewhat brittle material that has been used in MFCs for the anode due to its high conductivity and a high specific surface area. Values of RVC used in MFCs range widely from 6100 m^2/m^3 (Ringeisen *et al.* 2006) to 51 m^2/m^3 (He *et al.* 2005). These materials do offer the advantage over cloth and paper materials in that they provide high porosities, on the order of 97% (Logan 2008).

Another way to simultaneously achieve high porosities and high surface areas, without the use of a brittle material, is to use graphite fiber brush electrodes (Logan *et al.* 2007). These can be manufactured using common commercial equipment and are therefore easy to produce. They consist of graphite fibers wound into a brush design using a double core of wire. In laboratory reactors, titanium has been used due to its relatively inert nature (Figure 9.1). Other materials should be investigated that are non-corrosive and less expensive than titanium. Copper is highly conductive and inexpensive, but corrodes quickly and dissolved copper can be toxic to the microorganisms. Brush porosites and densities vary. Brush anodes used in single-bottle MFCs had a diameter of 5 cm and were 7 cm in length, with a total surface area of 1.06 m^2 and 7170 m^2/m^3-brush volume with 98% porosity (Logan *et al.* 2007). The final specific surface area depends on the volume of liquid around the brush, but as used in these tests in bottles with a capacity of 320 ml, the specific surface area for the liquid volume used was $A_s = 4200$ m^2/m^3. For smaller MFCs having a liquid volume of 26 ml, smaller brushes were used. These were 2.5 cm in outer diameter and 2.5 cm in length, with estimated surface area of 0.22 m^2 or 18,200 m^2/m^3-brush volume producing a 95% porosity. These porosities and surface areas are based on the individual filament diameter, which in these tests was 7.2 μm. Note that this differs from the way that surface area is calculated for carbon cloth, where projected surface area is used. These brushes may not have more surface area than carbon cloth, it is just that the porosity is very high and the methods used to calculate the surface areas are different. Long-term issues of biofouling of these electrodes have not been investigated, but a demonstration unit (see www.engr.psu.edu/ce/enve/logan/bioenergy/research_mfc.htm) has run for over one year with brush anodes without evidence of fouling.

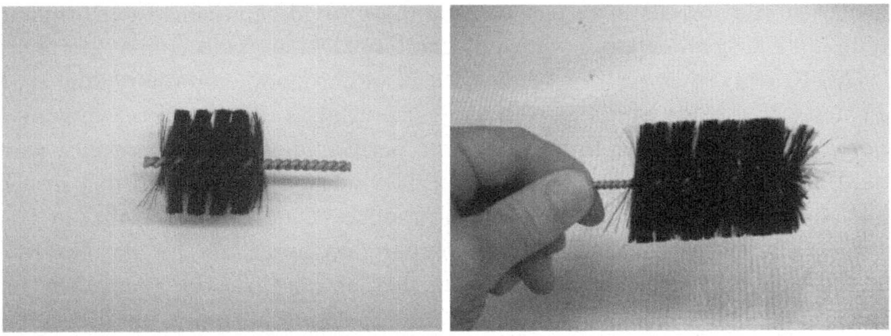

Figure 9.1. Brush electrodes used in MFCs. The core for these wires is made of titanium, although other materials can be used. The brush material is made from electrically conductive graphite fibers. From Logan *et al.* (2007). Reprinted with permission of the American Chemical Society

9.2.2 Cathode

The most challenging part of an MFC is the design of the cathode. In air-cathode MFCs, this electrode must allow a three-phase interaction: oxygen from air, protons from water, and electrons from the circuit. Most researchers use carbon paper or carbon cloth containing a catalyst on the water facing side for increasing the reaction rate of oxygen reduction. In other cases, a non-conductive material is made to be electrically conductive by applying a coating as described below.

No matter what type of material is used for the cathode, when a particulate catalyst is used it must be bound to the surface of the electrode, requiring a binder, such as Nafion or PTFE [poly(tetrafluoroethylene)], in an electrically conductive paste. The catalyst, such as Pt, is added as small particles minimizing the amount of this expensive material that must be added to the electrode. To make a Pt-catalyzed cathode, a pre-mixed paste containing Pt, for example 10 wt% Pt/C (E-TEK) is mixed with the binder, such as Nafion (5% solution) (Cheng *et al.* 2006c). This paste (7 μl of binder per mg of Pt/C catalyst) is applied to the carbon cloth and dried at room temperature for 24 h. If PTFE is used, the same procedure is followed with a PTFE suspension (2%) but it is dried at 350°C for 0.5 h. It has been found that the Nafion works better than the PTFE, but it is much more expensive (Cheng *et al.* 2006c). Thus, there is a trade-off in choosing the binder between performance and cost. Performance of one MFC design was not affected by the Pt loading over a range of 0.1 to 2 mg Pt/cm^2 with either Nafion or PTFE as the binder. The performance of the cathode is reduced when the catalyst was applied to the air-facing side of the electrode (Yang *et al.* 2009).

Alternative catalysts to Pt can be used that provide equivalent performance. Cheng and Logan (2006c) showed that cobalt tetramethylphenylporphyrin (CoTMPP) worked as well or better than Pt in chronopotentiometry studies. In MFC tests, CoTMPP works as well as Pt if a catalyst loading of 0.5 mg/cm^2 is used. Lower catalyst loading can affect performance. Similar results using CoTMPP were obtained by Zhao *et al.* (2005). They also showed that iron(II) phthalocyanine (FePc) could be used to produce results similar to that of Pt at current densities >0.2 mA/cm^2. The preparation procedure for the CoTMPP material is a bit more complicated than that of Pt. CoTMPP (Acros) is first dissolved in chloroform (0.001 mol/l) before being mixed with the carbon (Vulcan XC-72; 320 mg per 100 ml) and then the mixture is air-dried. The cathode must then be heated in an inert-gas atmosphere at 800°C for 1 h. The mixture is cooled, and then applied at a loading of 0.6 mg cm^{-2} to carbon cloth using a Nafion binder by the same method used for the Pt catalyst cathode.

Other catalysts and carbon materials have also been used. For example, in electrochemical tests Yu *et al.* (2007) examined tetramethoxyphenylporphyrin (TMPP), CoTMPP and FeCoTMPP, and metal phthalocyanine (Pc), FePc, CoPc and FeCuPc, supported on Ketjenblack (KJB) carbon. The best results were obtained using iron phthalocyanine supported on KJB carbon (FePc-KJB) for oxygen reduction in neutral pH solutions. The carbon used for the catalyst and binder also affects cathode performance. FePc on Ketjenblack carbon (FePc-KJB) had a slightly higher activity than the same catalyst on Vulcan XC carbon (FePcVC). Tests with some of these materials verified their effectiveness for oxygen reduction and power generation in MFCs. FePc-KJB, for example, produced 634 mW/m^2 (50 mmol/l phosphate buffer medium at pH 7) compared to 593 mW/m^2 with Pt. In MFC tests optimized for solution conductivity and anode surface area (carbon brush anode and 200 mmol/l phosphate buffer), a power density of 2011 mW/m^2 was obtained demonstrating that this combination of materials provided good performance (Yu *et al.* 2007).

Providing additional cathode surface area will usually result in more power generation (unless the anode surface area is much less than that of the cathode). For example, doubling the cathode surface area by placing a cathode on each side of a reactor increased power from 303 W/m^3 to 627 W/m^3 (Fan *et al.* 2007a).

Tubular cathodes. Flat cathode materials work fine in the laboratory for examining power generation using different substrates or materials, but the surface area that can be provided in the reactor is limited by the size of the reactor. The most cathode surface area that can be provided by this approach is when the cathode is wrapped completely around the reactor, forming a tubular reactor (Rabaey *et al.* 2005; Scott *et al.* 2007). However, this restricts the reactor to be a single tube. Alternatively, the specific surface area of the cathodes (cathode area per volume) can be controlled and even increased using tube-based cathodes (Figure 9.2). While the term 'tube' is used here, it refers to a general structure. The cathode can be circular (Zuo *et al.* 2007), oblong, or even flat forming a cassette-type of structure (Shimoyama *et al.* 2008). The idea for using tubular cathodes is that a series of electrodes, for example brush anodes and tubular cathodes, can be dropped into a tank turning it into an MFC. The specific surface areas of the tubes (i.e. tube area per volume of the tank) can be varied to increase performance.

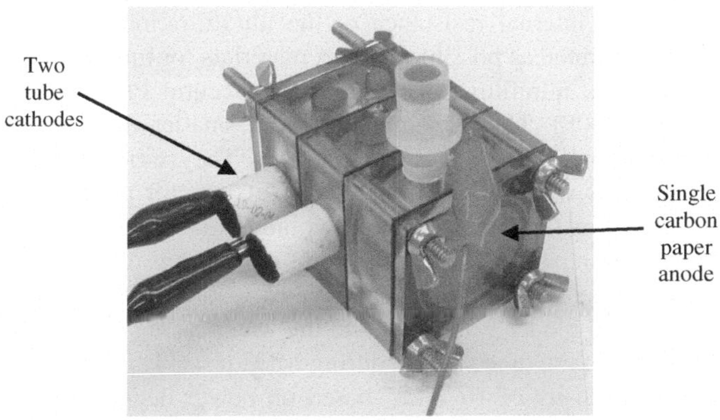

Figure 9.2. Example of tube cathodes in an MFC. From Zuo *et al.* (2007). Reprinted with permission of the American Chemical Society

The first cathode systems using tubes separated from the anode were developed by Zuo *et al.* (2007) using a hydrophilic tubular ultrafiltration membrane made of polysulfone on a composite polyester carrier. The tubes had an inner diameter of 14.4 mm (B0125, X-FLOW) and wall thickness of 0.6 mm, and were cut to a length of 3 cm. The tubes were made electrically conductive by applying a conductive graphite paint material (ELC E34 Semi-Colloidal, Superior Graphite Co.). CoTMPP was prepared as described above and used as the catalyst. The coating would not adhere to the water-facing side of the tube, so the coating was placed on the inside of the tube (air-facing side) even though this has been shown to reduce performance (Yang *et al.* 2009). Two tube cathodes were placed into an MFC reactor producing a specific surface area of $A_{cat,s} = 93\ m^2/m^3$, while the brush anode had a much higher specific surface area of $A_{an,s} = 7700\ m^2/m^3$. This combination of electrodes produced 18 W/m^3

with a coulombic efficiency of 70–74% (Zuo *et al.* 2007). The same performance for these electrodes was obtained under both batch and continuous flow operation. For comparison, the same type of reactor fitted with a flat anode produced 8.8 W/m^3, although the electrode surface areas of the anode was much lower (25 m^2/m^3). When a carbon paper cathode was used with a Pt catalyst, also with the same specific surface area, the power was 9.9 W/m^3. The main finding was that the tubular cathodes worked, but the surface areas of the cathodes needed to be large to offset their inherently lower efficiency compared to carbon cloth. This is due to the fact that the conductivity of the coating is not as high as carbon cloth or metals. Current collectors can be used to increase the efficiency of these tubular systems as discussed further below.

Due to the high internal resistance of the ultrafiltration membranes, other materials were examined as possible support materials for tubular cathode MFCs such as hydrophobic ultrafiltration membranes (Accurel PP V8/2, Membrana) (Zuo and Logan 2008). These membranes were smaller than the hydrophilic membranes, with each tube 0.70 cm in diameter. They were cut to a length of 3 cm and four tubes were placed in a small MFC reactor producing a specific surface area of 113 m^2/m^3 based on liquid volume. Flat anodes used were made of plain Toray carbon paper. These tubes were prepared as described above with CoTMPP and a Nafion binder, at a final catalyst loading of 0.5 mg/cm^2. In this case the catalyst and binder was applied to the water facing side of the tube and it remained stable on the tube surface during MFC tests. This system produces a maximum power density of 16 W/m^3 based on power density curves obtained with different external resistors for complete fed-batch cycles. The Coulombic efficiency of this system, however, was very low (6–12%). This was due to a high oxygen mass transfer coefficient for the hydrophobic membrane of $k = 2.2 \times 10^{-4}$ cm/s. for comparison, the hydrophilic membrane had an oxygen mass transfer coefficient of $k = 0.78 \times 10^{-4}$ cm/s.

Many types of ion exchange membranes have been used in MFCs (see below) as separators between the anode and cathode chambers. These are not available in tubular form in sizes that can be used in laboratory MFCs as separate tubes. Some researchers have rolled up these materials and bonded the ends together to form tubes (Rabaey *et al.* 2005). In studies in my lab, we have found that these materials crack when folded into small tube sizes (a few centimetres or smaller), limiting direct testing of the materials as separate tubes. However, these materials can be examined as support structures for conductive coatings in MFCs using flat materials. Anion (AEM) and cation (CEM) ion exchange membranes were cut into circular flat pieces having a projected surface area of 7 cm^2 (25 m^2/m^3), and made into cathodes by coating them with graphite paint and using a CoTMPP catalyst as described above. AEM cathodes produced better overall performance than the CEM membranes in tests using acetate in a 50 mmol/l phosphate buffer solution (PBS). The AEM cathodes achieved 449 mW/m^2 (projected cathode surface area) or 13.1 W/m^3 based on total reactor volume (Zuo *et al.* 2008). The CEM cathodes produced 286±30 mW/m^2 based on surface area, and 8.3±0.9 W/m^3 and 0.9 A/cm^2. The Coulombic efficiencies were similar for both membranes, and as high as 70%. The effect of the CoTMPP loading was also examined. It was found for the AEM cathode that decreasing the CoTMPP loading by 40–80% reduced power by 28–56%. In tests using a cathode with the conductive surface but no catalyst, power was reduced to only 16% (72 mW/m^2) of that generated with 0.5 mg/cm^2 of catalyst. The power was further increased using a current collector, as described below.

Brush cathodes. If dissolved oxygen is used as the catholytes, for example in MFCs in the ocean, then brush cathodes having a high surface area can be used. For example, brush cathodes were used by Reimers (2006) to power an MFC with the anodes in the sediment and the cathodes in the overlying water. Brush cathodes have also been used to power marine vehicles (Hasvold *et al.* 1997; Hasvold *et al.* 1999).

9.2.3 Membranes

In hydrogen fuel cells, membranes are used to transport the protons from the oxidized gas phase (hydrogen) to the oxidizer gas phase (oxygen). Thus, materials like Nafion are made to facilitate transport of protons, but the control of pH and water content are critical process variables that can affect performance. When the membrane becomes fully saturated (flooded), the performance of the fuel cell is adversely affected. The membrane also serves to prevent mixing of the two potentially explosive gases. In MFCs, membranes are mainly used as separators, or to keep the liquid containing the organic matter used by the bacteria from reaching the cathode. The membrane is always flooded with water in an MFC and, therefore, is always saturated. The pH in the membrane is likely that of the liquid, and so the proton concentration is not high. Thus, a membrane, such as Nafion under these conditions, is not functioning as it was meant to in terms of water content or pH.

The main challenge in choosing a membrane for *et al.* is to find one that is selective for the charge carrying species of interest. Protons are produced at the anode by bacteria, and consumed at the cathode through oxygen or proton reduction, but this does not guarantee that protons will be the main charge-carrying species through the membrane. Chemical cationic species, such as K^+, Na^+, Ca^{2+}, and Mg^{2+}, are present at orders of magnitude higher concentration than H^+, and are more likely to be exchanged across a cation membrane than protons. This transfer of charge rather than protons creates pH imbalances between the electrodes. Rozendal *et al.* (2006b) measured chemical species in a two-chamber MFC, and showed that cationic species, such as Na^{+} and Mg^{+}, increased over time in the cathode chamber. As shown in Figure 9.3, this resulted in an increase in the pH as the protons in the cathode chamber were consumed, but not equally replaced. The power produced decreased in response to this elevation of the pH due to the reduction of the cathode potential. In contrast, the anode potential was relatively unaffected.

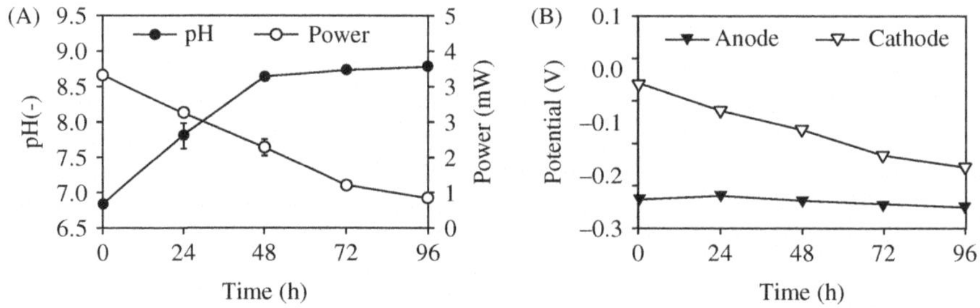

Figure 9.3. (A) Change in pH in the cathode and reduction in power in an MFC during a batch cycle (10 mmol/l phosphate buffer). (B) Electrode potentials showing that the main effect of pH was to reduce the performance of the cathode. From Rozendal *et al.* (2006b), with permission of the American Chemical Society

If protons do not readily cross the cation exchange membranes, then alternative methods are needed to balance pH in two-chambered systems. Kim *et al.* (2007) reasoned that one way to balance both charge and pH was to make use of negatively charged proton transporting phosphate buffers. They used an anion exchange membrane in an MFC to allow preferential transport of anions across the membrane. Because carbonate and phosphate buffers are present at higher concentrations than protons, these chemical species will be readily transported between the chambers. They found that an anion exchange membrane (AEM; Membranes International Inc.) produced 610 mW/m^2, which was more power than the 480 mW/m^2 produced by a similar cation exchange membrane (CEM; Membranes International Inc.) in an air-cathode MFC under otherwise identical conditions. Nafion produced slightly more power than the CEM, or 514 mW/m^2. In order to see what effect a membrane designed more for selection of molecules based on size might have on power generation, they tested a 1000 Dalton cut off ultrafiltration membrane. However, this UF membrane produced less power than these other membranes (462 mW/m^2)

mainly due to higher internal resistance of 98 Ω compared to the AEM (88 Ω) or the CEM or Nafion membranes (84 Ω).

Membranes used in BES not only exchange charged species, but they are also permeable to oxygen and other chemical species. Chae *et al.* (2007) examined the properties of Nafion 117 membranes when used in MFCs. They found that Nafion had a mass transfer coefficient of $\mathbf{k}_O = 2.80\times10^{-4}$ cm/s, and a diffusivity of $D_O = 5.35\times10^{-6}$ cm^2/s. These values were slightly different from Kim *et al.* (2007), who found that Nafion 117 had a mass transfer coefficient of $\mathbf{k}_O = 1.3\times10^{-4}$ cm/s and a $D_O = 2.4\times10^{-6}$ cm^2/s. Both AEM and CEM membranes have low oxygen mass transfer coefficients of $\mathbf{k}_O = 0.94\times10^{-4}$ cm/s. Substrate can also diffuse through the membrane, although at much lower rates, with mass transfer coefficients for these membranes for acetate of $\mathbf{k}_A = 4.3\times10^{-8}$ cm/s (Nafion), $\mathbf{k}_A = 1.4\times10^{-8}$ cm/s (CEM) and 5.5 $\mathbf{k}_A = 5.5\times10^{-8}$ cm/s (AEM) (Kim *et al.* 2007). Chae *et al.* found that 0.8% of the 67 mg/l of acetate in the anode chamber diffused through a Nafion 117 membrane within 71 h. Thus, this is a low flux of chemical lost through a membrane, although the substrate lost in an experiment depends on the area of the membrane and the time for a batch cycle.

Bipolar membranes have an advantage of transporting ions formed through the dissociation of water molecules. When current flows through the system water molecules dissociate to form H^+ and OH^-, balancing charge. Ter Heijne *et al.* (2006) used a bipolar membrane in an MFC with a ferrous/ferric iron couple to produce power. They successfully produced power using this approach, but iron fouling of the membrane was noted to be a problem. In a comparison of ion exchange and bipolar membranes it was concluded that the bipolar membranes were the least suitable for MFCs due to a high polarization and an inability to prevent pH splitting between the anode and cathode compartments (Harnisch *et al.* 2008).

Separators. While membranes have traditionally been used to separate the anode and cathode, Fan *et al.* (2007a) took a different approach to this situation by covering the cathode with a piece of cloth (J-cloth, commonly used as a kitchen wipe cloth). This material was different from those used in other studies in that the cloth has substantial hydraulic permeability. When placed on the cathode, the anode could be placed right next to, and on the opposite side of, the cathode (Figure 9.4). In previous tests when the anode was placed near the cathode in a membrane-less MFC, power density had been reduced due to oxygen diffusion through the cathode into the anode chamber liquid (Liu *et al.* 2005a). The effects of oxygen diffusion could be overcome by advective flow through the anode towards the cathode (Cheng *et al.* 2006b). When a membrane, such as Nafion, was placed on the cathode, power decreased (Liu and Logan

2004). Thus, the finding that power did not substantially decrease due to the cloth was a real advantage to system design. The actual mechanism by which the cloth increases power is not known. It is thought the cloth reduces oxygen diffusion into the anode chamber, allowing the anode to be placed near the cathode without being placed into an aerobic environment.

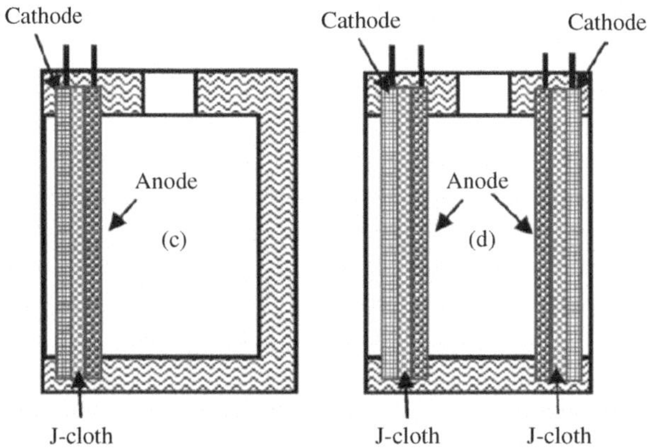

Figure 9.4. MFC designs using cloth separators: (A) cloth placed between the anode and cathode in a 3-ml liquid volume reactor; (B) two electrode and cloth assemblies in a single MFC. From Fan *et al.* (2007a). Reprinted with permission from Elsevier

The power density normalized to the electrode surface area decreases slightly when using a cloth separator, indicating there is some nominal effect of oxygen on the anode even with the cloth design. In an MFC with a 2-cm electrode spacing, a maximum power density of 1640 mW/m^2 (96 W/m^3) was obtained by Cheng and Logan (2006a), compared to 1460 mW/m^2 by Fan *et al.* (2007a) with a 1.7 cm spacing (liquid volume of 6 ml). When the J-cloth was used in a reactor with a 3 ml volume, the power per area further decreased to 1300 mW/m^2. However, the volumetric power density increased from 170 W/m^3 to 303 W/m^3 with the reduced reactor volume. By using two sets of electrodes in the same reactor (on opposite sides, with a 2.5 ml volume), the power density further decreased to 1120 mW/m^2 but the volumetric power density increased to 627 W/m^3. Thus, we see that there was a trade-off in power per area compared to power per volume. Clearly additional work is needed to understand the effect of the cloth on power generation, and other materials need to be investigated as separators in BES.

9.3 OTHER MATERIALS

9.3.1 Current collectors

Various other materials are used in an MFC, but current collectors are an extremely important aspect of MFC design relative to scale up of these systems. When small carbon paper electrodes are used in an MFC, for example, the electrical conductivity of the electrode over distances of a few centimetres does not contribute significant losses to the performance of the reactor. However, as the electrodes become larger, highly conductive materials are needed. In hydrogen fuel cells, the catalyst and a carbon layer are often directly applied to the membrane creating a highly catalytically active surface but one that is not relatively highly conductive. The plates that donate or accept electrons to these surfaces are highly conductive, allowing high current densities. As MFCs are scaled up, methods are needed for effectively conducting electrons from the electrode to the circuit.

In early BES designs using graphite plates, graphite rods were often inserted into the plate to ensure a good electrical connection and to avoid contact resistance (Rabaey *et al.* 2003). Later on when graphite granules were used, graphite rods similarly were inserted into the bed to make a good connection (Aelterman *et al.* 2006; Cheng and Logan 2007; Rabaey *et al.* 2005).

Brush electrodes provide a good example of how current collectors can be directly incorporated into the electrode design. Because the graphite fibers are wound directly into the metal core, the electrons have to travel over a relatively short distance before reaching the highly conductive metal. The distances that are optimum for the graphite fibers and the metal core have not yet been examined or optimized.

Relatively little work has been done on using current collectors for the cathode. However, in a recent study examining the use of ion exchange membranes as a support material for the cathode, it was found that a current collector could be used to increase power densities. Without a current collector, and in a low ionic strength solution, power production using an anion exchange membrane was 13.1 W/m^3 based on total reactor volume as noted above (Zuo *et al.* 2008). By adding a current collector, which in this case was a stainless steel mesh, that was pressed against the inside surface of the AEM cathode, and by using a higher concentration of buffer (200 mmol/l phosphate) a maximum volumetric power density was produced 21.2 W/m^3 (Zuo *et al.* 2008).

9.3.2 Wires, resistors and loads

Copper wires are often used in BES due to high electrical conductivity and low cost. However, copper will rapidly corrode, releasing copper ions which can be toxic to bacteria. When copper wires are used in reactors, all surfaces must be sealed with non-conductive surfaces. It is best within the solution to use relatively non-corrosive metals, such as stainless steel or titanium. Outside of the liquid, copper can be used to complete the circuit.

In laboratory MFCs a resistor is usually used to complete the circuit and to provide the circuit load. For larger systems, fans or lights can be used. Note that most light emitting diodes (LEDs) require a minimum voltage of 0.5 V, and may not function at the voltages produced by most MFCs. Thus, systems which can function on lower voltages and currents are needed.

9.4 MATERIALS FOR MICROBIAL ELECTROLYSIS CELLS

In a microbial electrolysis cell (MEC), oxygen is excluded from the reactor and a power source is used to boost the voltage produced by bacteria on the anode. When the overall cell potential is greater than 0.414 V, hydrogen can theoretically be produced (neutral pH conditions). The anode contributes approximately 0.2 to 0.25 V, so in practice more than ~0.2 V needs to be applied to get hydrogen gas evolution from the cathode. At low applied voltages of 0.3 V, or high organic loadings, cycle times can be long and this has resulted in high amounts of methane generation (Tartakovsky *et al.* 2008). Therefore, the applied voltages of 0.4 V and larger are generally used and organic loading is kept low to allow for rapid cycle times (Logan *et al.* 2008).

The materials used in MECs have generally followed the same evolutionary path as MFCs. The first two systems both used carbon paper electrodes suspended in an H-type reactor, with a plain carbon anode and a Pt-catalyzed cathode (Liu *et al.* 2005c; Rozendal *et al.* 2006a). Current densities, and as a consequence hydrogen gas production rates, were low (Table 9.1). The next MEC designs took two different paths. In some designs the cathode materials were altered, while in the others improvements in the anode were made. Rozendal *et al.* (2007) examined the use of a membrane electrode assembly with the cathode exposed to a gas phase instead of water. They used a platinum coated (5 mg/cm^2) titanium mesh as a cathode rather than Pt-coated carbon paper (0.5 mg/cm^2) and succeeded in increasing hydrogen production rates (0.3 m^3 H_2/m^3 d) relative to earlier studies. However, hydrogen recoveries were low and a high applied voltage was used (E_{ap} = 1.0 V) resulting in low overall energy recovery (Table 9.1). Tartakovsky *et al.* (2008) examined the used of a Pd/Pt coated carbon paper cathode (0.5 mg/cm^2), but hydrogen production rates were low and substantial methane was produced. It was also shown that the catalyst on the cathode could be eliminated in *et al.* where microorganisms were used to form a biocathode that enhanced hydrogen evolution at the cathode relative to a plain cathode (Rozendal *et al.* 2008). The applications of biocathodes for hydrogen evolution in MECs appears promising, but it needs to be further investigated and the microbes responsible for the hydrogen evolution need to be discovered and isolated. The use of stainless steel as MEC cathodes appears promising. Flat sheets of stainless steel A286 performed better than a sheet of Pt in a phosphate buffer solution, although total hydrogen production rates were lower than rates achieved with Pt catalysts in carbon cloth (Selembo *et al.* 2009). Recent tests with a stainless steel brush cathode showed hydrogen production rates that were comparable to those with tests conducted with Pt in carbon cloth (Call *et al.* 2009).

Table 9.1. Studies on hydrogen production in MECs evaluated in terms of applied potential and performance. All substrates are acetate, except as noted. Data from Logan *et al.* (2008) and Call *et al.* (2009)

Study	Applied Voltage (V)	H_2 production rate (m^3/m^3 day)	Overall H_2 Yield (%)	Energy input (kWh/m^3)
Liu *et al.* (2005c)	0.45	0.37	61	1.0
Rozendal *et al.* (2006a)	0.5	0.02	53	1.9
Ditzig *et al.* (2007)*	0.5	0.01	9.8	2.5
Cheng and Logan (2007)	0.6	1.1	88	1.3
Hu *et al.* (2008)	0.6	0.69	64	1.4
Call *et al.* (2009)	0.6	1.7	91	1.4
Call and Logan (2008)	0.8	3.12	93	1.7
Rozendal *et al.* (2007)	1	0.3	23	2.2

*Substrate was domestic wastewater.

The anode performance was addressed by using graphite granules instead of flat carbon paper, increasing anode surface area. When this anode was used with a Nafion membrane and wastewater, however, current densities, hydrogen production rates and hydrogen recoveries were all low. However, by using an anion exchange membrane, a graphite rod as the connector to the graphite anode, and using closely spaced electrodes, current densities were improved and hydrogen production rates increased substantially to 1.1 m^3/m^3 d at an applied voltage of 0.6 V (Cheng and Logan 2007).

Several MECs have been developed that do not use a membrane between the anode and cathode (Call and Logan 2008; Call *et al.* 2009; Hu *et al.* 2008). In water electrolyzers the gases produced at the electrodes (oxygen and hydrogen) must be kept separated. However, in et al. no oxygen gas is evolved. Thus, the membrane functions primarily to keep microorganisms from growing on the cathode and consuming the hydrogen gas, as well as increasing the hydrogen gas composition by reducing its contamination with carbon dioxide produced at the anode. Using a brush anode and a Pt-catalyzed cathode, a membraneless MEC achieved a hydrogen production rate of 3.12 m^3 H_2/m^3 d at an applied voltage of 0.8 V (Call and Logan 2008). Little methane was found in the product gas (1.9 %) at applied voltages of >0.3 V. There appeared to be little hydrogen consumption by bacteria in the system. Overall coulombic efficiencies were high (92%), and cathodic hydrogen recoveries ranged from 78% to 96% depending on the applied voltage (0.3–0.8 V). Carbon dioxide was typically present in the product gas at 7–8% as a result of the single-chamber design, compared to $<0.5\%$ in MEC reactors using membranes.

The development of MECs has not advanced as quickly as those of MFCs, but it is clear that these systems have great potential for the production of a useful energy product in BES.

9.5 CONCLUSIONS AND OUTLOOK

Over the past decade, there have been tremendous improvements in our understanding of the factors that affect current generation in BES. This has resulted in great increases in power densities in MFCs, and hydrogen production rates in MECs. Commercialization of these technologies, however, will require the development of systems based on the use of low-cost materials. We have already seen system architectures being developed that are inherently scalable, compared to the early designs based on H-type cells. These include the use of high surface area anodes, such as graphite fiber brushes, and tubular cathodes for MFCs. The long-term sustainability of the materials also needs to be better examined and understood. While many systems use Pt on the cathodes, the lifetime of these precious metals and non-precious metal replacements needs to be examined. The use of biocathodes in different BES is also a promising development as this would completely eliminate the need for metal catalysts.

The challenge remains to continue to improve these designs through additional research on different architectures and designs, and continued reduction on materials with substantially lower costs than those used today in the laboratory. BES reactors are biofilm systems that to date have required high surface areas and relatively large volumes compared to suspended growth systems. Reducing the volume of the reactor, through increased volumetric current densities, will lead to a reduction in the capital costs needed for the system. Reducing the materials costs will substantially affect the usefulness of these systems and in the long run will allow these systems to emerge from the laboratory into large scale, commercialized systems.

REFERENCES

Aelterman, P., Rabaey, K., Pham, T.H., Boon, N. and Verstraete, W. (2006) Continuous electricity generation at high voltages and currents using stacked microbial fuel cells. *Environ. Sci. Technol.* **40**, 3388–3394.

Bond, D.R. and Lovley, D.R. (2003) Electricity production by *Geobacter sulfurreducens* attached to electrodes. *Appl. Environ. Microbiol.* **69**(3), 1548–1555.

Call, D. and Logan, B.E. (2008) Hydrogen production in a single chamber microbial electrolysis cell (MEC) lacking a membrane. *Environ. Sci. Technol.* **42**(9), 3401–3406.

Call, D., Merrill, M.D. and Logan, B.E. (2009) High surface area stainless steel brushes as cathodes in microbial electrolysis cells (MECs). *Environ. Sci. Technol.* **43**(6), 2179–2183.

Chae, K.J., Choi, M., Ajayi, F.F., Park, W., Chang, I.S. and Kim, I.S. (2007) Mass transport through a proton exchange membrane (Nafion) in microbial fuel cells. *Energy Fuels* **22**, 169–176.

Chaudhuri, S.K. and Lovley, D.R. (2003) Electricity generation by direct oxidation of glucose in mediatorless microbial fuel cells. *Nat. Biotechnol.* **21**(10), 1229–1232.

Cheng, S. and Logan, B.E. (2007) Sustainable and efficient biohydrogen production via electrohydrogenesis. *Proc. Natl. Acad. Sci. USA* **104**(47), 18871–18873.

Cheng, S., Liu, H. and Logan, B.E. (2006a) Increased performance of single-chamber microbial fuel cells using an improved cathode structure. *Electrochem. Commun.* **8**, 489–494.

Cheng, S., Liu, H. and Logan, B.E. (2006b) Increased power generation in a continuous flow MFC with advective flow through the porous anode and reduced electrode spacing. *Environ. Sci. Technol.* **40**, 2426–2432.

Cheng, S., Liu, H. and Logan, B.E. (2006c) Power densities using different cathode catalysts (Pt and CoTMPP) and polymer binders (Nafion and PTFE) in single chamber microbial fuel cells. *Environ. Sci. Technol.* **40**, 364–369.

Ditzig, J., Liu, H. and Logan, B.E. (2007) Production of hydrogen from domestic wastewater using a bioelectrochemically assisted microbial reactor (BEAMR). *Int. J. Hydrogen Energy* **32**(13), 2296–2304.

Fan, Y., Hu, H. and Liu, H. (2007a) Enhanced coulombic efficiency and power density of air-cathode microbial fuel cells with an improved cell configuration. *J. Power Sources* **171**(2), 348–354.

Fan, Y., Hu, H. and Liu, H. (2007b) Sustainable power generation in microbial fuel cells using bicarbonate buffer and proton transfer mechanisms. *Environ. Sci. Technol.* **41**(23), 8154–8158.

Fan, Y., Sharbrough, E. and Liu, H. (2008) Quantification of the internal resistance distribution of microbial fuel cells. *Environ. Sci. Technol.* **42**(21), 8101–8107.

Harnisch, F., Schröder, U. and Scholz, F. (2008) The suitability of monopolar and bipolar ion exchange membranes as separators for biological fuel cells. *Environ. Sci. Technol.* **42**, 1740–1746.

Hasvold, Ø., Henriksen, H., Melvær, E., Citi, G., Johansen, B.Ø., Kjønigsen, T. and Galetti, R. (1997) Sea-water battery for subsea control systems. *J. Power Sources* **65**, 253–261.

Hasvold, Ø., Johansen, K.H., Mollestad, O., Forseth, S. and Størkersen, N. (1999) The alkaline aluminum/hydrogen peroxide power source in the Hugin II unmanned underwater vehicle. *J. Power Sources* **80**, 254–260.

He, Z., Minteer, S.D. and Angenent, L.T. (2005) Electricity generation from artificial wastewater using an upflow microbial fuel cell. *Environ. Sci. Technol.* **39**(14), 5262–5267.

Hu, H., Fan, Y. and Liu, H. (2008) Hydrogen production using single-chamber membrane-free microbial electrolysis cells. *Water Res.* **42**(15), 4172–4178.

Ishii, S., Watanabe, K., Yabuki, S., Logan, B.E. and Sekiguchi, Y. (2008) Characterization of electrode reducing rates of *Geobacter sulfurreducens* and an enriched electricity-generating mixed consortium in a microbial fuel cell. *Appl. Environ. Microbiol.* **74**(23), 7348–7355.

Kim, J.R., Cheng, S., Oh, S.-E. and Logan, B.E. (2007) Power generation using different cation, anion and ultrafiltration membranes in microbial fuel cells. *Environ. Sci. Technol.* **41**(3), 1004–1009.

Liu, H. and Logan, B.E. (2004) Electricity generation using an air-cathode single chamber microbial fuel cell in the presence and absence of a proton exchange membrane. *Environ. Sci. Technol.* **38**(14), 4040–4046.

Liu, H., Cheng, S. and Logan, B.E. (2005a) Power generation in fed-batch microbial fuel cells as a function of ionic strength, temperature, and reactor configuration. *Environ. Sci. Technol.* **39**(14), 5488–5493.

Liu, H., Cheng, S. and Logan, B.E. (2005b) Production of electricity from acetate or butyrate in a single chamber microbial fuel cell. *Environ. Sci. Technol.* **39**(2), 658–662.

Liu, H., Grot, S. and Logan, B.E. (2005c) Electrochemically assisted microbial production of hydrogen from acetate. *Environ. Sci. Technol.* **39**(11), 4317–4320.

Liu, H., Cheng, S., Huang, L. and Logan, B.E. (2008) Scale up of a single-chamber microbial fuel cell through optimization of the anode to cathode area ratio. *J. Power Sources* **179**, 274–279.

Logan, B.E. (1999) Environmental Transport Processes. Wiley, New York.

Logan, B.E. (2008) Microbial Fuel Cells. John Wiley & Sons, Inc., Hoboken, NJ.

Logan, B.E., Cheng, S., Watson, V. and Estadt, G. (2007) Graphite fiber brush anodes for increased power production in air-cathode microbial fuel cells. *Environ. Sci. Technol.* **41**(9), 3341–3346.

Logan, B.E., Call, D., Cheng, S., Hamelers, H.V.M., Sleutels, T.H.J.A., Jeremiasse, A.W. and Rozendal, R.A. (2008) Microbial electrolysis cells for high yield hydrogen gas production from organic matter. *Environ. Sci. Technol.* **42**(23), 8630–8640.

Logan, B.E. and Regan, J.M. (2006) Electricity-producing bacterial communities in microbial fuel cells. *Trends Microbiol.* **14**(12), 512–518.

Nevin, K.P., Richter, H., Covalla, S.F., Johnson, J.P., Woodard, T.L., Orloff, A.L., Jia, H., Zhang, M. and Lovley, D.R. (2008) Power output and columbic efficiencies from biofilms of *Geobacter sulfurreducens* comparable to mixed community microbial fuel cells. *Environ. Microbiol.* **10**(10), 2505–2514.

Oh, S. and Logan, B.E. (2006) Proton exchange membrane and electrode surface areas as factors that affect power generation in microbial fuel cells. *Appl. Microbiol. Biotechnol.* **70**(2), 162–169.

Oh, S., Min, B. and Logan, B.E. (2004) Cathode performance as a factor in electricity generation in microbial fuel cells. *Environ. Sci. Technol.* **38**(8), 4900–4904.

Rabaey, K., Boon, N., Siciliano, S.D., Verhaege, M. and Verstraete, W. (2004) Biofuel cells select for microbial consortia that self-mediate electron transfer. *Appl. Environ. Microbiol.* **70**(9), 5373–5382.

Rabaey, K., Clauwaert, P., Aelterman, P. and Verstraete, W. (2005) Tubular microbial fuel cells for efficient electricity generation. *Environ. Sci. Technol.* **39**(20), 8077–8082.

Rabaey, K., Lissens, G., Siciliano, S.D. and Verstraete, W. (2003) A microbial fuel cell capable of converting glucose to electricity at high rate and efficiency. *Biotechnol. Lett.* **25**(18), 1531–1535.

Reimers, C.E., Girguis, P., Stecher, H.A., Tender, L.M., Ryckelynck, N. and Whaling, P. (2006) Microbial fuel cell energy from an ocan cold seep. *Geobiology* **4**, 123–136.

Ringeisen, B.R., Henderson, E., Wu, P.K., Pietron, J., Ray, R., Little, B., Biffinger, J.C. and Jones-Meehan, J.M. (2006) High power density from a miniature microbial fuel cell using *Shewanella oneidensis* DSP10. *Environ. Sci. Technol.* **40**(8), 2629–2634.

Rismani-Yazdi, H., Carver, S.M., Christy, A.D. and Tuovinen, O.H. (2008) Cathodic limitations in microbial fuel cells: an overview. *J. Power Sour.* **180**, 683–694.

Rozendal, R., Hamelers, H.V.M., Molenkamp, R.J. and Buisman, C.J.N. (2007) Performance of single chamber biocatalyzed electrolysis with different types of ion exchange membranes. *Water Res.* **41**, 1984–1994.

Rozendal, R.A., Hamelers, H.V.M., Euverink, G.J.W., Metz, S.J. and Buisman, C.J.N. (2006a) Principle and perspectives of hydrogen production through biocatalyzed electrolysis. *Int. J. Hydrogen Energy* **31**(12), 1632–1640.

Rozendal, R.A., Hamelers, H.V.V. and Buisman, C.J.N. (2006b) Effects of membrane cation transport on pH and microbial fuel cell performance. *Environ. Sci. Technol.* **40**(17), 5206–5211.

Rozendal, R.A., Jeremiasse, A.W., Hamelers, H.V.M. and Buisman, C.J.N. (2008) Hydrogen production with a microbial biocathode. *Environ. Sci. Technol.* **42**(2), 629–634.

Scott, K., Murano, C. and Rimbu, G. (2007) A tubular microbial fuel cell. *J. Appl. Electrochem.* **37**, 1063–1068.

Selembo, P.A., Merrill, M.D. and Logan, B.E. (2009) The use of stainless steel and nickel alloys as low-cost cathodes in microbial electrolysis cells. *J. Power Sour*ces doi:10.1016/j.jpowsour.2008.1012.1144.

Shimoyama, T., Komukai, S., Yamazawa, A., Ueno, Y., Logan, B.E. and Watanabe, K. (2008) Electricity generation from model organic wastewater in a cassette-electrode microbial fuel cell. *Appl. Environ. Microbiol.* DOI 10.1007/s00253-00008-01516-00250.

Tartakovsky, B., Manuel, M.-F., Neburchilov, V., Wang, H. and Guiot, S.R. (2008) Biocatalyzed hydrogen production in a continuous flow microbial fuel cell with a gas phase cathode. *J. Power Sources* **182**, 291–297.

ter Heijne, A., Hamelers, H.V.M., de Wilde, V., Rozendal, R.R. and Buisman, C.J.N. (2006) Ferric iron reduction as an alternative for platinum-based cathodes in microbial fuel cells. *Environ. Sci. Technol.* **40**, 5200–5205.

Yang, S., Jia, B. and Liu, H. (2009) Effects of the Pt loading side and cathode-biofilm on the performance of a membrane-less and single chamber microbial fuel cell. *Bioresource Technol.* **100**, 1197–1202.

Yu, E.H., Cheng, S.C., Scott, K. and Logan, B.E. (2007) Microbial fuel cell performance with non-Pt cathode catalysts. *J. Power Sources* **171**(2), 275–281.

Zhao, F., Harnisch, F., Schröder, U., Scholz, F., Bogdanoff, P. and Herrmann, I. (2005) Application of pyrolysed iron (II) phthalocyanine and CoTMPP based oxygen reduction catalysts as cathode materials in microbial fuel cells. *Electrochem. Commun.* **7**, 1405–1410.

Zuo, Y., Cheng, S., Call, D. and Logan, B.E. (2007) Tubular membrane cathodes for scalable power generation in microbial fuel cells. *Environ. Sci. Technol.* **41**(9), 3347–3353.

Zuo, Y., Cheng, S. and Logan, B.E. (2008) Ion exchange membrane cathodes for scalable microbial fuel cells. *Environ. Sci. Technol.* **42**(18), 6967–6972.

Zuo, Y. and Logan, B.E. (2008) Hydrophobic membranes improve power production in tubular cathode microbial fuel cells (MFCs). *Wat. Sci. Technol.*, submitted.

10

Technological factors affecting BES performance and bottlenecks towards scale up

Bert Hamelers, Tom H.J.A. Sleutels, Adriaan W. Jeremiasse, Jan W. Post, David P.B.T.B. Strik and René A. Rozendal

10.1 INTRODUCTION

Over the past few years, laboratory-scale BESs have significantly improved and are already capable of producing current densities of over 10 A/m^2 anode surface area (Fan *et al.* 2007a; Torres *et al.* 2007). With these current densities the technology has arrived in the next stage, which is the stage of scaling-up and practical implementation (Rozendal *et al.* 2008a). Generally, engineering and operation are regarded as straightforward and not primarily a scientific challenge. However, several researchers have aimed at developing scalable designs (Liu *et al.* 2008; Logan *et al.* 2007; Rabaey *et al.* 2005), but so far no study has demonstrated that the respective designs can indeed be operated satisfactorily beyond the litre-scale. Moreover, pilot scale MFC research at the University of Queensland (Australia, Keller and Rabaey 2008) and Wageningen University (The Netherlands; unpublished) has shown that construction and maintenance can be demanding. Possible problems such as leakages, clogging of electrodes, membrane damage and short circuits call for flexible designs.

In this chapter, we discuss these and other design challenges that we will be facing when scaling-up BESs for wastewater application. Next to identifying these design constraints, we will indicate directions for meeting these challenges. Given the little information available on scale-up, this chapter must be considered as provisional, as it is at this point of time not possible to be complete on constraints and certainly not on solutions.

In this chapter we will first discuss those design constraints that arise from application of BESs for wastewater treatment. A first constraint comes from the required energetic efficiency and footprint that are necessary to compete with the digestion technology. The footprint of BES should be comparable to a digestion system while the energetic efficiency of especially MFCs should preferably be higher. Secondly, as for any biological treatment process, successful BES application will need wastewater with a suitable composition, i.e. factors as toxicity, temperature should be in the right range. Here we will focus on conductivity and buffer capacity, two crucial wastewater properties that are of specific concern in the case of BESs. Alkalinity and conductivity levels can not be increased by addition of salts as wastewater discharge legislation and economics make this undesirable. Therefore conductivity and alkalinity are important parameters for design.

In a further section we will discuss the technological constraints that arise from scale-up. Issues that might not be important in lab scale can gain importance at large scale. We will discuss the effects of increased current, the membrane choice and the hydrodynamics. Just scaling up without paying attention to these factors will inevitably lead to additional energy losses. The design of a BES must, thus, accommodate these different factors.

A separate section deals with the cost of the system in relation to material choice. A BES system has much more internals than an anaerobic digestion system. The cost of these internals will determine a large part of the economics of the system and merits a separate section.

In the final section a wrap up will be given of the constraints identified in this chapter. Although this listing might seem discouraging for technology development, it must be faced, as neglect of some issues will lead to inadequate designs. For each of the constraints, possible solutions are proposed. This shows that the constraints are not absolute but can be met with clever design(s).

This chapter thus aims at showing that in scale up and applications, real challenges lie ahead. These challenges can be met with clever designs based on insight in process and materials. To date, little information is available on scale up, however with our increasing knowledge on processes as obtained from lab research and building on an increasing basis of scale-up research and practical experiences, it must be possible to come up with competitive designs. It must be kept in mind that BESs have advantages that will be more and more important for a sustainable wastewater treatment and that practical bottlenecks will be resolved to bring the emerging BESs to function in practice. Scale-up research is thus, of, utmost importance to realise these advantages in practice.

10.2 DESIGN CONSTRAINTS AS DETERMINED BY WASTEWATER APPLICATION

10.2.1 Footprint and energetic efficiency

A BES consists of three principle elements: the anode where the oxidation takes place and the electrons enter the electrical circuit; the cathode where the reduction reaction takes place and the electrons leave the electrical circuit and a separator. This separator separates the anode and the cathode in such a way that electronic short-circuiting is prevented while ion transport is enhanced. The separator can be some kind of membrane or a simple layer of water containing salts.

To become an attractive option for wastewater treatment besides the economics, the footprint and energetic efficiency of the system are important measures for evaluation. With activated sludge systems a treatment capacity of ~0.5–2 kg COD/m^3 reactor volume per day could be achieved and with anaerobic systems ~8–20 kg COD/m^3 (Logan *et al.* 2006). With a current density of 10 A/m^2, a treatment capacity of BESs could be achieved of ~7 kg COD/m^3 reactor volume per day (1000 A/m^3) if a surface area of 100 m^2/m^3 of reactor volume can be realized (Rozendal *et al.* 2008a).

There are two types of design that can realize such a high specific surface area, the planar design (Figure 10.4) and the tubular design (Figure 10.1). In the planar design the anode, separator and cathode are planar and lie parallel to each other within one cell. The distance between the plates is small and if a cell thickness of 1 cm can be realised the 100 m^2/m^3 can be realized. In the tubular design, the separator is tubular shaped, on the inside the anode is present and on the outside the cathode or vice versa. A high surface density can be realized by choosing a small diameter tubing and a high number of tubes within one cell. Assuming an equal volume needed for the anode and cathode, it can be calculated that a 2 cm diameter tubing is needed. Experience from membrane bioreactors would indicate that tubular designs achieve higher specific surface areas.

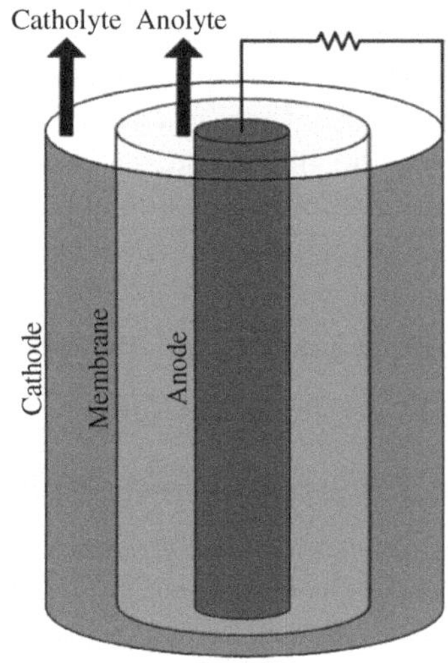

Figure 10.1. Basic lay-out of a tubular BES. Although here an MFC is depicted, the same lay-out is possible for an MEC. Variations exist, such as a cathode attached to the membrane in a membrane electrode assembly

The principal advantage of BESs is that they produce energy from the organic material in the wastewater. However, by doing this, BESs directly compete with anaerobic digesters, which is particularly the case for MFCs. In combination with a gas combustion engine, anaerobic digesters can produce electricity from organic matter with an efficiency of 30–35%. The energetic efficiency of MFCs, on the other hand, depends on the product of coulombic efficiency and voltage efficiency.

The coulombic efficiency is the fraction of the electrons from converted organic matter that ends up in the electrical circuit. Competing microbial processes such as methanogenesis and biomass growth lower the coulombic efficiency as electrons and thus energy is lost for electricity production. Coulombic efficiency values reported in the literature vary for different reactor designs, types of wastewaters, temperatures etc., but typically range from 5% (Min and Logan 2004) to 38% (Wang *et al.* 2008). For synthetic wastewaters however, much higher efficiencies have been reported, up to 100% (Aelterman *et al.* 2008).

The voltage efficiency describes how much of the chemical energy from the converted organic matter ends up as electrical energy. In theory, the electrons can have a voltage equal to the thermodynamically determined energy level of the electrons. In practice this value will be lower due to all kinds of internal losses in the MFC. For example, an MFC operated on acetate can theoretically generate about 1.1 V (Cheng *et al.* 2006a); operating this system at 0.55 V would mean it operates at 50% voltage efficiency. The voltage efficiency is a function of the power density of the system. It is generally accepted that a system with a good catalyst has a 50% voltage efficiency at maximum power density. It is clear that at maximum power density, the coulombic efficiency should be at least 80% to get an energetic efficiency of 40% to be at least competitive with anaerobic digestion. In more general terms this is shown in Figure 10.2, which illustrates that both the voltage efficiency and coulombic efficiency should be high to compete with an anaerobic digester.

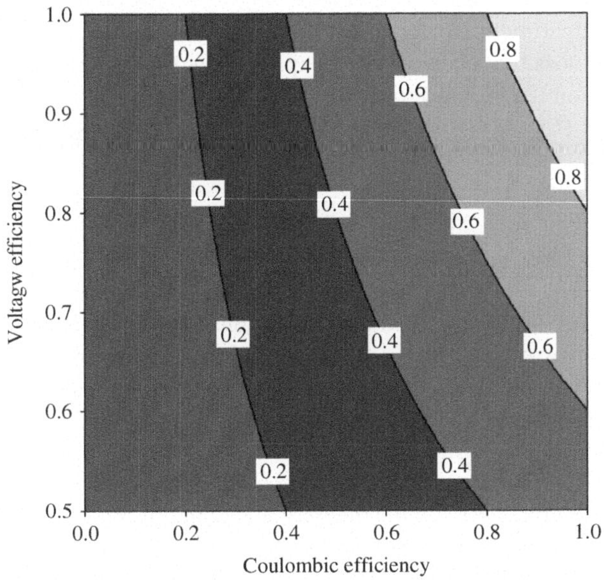

Figure 10.2. Contour plot of energetic efficiency as a function of coulombic efficiency and voltage efficiency. Only BESs with energetic efficiencies above 0.4 are starting to get competitive with anaerobic digestion. If for example a minimal energetic efficiency of 60% is required, the green and yellow area is the operational bandwidth

Increasing the coulombic efficiency is mainly a matter of reducing the flow of electrons going to the methanogens and biomass growth. How to steer this competition is an important aspect of the biofilm microbiology and is discussed in Chapter 11. The voltage efficiency, however, is strongly depending on design and is, thus, a crucial aspect of scale up.

10.2.2 Effect of conductivity

In BESs two flows of charge take place, a negatively charged flow of electrons from the anode to the cathode and an ionic charge flow to compensate for the negative charge flow of the electrons. Therefore in BESs two types of ohmic losses can be distinguished (Logan *et al.* 2006; Rozendal *et al.* 2008a): (i) ohmic losses due to resistance of the flow of electrons through electrodes, wiring and interconnections, and (ii) ohmic losses due to resistance of ions moving through membrane, anolyte and catholyte.

In a BES treating wastewater, the anolyte consists of the wastewater that is treated. The conductivity of the anolyte will thus be determined by the conductivity of the wastewater. In this section we focus on the anolyte losses as this is the compartment in which the wastewater is treated.

The energy loss that occurs can be expressed as the voltage loss. The voltage loss in a solution (ΔE_s) is determined by the current density (I) and the distance the ions travel (L) and the solution conductivity ρ_s. This loss can be calculated from Ohm's law as:

$$\Delta E_s = I \cdot \frac{L}{\rho_s} \tag{10.1}$$

To achieve an acceptable volumetric capacity, current densities should be in the range of 10 A/m^2 and preferably higher. At these current densities, voltage losses should be limited. As the conductivity of the anolyte is determined by the wastewater, this leaves the transport distance of ions as the only technological factor under control.

We investigate the importance of the conductivity by analyzing a situation where a voltage loss of 20% of the theoretical maximum is considered acceptable. Using Eq. (10.1) the maximum transport distance at a given conductivity can be calculated. Figure 10.3 shows this maximum distance as a function of the wastewater conductivity.

These requirements determine the acceptable ionic transport distance. This transport will also take place in membrane and cathode. However in the anode the biggest resistance is expected, as the conductivity is not under control.

The anolyte resistance for ions depends on the type of wastewater used. In laboratory BESs, synthetic wastewaters with relatively high conductivities

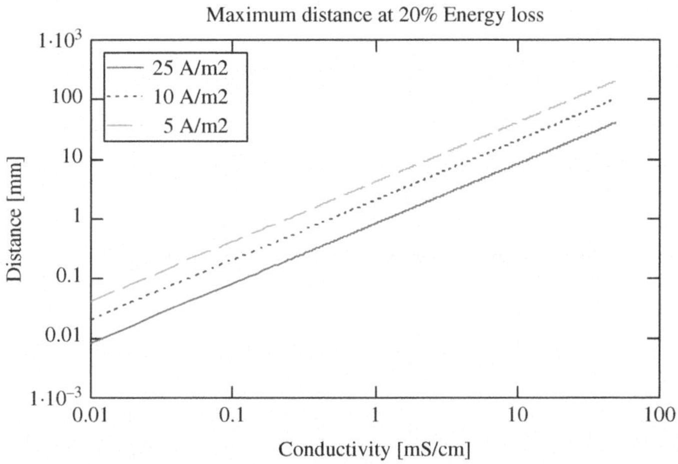

Figure 10.3. Electrolyte ohmic loss in BESs. Maximum allowable ionic transport distance (mm) to reduce the voltage loss less than 20% of the theoretical voltage as a function of the wastewater conductivity.

($>$5 mS/cm) have been used (Liu *et al.* 2005). And thus the accompanied voltage losses are relatively low.

Wastewaters can have a wide range of conductivities depending on their origin, for domestic wastewaters generally low ranges ($\sim$1 mS/cm) are reported (Rozendal *et al.* 2008a; Taylor and Gardner 2007) while some industrial wastewaters have very high conductivities. Figure 10.3 clearly shows that at relevant current densities in the low conductivity range ($<$1 mS/cm) only very small distances are allowed ($<$1mm). Because for a full-scale BES it would be too expensive to increase the conductivity of real wastewater by the addition of salt, it is important to minimize the electrode spacing. In low conductivity wastewaters the required small distances could be realised by using a Membrane Electrode Assembly (MEA) (Rozendal *et al.* 2007) In an MEA, the cathode is directly coated onto the membrane causing the transported ions to react directly on the membrane surface without having to travel through a boundary layer of electrolyte. In a BES without membrane, the resistance of anolyte and catholyte are the same. In a BES with membrane however, the catholyte resistance depends on the mode of operation of anode and cathode (batch or continuous) and the ion exchange membrane used. For instance, if the anode is operated in continuous-mode, while the cathode is operated in batch-mode, salts accumulate in the catholyte (Rozendal *et al.* 2008c). Consequently, the catholyte conductivity increases, resulting in a decrease in resistance.

10.2.3 Effect of buffer capacity

Oxidation of organic material by electrochemical active bacteria is an acid producing reaction. For each electron produced one proton is generally produced. As four electrons are accepted by molecular oxygen in its reduction to water, it can be calculated that 4 mmol/l H^+ are produced per mmol/l COD (32 mg COD). This means that the electric current is associated with an appreciable flux of protons from the bio-anode to the solution. To prevent local acidification, the protons need to be transported to the bulk solution. Therefore buffers are often added to the anolyte and catholyte to maintain a neutral pH, which is favourable for growth of electrochemically active microorganisms on the bioanode.

Furthermore, the addition of buffer increases the electrolyte conductivity. In laboratory BESs, often phosphate buffer of pH 7 is used, typically with concentrations of 10 to 50 mmol/l, but even concentrations as high as 320 mmol/lhave been reported (Hu *et al.* 2008). Also, other types of buffer have been used. For example, Fan *et al.* (2007b) investigated the use of 0.2 mol/l bicarbonate buffer of pH 9 in an MFC without membrane. They found that power outputs were higher than with 0.2 mol/l phosphate buffer at pH 7, and suggested that this was caused by the higher proton transfer rate via bicarbonate than via phosphate.

Nevertheless, in batch systems like cathodes, buffers maintain the pH neutrality only for a period of time. For example, in BESs containing a membrane, the pH in the cathode chamber eventually increases to values well over pH 12 (Rozendal *et al.* 2007, 2008c), which is an effect inherent in the use of ion exchange membranes (paragraph 10.2.4). Furthermore, addition of buffers to wastewaters in full-scale BESs is not practical because of the high costs and discharge regulations. The buffer concentration in full-scale BESs therefore, will be limited to the concentration of alkalinity already present in wastewater, and the concentration of bicarbonate alkalinity that is produced inside the BES through the oxidation of organic matter by electrochemically active micro-organisms (Fan *et al.* 2007b).

10.2.4 Membrane separator or not

Although BESs without membrane are increasingly being applied (Call and Logan 2008; Liu and Logan 2004), still a major part of BESs use a membrane to separate the anode from the cathode compartment. When a membrane is applied, charge will move through the membrane between anode and cathode to compensate for the flow of electrons outside the system. The first BESs used cation exchange membranes (CEM), such as Nafion 117 (Dupont Co., USA). In wastewater however, cations are present in much higher concentrations than protons and will therefore be transported through the membrane instead of protons (Gil *et al.* 2003; Kim *et al.* 2007; Rozendal *et al.* 2006). Because the protons that are consumed in the cathode are not replenished with protons from the anode, the pH will increase in the cathode and decrease in the anode. As a result of this membrane pH gradient a voltage loss will be suffered in the system (Rozendal *et al.* 2006). The voltage loss associated with the membrane pH gradient will become a significant problem especially in continuous full scale systems where the pH gradient has time to develop and the use of high concentrations of buffer is not economically feasible. To prevent this pH from rising, other membranes have been tested for BES application. These membranes include anion exchange membranes (AEM), charged mosaic membranes, bipolar membranes and ultra filtration membranes (Cheng and Logan 2007; Kim *et al.* 2007; Rozendal *et al.* 2008c; Ter Heijne *et al.* 2006). None of these membranes however were able to prevent the cathode pH from rising.

An advantage of applying a membrane is that it creates a physical barrier between anode and cathode. When a membrane is applied as a physical barrier in, for example, an MFC, it prevents oxygen moving from the cathode to the anode. Oxygen in the anode compartment is preferred by microorganisms as an electron acceptor over the electrode and this will decrease the coulombic efficiency (Liu and Logan 2004). In MECs, the membrane prevents the produced hydrogen moving to the anode where it is consumed by methanogens (Tartakovsky *et al.* 2008). Furthermore, it also prevents the produced carbon dioxide in the anode chamber mixing with the hydrogen in the cathode chamber, keeping the hydrogen pure. Still, membranes are never 100% selective and do not prevent the mixing of anolyte and catholyte entirely. When a flux of ions moves through a membrane, some exchange of anolyte and catholyte will occur. Especially in the future when BESs may be applied to produce specialty products at the cathode, this mixing should be kept in mind. The choice of using a membrane or not may depend on keeping the product in the cathode or preventing the mixing of anolyte and catholyte.

10.3 DESIGN CONSTRAINTS AS DETERMINED BY SCALE UP

10.3.1 Scale up and voltage losses

The obvious parameter that changes upon scale-up is the size of the current. Current increase can be a result of the increase of size of a single cell and/or the number of cells electrically connected in parallel.

A voltage loss will occur as a result of the resistance for the electron flow of the electrode material itself. This voltage loss is, however, scale dependent, as both the resistance of the electrode itself and the average current density increase with size. The exact relationship between scale and voltage loss depends also on the geometry of the electrode. We consider here a rectangular electrode whose width is much larger than the height. We assume further that along the whole width of the anode current is collected. As the resistance is linearly related to the path length for the electron flow, it can be safely assumed that the resistance increases linearly with the height of the electrode. A second increase is the increase of the average current through the anode, again it can be expected that this average current increases with the height of the electrode. Consider a planar anode, which has in laboratory reactors a typical size of a few centimetres. A scaled-up planar electrode might be a few metres big. This means that a height increase with a factor 100 will lead to a voltage loss increase with a factor 10000.

Also the electrical circuitry of the BES should be considered for a scaled-up system, e.g. whether individual cells should be serial or parallel electrically connected. To limit the voltage loss caused by increase in size, cells may be connected either in series or in parallel. Parallel connection of cells (for instance in tubular design as bundles of tubes) need not only be hydraulically sealed at the end of the reactor but also electrically connected.

In the most straightforward configuration, a BES reactor would be just a pile of cells with a thickness of ~1 cm each. These cells can be operated in parallel, i.e. keeping the cells independent from each other (Shimoyama *et al.* 2008). The advantages of this configuration are evident. Firstly, single-cell experiments may be regarded as representative and predictive for the stack performance. Secondly, in a properly designed stack, the failure and or cleaning of a single-cell would not affect the performance of the other cells. There may be a few challenges in this configuration. A major concern for this configuration mentioned in the literature (Rozendal *et al.* 2008a) is the long travel distance of the electrons that have to flow all the way through electrodes. This is particularly challenging for carbon or graphite electrodes (see Section 10.3.2). In a scaled-up system, the associated ohmic voltage drop could amount up to several hundreds of millivolts due to higher absolute currents when compared to laboratory setups.

Alternatively, a stack configuration with bipolar plates between the cells could be a viable design as the ohmic loss due to the low conductivity of electrodes is minimized due to the short electron transport distances. Bipolar plate stack design, however, poses other challenges that need to be resolved, such as cell reversal (Section 10.5.1).

10.3.2 Hydrodynamics and mechanics

Whatever the type of scaled up configuration might be used in future, attention should be placed on hydrodynamics and mechanics. For instance, a well-engineered hydrodynamic design is needed to obtain and maintain a good distribution of anolyte and catholyte to the cells from a shared manifold. The fluid dynamics can especially be critical in case changes in pressure drop would occur during operation due to clogging or other failures of one cell. For up-scaled systems, the crossflow velocity and path length for the fluids would become critical. In a single-pass system, the crossflow velocity would become too low to provide good mixing causing pH gradients (perpendicular to the biofilm), and the path length would become too long causing acidification of the anodic compartment and depletion of biodegradable organic matter (in a longitudinal direction). These problems may be encountered by applying serial staged stacks or by applying recycling. Energy loss due to pumping should be minimized. It might be expected that this loss is larger than in the case of digestion, where the gas can be used to mix the system.

The mechanical strength and dimensional stability of the BES, the individual cells, and its components, should also be considered. In single-cell experiments, the mechanic strength of the cell and its components is relatively uncomplicated as applied areas are limited and dense stiff endplates are used to keep the components together and shaped. With an increased area and number of cells in a stack, the degree of deformation would increase due to a piling of (flexible) layers.

Most of the topics related to hydrodynamics and mechanics are not new and have already been explained by, for example, Tanaka (2007).

10.4 COSTS AND CHOICE OF MATERIALS

10.4.1 Material properties and costs

Due to the intrinsic complexity of BESs it is likely that the capital costs will always remain higher than for conventional wastewater treatment systems (Rozendal *et al.* 2008a). This is the direct result of the large amounts of internals (i.e. electrodes and membranes) that are required in BESs for achieving reasonable organics conversion rates. For example, at an assumed volumetric current density of 1000 A/m^3 (~7 kg COD/m^3/day), a surface area-based current density of 10 A/m^2, and an electrochemical cell thickness of 1 cm (Rozendal *et al.* 2008a), about 100 m^2 of anode, cathode, and membrane are required per m^3. It is clear that this puts considerable constraints on the cost and choice of materials, and the electrochemical cell design.

10.4.2 Anode

In nearly all BES studies to date, researchers have been using graphite (or carbon) as the anode material of choice (Logan *et al.* 2006). The reasons for this are that graphite is (i) well compatible with electrochemically active biofilms, (ii) widely available in high specific area structures (e.g. graphite felt), and (iii) relatively inexpensive (Table 10.1). One major disadvantage of graphite materials, however, is that they are only moderately conductive (i.e. relatively high electrical resistivity) compared to other materials, particularly metals (Table 10.1). This is not really an issue in small-scale BESs, as the currents and the travel distances of the electrons are relatively small. However, in large-scale systems the use of graphite can become problematic, as the electrochemical losses can become quite significant if the system is not properly designed (see below).

Table 10.1. Electrical resistivity and material costs of various electrode materials

Material	Electrical resistivity[a] (μΩcm)	Material costs[b] ($/kg)
Graphite	1.47×10^3–1.90×10^3	0.24–10[c]
Titanium	39[d]	13–28
Palladium	10.73[e]	1.04×10^4
Platinum	10.7[e]	3.68×10^4
Stainless steel	9.87[e]	2.1[f]
Nickel	6.93[e]	24.2
Tungsten	5.39[e]	27–36
Gold	2.255[e]	1.95×10^4
Copper	1.712[e]	0.69
Silver	1.617[e]	3.73×10^2

[a]From (*Lide* 2008–2009)
[b]2006 values from http://minerals.usgs.gov/minerals/pubs/mcs/index.html
[c]Natural graphite (<97% carbon): 0.24–0.95 $/kg
Synthetic graphite (99.95% carbon): 3–10 $/kg
[d]At 273 K
[e]At 298 K
[f]Based on 18/8 stainless steel scrap (18% chromium, 8% Nickel).

Electrochemically active microorganisms have been shown to be able to interact well with other, more conductive materials (i.e. relatively low electrical resistivity), such as titanium-based anodes (Ter Heijne *et al.* 2008). However, these materials are much more expensive than graphite and it is unlikely that these large costs can be justified for practical application of bioelectrochemical wastewater treatment (Rozendal *et al.* 2008c).

10.4.3 Cathode

Platinum had been the cathode electrocatalyst of choice in most early BES studies (see Chapter 14), but soon researchers started to realize that platinum is too expensive for application in BESs (Rozendal *et al.* 2008a). Moreover, most platinum catalysis does not work as well under typical BES conditions as it does in PEM fuel cells and electrolyzers. The cause of this low performance is not well understood, but the relatively mild conditions in BESs (low *T*, neutral bulk pH) compared to those in PEM fuel cells and electrolyzers (323–373 K, $pH < 1$) are probably part of the explanation. In addition, under typical BES conditions, platinum electrocatalysts are likely to suffer from sulfide poisoning as a result of the presence of reduced sulfur compounds in feed streams (e.g. wastewater) (Protopopoff and Marcus 1988).

In view of the above, there is a strong demand for novel cathode electrocatalysts that show high catalytic activity at relatively low costs. The only way to realistically achieve this is to use an inexpensive base material, such as graphite (Table 10.1), in combination with a low-cost electrocatalyst (see Chapter 14). For example, for MFCs, researchers have demonstrated that transition metal macrocycle based catalysts can efficiently be used to catalyze the oxygen reduction reaction (see Chapter 14 (Cheng *et al.* 2006b; Zhao *et al.* 2005). In addition, a lot can also be expected from the development of microbial biocathodes (Chapter 15) (Clauwaert *et al.* 2007; He and Angenent 2006; Rozendal *et al.* 2008b). Microbial biocathodes have the advantage that they use a self-regenerating catalyst, which reduces cathode costs to that of the biocathode base material (e.g. graphite).

10.4.4 Membranes

In addition to the operational problems associated with the use of ion exchange membranes in BESs (i.e. membrane pH gradients; see Paragraph 10.2.4), they also represent a large capital cost. Nafion, for instance, which was used in most early MFC experiments, costs about 1400 $\$/m^2$ (Logan 2008). In a full-scale BES, requiring an estimated amount of 100 m^2 membrane per cubic metre of reactor volume, the associated capital investment is not justifiable for commercial applications (Rozendal *et al.* 2008a).

Nafion is regarded as one of the best cation exchange membranes available on the market, as it is designed for electrochemical applications with much higher current densities than BESs, such as the choralkali process and PEM fuel cells. However, BESs, which exhibit current densities that are about three orders of magnitude lower, can do with much cheaper membranes, such as CMI-7000 (Membranes International; 80 $\$/m^2$) (Rabaey *et al.* 2003, 2005). These cheaper membranes might have a slightly higher resistance, but the associated increase in resistance is almost negligible in a BES context. Moreover, it can be expected that the price of ion exchange membranes will continue to decrease with increasing demands, so it is not unlikely that ion exchange membranes might one day be available for less then 10 $\$/m^2$.

Ultimately, membrane-less approaches (Call and Logan 2008; Liu and Logan 2004) are preferred, as this would completely eliminate the capital expense required for membranes. However, membranes-less approaches are not straightforward as they generally lead to loss of process efficiency, as a result of fuel/oxidant cross-over in MFCs, and hydrogen loss due to anodic oxidation and/or hydrogenotrophic methanogenesis in MECs (Rozendal *et al.* 2008a).

10.5 OVERCOMING DESIGN CONSTRAINTS

10.5.1 Constraints and solutions

In this chapter we have so far identified a number of constraints originating from application and scale-up. To make BESs successful for wastewater treatment, a number of constraints need to be met. Table 10.2 summarizes in the first column the identified constraints. Although this list might seem disheartening, it does not imply that the BES system will not become a success. On the contrary, neglecting these aspects will without doubt lead to a failure. Solving these constraints will be a key to success.

Directions to solve these constraints are given in the second column of Table 10.2. For each constraint, several possible solutions are listed. It shows that for each of the constraints several solutions are available. This list of solutions is definitely far from being exhaustive as research into scale-up has hardly started.

The listed solutions are useful approaches for a specific constraint. A BES designed for a certain application in a specific situation must of course meet all relevant constraints.

For example a BES treating liquid manure has fewer constraints because of the nature of the substrate, i.e. high conductivity and alkalinity. However this does not solve the constraint of the resistivity of the electrode material. This is no issue in small-scale laboratory BES experiments but can become critical at large-scale (Table 10.1) (Rozendal *et al.* 2008a). Graphite, for example, is a perfect electrode material at lab-scale, but can exhibit large ohmic losses at large-scale. These large ohmic losses can particularly emerge in large-scale single cell BES designs (i.e. is the typical design of a laboratory BES; Figure 10.4A), as large currents need to be transported though the graphite over large distances. To eliminate these ohmic losses two strategies can be adopted: (i) a current collector can be used to support the electrode(s), or (ii) a bipolar plate stack design (Figure 10.4B) can be adopted.

Table 10.2. Constraints and solutions for scale-up of BESs

Constraint	Solution
Sufficient conductivity	Use of membrane electrode assembly Lower I via increase surface area Aim at high conductivity wastewater
Sufficient buffer capacity	Lower I via increase surface area Aim at high alkalinity wastewater Use anode effluent as cathode influent and vice versa (Freguia *et al.* 2008)
Reactor footprint	Aim at high COD wastewater Use of a high surface area
Cost	Use of cheap, appropriate materials Use of thinner materials Production high value products
Membrane	More selective membranes Membrane-less design
Cathode	Cheap chemical catalyst Biocathodes
High current	High surface area Current collector Bipolar stack designs

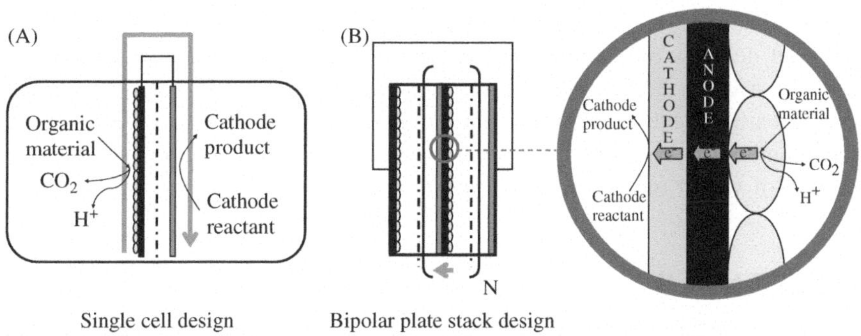

Figure 10.4. Single cell design versus bipolar plate stack design. The orange arrow indicates the travel distance of the electrons

With respect to the first strategy it is important to use a highly conductive (i.e. low resistivity), cost-effective current collector that at the same time is not susceptible to corrosion. This immediately eliminates copper as a suitable current collector, which, despite its low cost and low resistivity, is highly susceptible to corrosion. Stainless steel, on the other hand, is not very susceptible to corrosion, has a reasonably low resistivity, and can still be quite cost-effective.

Adopting a bipolar plate stack design (i.e. the second strategy) can eliminate the need for a current collector, as the travel distance of electrons can be reduced to an absolute minimum (Figure 10.4B). Bipolar plate stack designs, however, have hardly been investigated for BES applications (Shin *et al.* 2006). Moreover, issues such as the susceptibility to cell reversal (Aelterman *et al.* 2006; Oh and Logan 2007) still need to be resolved before the practical implementation of bipolar plate stack BES designs becomes reality.

Costs will always be an issue and it would be sensible to consider which product, e.g. electricity or hydrogen, is attractive. Hydrogen gives a much higher revenue than electricity. Of course one needs to consider whether hydrogen can be used in the specific situation.

To obtain a small footprint a high volumetric surface area is required. If one constructs cells of 1 cm thickness, one would achieve, a 100 m^2/m^3 surface area and a sufficiently high volumetric production might be obtained. Care should however be taken that the cell does not get clogged; thus some pre-treatment such as filtering might be required.

The example of the BES treating liquid manure illustrates that there are real challenges ahead when scaling up and applying BESs. Designs must be developed that meet these challenges. It should be possible to come up with designs that meet the identified design constraints. Which constraints are important depends on the specific situation, and specific designs are, therefore, probably needed. Up to date, little scale-up research has been performed, most research is lab scale. Research at lab scale is needed to investigate the fundamentals of processes and concepts. It cannot, however, replace scale-up research of which more is urgently needed.

REFERENCES

Aelterman, P., Rabaey, K., Pham, T.H., Boon, N. and Verstraete, W. (2006) Continuous electricity generation at high voltages and currents using stacked microbial fuel cells. *Environ. Sci. Technol.* **40**, 3388–3394.

Aelterman, P., Freguia, S., Keller, J., Verstraete, W. and Rabaey, K. (2008) The anode potential regulates bacterial activity in microbial fuel cells. *Appl. Microbiol. Biotechnol.* **78**(3), 409–418.

Call, D. and Logan, B.E. (2008) Hydrogen production in a single chamber microbial electrolysis cell lacking a membrane. *Environ. Sci. Technol.* **42**(9), 3401–3406.

Cheng, S., H., L. and Logan., B.E. (2006a) Increased power generation in a continuous flow MFC with advective flow through the porous anode and reduced electrode spacing. *Environ. Sci. Technol.* **40**, 2426–2432.

Cheng, S., Liu, H. and Logan, B.E. (2006b) Power densities using different cathode catalysts (Pt and CoTMPP) and polymer binders (Nafion and PTFE) in single chamber microbial fuel cells. *Environ. Sci. Technol.* **40**(1), 364–369.

Cheng, S. and Logan, E. (2007) Sustainable and efficient biohydrogen production via electrohydrogenesis. *PNAS* **104**(47), 18871–18873.

Clauwaert, P., Ha, H.V.D., Boon, N., Verbeken, K., Verhaege, M., Rabay, K. and Verstraete, W. (2007) Open air biocathode enables effective electricity generation with microbial fuel cells. *Environ. Sci. Technol.* **41**(21), 7564–7569.

Fan, Y., Hu, H. and Liu, H. (2007a) Enhanced Coulombic efficiency and power density of air-cathode microbial fuel cells with an improved cell configuration. *J. Power Sources* **171**(2), 348–354.

Fan, Y., Hu, H. and Liu, H. (2007b) Sustainable power generation in microbial fuel cells using bicarbonate buffer and proton transfer mechanisms. *Environ. Sci. Technol.* **41**(23), 8154–8158.

Freguia, S., Rabaey, K., Yuan, Z. and Keller, J. (2008) Sequential anode-cathode configuration improves cathodic oxygen reduction and effluent quality of microbial fuel cells. *Water Res.* **42**(6–7), 1387–1396.

Gil, G.C., Chang, I.S., Kim, B.H., Kim, M., Jang, J.K., Park, H.S. and Kim, H.J. (2003) Operational parameters affecting the performance of a mediator-less microbial fuel cell. *Biosensors Bioelectron.* **18**(4), 327–334.

He, Z. and Angenent, L.T. (2006) Application of bacterial biocathodes in microbial fuel cells. *Electroanalysis* **18**(19–20), 2009–2015.

Hu, H., Fan, Y. and Liu, H. (2008) Hydrogen production using single-chamber membrane-free microbial electrolysis cells. *Water Res.* **42**, 4172–4178.

Keller, J. and Rabaey, K. (2008) *Experiences from MFC Pilot Plant Operation.* Penn State University, USA.

Kim, J.R., Cheng, S., Oh, S.E. and Logan, B.E. (2007) Power generation using different cation, anion, and ultrafiltration membranes in microbial fuel cells. *Environ. Sci. Technol.* **41**(3), 1004–1009.

Lide, R.L. (ed.) (2008–2009) *CRC Handbook of Chemistry and Physics.* CRC Press, Boca Raton.

Liu, H. and Logan, B.E. (2004) Electricity generation using an air-cathode single chamber microbial fuel cell in the presence and absence of a proton exchange membrane. *Environ. Sci. Technol.* **38**(14), 4040–4046.

Liu, H., Cheng, S. and Logan, B.E. (2005) Power generation in fed-batch microbial fuel cells as a function of ionic strength, temperature, and reactor configuration. *Environ. Sci. Technol.* **39**(14), 5488–5493

Liu, H., Cheng, S., Huang, L. and Logan, B.E. (2008) Scale-up of membrane-free single-chamber microbial fuel cells. *J. Power Sources* **179**(1), 274–279.

Logan, B.E. (2008) *Microbial Fuel Cells.* John Wiley and Sons, Inc., Hoboken, New Jersey.

Logan, B.E., Hamelers, B., Rozendal, R., Schröder, U., Keller, J., Freguia, S., Aelterman, P., Verstraete, W. and Rabaey, K. (2006) Microbial fuel cells: methodology and technology. *Environ. Sci. Technol.* **40**(17), 5181–5192.

Logan, B.E., Cheng, S., Watson, V. and Estadt, G. (2007) Graphite fiber brush anodes for increased power production in air-cathode microbial fuel cells. *Environ. Sci. Technol.* **41**(9), 3341–3346.

Min, B. and Logan, B.E. (2004) Continuous electricity generation from domestic wastewater and organic substrates in a flat plate microbial fuel cell. *Environ. Sci. Technol.* **38**(21), 5809–5814.

Oh, S.E. and Logan, B.E. (2007) Voltage reversal during microbial fuel cell stack operation. *J. Power Sources* **167**(1), 11–17.

Protopopoff, E. and Marcus, P. (1988) Poisoning of the cathodic hydrogen evolution reaction by sulphur chemisorbed on platinum (110). *J. Electrochem. Soc.* **135**(12), 3073–3075.

Rabaey, K., Lissens, G., Siciliano, S.D. and Verstraete, W. (2003) A microbial fuel cell capable of converting glucose to electricity at high rate and efficiency. *Biotechnol. Lett.* **25**(18), 1531–1535.

Rabaey, K., Clauwaert, P., Aelterman, P. and Verstraete, W. (2005) Tubular microbial fuel cells for efficient electricity generation. *Environ. Sci. Technol.* **39**(20), 8077–8082.

Rozendal, R.A., Hamelers, H.V.M. and Buisman, C.J.N. (2006) Effects of membrane cation transport on pH and microbial fuel cell performance. *Environ. Sci. Technol.* **40**(17), 5206–5211.

Rozendal, R.A., Hamelers, H.V.M., Molenkamp, R.J. and Buisman, C.J.N. (2007) Performance of single chamber biocatalyzed electrolysis with different types of ion exchange membranes. *Water Res.* **41**, 1984–1994.

Rozendal, R.A., Hamelers, H.V.M., Rabaey, K., Keller, J. and Buisman, C.J.N. (2008a) Towards practical implementation of bioelectrochemical wastewater treatment. *Trends Biotechnol.* **26**(8), 450–459.

Rozendal, R.A., Jeremiasse, A.W., Hamelers, H.V.M. and Buisman, C.J.N. (2008b) Hydrogen production with a microbial biocathode. *Environ. Sci. Technol.* **42**, 629–634.

Rozendal, R.A., Sleutels, T.H.J.A., Hamelers, H.V.M. and Buisman, C.J.N. (2008c) Effect of the type of ion exchange membrane on performance, ion transport, and pH in biocatalyzed electrolysis of wastewater. *Water Sci. Technol.* **57**(11), 1757–1762.

Shimoyama, T., Komukai, S., Yamazawa, A., Ueno, Y., Logan, B.E. and Watanabe, K. (2008) Electricity generation from model organic wastewater in a cassette-electrode microbial fuel cell. *Appl. Microbiol. Biotechnol.* **80**(2), 325–330.

Shin, S.H., Choi, Y.J., Na, S.H., Jung, S.H. and Kim, S. (2006) Development of bipolar plate stack type microbial fuel cells. *Bull. Korean Chem. Soc.* **27**(2), 281–285.

Tanaka, Y. (2007) Ion exchange membranes fundamentals and applications, Elsevier.

Tartakovsky, B., Manuel, M.-F., Neburchilov, V., Wang, H. and Guiot, S.R. (2008) Biocatalyzed hydrogen production in a continuous flow microbial fuel cell with a gas phase cathode. *J. Power Sources* **182**(1), 291–297.

Taylor, B. and Gardner, T. (2007) Southeast Queensland recycled water aspects and soil impacts. Australian Water Association, Sunshine Coast, Australia.

Ter Heijne, A., Hamelers, H.V.M., De Wilde, V., Rozendal, R.A. and Buisman, C.J.N. (2006) A bipolar membrane combined with ferric iron reduction as an efficient cathode system in microbial fuel cells. *Environ. Sci. Technol.* **40**(17), 5200–5205.

Ter Heijne, A., Hamelers, H.V.M., Saakes, M. and Buisman, C.J.N. (2008) Performance of non-porous graphite and titanium-based anodes in microbial fuel cells. *Electrochim. Acta* **53**(18), 5697–5703.

Torres, C.I., Marcus, A.K. and Rittman, B.E. (2007) Kinetics of consumption of fermentation products by anode-respiring bacteria. *Appl. Microbiol. Biotechnol.* **77**(3), 689–697.

Wang, X., Feng, Y. and Lee, J.H. (2008) Electricity production from beer brewery wastewater using single chamber microbial fuel cell. *Water Sci. Technol.* **57**(7), 1117–1121.

Zhao, F., Harnisch, F., Schröder, U., Scholz, F., Bogdanoff, P. and Herrmann, I. (2005) Application of pyrolysed iron(II) phthalocyanine and CoTMPP based oxygen reduction catalysts as cathode materials in microbial fuel cells. *Electrochem. Comm.* **7**(12), 1405–1410.

11
Organics oxidation

Stefano Freguia

11.1 INTRODUCTION

The anodic oxidation of organic compounds was the first application to be envisioned for bioelectrochemical systems. Indeed, the ability of microorganisms to generate electricity via direct electron transfer to electrodes was first proven using organic carbon as the energy source (Kim *et al.* 2002). Oxidation of organics is coupled to a cathodic process, which may be oxygen reduction (microbial fuel cells, MFCs), hydrogen production (microbial electrolysis cells, MECs), or any other reductive process (more generally, bioelectrochemical systems, BESs). The abundance of organic waste in the form of wastewaters, domestic solid waste and agricultural waste matter makes this application attractive for large scale energy generation. It is estimated that the energy that could be yielded by the oxidation of organic matter in domestic sewage is approximately 10% of the residential electricity consumption (worldwide average, calculated from data made available by World Resources Institute).

 Bioelectrochemical Systems: From Extracellular Electron Transfer to Biotechnological Application. Edited by Korneel Rabaey, Largus Angenent, Uwe Schröder and Jürg Keller. ISBN: 9781843392330. Published by IWA Publishing, London, UK.

This is the main reason that has triggered wide interest in microbial fuel cells as renewable energy generation systems. In this application, energy generation is the primary goal and it is often associated with the need to treat or purify the organics source. In particular, wastewater treatment is the most studied application, as well as the most advanced towards full scale implementation. As a technology for wastewater treatment, microbial fuel cells compete with established processes, in particular activated sludge and anaerobic digestion (AD). While still suffering from higher installation cost (Rozendal *et al.* 2008), several advantages make MFCs an attractive alternative to AD for energy generation from wastewaters, including better effluent quality, ability to treat dilute influents, no heating requirement, no need for gas treating and handling and single step conversion of chemical energy to electricity. Figure 11.1 represents a scheme of a microbial fuel cell for organics oxidation.

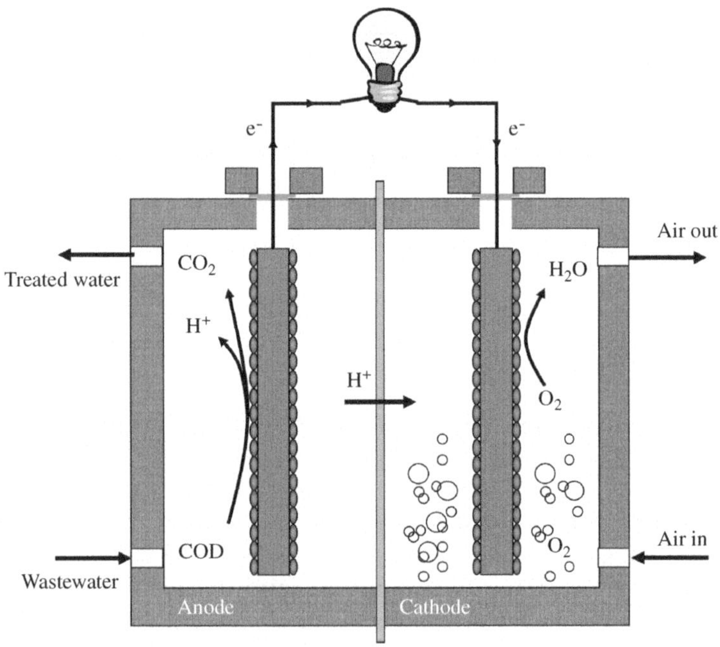

Figure 11.1. Schematic representation of a microbial fuel cell for wastewater treatment

For other BESs, including MECs, the anodic process is unchanged, while the cathode reaction is replaced by a useful product generating reduction. As the focus of this chapter is on anodic organics oxidation, the information presented herein is applicable to all types of BES, as long as the anodic electron donor is organic carbonaceous matter.

Bioanode-dwelling microbial communities carry out the full oxidation of organic substances to carbon dioxide (CO_2), with quantitative transfer of the substrate reducing equivalents to an anode in the form of electrical current. If we take glucose as a sample organic electron donor, the net reaction is shown in Eq. (11.1).

$$C_6H_{12}O_6 + 6H_2O \rightarrow 6CO_2 + 24H^+ + 24e^- \quad (11.1)$$

The range of substrates that can undergo this oxidation is extremely wide and so is the phylogenetic diversity of the microorganisms that are able to carry it out. Single substrate studies have shown that microorganisms in pure or mixed culture are able to fully oxidize to CO_2 acetate, propionate, butyrates, valerates, glucose, lactate, xylose, formate and ethanol, with coulombic efficiencies as high as 100% (Freguia *et al.* 2007b) and current densities up to 0.6 mA cm^{-2} (Fan *et al.* 2007). This list of organic compounds includes substances that are very common in wastewaters and polluted waterways, and make up the easily biodegradable fraction of their organic content. However, wastewaters contain also a wide variety of slowly biodegradable matter as well as recalcitrant compounds. These are responsible for the lower coulombic efficiencies of microbial fuel cells when real wastewaters are used, as often these compounds are hardly converted to current by anode dwelling microorganisms. However, COD removals up to 90% have been observed at MFC anodes fed with real wastewaters (Ghangrekar and Shinde 2008), thus implying that hydrolysis of large organic molecules does occur in these environments, in a fashion that must be entirely analogous to the first (hydrolytic) step of anaerobic digestion.

This chapter addresses the mechanisms of microbial carbon oxidation at microbial fuel cell anodes, with particular emphasis on metabolic pathways and syntrophic interactions in mixed anodic biofilms. First, an overview of respiratory metabolism will be provided, with a special focus on the aspects of this metabolism which are important when an anode is the electron acceptor in anaerobic conditions. An overview of fermentative metabolism will follow, which will lead to a section on the important syntrophic interactions between fermenters and anode respirers. The competing and detrimental effect of methanogenesis will be also introduced to complete the overview of the three major functional populations at MFC anodes. Section 11.6 describes a technological twist to the normal biofilm situation, where improvements can be achieved by physically separating fermentation and oxidation of fermentation products.

11.2 RESPIRATORY OXIDATION TO CARBON DIOXIDE

Organic substrates can undergo a variety of bioreactions in a mixed bacterial community. These can be divided into two categories based on the availability of electron acceptors: respirations and fermentations.

Direct anodic oxidation is a form of respiration, where the terminal electron acceptor is an insoluble electrode. This mechanism of energy generation is not different from that of any other respiration, such as oxygen respiration or nitrate respiration. The net products of respiration are H_2O, CO_2 and ATP. Like other forms of respiration, anodic oxidation avails itself of the tricarboxylic acid (TCA) cycle (Figure 11.2) and membrane bound electron transfer chain. Acetate, lactic acid and ethanol can enter the TCA cycle through pyruvate or acetyl-CoA via one or two enzyme steps (Figure 11.2), thus numerous species of anodophilic bacteria are able to take them up and utilize them directly as electron donors (Bond and Lovley 2003; Kim *et al.* 1999; Kim *et al.* 2007). On the contrary, carbohydrates such as glucose require a more involved enzymatic process called glycolysis (Figure 11.2). Fermentative bacteria are higly specialized in glycolysis and can carry it out at high rates, thus often successfully competing with anodophilic bacteria for carbohydrates as substrates. While with pure cultures anodic glucose oxidation has been observed (Chaudhuri and Lovley 2003), in mixed cultures the vast majority of the glucose is fermented (Freguia *et al.* 2008).

Bacteria generate NADH, NADPH and $FADH_2$ from the TCA cycle. These are the starting points of the electron transfer chain, with electrons travelling through flavoproteins, iron-sulfur proteins, quinone pool and a series of cytochromes. The electron transfer chain generates a proton motive force across the membrane, by extruding approximately 10 protons to the periplasm for each 2 electrons (derived from 1 NADH) being transported. This proton translocation is not a continuous process, but it occurs in steps. Indeed, there are only 3 steps which have been shown to be associated with the release of protons: complex I, III and IV. The first 4 protons are extruded by Complex I, a membrane-spanning proteic complex which makes up the first step of electron transfer from NADH and comprises flavoproteins (hydrogen atom carriers) and Fe-S proteins (electron carriers). Another $4H^+$ are translocated through complex III, comprising cytochrome b, Fe-S protein and cytochrome c1. This complex oxidizes and reduces quinones from the quinone pool and exploits their ability to diffuse across the cytoplasmic membrane in order to generate proton motive force (Q cycle). Complex IV is responsible for the extrusion of the last 2 protons by the cytochrome oxidase protein.

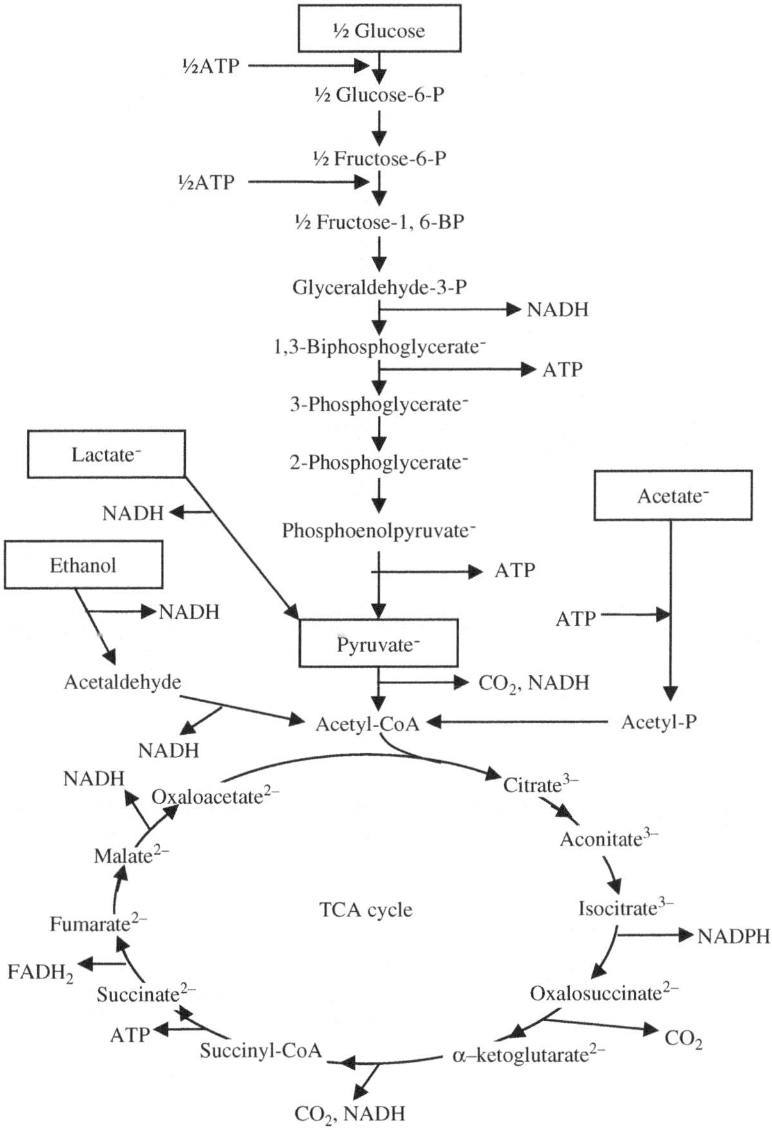

Figure 11.2. Pathways of oxidation of some common organic substances in respiratory metabolism. Each step is catalyzed by one enzyme. Acetyl-CoA is the gateway to the TCA cycle, which is responsible for the generation of 2 NADH, 1 NADPH and 1 $FADH_2$. Glucose requires 10 enzymatic steps to acetyl-CoA (glycolysis), whereas lactate, acetate and ethanol each require only 2 steps

The proton motive force drives ATP generation through the membrane-spanning ATP-synthase enzyme, which exploits the electrochemical potential liberated by the protons as they return to the cytoplasm. This process is named oxidative phosphorylation. Roughly three protons are required to generate one ATP. As these numbers are all approximate and the mechanisms not entirely understood, it is generally accepted that the maximum theoretical ATP yield for prokaryotes is 3 ATP per NADH (White 2007). This ratio is known as respiration ratio and its actual value is reduced by the utilization of the proton motive force for an array of other cell activities, including motility and solute transport, as well by any 'shortcut' in the electron transfer chain.

As previously described (Chapters 5 and 6), the last electron transfer step towards an anode can occur directly via cytochromes and conductive pili (nanowires) or via soluble endogenous redox mediators. The latter can enter the electron transfer chain at different stages, depending on their redox potential. Riboflavin (as well as its derivatives FAD and FMN), which is a common bacterial excreted flavoprotein (Marsili *et al.* 2008), transfers electrons at a potential around −220 mV NHE, while soluble excreted quinones deliver electrons at higher potential: 2-amino-3-carboxy-1,4-naphthoquinone (ACNQ, $E^{0'} = -71$ mV NHE, Yamazaki *et al.* (1999)) and 1,4-dihydroxy-2-naphthoic acid (DHNA, $E^{0'} = +300$ mV NHE approximately, Wang *et al.* 2008) are commonly excreted quinones; phenazines such as pyocyanine have also been detected as endogenous redox shuttles, their standard redox potential being approximately −100 mV NHE (Rabaey *et al.* 2005). When bacteria establish electrical contact with an electrode, the electron transfer chain is not exploited in its entirety, as the potentials of the soluble mediators or cytochromes used for the terminal electron transfer step are not high enough to receive electrons from the further steps in the electron transfer chain. The maximum ATP yield is reduced according to where the electron transfer chain is interrupted to convey electrons to the anode. Table 11.1 displays the maximum theoretical ATP yields when different electron carriers are responsible for the terminal electron transfer step. Here these yields are calculated both based on thermodynamic potential values (i.e. yield proportional to the difference in redox potential between NADH and terminal step electron carrier) and on the number of protons extruded by each step of the electron transfer chain. For instance, if cytochrome c is the last step of electron transfer, eight protons out of the theoretical maximum ten can be translocated via complex I and complex III; if a quinone is the electron transfer mediator to the anode, only four out of ten protons can be translocated by complex I.

As the number of ATPs generated via ATP synthase is proportional to the total number of protons extruded, ATP yields are also accordingly proportional.

As the ability of bacteria to grow is directly affected by the availability of intracellular ATP, bacterial growth yields depend on the mechanism of electron transfer. The selection by bacteria of which compound/complex to exploit as extracellular electron transfer mediator depends on the potential of the anode. For example, if the anode potential is less than −200 mV NHE, cytochromes and quinones cannot be exploited for the terminal electron transfer step. Ultimately, the anode potential affects the growth yields by limiting the ATP that can be produced per electron-equivalent being oxidized. In a mixed culture growing on acetate, growth yields were found to fall from 30% to 0% as the anodic potential was lowered from −20 mV to −220 mV NHE (Freguia *et al.* 2007b).

Table 11.1. ATP yields for different terminal electron transfer steps during anodic respiration

Terminal electron carrier	Redox potential $E^{0\prime}$(V NHE)	Proton extrusion H+/NADH[1]	ATP yield (E based) (ATP/NADH)	ATP yield (H+ based) (ATP/NADH)	ATP yield[2] (E based) (ATP/glucose)
Cyt a3 (full respiration)	+385	10	3.0	3.0	38
Cyt c	+250	8	2.4	2.4	31
Quinone (ACNQ)	−71	4	1.1	1.2	16
Riboflavin	−220	0	0.4	0.0	8
NADH	−320	0	0.0	0.0	3

[1]Theoretical number of protons extruded per NADH, based on the known proton pumps and Q cycle.
[2]ATP yields from glucose account for 2 ATPs generated from glycolysis and 1 ATP produced by the TCA cycle.

11.3 FERMENTATION AT MICROBIAL FUEL CELL ANODES

Fermentation is the bacterial generation of ATP in the absence of electron acceptors. Substrate is first oxidized to intermediate metabolites which are subsequently reduced to fermentation products, with net generation of energy. As MFC anodes are anaerobic and lack significant amounts of soluble electron acceptors, fermentations are common in these environments. While an MFC anode acts as extracellular insoluble electron acceptor, the fact that not all bacterial species can directly exploit it as electron sink makes fermentative processes extremely competitive in a mixed community MFC anode environment, even more so if we consider that fermenters often have much higher substrate turnover rates compared to respiring organisms. Some common fermentations of glucose are shown in Figure 11.3.

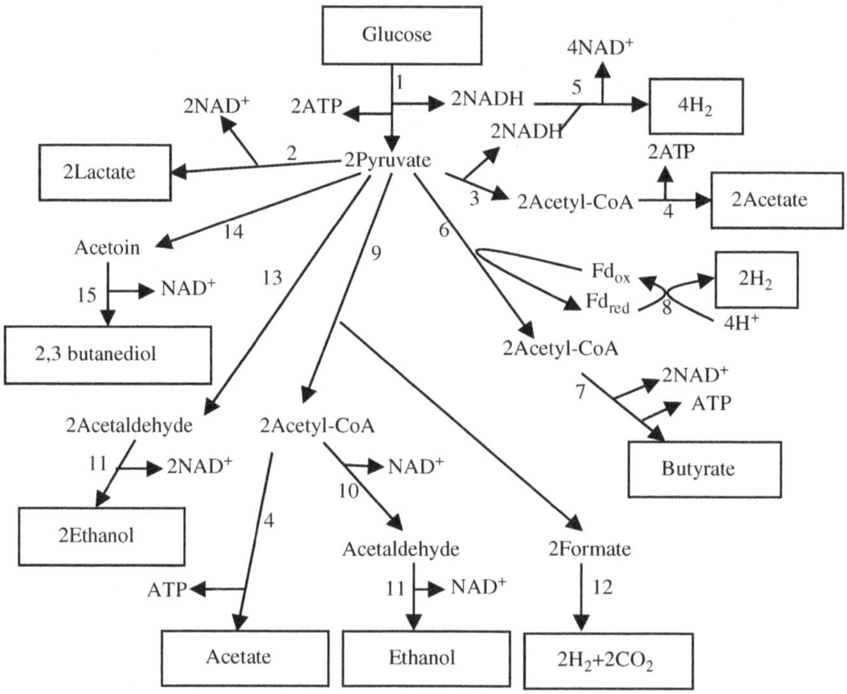

Figure 11.3. Some fermentative pathways from glucose. Except when acetate or butyrate are produced, the only source of ATP is glycolysis. Reactions involved, with the enzymes which catalyze them: (1) glycolysis (9 enzymatic steps, see Figure 11.2); (2) lactate dehydrogenase; (3) pyruvate dehydrogenase; (4) 2 enzymatic steps, phosphotransacetylase and acetate kinase; (5) NADH dehydrogenase; (6) pyruvate-ferredoxin oxidoreductase; (7) 6 enzymatic steps to butyrate, including the generation of 1 ATP; (8) hydrogenase; (9) pyruvate-formate lyase; (10) acetaldehyde dehydrogenase; (11) alcohol dehydrogenase; (12) formate-hydrogen lyase; (13) pyruvate decarboxylase; (14) 3 enzymatic steps to acetoin (2-hydroxy-3-butanone); (15) 2,3 butanediol dehydrogenase

The generation of reduced by-products is linked to the need to consume the reducing power generated during glycolysis. This means that for every NADH produced during the oxidation of glucose to pyruvate, equimolar amounts of NAD^+ need to be produced by the reduction of pyruvate to other products, e.g. ethanol, lactate and 2,3-butanediol. Alternatively, NADH may be directly reoxidized with generation of molecular hydrogen, via the NADH dehydrogenase or the ferredoxin pathway (respectively, steps 5 and 8 in Figure 11.3). Lacking a high potential electron acceptor, fermentations can yield only a limited amount of energy compared to most respirations. Table 11.2 displays several fermentation stoichiometries as well as the net Gibbs free energy and ATP yields for the reactions. Acetogenic fermentation can yield up to 4 ATPs per glucose, whereas aerobic respiration can yield up to 38 (Table 11.1). Most other fermentations (such as homolactic fermentation) are limited to 2 ATPs per glucose. For this reason fermenters require high substrate turnover rates in order to make sufficient amounts of ATP to support growth. In turn, high substrate turnover rates make fermenters highly competitive with respirers for fermentable substrates. Fermenters exhibit high turnover rates thanks to the high expression of glycolytic enzymes, as well as increased transport of substrates into the cells via expression of the substrate specific phosphotransferase systems.

Table 11.2. Stoichiometries, Gibbs free energy of reaction and ATP yields of some common bacterial fermentations and archaeal methanogenesis

Fermentations	Stoichiometry	$\Delta G^{0'}$ (i) (kJ/mol)	ATP/ Glucose
Homolactic	Glucose → 2 Lactate-$+2H^+$	−197	2
Alcoholic	Glucose → 2 Ethanol$+2CO_2$	−235	2
Acetogenesis	Glucose$+2H_2O$ → 2Acetate$^-+2CO_2$ $+4H_2+2H^+$	−216	4
Butanediol	Glucose → Butanediol$+2CO_2+H_2$	−194	2
Butyric	Glucose → Butyrate$+2CO_2+2H_2+H^+$	−264	3
Methanogenesis			
Acetoclastic	Acetate-+H+ → CH_4+CO_2	−36	
Hydrogenotrophic	$4H_2+CO_2$ → CH_4+2H_2O	−131	

(i) Source of Gibbs free energy data: Heijnen (1999)

When hydrogen production occurs (such as in acetogenic fermentation), hydrogen accumulation may easily bring fermentation to a stop if H_2 is generated via NADH dehydrogenase (step 5 in Figure 11.3). This is due to the fact that the standard potential of the NAD^+/NADH couple ($E^{0'} = -320$ mV NHE) is higher than that of the H^+/H_2 couple ($E^{0'} = -414$ mV NHE). As the NAD^+/NADH ratio is normally close to 1 in bacteria (Berríos-Rivera *et al.* 2002), thermodynamics imposes an upper limit of approximately 60 Pa for H_2 partial pressure (Angenent *et al.* 2004). If this limit is exceeded, the bioreaction becomes thermodynamically not feasible. In mixed communities, this problem is often solved by the presence of hydrogenotrophic organisms, especially methanogens but also homoacetogens, which effectively scavange H_2 and enable the fermentation to proceed.

In pure culture or in the absence of hydrogenotrophic populations, bacteria resort to other solutions. Hydrogen can be produced directly from pyruvate via the ferredoxin pathway (step 8 in Figure 11.3). This pathway does not carry tight thermodynamic limitations as the potential of ferredoxin is −410 mV NHE; thus up to 30 000 Pa H_2 can be released. However, this fermentation produces only 2 ATPs per glucose instead of the 4 ATP per glucose when the NADH pathway is used. Moreover, the ferredoxin pathway does not consume NADH, thus other reduced fermentation products such as butyrate or ethanol must be produced to close the NADH balance.

More recently it was discovered that in the presence of a polarized anode or other extracellular electron acceptors, bacteria can resort to another way to dispose of the extra reducing equivalents. Instead of making hydrogen, the NADH electrons can be conveyed directly to the anode via extracellular electron transfer. This has been shown thus far for a gram-positive fermenter, *Propionibacterium freudenreichii* (Wang *et al.* 2008), which has displayed the ability to produce and secrete soluble quinones to carry out the extracellular process. This modified fermentation is equivalent to step 5 in Figure 11.3, with the difference that an anode replaces H^+ as the terminal electron acceptor and the proteins/electron carriers involved in this pathway are those of the membrane-bound electron transfer chain. In light of this, fermenters may be able to take on a very important role in MFC anodic biofilms: they provide a way to initiate carbohydrate oxidation, they produce fermentation by-products that are good substrates for anodophilic respirers, and they contribute to the overall electrochemical activity of the biofilm by excreting redox active compounds.

11.4 SYNTROPHY BETWEEN FERMENTERS AND ANODOPHILES

We have so far described the processes of respiration and fermentation separately. However, in a complex mixed bacterial community growing as a biofilm on an anode, populations of respiring and fermenting microorganisms inevitably compete for substrates and interact with each other by commensal and bioelectrochemical means. Recent extensive studies by Freguia *et al.* (2008) and Lee *et al.* (2008) on the matter have unravelled the complex food web of anodic biofilms. The following are the governing principles of such ecosystems. Figure 11.4 is a schematic representation of this food web.

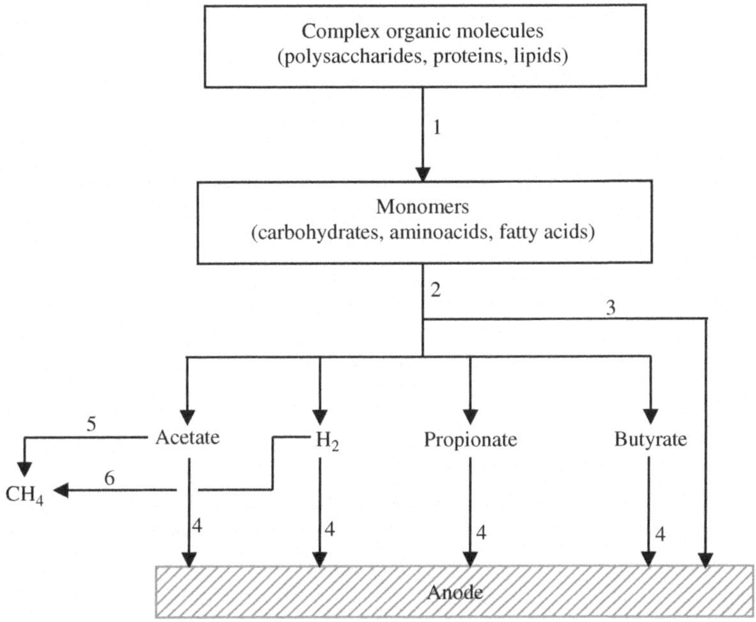

Figure 11.4. Food web of microbial fuel cell anodic communities. Polymeric organic molecules such as polysaccharides, proteins and lipids are hydrolyzed to monomeric carbohydrates, aminoacids and fatty acids by fermentative bacteria (1), which subsequently ferment them to acetate, propionate, butyrate (2), with the reducing equivalents being released as molecular hydrogen or directly as electrons towards the anode via redox mediators (3); step 3 may involve respiring bacteria which harvest some of the potential difference between the redox mediators excreted by the fermenters and the anode. Step 4 is the respiratory anodic oxidation of fermentation products by anodophilic bacteria: the products of this oxidation are CO_2 and electrons. Hydrogen and acetate can also be taken up by methanogens, with generation of CH_4 via acetogenic (5) and hydrogenotrophic (6) methanogenesis

Competition for carbohydrates is effectively won by fermentative microorganisms. At least 60% and in most cases over 90% of sugars are converted to fermentation products rather than electricity. Being the highest ATP yielding fermentation (Table 11.2), acetogenic fermentation strongly prevails over other fermentations. Acetate is always the main fermentation product, with propionate and butyrate acquiring importance as overflow metabolites at low anodic potential (less than −100 mV NHE) or high organic loading rates (greater than 3 $kg_{COD}m^{-3}d^{-1}$). Lactic and alcoholic fermentations are generally barely competitive in mixed anodic biofilms.

All fermentation products are substrates for anodophilic respirers. As most carbohydrates are first fermented in mixed anodic communities, electricity generation from carbohydrates is thus almost entirely a syntrophic process. While acetate is possibly the fastest organic compound to be converted to electricity by bacteria, propionate (Bond and Lovley 2005), butyrate (Liu *et al.* 2005) and other common fermentation products such as lactate (Park and Zeikus 2003), ethanol (Kim *et al.* 2007) and formate (Ha *et al.* 2008) have also been shown to work well as substrates for anodophiles. The question is on the fate of hydrogen, as all fermentations that produce volatile fatty acids also generate hydrogen gas to some extent. Indeed, some species of anodophilic bacteria are able to take up H_2 very effectively and to convert it to electrical current lithoheterotrophically, with up to 15% of the current deriving from H_2 oxidation in some biofilms (Freguia *et al.* 2008).This situation is known as interspecies hydrogen transfer. Moreover, as described above, some fermenters are able to skip hydrogen and dump the electrons directly onto the anode via excreted soluble redox mediators such as hydroquinones (Wang *et al.* 2008). In a mixed anodic community, these soluble mediators may also be taken up as electron donors by other anodophiles, where they could insert into the electron transfer chain and drive a proton motive force with the anode as terminal electron acceptor. Other means to dispose of NADH reducing equivalents involve the expression of nanowires, conductive pili that can lead electrons from fermenters directly to the electrode or to anodophilic bacteria (Gorby *et al.* 2006).

As all the reducing equivalents of carbohydrates can ultimately be transferred to the anode, fermentations are not intrinsically detrimental to the anodic process. In fact they are essential as they very rapidly convert carbohydrates into ideal substrates for anodophiles, as well as enriching the biofilm with electroactive compounds. The only true coulombic loss occurs when methanogens develop.

11.5 METHANOGENS COMPETE FOR FERMENTATION PRODUCTS

Methanogenesis is a type of bioreaction carried out exclusively by Archaea, where the substrate is converted to methane gas (CH_4). The possible substrates of methanogenesis are very few and, for what concerns microbial fuel cell anodes, are limited to acetate (acetoclastic methanogenesis) and H_2 (hydrogenotrophic methanogenesis, see Table 11.2). In any strictly anaerobic environment where acetate and/or H_2 are present, methanogens are doomed to develop. Thus, it is not surprising that methanogens are routinely isolated from MFC anodic communities. Any substrate converted to methane is lost to the electrode and represents an irreversible coulombic loss for the fuel cell. Inevitably, competition arises between methanogens and anodophiles, as both feed on the same two substrates. Methanogens are at an advantage over anodophilic bacteria as they are not constrained to maintain electrical contact with the electrode and thus they can thrive at large distances from the anode as well as in planktonic form. The methane production of an anodic community appears to be affected by the long-term operation of the MFC and not by the anodic potential, as would instead be expected if true competition were occurring. This fact suggests that anodophiles and methanogens are not strictly in competition with each other, but rather both thriving in different niches of the biofilm. At optimal loading conditions (as much COD is fed as can be converted based on the biological activity) methane production in continuous glucose-fed MFCs settle at around 15–25% of the converted glucose (Freguia *et al.* 2007a). At these conditions methanogenesis is only hydrogenotrophic. Acetoclastic methanogens are slower growers, thus they appear to develop only when substrate in excess is present, that is when the organic loading rate largely exceeds the optimal value during long-term operation. Acetoclastic methanogenesis, unlike hydrogenotrophic methanogenesis, appears to be dependent on the anodic potential. At a potential of −150 mV NHE, over 50% of the electrons of a glucose fed anode may be converted to methane (Freguia *et al.* 2007a).

Methanogenesis is the only true coulombic loss of microbial fuel cell anodes. It is ubiquitous and very hard to root out. Attempts have been made with periodic aeration of the anode compartment. Most anodophilic respirers are also aerophilic (as they thrive on the respiratory electron transfer chain) thus they are not affected by oxygen. Some fermenters are aerotolerant and their electrochemical activity may even be triggered by temporary aerobic conditions (Yamazaki *et al.* 2002). On the contrary, methanogens are strict anaerobes and as such are rapidly killed by even small amounts of O_2. Unfortunately, if the biofilm grows thick, O_2 may not be able to reach the deepest layers where methanogens continue to thrive. Frequent aeration from the start of the bioanode may be the solution to the problem of methanogenesis at MFC anodes.

11.6 ELECTROCATALYTIC OXIDATION OF FERMENTATION PRODUCTS

As an alternative to syntrophic oxidation of fermentation products in mixed community bioanodes, the location where fermentation and oxidation occur may be physically separated. In that case, the oxidation may be chemically catalysed rather than biocatalysed. The separation between the two steps may be carried out via anode coatings which are permeable to fermentation products but not to bacteria. Fermentation of complex organic matter produces simple energy-rich reduced fermentation products (such as hydrogen, ethanol or formate) which can be considered as secondary fuels – accessible for direct oxidation and current generation. Yet, such an oxidation requires the use of electrocatalytic anodes to decrease the normally too large oxidation overpotential. The uncoupling of biological fermentation and electrochemical (abiotic) oxidation may curb the problem of methane generation as by-product. First reports on the use of fermentation products for direct electricity generation were made by Karube and coworkers (Karube *et al.* 1977; Suzuki *et al.* 1980) who utilized immobilized hydrogen producing cultures as biocatalysts and platinum as electrocatalyst for hydrogen oxidation. Due to the liability of the platinum electrodes to poisoning and deactivation the reported power densities were rather low. Considerable progress was achieved by the development of sandwich electrodes consisting of platinum, protected from poisoning reactions by an overlay of conductive polymers such as polyaniline or its fluorinated forms. With current densities of up to 1.5 mA cm^{-2} these electrodes considerably improved the performance of microbial fuel cells (Nießen *et al.* 2004a; Schröder *et al.* 2003). Further, they offered access to exploit a great number of heterotrophic, photoheterotrophic (Rosenbaum *et al.* 2005a) and even purely photosynthetic (Rosenbaum *et al.* 2005b) microorganisms and access to complex carbohydrates such as starch (Nießen *et al.* 2004b, 2006) and cellulose (Nießen *et al.* 2005) for current generation in microbial fuel cells.

Based on its high market price, the use of platinum (both for the cathode and for the anode) is not an option for microbial fuel cells. This circumstance entails the requirement for the development of alternative, noble-metal free electrocatalysts. Tungsten carbide, an inexpensive and yet effective electrocatalyst, known since the end of the 1960s (Binder *et al.* 1969a, b) may represent a promising candidate, since it possesses an excellent electrocatalytic oxidation performance at low costs. Current performance data (3 mA cm^{-2} maximum current density, and a maximum power density of 586 μW cm^{-2}) mark the top values of present microbial fuel cell technology. The electrocatalyst not only shows sound activity towards the oxidation of hydrogen, but also indicates good activity for the oxidation of formate (1.9 mA cm^{-2}) and lactate (Rosenbaum *et al.* 2006, 2007).

11.7 SUMMARY

Organic carbon oxidation at BES anodes is a fairly well understood process. Bacteria generate electricity via respiratory metabolism, where the terminal electron acceptor is the electrode. This metabolism enables the anodophilic bacteria to generate ATP and thus grow. As the energy gain is limited by the anodic potential, ATP yields and hence growth yields are far smaller than those observed during aerobic metabolism. As smaller growth yields mean less waste sludge and larger coulombic efficiencies, this feature of BES makes them attractive as wastewater treatment technologies. Besides, when compared with currently established technologies (activated sludge and anaerobic digestion), BES anodes present a plethora of advantages which justify extensive research in the field.

When the organic substrates to be treated are fermentable (as most often is the case), fermenting bacteria often play a crucial role in carbon conversion in anodic communities, by quickly metabolizing substrates and supplying anode respiring organisms with easily biodegradable compounds. A disadvantage of fermentation is that any hydrogen thus produced will support the growth of populations of hydrogenotrophic methanogens, which are normally responsible for coulombic losses of 15–25% as methane in these systems. This problem represents the main drawback of organic carbon oxidation at BES anodes and demands further research. Possible solutions may involve periodic aeration of the anode compartment to eliminate the strictly anaerobic methanogens.

In terms of current output, at values up to 0.6 mA cm^{-2} for simple biofilm anodes and 3 mA cm^{-2} for coated anodes, anodes do not currently represent the limiting step to electron transfer in bioelectrochemical systems, with cathodes and electrolyte resistance instead accounting for most of the loss.

REFERENCES

Angenent, L.T., Karim, K., Al-Dahhan, M.H., Wrenn, B.A. and Domiguez-Espinosa, R. (2004) Production of bioenergy and biochemicals from industrial and agricultural wastewater. *Trends Biotechnol.* **22**(9), 477–485.

Berríos-Rivera, S.J., Bennett, G.N. and San, K.-Y. (2002) The effect of increasing NADH availability on the redistribution of metabolic fluxes in escherichia coli chemostat cultures. *Metab. Engng.* **4**(3), 230–237.

Binder, H., Köhling, A., Kuhn, W., Lindner, W. and Sandstede, G. (1969a) Das Verhalten von Wolframcarbid in Elektroden für Brennstoffzellen mit saurem Elektrolyten. Pergamon Press, Vol. 10, pp. 25–28.

Binder, H., Köhling, A., Kuhn, W. and Sandstede, G. (1969b) Electrochemical oxidation of aldehydes, formic acid, and carbon monoxide at tungsten carbide electrodes in 2 N sulfuric acid. *Angew. Chem. Int. Edn.* **8**(10), 757–758.

Bond, D.R. and Lovley, D.R. (2003) Electricity production by Geobacter sulfurreducens attached to electrodes. *Appl. Environ. Microbiol.* **69**(3), 1548–1555.

Bond, D.R. and Lovley, D.R. (2005) Evidence for involvement of an electron shuttle in electricity generation by Geothrix fermentans. *Appl. Environ. Microbiol.* **71**(4), 2186–2189.

Chaudhuri, S.K. and Lovley, D.R. (2003) Electricity generation by direct oxidation of glucose in mediatorless microbial fuel cells. *Nature Biotechnol.* **21**(10), 1229–1232.

Fan, Y., Hu, H. and Liu, H. (2007) Enhanced Coulombic efficiency and power density of air-cathode microbial fuel cells with an improved cell configuration. *J. Power Sources* **171**(2), 348–354.

Freguia, S., Rabaey, K. and Keller, J. (2007a) Competition and interaction between fermentation and electricity generation at microbial fuel cell anodes. *Proceedings of the 11th Anaerobic Digestion Congress*, 23–27 September 2007, Brisbane, Australia.

Freguia, S., Rabaey, K., Yuan, Z. and Keller, J. (2007b) Electron and carbon balances in microbial fuel cells reveal temporary bacterial storage behavior during electricity generation. *Environ. Sci. Technol.* **41**(8), 2915–2921.

Freguia, S., Rabaey, K., Yuan, Z. and Keller, J. (2008) Syntrophic processes drive the conversion of glucose in microbial fuel cell anodes. *Environ. Sci. Technol.* **42**(21), 7937–7943.

Ghangrekar, M.M. and Shinde, V.B. (2008) Simultaneous sewage treatment and electricity generation in membrane-less microbial fuel cell. *Water Sci. Technol.* **58**(1), 37–43.

Gorby, Y.A., Yanina, S., McLean, J.S., Rosso, K.M., Moyles, D., Dohnalkova, A., Beveridge, T.J., Chang, I.S., Kim, B.H., Kim, K.S., Culley, D.E., Reed, S.B., Romine, M.F., Saffarini, D.A., Hill, E.A., Shi, L., Elias, D.A., Kennedy, D.W., Pinchuk, G., Watanabe, K., Ishii, S., Logan, B., Nealson, K.H. and Fredrickson, J.K. (2006) Electrically conductive bacterial nanowires produced by Shewanella oneidensis strain MR-1 and other microorganisms. *Proc. Natl. Acad. Sci. USA* **103**(30), 11358–11363.

Ha, P.T., Tae, B. and Chang, I.S. (2008) Performance and bacterial consortium of microbial fuel cell fed with formate. *Amer. Chemical Soc.* pp. 164–168

Heijnen, J.J. (1999) Bioenergetics of microbial growth. In: *Bioprocess Technology: fermentation, biocatalysis, bioseparation*. Flickinger, M.C. and Drew, S.W. (eds). John Wiley and Sons, pp. 267–291.

Karube, I., Matsunaga, T., Tsuru, S. and Suzuki, S. (1977) Biochemical fuel cell utilizing immobilized cells of Clostridium butyricum. *Biotechnol. Bioeng.* **19**(11), 1727–1733.

Kim, B.H., Ikeda, T., Park, H.S., Kim, H.J., Hyun, M.S., Kano, K., Takagi, K. and Tatsumi, H. (1999) Electrochemical activity of an Fe(III)-reducing bacterium, Shewanella putrefaciens IR-1, in the presence of alternative electron acceptors. *Biotechnol. Tech.* **13**(7), 475–478.

Kim, H.J., Park, H.S., Hyun, M.S., Chang, I.S., Kim, M. and Kim, B.H. (2002) A mediator-less microbial fuel cell using a metal reducing bacterium, Shewanella putrefaciens. *Enzyme Microb. Technol.* **30**(2), 145–152.

Kim, J.R., Jung, S.H., Regan, J.M. and Logan, B.E. (2007) Electricity generation and microbial community analysis of alcohol powered microbial fuel cells. *Bioresour. Technol.* **98**(13), 2568–2577.

Lee, H.S., Parameswaran, P., Kato-Marcus, A., Torres, C.I. and Rittmann, B.E. (2008) Evaluation of energy-conversion efficiencies in microbial fuel cells (MFCs) utilizing fermentable and non-fermentable substrates. *Water Res.* **42**(6–7), 1501–1510.

Liu, H., Cheng, S.A. and Logan, B.E. (2005) Production of electricity from acetate or butyrate using a single-chamber microbial fuel cell. *Environ. Sci. Technol.* **39**(2), 658–662.

Marsili, E., Baron, D.B., Shikhare, I.D., Coursolle, D., Gralnick, J.A. and Bond, D.R. (2008) Shewanella Secretes flavins that mediate extracellular electron transfer. *Proc. Natl. Acad. Sci. USA* **105**(10), 3968–3973.

Nießen, J., Schröder, U., Rosenbaum, M. and Scholz, F. (2004a) Fluorinated polyanilines as superior materials for electrocatalytic anodes in bacterial batteries. *Electrochem. Commun.* **6**, 571–575.

Nießen, J., Schröder, U. and Scholz, F. (2004b) Exploiting complex carbohydrates for microbial electricity generation – a bacterial fuel cell operating on starch. *Electrochem. Commun.* **6**, 955–958.

Nießen, J., Schröder, U., Harnisch, F. and Scholz, F. (2005) Gaining electricity from *in situ* oxidation of hydrogen produced by fermentative cellulose degradation. *Lett. Appl. Microbiol.* **41**, 286–290.

Nießen, J., Harnisch, F., Rosenbaum, M., Schröder, U. and Scholz, F. (2006) Heat treated soil as convenient and versatile source of bacterial communities for microbial electricity generation. *Electrochem. Commun.* **8**, 869–873.

Park, D.H. and Zeikus, J.G. (2003) Improved fuel cell and electrode designs for producing electricity from microbial degradation. *Biotechnol. Bioengng.* **81**(3), 348–355.

Rabaey, K., Boon, N., Hofte, M. and Verstraete, W. (2005) Microbial phenazine production enhances electron transfer in biofuel cells. *Environ. Sci. Technol.* **39**(9), 3401–3408.

Rosenbaum, M., Schröder, U. and Scholz, F. (2005a) *In situ* Electrooxidation of photobiological hydrogen in a photobiological fuel cell based on *Rhodobacter sphaeroides*. *Environ. Sci. Technol.* **39**(16), 6328–6333.

Rosenbaum, M., Schröder, U. and Scholz, F. (2005b) Utilizing the green alga *Chlamydomonas reinhardtii* for microbial electricity generation: a living solar cell. *Appl. Microbiol. Biot.* **68**, 753–756.

Rosenbaum, M., Zhao, F., Schröder, U. and Scholz, F. (2006) Interfacing electrocatalysis and biocatalysis using tungsten carbide: a high performance noble-metal-free microbial fuel cell. *Angew. Chem. Int. Edn* **45**, 6658–6661.

Rosenbaum, M., Zhao, F., Quaas, M., Wulff, H., Schröder, U. and Scholz, F. (2007) Evaluation of catalytic properties of tungsten carbide for the anode of microbial fuel cells. *Appl. Catal. B: Environ.* **74**, 262–270.

Rozendal, R.A., Hamelers, H.V.M., Rabaey, K., Keller, J. and Buisman, C.J.N. (2008) Towards practical implementation of bioelectrochemical wastewater treatment. *Trends Biotechnol.* **26**(8), 450–459.

Schröder, U., Nießen, J. and Scholz, F. (2003) A generation of microbial fuel cell with current outputs boosted by more than one order of magnitude. *Angew. Chem. Int. Ed.* **115**, 2986–2989.

Suzuki, S., Karube, I. and Matsunaga, T. (1980) Biochemical energy conversion using immobilized whole cells of Clostridium butyricum. *Biochem.* **62**(5–6), 353–358.

Wang, Y.F., Masuda, M., Tsujimura, S. and Kano, K. (2008) Electrochemical regulation of the end-product profile in Propionibacterium freudenreichii ET-3 with an endogenous mediator. *Biotechnol. Bioengng.* **101**(3), 579–586.

White, D. (2007) *The Physiology and Biochemistry of Prokaryotes*. Oxford University Press.

Yamazaki, S.-i., Kano, K., Ikeda, T., Isawa, K. and Kaneko, T. (1999) Role of 2-amino-3-carboxy-1,4-naphthoquinone, a strong growth stimulator for bifidobacteria, as an electron transfer mediator for NAD(P)+ regeneration in Bifidobacterium longum. *Biochim. Biophys. Acta (BBA) – General Subjects* **1428**(2–3), 241–250.

Yamazaki, S., Kaneko, T., Taketomo, N., Kano, K. and Ikeda, T. (2002) Glucose metabolism of lactic acid bacteria changed by quinone-mediated extracellular electron transfer. *Biosci., Biotechnol., Biochem.* **66**(10), 2100–2106.

12

Conversion of sulfur species in bioelectrochemical systems

Paritam K. Dutta, Jürg Keller, Zhiguo Yuan, René A. Rozendal and Korneel Rabaey

12.1 INTRODUCTION

Sulfur is one of the most abundant elements on earth and an essential element for life. Sulfur can prevail in many oxidation states and it exists in a great number of different chemical forms within its oxidation states. Most of them play a very important role in aqueous systems, where redox reactions occur either microbially or chemically following the thermodynamics of the system. Sulfur is highly redox sensitive and the many possible oxidation/reduction reactions make the sulfur chemistry quite complex. Sulfur species are also constituents of a large number of industrial products (for instance, sulfuric acid, H_2SO_4 and sulfite, SO_3^{2-}), and SO_2 and sulfides represent major environmental pollutants (Levillain *et al.*, 2007).

Additionally, hydrogen sulfide (H_2S) is a important human health hazard that causes deaths through asphyxiation. Therefore, a number of remediation strategies for polluting or dangerous sulfur species have been developed and implemented in the past decades. These strategies involve physicochemical processes (i.e. air stripping, chemical oxidation and precipitation) and biotechnological processes. Recently, the conversion of sulfur species in bioelectrochemical systems received considerable attention, mainly for the removal of sulfide from wastewater and the impact of sulfur species on the performance of BESs.

In this chapter, the conversion of sulfur species in abiotic and biotic electrochemical cells is described. First, the chemical and electrochemical properties of elemental sulfur, sulfide, polysulfides and sulfate are briefly described. Second, existing sulfide and sulfate removal technologies are outlined. Third, sulfide oxidation in abiotic and biotic electrochemical cells and the interactions of oxidation products both in abiotic and biotic media are overviewed.

12.2 PROPERTIES OF SULFUR SPECIES

12.2.1 Elemental sulfur

At ambient pressure and temperature, elemental sulfur (S^0) exists as rings of different sizes (S_n) or as polymeric chains of high molecular mass (S_μ). Sulfur has many allotropes. Out of these, the greenish-yellow coloured orthorhombic α-S_8 is thermodynamically most stable and is the main constituent of commercial sulfur (Steudel 2000). It contains crown shaped S_8 molecules which are stacked in a complex array. Elemental sulfur is weakly soluble in water, and this low solubility does not allow notable reduction in aqueous media. In contrast to chemically produced elemental sulfur, biologically produced elemental sulfur particles have typically more hydrophilic properties (Janssen *et al.* 1999).

12.2.2 Sulfide and polysulfides

Hydrogen sulfide (H_2S) is ubiquitously present in certain types of natural gas, waste gases and wastewaters. Sulfide is toxic, corrosive and odorous. Sulfide in aqueous solution is a weak acid and dissociates as follows (Steudel 2000):

$$H_2S(aq) \rightleftharpoons HS^- + H^+ \qquad pK_1 = 7(20^\circ C) \tag{12.1}$$

$$HS^- \rightleftharpoons S^{2-} + H^+ \qquad pK_2 = 17(20^\circ C) \tag{12.2}$$

The extremely high value of pK_2 rules out the formation of S^{2-} in almost all solutions except at pH values near 14, which is the previously determined value for pK_2 according to the literature (Steudel 2000). At neutral pH, only the first dissociation of H_2S is of importance. H_2S is well soluble in water, ranging between 3,000 and 4,000 mg l^{-1} at atmospheric pressure and normal temperature. The H_2S solubility in water decreases with increasing temperatures (Bowker *et al.* 1989).

Polysulfide anions (S_n^{2-}) are unbranched chains of sulfur atoms. Each polysulfide anion (S_n^{2-}) can be represented by the formula $[S_{n-1}^{(0)}S^{2-}]^{2-}$which implies that anions contain ($n-1$) zerovalent sulfur atoms (contribute to $S_{dissolved}$) and one sulfide atom of divalent state. (Levillain *et al.* 2007; Kamyshny Jr *et al.* 2004). Under acidic conditions polysulfides are unstable. Polysulfide anions can therefore only exist to a significant level in water at pH values above 6 (Steudel 2000). The chain length of the polysulfide anion strongly depends on the pH value, temperature, and on the total sulfur concentration at a certain temperature.

12.2.3 Sulfate and other oxyanions

Sulfate (SO_4^{2-}) is one of the most common forms of sulfur species. It is generally present as a dissolved compound in marine and freshwater environments or as an insoluble salt in minerals, e.g. gypsum (Lens *et al.* 1998). The sulfate anion is chemically rather inert, stable in aqueous solution at all pH values both in the presence and absence of oxidizing agents. It needs to be activated for biological reduction at ambient temperature. Many industrial wastewaters and even domestic sewage (particularly when seawater infiltration is occurring) can contain high sulfate concentrations. Besides sulfate (SO_4^{2-}), thiosulfate ($S_2O_3^{2-}$), sulfite (SO_3^{2-}), dithionate ($S_2O_4^{2-}$) represent other commonly encountered sulfur oxyanions in aqueous solution.

12.2.4 Relationship of electrochemical potential and pH for sulfur species in aqueous systems

The thermodynamic stability area of different sulfur species in the aqueous system can be visualized in an E_h–pH diagram (known as a Pourbiax Diagram). Stability areas are represented as a function of pH and electrochemical potential E_h (potential versus Standard Hydrogen Electrode (SHE)). The E_h–pH diagram enables us to see at a glance the stability region of aqueous sulfur species and solid compounds at particular E_h and pH values in relation to temperature and the overall sulfur concentration at equilibrium conditions. It has to be noted though that this does not provide any information regarding the reaction kinetics, which may control the transitions between different states, even though the thermodynamics would favour such reactions.

The Nernst equation is used to draw the E_h–pH diagram for a chemical system containing one or a combination of elements in an aqueous solution for a particular temperature and a known concentration. Figure 12.1 shows the E_h–pH diagram of the S–H_2O system at 25°C. Table 12.1 provides details on important half cell reactions of sulfur species in aqueous media with their theoretical potential values. In Figure 12.1, the relative stability areas of sulfide (H_2S, HS^- and S^{2-}), elemental sulfur (S^0) and sulfate (SO_4^{2-}) are shown. The stability domain of elemental sulfur is triangular, very narrow and falls within the stability domain of water as shown in Figure 12.1. Elemental sulfur is considered to be stable in the presence of water and acid solutions free of oxidizing agents. At alkaline and lower potential solutions, S^0 disproportionates to H_2S, HS^-, S_n^{2-} and SO_4^{2-} and other oxidation products but in practice these reactions are slow and take place only in hot and very alkaline media (Pourbaix 1966).

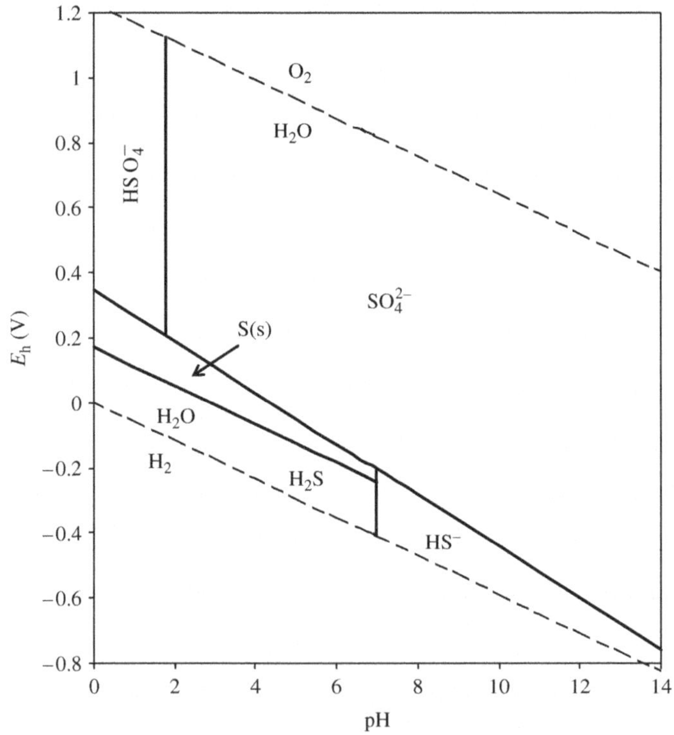

Figure 12.1. Potential (E_h) – pH diagram for the S-H_2O system at 25°C, pH_2 and pO_2 1 bar, $\Sigma S = 0.1M$ (adapted from Pourbaix 1966)

However, Ateya *et al.* (2003) reports that the stability of elemental sulfur is controlled by reaction kinetics rather than thermodynamic considerations because sulfur is found to be formed at potentials where the Pourbaix diagram predicts it to be unstable. Thiosulfate ($S_2O_3^{2-}$), sulfite (SO_3^{2-}), dithionate ($S_2O_4^{2-}$), tetrathionate ($S_4O_6^{2-}$), polysulfides (S_n^{2-}) and other important sulfoxy species are considered to form so-called 'false' equilibria in aqueous solution and were not shown in the Pourbaix diagram, Figure 12.1 (Pourbaix 1966). Different useful metastable E_h–pH diagrams showing metastable sulfoxy species (species are in equilibrium, but are susceptible to change to other forms with only slight interaction) and polysulfides for S–H_2O system are available in the literature (Hamilton and Woods 1983; Kelsall and Thompson 1993).

Table 12.1. Standard electrode potentials (E_0) and theoretical potentials (E_0') at pH 7 calculated according to the Nernst equation for important sulfur half reactions in aqueous media (concentrations 1 mol/l) (Pourbaix 1966; Zhdanov 1975)

Redox Reaction	E_0 (V)	E_0' (V) [pH 7]
$S(s) + 2e^- + 2H^+ \rightarrow H_2S$ (aq)	0.142	−0.272
$S(s) + 2e^- + H^+ \rightarrow HS^-$	−0.065	−0.271
$S(s) + 2e^- \rightarrow S^{2-}$	−0.476	−0.476
$S_2O_3^{2-} + 8e^- + 8H^+ \rightarrow 2HS^- + 3H_2O$	0.200	−0.213
$SO_4^{2-} + 8e^- + 9H^+ \rightarrow HS^- + 4H_2O$	0.252	−0.213
$SO_4^{2-} + 8e^- + 10H^+ \rightarrow H_2S(aq) + 4H_2O$	0.303	−0.214
$SO_4^{2-} + 6e^- + 8H^+ \rightarrow S(s) + 4H_2O$	0.357	−0.194
$S_5^{2-} + 8e^- + 5H^+ \rightarrow 5HS^-$	0.003	−0.255
$5S(s) + 2e^- \rightarrow S_5^{2-}$	−0.319	−0.319

12.3 EXISTING SULFIDE AND SULFATE REMOVAL TECHNOLOGIES

12.3.1 Sulfide removal technologies

Sulfide is present in a plethora of wastewaters, including domestic sewage and in a considerable number of gas streams, such as biogas, natural gas, tail gas, syngas, and off-gases from wastewater treatment plants. It has to be removed because of its toxicity, corrosive properties and unpleasant odour. A number of physicochemical and biological removal processes are commonplace for aqueous and gaseous sulfide. Recently, Zhang *et al.* (2008) reviewed chemical and biological technologies for hydrogen sulfide emission control in sewer systems.

12.3.1.1 Physicochemical processes

Common physicochemical processes involve air stripping, chemical oxidation and precipitation. The removal process is strongly dependent on the source of the sulfide. The kinetics of air/oxygen oxidation of aqueous sulfide and air stripping for gaseous sulfide oxidation are well described and documented in the literature (Bowker *et al.* 1989; Chen and Morris 1972; Kuhn *et al.* 1983). Various physicochemical processes for the removal of sulfide from gases such as natural gas and the desulfurization of petroleum stocks have been extensively reviewed (Eow 2002). Several chemical oxidants such as hydrogen peroxide, chlorine, potassium permanganate and ozone have also been used for sulfide control in wastewaters (Bowker *et al.* 1989; Millero *et al.* 1989). Iron salts have been used in many cases for sulfide control by precipitating out sulfide as insoluble iron-sulfide (Padival *et al.* 1995). For gas streams, the Claus process can be used. It uses oxygen for recovering elemental sulfur from H_2S containing gases, particularly natural gas.

12.3.1.2 Biological technologies

Microbial sulfide oxidation is a key process in the biogeochemical sulfur cycle. Sulfur conversions are intimately linked with the carbon, nitrogen and oxygen cycle. Biotechnological or microbiological sulfide removal processes operate at ambient temperatures and atmospheric pressures, which cuts the energy costs and allows high removal efficiencies even at low sulfide concentrations. Biological sulfide oxidation rates in cell suspensions were up to 75 times faster than the chemical non-catalyzed oxidation rates at concentrations of 10 mg l^{-1} and 7 times faster at sulfide concentrations of 100 mg l^{-1} (Buisman *et al.* 1990). The organisms responsible for this oxidation, phototrophic green and purple sulfur bacteria and aerobic colourless sulfur bacteria are well described for their ability to oxidize H_2S. Basu *et al.* (1996) classified the bacteria capable of metabolizing H_2S into three major groups: (i) aerobic sulfate producers; (ii) anaerobic sulfate producers (nitrate as electron acceptor); and (iii) anaerobic elemental sulfur producers (photosynthetic bacteria).

Phototrophic, heterotrophic, chemotrophic, and autotrophic microorganisms have been utilized for the treatment of H_2S-containing gases and wastewaters based on the inherent difference in nutritional requirements and different process schemes. The biological processes for removal of hydrogen sulfide from gas streams were recently reviewed by Syed *et al.* (2006). The authors particularly highlighted the use of phototrophic fixed film reactors as an interesting concept for cost-effective anaerobic H_2S removal from biogas, due to the ability to operate the system over extended time periods without requiring a biomass separation step and the ability to operate at high and variable loadings. However, Hurse and Keller (2004) reported the necessity to periodically harvest biofilm material to recover elemental sulfur that accumulates in phototrophic fixed film reactors. On the other hand, light supply and distribution are considered to be major constraints for anaerobic methods involving phototrophic bacteria.

12.3.2 Sulfate removal technologies

Likewise several chemical, physicochemical and biological processes are currently applied for sulfate removal from wastewater. They include energy intensive processes such as reverse osmosis, electrodialysis, and nanofiltration. Sulfate can be removed by precipitating out as barium sulfate ($BaSO_4$) and calcium sulfate ($CaSO_4$) (Lens *et al.* 1998). For high strength wastewaters, biological sulfate removal is a cost-effective alternative. Biological sulfate removal generally consists of two steps, first sulfate is reduced to sulfide, which is subsequently oxidized biologically to elemental sulfur (S^0) or removed as metal sulfide (Lens *et al.* 1998).

12.3.3 Evaluation of existing technologies

The high cost (ranging from € 1.9 to 7.2 kg^{-1} S removed) is the main problem of the existing sulfide removal technologies in sewage systems (Zhang *et al.* 2008). Biotechnological or microbiological sulfide removal processes, which are considered as environmentally benign processes also involve high aeration costs and need stringent control to avoid stripping of sulfides if air is used as oxidant (Syed *et al.* 2006). Furthermore, the commonly used oxidants, oxygen, and nitrate also oxidize organic compounds, which need to be preserved for nutrient removal in a downstream wastewater treatment plant. While several sulfide removal technologies are clearly available, there is still a need for more cost-effective and efficient methods to control sulfide in sewage systems (Zhang *et al.* 2008). Sulfide removal using an electrochemical or bioelectrochemical system might have the potential to overcome many of the shortcomings of the existing processes, if successfully developed. In addition, where abiotic electrochemical systems can enable selective aqueous sulfide removal, a biotic electrochemical system can remove sulfide and organics simultaneously without the expense of energy and with the scope of harvesting solid sulfur in both cases.

12.4 ABIOTIC ELECTROCHEMICAL REMOVAL OF AQUEOUS SULFIDE

12.4.1 Introduction

Electrochemical remediation is often attractive due to the low cost of electricity compared to chemical reagents. Moreover, there is no risk of residual pollution of unreacted reagents from electrochemical processes. A considerable number of electrochemical sulfide oxidation studies involved a wide range of sulfide sources such as H_2S gas, sulfide in aqueous and non-aqueous solvents, such as DMSO (dimethyl sulfoxide), DMF (dimethyl formamide), CS_2 and methanol (Levillain *et al.* 2007). Aspects, such as the type of electrode materials, the geometry, different reactor configurations, altered potential, and current densities were investigated. However, the electrochemical behaviour of sulfur, sulfide, and polysulfide ions in aqueous solutions is not well documented compared to non-aqueous solvents (Levillain *et al.* 2007).

Most of the studies in aqueous solutions were done in alkaline media by electrolysis in single chamber reactors, over a wide temperature range. Table 12.2 summarizes a number of electrochemical aqueous sulfide oxidation studies via electrolysis. Electrochemical oxidation of pyrite (FeS_2), the most common sulfide mineral, is reported in several studies (Hamilton and Woods 1981; Holmes and Crundwell 2000; Kelsall *et al.* 1999; Zhu *et al.* 1994). The electrochemical oxidation of hydrogen sulfide is attractive as it allows us to extract two valuable products: hydrogen and sulfur (Anani *et al.* 1990; Kameyama *et al.* 1981; Zaman and Chakma 1995).

The studies discussed above and tabulated in Table 12.2 did not examine spontaneous electrochemical sulfide oxidation coupled to electricity production. Chuang *et al.* (2000) reported a proton conducting solid-state H_2S–O_2 electrochemical cell to oxidize H_2S gas in both fuel cell and electrolysis modes (120–145°C and 235–510 kPa). The authors used a two-chambered fuel cell system (Nafion membrane) using Pd, Pt and MoS_2 catalysts with carbon as anode materials. They indicated that long-term operation of H_2S fuel cells is feasible at temperatures above the melting point of sulfur. Different research groups working on seafloor or marine sediment fuel cells reported that a significant portion of the anodic current of these fuel cells resulted from the electrochemical oxidation of dissolved and mineral sulfide (DeLong and Chandler 2002; Reimers *et al.* 2006; Ryckelynck *et al.* 2005; Tender *et al.* 2002).

Table 12.2. Summary of studies on electrochemical sulfide oxidation in electrolysis mode

Cell configuration	Anode material	Source of sulfide	Applied anode potential E_{AN} (SHE) or current	References
Pyrex beaker & glass cell	Boron doped diamond	Synthetic	0.1–0.8 A	(Waterston *et al.* 2007)
Single chamber	Porous carbon felt	Polluted brine	−0.95 to 2.5 V	(Ateya *et al.* 2005)
Single chamber (rotating anode)	Graphite disk	Polluted brine	−0.8 to 1.4 V	(Ateya *et al.* 2003)
Single chamber (parallel plate)	Platinized Ti and Ti–MnO_2	Tannery waste	Current density 20 to 200 A/m^2	(Rajalo and Petrovskaya 1996)
Two chamber (separated by Nafion membrane)	Graphite, Ni, porous Ni–Cr alloy, Ti	Alkaline aqueous solution	Current density 550–1050 A/m^2	(Anani *et al.* 1990)
Single chamber (rotating and stationary anode)	Pt	Synthetic (pH 9.0)	−0.9 to 0.4 V	(Szynkarczuk *et al.* 1994)
Continuous stirred tank reactor[1]	Pt	Synthetic alkaline sulfide	1.8 to 4.8 V	(Shih and Lee 1986)
Rotating electrode assembly	Au	Synthetic (pH 6.8 and 9.2)	−0.9 to 0.25 V	(Hamilton and Woods 1983)
Single chamber (rotating tripolar cell with wiper blade)	Pt	Synthetic	0.04 to 1.5 V	(Farooque and Fahidy 1977)

[1]Toluene to dissolve product sulfur.

12.4.2 Spontaneous sulfide oxidation and electricity generation

Sulfide is an electrochemically active compound and can be spontaneously oxidized at an anode without any catalyst if the anode potential is sufficiently high to overcome the overpotential for the oxidation. In 1991, Habermann and Pommer developed a process in which bacteria were used to reduce sulfate to sulfide, using organic electron donors. The sulfide was subsequently oxidized in a fuel cell type reactor, allowing indirect generation of electrical current from organics oxidation. Recently, Dutta *et al.* (2008) described aqueous sulfide removal by spontaneous electrochemical oxidation with concomitant electricity generation at ambient temperature, pressure and neutral pH using low cost graphite granule electrodes and a two-chambered fuel cell separated by a membrane. Figure 12.2 illustrates spontaneous sulfide oxidation with electricity generation; a much higher sulfide oxidation rate and current was found at 10 Ω in comparison to 50 Ω in a lab-scale batch experiment. During continuous operation, on average, the power generated was 5 (±) 1 W m^{-3} of total anodic compartment (TAC), with a maximum of 78 W m^{-3} TAC using a ferricyanide cathode. Reimers *et al.* (2006) attributed (part of the) current generation from a benthic MFC to abiotic sulfide oxidation by uncoupling biological factors. Abiotic electron transfer to seafloor fuel cells is the result of the oxidation of dissolved and solid phase forms of reduced sulfur as reported by Ryckelynck *et al.* (2005). Tender *et al.* (2002) attributed about 40% of the current density to oxidation of sediment sulfide in marine sediment MFCs. Sulfur has many oxidation states and their calculated theoretical potentials under particular conditions are closely matched. Therefore, a number of aqueous sulfur species may have evolved spontaneously in parallel or consecutive redox reactions. But the reaction $S(s) + 2e^- + 2H^+/H^+ \rightarrow H_2S/HS^-$ is shown to be primarily responsible for aqueous sulfide oxidation in the system described by Dutta *et al.* (2008).

12.4.3 Final product of sulfide oxidation

Elemental sulfur (S^0), polysulfides (S_n^{2-}), sulfate (SO_4^{2-}), sulfite (SO_3^{2-}) dithionate ($S_2O_4^{2-}$) and thiosulfate $S_2O_3^{2-}$ may be produced during oxidation of sulfide depending on the experimental conditions. Elemental sulfur was reported to be the only final oxidation product in a number of studies (Ateya *et al.* 2003; Szynkarczuk *et al.* 1994). Based on a mass and charge balance, approximately 95% of the sulfide (by mass of sulfur) could be recovered as elemental sulfur during spontaneous sulfide oxidation and current generation (Dutta *et al.* 2008). The remaining 5% of the sulfide is further oxidized to dissolved sulfur oxyanions, mainly thiosulfate, which is probably due to auto-oxidation (oxidation by O_2) of sulfide. Solid elemental sulfur was the only final oxidation product in wide potential range for various anode materials.

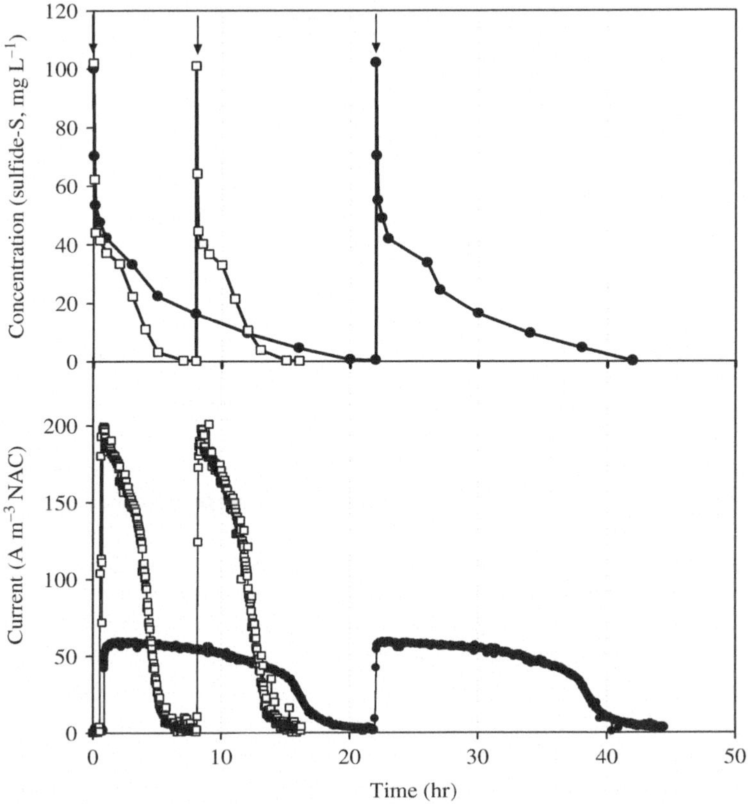

Figure 12.2. Profiles of aqueous sulfide concentration and corresponding current in A m^{-3} of NAC (net anodic compartment) at different resistances in batch mode (□) 10 Ω, (•) 50 Ω, (pH 6.75), arrows indicate sulfide spiking. The cell voltage in the start up phase was 0.59 V (anode potential −0.250 V, SHE) for 50 Ω and 0.4 V (anode potential −0.082 V, SHE) for 10 Ω (Dutta *et al.* 2008). (Published with the permission of Elsevier)

There appears to be negligible transfer of H_2S gas from the anode to the cathode through the membrane, in case a two chamber cell was used. Chuang *et al.* (2000) ruled out possible cross over of H_2S gas through Nafion membranes as used in their study. However, this H_2S crossover may depend on the concentration and the fuel cell geometry, as Zhao *et al.* (2008) reported.

Polysulfides have been identified as an intermediate when sulfide oxidation experiments were performed at alkaline media (Szynkarczuk *et al.* 1994). Polysulfide could have been formed either by chemical dissolution (Eq. 12.3) of elemental sulfur by sulfide or by the two-step reactions (Eqs. 12.4–12.5) proposed for alkaline media (Szynkarczuk *et al.* 1994):

$$HS^- + \frac{n-1}{8}S_8 \leftrightarrow S_n^{2-} + H^+ \qquad (12.3)$$

$$S_n^{2-} + nH_2O + 2(n-1)e^- \rightarrow nHS^- + nOH^- \qquad (12.4)$$

$$nS^0 + 2e^- \rightarrow S_n^{2-} \qquad (12.5)$$

The concentration of polysulfide depends on the anode potential, pH, sulfide concentration, and the amount of elemental sulfur present on the electrode surface. The polysulfide formed eventually oxidizes spontaneously. As indicated earlier, polysulfide anions can exist in water only at relatively high pH levels (>6) (Kleinjan *et al.* 2005; Steudel 2000).

When boron doped diamond (BDD) anodes were used for sulfide oxidation, sulfate was reported as the only final oxidation product (Waterston *et al.* 2007). For gold electrodes, oxidation of sulfide to sulfate was reported in parallel with sulfur deposition (Hamilton and Woods 1983). Sulfide conversion to sulfate may appear suitable for practical applications, as sulfide discharge limits are more stringent than those for sulfate, but sulfide to sulfate conversion needs higher currents as this conversion requires eight electrons, whereas sulfide to sulfur requires only two. This sulfide oxidation to sulfate can also be achieved by aerating water. However, forming sulfate creates a risk of downstream reduction to sulfide by microorganisms, while an effective removal of sulfide as elemental sulfur alleviates this risk.

12.4.4 Properties of electrodeposited sulfur

There is notable contradiction and speculation in the literature on the properties of electrodeposited sulfur. The sole agreement is that sulfur deposit passivates the electrode surface. Elemental sulfur has been considered as surface poison in several studies because of this passivation (Farooque and Fahidy 1977).

Passivation or deactivation of the anode surface and finally termination of the reaction due to the sulfur deposition was also reported in several studies (Ateya *et al.* 2003; Reimers *et al.* 2006). Dutta *et al.* (2008) observed that the percentage sulfide removal decreased from an initial 80% to 10% due to sulfur passivation after 3 months of operation in a continuous flow two-chambered electrochemical reactor. The capability of activated carbon cloth anodes for sulfide oxidation was also reported to decrease over time due to elemental sulfur deposition (Zhao *et al.* 2008). On the other hand, Turcotte *et al.* (1993) stated that electrodeposited sulfur does not behave as an electron insulator as much as bulk elemental sulfur does in its orthorhombic state. They reported sulfur film conducting properties more similar to semiconductors. Hamilton and Woods (1983) reported sub-monolayer coverage of the anode electrode by sulfur at under potentials and a multilayered sulfur coverage at more positive potentials for gold electrodes. Chemical images of the distribution of elemental sulfur on oxidized pyrite and arsenopyrite surfaces by near Raman imaging microscopy revealed that sulfur did not form a continuous film, but deposits as discrete patches, dispersed randomly across the sample surface (McGuire *et al.* 2001). At present it is not clear why elemental sulfur is found in these specific, localized patches at the mineral surface. Ryckelynck *et al.* (2005) also reported irregular distribution of electrochemically formed elemental sulfur on the anode of a sediment microbial fuel cell. Crystalline sulfur on a carbon anode (Figure 12.3A) has been observed in scanning electron microscopic (SEM) images. An elemental sulfur layer originating from pyrite oxidation was an active form of sulfur (S^0_{act}), which implies that such S^0 could be further oxidized or reduced (Zhu *et al.* 1994). Electrochemically converted elemental sulfur might not be S_8, which has limited electrochemical reactivity (Szynkarczuk *et al.* 1994). Farooque and Fahidy (1977) reported that anode passivation and related problems have hindered the development of electrochemical sulfide removal. Mao *et al.* (1991) proposed an electrochemical process which delayed the electrode passivation by sulfur dissolution. However, Tender *et al.* (2002) reported that the long-term sustained power generation in marine sediments was limited by the sulfide and organics influx rather than anode fouling. At potentials below the open circuit potential (below the equilibrium potential of the sulfide/S^0 couple), elemental sulfur has been reported to reduce back to sulfide and polysulfide at graphite granules (Dutta *et al.* 2008). The reactivity of electrodeposited sulfur might help to regenerate the sulfur loaded electrode. Moreover, concentrated sulfide solutions could be obtained from low concentration sulfide wastewaters by trapping sulfide as sulfur and then reducing the deposited sulfur back to sulfide.

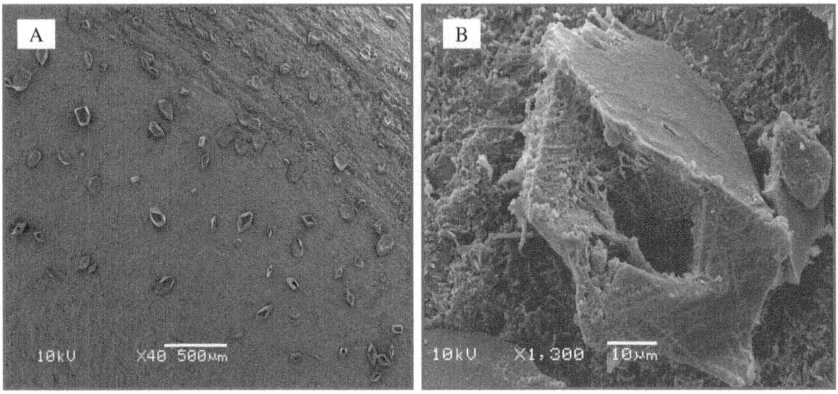

Figure 12.3. SEM images of (A) sulfur crystals on a carbon anode (×40); (B) unspecified structures grown on a sulfur crystal of a BES (×1,300).

12.5 REMOVAL OF AQUEOUS SULFIDE IN BES

12.5.1 Introduction

The conversion of inorganic sulfur compounds in a bioelectrochemical system (BES) may considerably influence its operation. Sulfate can function as an alternative electron acceptor for the oxidation of organic carbon compounds in BESs. Indeed, sulfate reduction to sulfide has been demonstrated in a glucose fed MFC (Rabaey *et al.* 2005).

In anaerobic conditions, sulfate reduction to sulfide by sulfate reducing bacteria (SRB) using organics as electron donor is a very common phenomenon. Habermann and Pommer (1991) reported that this sulfide could be oxidized or stored as fuel in the anode. Here, sulfide forms an intermediate for the oxidation of organic carbon compounds. When sulfate reduction and sulfide oxidation occur simultaneously at the anode, simultaneous removal of organics, sulfide and sulfate from wastewaters can be achieved in MFCs or BESs (Rabaey *et al.* 2006). However, sulfate reduction can also occur externally in a separate vessel, while the produced sulfide could be oxidized in an abiotic fuel cell (Zhao *et al.* 2008). In addition to sulfate, some SRB could also reduce thiosulfate and sulfite to sulfide (Bruser *et al.* 2000; Zhao *et al.* 2009).

Bacteria growing in the anode compartment of a BES appear only able to reduce sulfate with fermentable electron donors. Sulfate reduction was not observed when acetate was supplied, during both open and closed circuit conditions (Dutta *et al.* 2009). SRB, which can use only acetate as substrate, might not develop in an acetate-fed BES environment. The reason is not yet clear, as many SRB are known for their ability to use acetate.

12.5.2 Sulfide oxidation in a biotic cell

Rabaey *et al.* (2006) reported sulfide removal in the presence of microorganisms and residual organics in an MFC. Bacterial isolates from this MFC were identified as either *Alcaligenes* sp. or *Paracoccus* sp.; the latter was capable of oxidizing sulfide using nitrate as electron acceptor. However, biological factors were not uncoupled from abiotic spontaneous electrochemical sulfide oxidation in this study. The study predicted that sulfide oxidizing bacteria could donate electrons directly to the electrode, but could not confirm it. The processes described by Rabaey *et al.* (2006) were likely mainly abiotic as aqueous sulfide oxidizes spontaneously at the anode. Tender *et al.* (2002) reported power generation from marine sediments via at least two anode reactions: (i) oxidation of sediment sulfide – a by-product of microbial heterotrophy; and (ii) microbially catalyzed organics oxidation. They also predicted that sulfur oxidizing bacteria might also oxidize sulfur to sulfate with the anode electrode serving as electron acceptor in marine sediments. As a result of this sulfur oxidation, no fouling of the anode was observed. However, in another study the presence of biofilms did not impact on the sulfide oxidation products and rates in comparison to abiotic systems (Dutta *et al.* 2009). The fact that sulfide removal rates do not change due to the presence of a biofilm does not necessarily imply that biological sulfide oxidation does not happen. In an MFC system, the abiotic and biotic processes might proceed at the same time, but the combined sulfide oxidation rate would remain unaltered in comparison to abiotic oxidation if the reaction rate is limited by the electrode or the fuel cell operation.

Elemental sulfur deposited on an anode can be quite readily reduced again to sulfide if a favourable potential is maintained. The formed elemental sulfur can be a preferred electron acceptor over the anode for heterotrophic bacteria. The formed sulfide would subsequently be oxidized again at the anode. Hence, the deposited elemental sulfur and sulfide would form mediators, or electron shuttles, for carbon oxidation. Bacterial mediated elemental sulfur reduction has been reviewed in the past (Schauder and Kröger 1993). For instance, dissimilatory sulfur-reducers from the genus *Desulfuromonas* are known to grow anaerobically by coupling the oxidation of acetate, ethanol and several other donors to the reduction of elemental sulfur to sulfide, while being unable to reduce sulfate (Madigan *et al.* 2003).

A recent study reported sulfide generation during acetate oxidation in a BES in the presence of electrodeposited sulfur on the anode (Dutta *et al.* 2009).

Microorganisms used electrodeposited sulfur as preferable electron acceptor over the solid anode and soluble sulfate, and produced sulfide irrespective of the electrochemical conditions. Bacteria might use sulfur as electron acceptor/sink while a potentially biological structure was observed on top of a sulfur crystal SEM image (Figure 12.3B). The direct usage of electrodeposited sulfur as an electron acceptor by the bacteria indicated a metabolism similar to *Desulfuromonas acetoxidans.* The latter is an abundant organism in the seafloor and at benthic MFC anodes (Bond *et al.* 2002; Reimers *et al.* 2006; Tender *et al.* 2002). A model was proposed for acetate oxidation at an anode in the presence of electrodeposited sulfur (Figure 12.4). More research is needed to identify whether bacteria reduce electrochemically active deposited sulfur either in the form of chemisorbed sulfur, elemental sulfur (amorphous/crystalline), or both, or in the form of soluble polysulfide, known as electron acceptor for a hyperthermophilic archaea (Blumentals *et al.* 1990) .

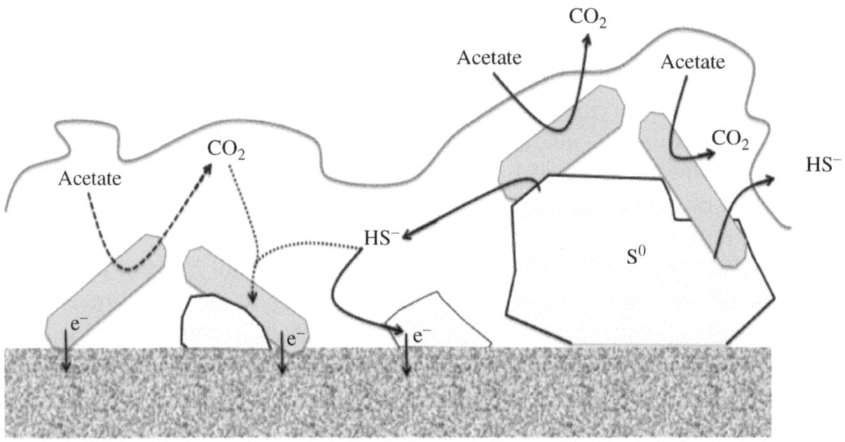

Figure 12.4. Acetate oxidation pathways in a BES in the presence of electrodeposited sulfur at an anode. Bacteria can oxidize acetate to CO_2 by donating electrons directly to the electrode or to elemental sulfur. In the latter case, the produced sulfide migrates to the bulk solution or oxidizes again to elemental sulfur depending on the anode potential. Sulfide oxidation occurs either abiotically or biotically or via a combination of both. (The model diagram is adapted from Dutta *et al.* 2009)

12.6 OUTLOOK

Sulfide removal from sewage systems could be one of the first applications of the abovementioned electrochemical process as the process offers particular advantages. The spontaneity of the abiotic sulfide oxidation reaction remains unaffected by the presence of organic carbon compounds (acetate), while acetate or other organics are not oxidized in the absence of bacteria. These features make the process suitable for selective sulfide removal from sewage without oxidizing organics. It appears unlikely that electricity generation by abiotic sulfide oxidation could become a significant goal, considering the often low sulfide concentrations in wastewater. But, at least spontaneous abiotic electrochemical sulfide oxidation could be an energy efficient sulfide removal process from wastewaters.

Polishing of anaerobic digester effluents could be another application of BESs for sulfide removal. In such a case biofilm formation on the anode would help to remove residual organics concomitantly with sulfide. BESs might also have potential applications for wastewaters having dissolved sulfate and fermentable substrates such as wastewaters from food processing industries (Lens *et al.* 1998). Dissolved sulfate can be removed via the sulfide pathway with the generation of electricity. Sulfur conversion and subsequent power generation in marine sediment based BESs is promising to power marine oceanographic instruments such as integrated conductivity, temperature and depth sensors, hydrophones, etc. (Tender *et al.* 2002). However, an appropriate anode potential and loading rate have to be maintained for efficient sulfur conversions and organics removal in continuous BESs.

REFERENCES

Anani, A.A., Mao, Z., White, R.E., Srinivasan, S. and Appleby, A.J. (1990) Electrochemical production of hydrogen and sulfur by low-temperature decomposition of hydrogen sulfide in an aqueous alkaline solution. *J. Electrochem. Soc.* **137**(9), 2703–2709.

Ateya, B.G., AlKharafi, F.M. and Al-Azab, A.S. (2003) Electrodeposition of sulfur from sulfide contaminated brines. *Electrochem. Solid-State Lett.* **6**(9), C137–C140.

Ateya, B.G., AlKharafi, F.M., Abdullah, R.M. and Al-Azab, A.S. (2005) Electrochemical removal of hydrogen sulfide from polluted brines using porous flow through electrodes. *J. Appl. Electrochem.* **35**, 297–303.

Basu, R., Clausen, E.C. and Gaddy, J.L. (1996) Bilogical conversion of hydrogen sulfide into elemental sulfur. *Bioengng. Resources* **15**(4), 234–238.

Blumentals, I.I., Itoh, M., Olson, G.J. and Kelly, R.M. (1990) Role of polysulfides in reduction of elemental sulfur by the hyperthermophilic archaebacterium pyrococcus furiosus. *Appl. Environ. Microbiol.* **56**(5), 1255–1262.

Bond, D.R., Holmes, D.E., Tender, L.M. and Lovley, D.R., 2002. Electrode-Reducing microorganisms that harvest energy from marine sediments. *Science* **295**(5554), 483–485.

Bowker, R.P.G., Smith, J.M. and Webster, N.A. (1989) *Odor and Corrosion Control in Sanitary Sewerage Systems and Treatment Plants*. Park Ridge, New Jersey, Noyes Data Corporation.

Bruser, T., Lens, P.N.L. and Truper, H.G. (2000) The biological sulfur cycle, In *Environmental Technologies to Treat Sulfur Pollution*. Lens, P. and Hulshoff Pol, L. (eds). IWA, London.

Buisman, C., Uspeert, P., Janssen, A. and Lettinga, G. (1990) Kinetics of chemical and biological sulphide oxidation in aqueous solutions. *Water Res.* **24**(5), 667–671.

Chen, K.Y. and Morris, J.C., 1972. Kinetics of oxidation of aqueous sulfide by O_2. *Environ. Sci. Technol.* **6**(6), 529–537.

Chuang, K.T., Donini, J.C., Sanger, A.R. and Slavov, S.V. (2000) A proton-conducting solid state H2S-O2 fuel cell: 2. Production of liquid sulfur at 120–145°C. *Int. J. Hydrogen Energy* **25**(9), 887–894.

DeLong, E.F. and Chandler, P. (2002) Power from the deep. *Nature Biotechnol.* **20**, 788–789.

Dutta, P.K., Keller, J., Yuan, Z., Rozendal, R.A. and Korneel, R. (2009) The role of sulfur during acetate oxidation in biological anodes. *Environ. Sci. Technol.* Accepted.

Dutta, P.K., Rabaey, K., Yuan, Z. and Keller, J. (2008) Spontaneous electrochemical removal of aqueous sulfide. *Water Res.* **42**(20), 4965–4975.

Eow, J.S. (2002) Recovery of sulfur from sour acid gas: a review of the technology. *Environmental Progress* **21**(3), 143–162.

Farooque, M. and Fahidy, T.Z. (1977) Low potential oxidation of hydrogen sulfide on a rotating tripolar wiper-blade electrode via continuous anode reactivation. *Electrochem. Sci. Technol.* **124**(8), 1191–1195.

Habermann, W. and Pommer, E.H. (1991) Biological fuel cells with sulphide storage capacity. *Appl. Microbiol. Biotechnol.* **35**(1), 128–133.

Hamilton, I.C. and Woods, R. (1981) An investigation of surface oxidation of pyrite and pyrrhotite by linear potential sweep voltammetry. *J. Electroanal. Chem.* **118**(FEB), 327–343.

Hamilton, I.C. and Woods, R. (1983) An investigation of the deposition and reactions of sulphur on gold electrodes. *J. Appl. Electrochem.* **13**(6), 783–794.

Holmes, P.R. and Crundwell, F.K. (2000) The kinetics of the oxidation of pyrite by ferric ions and dissolved oxygen: an electrochemical study. *Geochim. Cosmochim. Acta* **64**(2), 263–274.

Hurse, T.J. and Keller, J. (2004) Performance of a substratum-irradiated photosynthetic biofilm reactor for the removal of sulfide from wastewater. *Biotechnol. Bioeng.* **87**(1), 14–23.

Janssen, A.J.H., Lettinga, G. and de Keizer, A. (1999) Removal of hydrogen sulphide from wastewater and waste gases by biological conversion to elemental sulphur: Colloidal and interfacial aspects of biologically produced sulphur particles. *Colloids Surfaces A: Physicochem. Engng. Aspects* **151**(1–2), 389–397.

Kameyama, T., Dokiya, M., Fujishige, M., Yokokawa, H. and Fukuda, K. (1981) possibility for effective production of hydrogen sulfide by means of a porous vycor glass membrane. *Indust. Engng. Chem. Fundam.* **20**(1), 97–99.

Kamyshny, Jr., A., Goifman, A., Gun, J., Rizkov, D. and Lev, O. (2004) Equilibrium distribution of polysulfide ions in aqueous solutions at 25°C: a new approach for the study of polysulfides' equilibria. *Environ. Sci. Technol.* **38**(24), 6633–6644.

Kelsall, G.H. and Thompson, I. (1993) Redox chemistry of H_2S oxidation in the British-gas stretford process. 1. Thermodynamics of sulfur-water systems at 298k. *J. Appl. Electrochem.* **23**(4), 279–286.

Kelsall, G.H., Yin, Q., Vaughan, D.J., England, K.E.R. and Brandon, N.P. (1999) Electrochemical oxidation of pyrite (FeS_2) in aqueous electrolytes. *J. Electroanal. Chem.* **471**(2), 116–125.

Kleinjan, W.E., Keizer, A.D. and Janssen, A.J.H. (2005) Kinetics of the chemical oxidation of polysulfide anions in aqueous solution. *Water Res.* **39**(17), 4093–4100.

Kuhn, A.T., Kelsall, G.H. and Chana, M.S. (1983) A review of the air oxidation of aqueous sulphide solutions. *J. Chem. Technol. Biotechnol.* **33A**, 406–414.

Lens, P.N.L., Visser, A., Janssen, A.J.H., Pol, L.W.H. and Lettinga, G. (1998) Biotechnological treatment of sulfate-rich wastewaters. *Critical Rev. Environ. Sci. Technol.* **28**(1), 41–88.

Levillain, E., Demortier, A. and Lelieur, J.P. (2007) Sulfur. In: *Encyclopedia of Electrochemistry*. Bard, A.J., Stratmann, M., Scholz, F. and Pickett, C.J. (eds) Wiley-VCH Verlag GmbH & Co. KGaA.

Madigan, M.T., Martinko, J.M. and Parker, J. (2003) *Brock Biology of Microorganisms*. Prentice Hall International, Inc.

Mao, Z., Anani, A., White, R.E., Srinivasan, S. and Appleby, A.J. (1991) A modified electrochemical process for the decomposition of hydrogen sulfide in an aqueous alkaline solution. *J. Electrochem. Soc.* **138**(5), 1299–1303.

McGuire, M.M., Jallad, K.N., Ben-Amotz, D. and Hamers, R.J. (2001) Chemical mapping of elemental sulfur on pyrite and arsenopyrite surfaces using near-infrared Raman imaging microscopy. *Appl. Surf. Sci.* **178**(1–4), 105–115.

Millero, F.J., LeFerriere, A., Fernandez, M., Hubinger, S. and Hershey, P. (1989) Oxidation of H_2S with H_2O_2 in natural waters. *Environ. Sci. Technol.* **23**(2), 209–213.

Padival, N.A., Kimbell, W.A. and Redner, J.A. (1995) Use of Iron salts to control dissolved sulfide in trunk systems. *J. Environ. Engng.* **121**(11), 824–829.

Pourbaix, M. (1966) *Atlas of Electrochemical Equilibrium in Aqueous Solutions*. Pergamon, Oxford.

Rabaey, K., Clauwaert, P., Aelterman, P. and Verstraete, W. (2005) Tubular microbial fuel cells for efficient electricity generation. *Environ. Sci. Technol.* **39**(20), 8077–8082.

Rabaey, K., VandeSompel, K., Maignien, L., Boon, N., Aelterman, P., Clauwaert, P., DeSchamphelaire, L., Pham, H.T., Vermeulen, J., Verhaege, M., Lens, P. and Verstraete, W. (2006) Microbial fuel cells for sulfide removal. *Environ. Sci. Technol.* **40**(17), 5218–5224.

Rajalo, G. and Petrovskaya, T., 1996. Selective electrochemical oxidation of sulphides in tannery wastewater. *Environ. Technol.* **17**, 605–612.

Reimers, C.E., Girguis, P., Stecher, H.A., Tender, L.M., Ryckelynck, N. and Whaling, P., 2006. Microbial fuel cell energy from an ocean cold seep. *Geobiology* **4**(2), 123–136.

Ryckelynck, N., Stecher, H.A. and Reimers, C.E. (2005) Understanding the anodic mechanism of a seafloor fuel cell: interactions between geochemistry and microbial activity. *Biogeochem.* **76**(1), 113–139.

Schauder, R. and Kröger, A. (1993) Bacterial sulphur respiration. *Arch. Microbiol.* **159**(6), 491–497.

Shih, Y.S. and Lee, J.L. (1986) Continuous solvent extraction of sulfur from the electrochemical oxidation of a basic sulfide solution in the CSTER system. *Ind. Eng. Chem. Proc. Des. Dev.* **25**(3), 834–836.

Steudel, R. (2000) The chemical sulfur cycle. In *Environmental Technologies to Treat Sulfur Pollution*. Lens, P.N.L. and Hulshoff Pol, L. (eds). IWA Publishing, London.

Syed, M., Soreanu, G., Falletta, P. and Beland, M. (2006) Removal of hydrogen sulfide from gas streams using biological processes – a review. *Canadian Biosystems Engineering/Le Genie des biosystems au Canada*. **48**, 2.1–2.14.

Szynkarczuk, J., Komorowski, P.G. and Donini, J.C. (1994) Redox reactions of hydrosulphide ions on the platinum electrode–I. The presence of intermediate polysulphide ions and sulphur layers. *Electrochim. Acta* **39**(15), 2285–2289.

Tender, L.M., Reimers, C.E., Stecher, H.A., Holmes, D.E., Bond, D.R., Lowy, D.A., Pilobello, K., Fertig, S.J. and Lovley, D.R. (2002) Harnessing microbially generated power on the seafloor. *Nature Biotech.* **20**(8), 821–825.

Turcotte, S.B., Benner, R.E., Riley, A.M., Li, J., Wadsworth, M.E. and Bodily, D.M. (1993) Surface analysis of electrochemically oxidized metal sulfides using Raman spectroscopy. *J. Electroanal. Chem.* **347**(1–2), 195–205.

Waterston, K., Bejan, D. and Bunce, N. (2007) Electrochemical oxidation of sulfide ion at a boron-doped diamond anode. *J. Appl. Electrochem.* **37**(3), 367–373.

Zaman, J. and Chakma, A. (1995) Production of hydrogen and sulfur from hydrogen sulfide. *Fuel Process Technol.* **41**(9), 159–198.

Zhang, L., De Schryver, P., De Gusseme, B., De Muynck, W., Boon, N. and Verstraete, W. (2008) Chemical and biological technologies for hydrogen sulfide emission control in sewer systems: a review. *Water Res.* **42**(1–2), 1–12.

Zhao, F., Rahunen, N., Varcoe, J.R., Chandra, A., Avignone-Rossa, C., Thumser, A.E. and Slade, R.C.T. (2008) Activated carbon cloth as anode for sulfate removal in a microbial fuel cell. *Environ. Sci. Technol.* **42**(13), 4971–4976.

Zhao, F., Rahunen, N., Varcoe, J.R., Roberts, A.J., Avignone-Rossa, C., Thumser, A.E. and Slade, R.C.T. (2009) Factors affecting the performance of microbial fuel cells for sulfur pollutants removal. *Biosensors Bioelectron.* **24**(7), 1931–1936.

Zhdanov, S.I. (1975) Sulfur, In *Encyclopedia of Electrochemistry of the Elements*. Bard, A.J. (ed.) Macel Dekker, New York and Basel.

Zhu, X., Li, J. and Wadsworth, M.E. (1994) Characterization of surface layers formed during pyrite oxidation. *Colloids Surfaces A: Physicochem. Eng. Aspects* **93**, 201–210.

13

Chemically catalyzed cathodes in bioelectrochemical systems

René A. Rozendal, Falk Harnisch, Adriaan W. Jeremiasse and Uwe Schröder

13.1 INTRODUCTION

The recent progress especially in the anodic performance of BESs causes a considerable demand for the development of adequate cathode technologies. Whereas cathode research was rather neglected at first, it has now become evident that the cathodic reactions represent a major bottleneck in bioelectrochemical devices. Table 13.1 provides an overview of the relevant cathode reactions used in bioelectrochemical systems.

In theory, a wide range of cathode reactions are possible, but in the early days of BESs only cathode reactions suitable for electricity production (i.e. MFC application) were considered (Logan *et al.* 2006). Experimental cathode systems

 Bioelectrochemical Systems: From Extracellular Electron Transfer to Biotechnological Application. Edited by Korneel Rabaey, Largus Angenent, Uwe Schröder and Jürg Keller. ISBN: 9781843392330. Published by IWA Publishing, London, UK.

that are easy to handle, such as hexacyanoferrate (III/II) (ferri/ferrocyanide) have been used widely, but these systems are not sustainable and have to be replenished regularly (Rabaey *et al.* 2005). It has thus been generally acknowledged that the only cathode reaction practically relevant for electricity production in microbial fuel cells is the oxygen reduction reaction (ORR).

Table 13.1. Overview of the cathode reactions used in bioelectrochemical systems

Cathodic reactions	Conditions	$E^{0'}$ at pH 7 (V)	References
Microbial fuel cells			
Oxygen/ H_2O	Theoretical	0.82	(Logan *et al.* 2006)
Oxygen/ H_2O	Effective (open circuit)	0.51	(Cheng *et al.* 2006b; Zhao *et al.* 2006)
Hexacyanoferrate (III/II) (ferro/ferricyanide)	In solution (1:1 ratio)	0.36	(Bard *et al.* 1985)
Microbial electrolysis cells			
$2H^+/H_2$	Theoretical	−0.41	(*Logan et al.* 2008)
$2H^+/H_2$	Effective (open circuit)	−0.41	(Call and Logan 2008; Cheng and Logan 2007; Rozendal *et al.* 2006)

In 2003, it was realized that MFCs can be modified to become MECs by replacing the ORR at the cathode by the hydrogen evolution reaction (HER) (Liu *et al.* 2005; Logan *et al.* 2008; Rozendal and Buisman 2005; Rozendal *et al.* 2006). In an MEC the anodic oxidation of organic material is coupled to the HER in one device, which is energetically favorable compared to early concepts that coupled a conventional electrolyzer to microbial fuel cells (Kreysa *et al.* 1994). Due to the low energy requirements of MECs (~1 to 2 kWh/m^3 H_2) compared to water electrolysis cells (~4 to 5 kWh/m^3 H_2) and due to the high conversion rates compared to fermentation (Logan *et al.* 2008), microbial electrolysis systems hold great promise as a hydrogen production device in a future hydrogen economy. However, to fulfill this prospect as an interesting system a cost-effective HER cathode is required.

At pH 7 and a partial oxygen pressure of 0.2 bar, the theoretical redox potential of the ORR is 0.81V (Logan *et al.* 2006). However, under open circuit conditions the potential is generally below 0.51 V at conventional electrocatalysts (Cheng *et al.* 2006b; Zhao *et al.* 2006), which is due to the formation of a mixed potential at the electrode surface (Bard *et al.* 2008) (also see Chapter 8). Moreover, several hundreds of millivolts more are typically lost under practical working conditions of MFCs, i.e. when current is flowing.

Also, the HER can suffer significant losses. The theoretical redox potential of the HER at pH 7 is −0.41 V (Logan *et al.* 2008), which can be approached closely at low current densities. Yet, under practical working conditions of MECs, several hundred of millivolts of potential loss are also encountered for this reaction (Call and Logan 2008; Cheng and Logan 2007; Rozendal *et al.* 2006).

As both the ORR and HER are proton consuming (or hydroxyl ion producing) reactions (see below), the redox potentials of these reaction are directly linked to the pH of the electrolyte solution and at the electrode surface. Therefore, any changes in pH cause strong shifts of the redox potential of these reactions. For cathode reactions like the HER and ORR in bioelectrochemical systems, two major pH based issues are observed:

(i) Shift of the cathode potential towards negative values due to alkalinization of the cathode compartment;
(ii) Shift of the cathode potential towards negative values due to pH gradients occurring at the electrode/electrolyte solution interface at higher current densities.

Both effects (which are discussed in more detail in Chapter 8 – Electrochemical Losses) lead to a strong decrease of the cell voltage, and thus of the overall performance of the bioelectrochemical device.

This chapter discusses the fundamentals of the two most studied cathode reactions in bioelectrochemical systems: (i) the ORR; and (ii) the HER. Also it discusses the BES configurations in which those cathode reactions have been implemented. The ORR is extensively discussed in Section 13.2; whereas the following Section 13.3 focuses on the HER. Finally, this chapter provides an outlook on future cathode systems for BES application (Section 13.3). This chapter limits itself to chemically catalyzed cathode systems; biologically catalyzed cathode systems are reviewed in Chapter 15.

13.2 OXYGEN REDUCTION REACTION (ORR)

13.2.1 Introduction

The reduction of oxygen is doubtlessly the best choice for the cathodic reaction, in microbial fuel cells. The direct and virtually inexhaustible availability of oxygen as well as its positive redox potential and the harmlessness of the reduction product (water), are three major reasons for the superiority of this element as the oxidant for microbial fuel cells.

For conventional chemical fuel cells, the improvement of the performance of the oxygen reduction cathode at lowest possible costs and at high long-term stability of the cathode material represents one of the most important research issues, since it is often the cathodic reaction that limits the performance of chemical fuel cells. Microbial fuel cells with their comparably low current and power output, which are orders of magnitude below that of chemical fuel cells, on a first glance would not appear to share the same problem. Yet, it has become obvious that this exactly is the case: the conditions for an efficient oxygen reduction in microbial fuel cells are far from ideal, and they differ significantly from those present in conventional fuel cells (Zhao *et al.* 2006). Ambient temperature, low concentrations of the supporting electrolytes as well as the mostly near to neutral pH put thermodynamic as well as kinetic constraints on the performance of the MFC cathode.

With well-catalyzed systems, the ORR generally proceeds as a direct, four-electron reduction:

$$O_2 + 4e^- + 4\,H^+ \rightarrow 2\,H_2O \tag{13.1a}$$

under acid conditions, or

$$O_2 + 4e^- + 2\,H_2O \rightarrow 4\,OH^- \tag{13.1b}$$

under alkaline conditions.

Alternatively, the ORR can also proceed as two, consecutive, two-electron reaction steps, with hydrogen peroxide as the intermediate product. This is also the typical mechanism when carbon is used as cathode material:

$$\text{Step 1: } O_2 + 2e^- + 2\,H^+ \rightarrow H_2O_2 \tag{13.2a}$$

$$\text{Step 2: } H_2O_2 + 2e^- + 2\,H^+ \rightarrow 2\,H_2O \tag{13.3a}$$

under acid conditions, or

$$\text{Step 1: } O_2 + 2e^- + H_2O \rightarrow HO_2^- + OH^- \tag{13.2b}$$

$$\text{Step 2: } HO_2^- + 2e^- + H_2O \rightarrow 3\,OH^- \tag{13.3b}$$

under alkaline conditions.

The reaction mechanism of the ORR depends on several parameters, such as the nature of the electrode material, the pH and the current density. Generally, three main reaction steps can be distinguished, each being potentially rate limiting: (i) the initial chemisorption of the oxygen molecule at a suitable reaction site on the electrode surface, (ii) the dissociation of the oxygen molecule and (iii) the electron transfer step.

At bare, unmodified electrode materials like carbon or steel the ORR proceeds only at very high overpotentials (Zhao *et al.* 2006). Therefore, in any study in which plain carbon electrodes are used as cathode material, the achieved current densities are extremely low. A comparably effective approach to circumvent the strong polarization of carbon is the considerable enlargement of the active electrode surface area (e.g. via the use of packed granule cathodes (Freguia *et al.* 2007), which effectively decreases the surface specific current density and thus the polarization). Moreover, it is often observed, that the performance of carbon based MFC cathodes (e.g. in benthic fuel cells) improves over time. The reason for this phenomenon lies in a colonization of the electrode with bacteria that can assist the electrochemical oxygen reduction (see Chapter 15).

13.2.2 Oxygen reduction catalysts

13.2.2.1 Platinum

Platinum is the best known oxygen reduction catalyst and has long been used as a benchmark material in chemical and microbial fuel cell research. However, facing its increasing price in combination with the low current densities achievable in BESs, its application therein is obviously not rational.

A possible approach to lower the costs of Pt based cathodes is the drastic decrease of the Pt load as suggested by Cheng *et al.* (2006b). Their study showed that in their experimental setup the reduction of the Pt load had only a small impact on the electrode performance. In this example, however, the high internal resistance of the used MFC system may have been determining the overall fuel cell performance. Moreover, lowering the Pt load may lower the costs for a respective electrode, but it does not solve problems associated with the susceptibility of Pt to (often irreversible) poisoning in 'dirty', non-sterile biological environments.

The problems associated with the use of Pt as an electrocatalyst in microbial environments urges the development of non-noble metal based oxygen reduction electrocatalysts. In the following paragraphs, three classes of catalysts – transition metal macrocycle based catalysts, metal oxides, and isolated enzymes, are discussed in the light of application as cathode material in MFCs.

13.2.2.2 Transition metal macrocycle based catalysts

The development of fuel cell oxygen reduction electrocatalysts represents a major research field, and increasing efforts have been made to search for inexpensive non-noble metal materials. An emphasis is on biomimetic approaches utilizing transition metal porphyrines and phthalocyanines (e.g. Fukuzumi and Imahori 2001). These materials are known to be good oxygen reduction catalysts that represent inexpensive alternatives to platinum (e.g. Baranton *et al.* 2005). Jasinski was the first to report that transition metal porphyrines and phthalocyanines show an electrochemical activity towards the oxygen reduction reaction and can be used as cathode catalyst in fuel cells (Jasinski 1964). The molecules have planar structures with the metal ion symmetrically surrounded by four nitrogen atoms, which leads to the common term transition metal N_4 macrocycles. Jahnke *et al.* improved the stability as well as the electrochemical activity of transition metal porphyrins deposited on a carbon support by a pyrolytic heat treatment step in the range from 450 to 900°C in an inert atmosphere (Jahnke *et al.* 1976).

For microbial fuel cells, pyrolysed phthalocyanines and porphyrines of iron(II) and cobalt have recently been introduced (Zhao *et al.* 2005). Subsequent investigations demonstrated that, except for the cathode material itself, the electrode manufacturing, e.g. the electrode binder (Cheng *et al.* 2006b) and the catalyst load (Zhao *et al.* 2006) greatly influence the electrode performance.

A major research issue is the improvement of the catalysts' long-term stability. Especially the problems of leaching of the transition metal ions (the catalytic centres) out of the catalyst and the oxidative degradation by H_2O_2, the intermediate product of the oxygen reduction (see Eq. 13.2) have to be solved. For the latter issue, the incorporation of MnO_x may represent a feasible solution (Roche *et al.* 2007). Here, the metal oxide increases the rate of the reductive hydrogen peroxide decomposition (and thus the life time of this compound) and also increases the average number of electrons per reduced oxygen molecule.

13.2.2.3 Metal oxides

Metal oxides are also interesting candidates for ORR electrocatalysts, but so far only two preliminary studies demonstrated their principle applicability. Lead dioxide (PbO_2) has been shown to outperform platinum in a salt-bridge based two chamber MFC system (Morris *et al.* 2007). Yet, due to this system's inherent constraints, this information may be of only qualitative nature.

MnO_x has been shown to enhance the ORR-activity of transition metal macrocycle based catalyst materials (Roche *et al.* 2007), firstly (as described above) by promoting the reductive H_2O_2 decomposition, thus by shifting the ORR mechanism from a pure two-electron reaction to a four-electron or two consecutive two-electron step pathway. Secondly, the Mn(III)/Mn(IV) redox couple was proposed to directly catalyse the oxygen reduction reaction. The activity was more pronounced if the MnO_x was doped by additional divalent metal cations (e.g. Ni(II), Mg(II)).

Both studies showed promising but preliminary results, and considerable research efforts, e.g. concerning the long-term stability of the metal oxides in pH neutral electrolyte solutions are required.

13.2.2.4 Enzymes

Potentially attractive is the application of enzyme based cathodes, because due to a higher selectivity and lower ORR overpotentials, redox enzymes (e.g. oxidases, such as cytochrome oxidase and bilirubin oxidase), laccases, and peroxidases (e.g. horseradish peroxidase) may allow a considerable increase of the MFC cell voltage (Miura *et al.* 2007; Tsujimura *et al.* 2003). So far, only one study (Schaetzle *et al.* 2009) on this type of MFC cathode is available. In this study, in which laccase (from *Trametes versicolor*) was used as oxygen reduction catalyst, an open circuit potential of 1.1 V was achieved, which underlines the feasibility of such an approach. Yet, as also mentioned in the above sections, the long-term stability of the catalyst may represent an important issue, which needs to be addressed in further studies.

13.2.3 MFC cathode configurations

13.2.3.1 Aqueous cathodes

Aqueous cathodes are generally used in classical, two chamber MFC (often H-shape) systems with the anode and the cathode being immersed in aqueous electrolyte solutions, separated by a diaphragm. Such systems are usually simple to handle and are thus ideal for experimental studies. They also allow the temporary use of experimental cathodes like hexacyanoferrate. Yet, the major bottleneck of aqueous cathodes is the limited oxygen solubility (6 mg l^{-1} O_2 in water at 25°C (Lide 2004)) and thus availability in aqueous solutions, which considerably limits the cathode performance. Solutions have been proposed, like the pressurization (Fornero *et al.* 2008) of the cathode chamber. However, such approaches may be energy demanding (lowering the overall efficiency) and require a more complex MFC design.

13.2.3.2 Air cathodes

In these systems, first introduced for MFCs by Sell and co-workers (Sell *et al.* 1989), the cathode electrocatalyst directly faces the ambient air and thus allows virtually unlimited access to the oxidant. However, as the ORR is dependent on the ability of ions to migrate towards (i.e. protons; Eq. 13.1a) or away (i.e. hydroxyl ions; Eq. 13.1b) from the electrocatalyst, a layer of liquid is always required at the cathode surface. As it can be derived from Eq. (13.1a), the ORR itself can produce water. Larger amounts of water, however, are generally also transferred across the separator, due to the hydrodynamic pressure gradient and between anode chamber and cathode. Especially in so-called membrane-less systems in which, e.g. only a J-cloth separates the anode chamber from the cathode (Fan *et al.* 2007), the water penetration (leaking) can be considerable. Here, however, the transfer of water from the anode chamber to the cathode can be accompanied by a significant crossover of organic compounds, leading to a strong decrease of the cathode performance. At the same time, the penetration of oxygen into the anode chamber represents an important issue that leads to a limitation of the coulombic efficiency of one-chamber cells when compared to two-chamber systems (Cheng *et al.* 2006a).

Thus, except for the minute volume of electrolyte solution, air cathodes are not dissimilar to aqueous cathodes, and they do share similar restrictions. The major advantage lies undoubtedly in the fast transfer of oxygen to the reaction site, which does not require any mechanical mass transfer enhancement (stirring, bubbling) as is the case for aqueous cathodes.

13.3 HYDROGEN EVOLUTION REACTION (HER)

13.3.1 Introduction

The hydrogen evolution reaction (HER) is by far the most studied electrochemical reaction (Vetter 1967). The reason for this is twofold: (i) it is a highly relevant reaction for society, as hydrogen might someday replace fossil fuels as transportation fuel, and (ii) hydrogen evolution was once thought to be one of the simplest electrochemical reactions, due to its straightforward reaction scheme:

$$2\,H^+ + 2e^- \rightarrow H_2 \tag{13.4a}$$

under acid conditions, or

$$2\,H_2O + 2e^- \rightarrow H_2 + 2\,OH^- \tag{13.4b}$$

under alkaline conditions.

Since protons (or hydroxyl ions) are involved in the hydrogen evolution reaction, the equilibrium potential of the reaction is highly dependent on the pH at the cathode, according to:

$$E^0_{H2,pH} = -0.059 \times pH \tag{13.5}$$

In which $E^0_{H2,pH}$ (in Volts) refers to the equilibrium potential of the hydrogen evolution as a function of the pH at standard temperature (298 K) and 1 atm partial hydrogen pressure. It is important to note that this pH is not the bulk pH of the solution, but refers to the pH at the electrode surface (Andersen *et al.* 1993). In MECs with near neutral bulk pH, the pH near the surface of the electrode is expected to be alkaline, as the HER is a proton-consuming or hydroxyl ion producing reaction (Eq. 13.4). This can negatively affect the cathode potential and MEC performance.

Research has shown that the hydrogen evolution reaction is not as simple as initially assumed from its straightforward reaction scheme (Vetter 1967), but it is now generally agreed upon that the HER mechanism involves three possible reaction steps (Kibler 2006; Saraby-Reintjes 1986; Vetter 1967):

$$\text{Step 1:} \quad H^+ + e^- \rightarrow H_{ad} \tag{13.6a}$$

$$\text{Step 2:} \quad 2\,H_{ad} \rightarrow H_{2,ad} \tag{13.7a}$$

$$\text{Step 3:} \quad H^+ + H_{ad} + e^- \rightarrow H_{2,ad} \tag{13.8a}$$

under acid conditions, or

$$\text{Step 1:} \quad H_2O + e^- \rightarrow H_{ad} + OH^- \tag{13.6b}$$

$$\text{Step 2:} \quad 2\,H_{ad} \rightarrow H_{2,ad} \tag{13.7b}$$

$$\text{Step 3:} \quad H_2O + H_{ad} + e^- \rightarrow H_{2,ad} + OH^- \tag{13.8b}$$

under alkaline conditions. H_{ad}, $H_{2,ad}$ refer to electrode adsorbed atomic and molecular hydrogen respectively.

Step 1 is referred to as the Volmer (or discharge) reaction, Step 2 is referred to as the Tafel (or combination) reaction, and Step 3 is referred to as the Heyrovski (or ion+atom reaction). In general, one of these reaction steps is rate determining (Saraby-Reintjes 1986). In some cases, also desorption and diffusion of the produced molecular hydrogen can be rate-determining steps (Vetter 1967).

For most electrode materials suitable for catalyzing the HER, cathodic overpotentials can be described with the Tafel equation over a wide range of current densities (10^{-9} to 100 A/cm^2 – Figure 13.1 (Vetter 1967)):

$$\eta = a + b \cdot \log(i) \tag{13.9}$$

With η the cathodic overpotential of HER, i the current density, and a and b constants. Constant a strongly depends on the current exchange density i_0. Constant b is referred to as the Tafel slope. High performance HER electrocatalysts should exhibit high current exchange density and low Tafel slopes (Tilak *et al.* 1986).

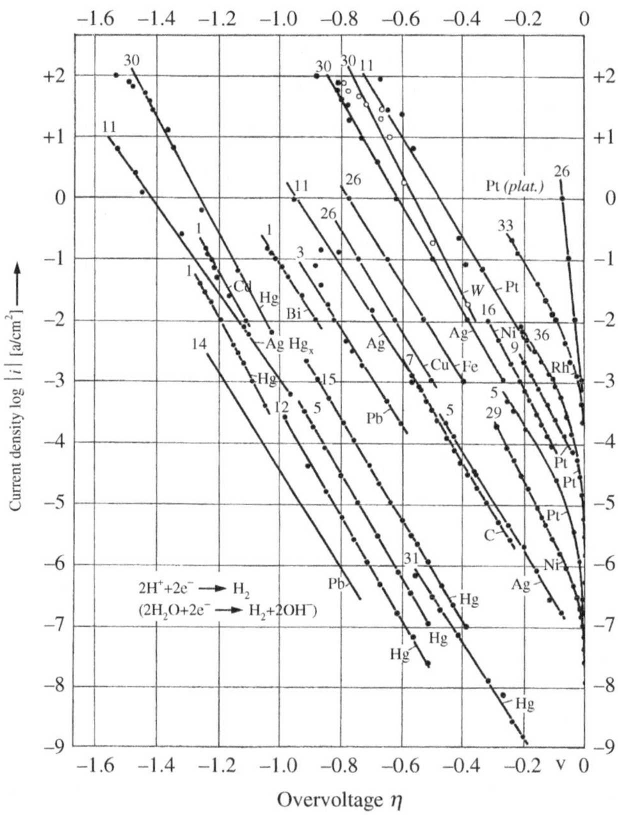

Figure 13.1. Tafel plot of the HER for various metals and electrolytes – from Vetter (1967). The numbers refer to literature references as listed in Vetter (1967). Copyright Academic Press, reproduced by permission

Several researchers realized that there is a direct relationship between the exchange current density for the HER for a certain material and the hydrogen atom chemisorption energy for this material (ΔE_H), which is an indicator of the metal-to-H bond strength (Conway and Jerkiewicz 2000; Kibler 2006; Norskov *et al.* 2005; Trasatti 1972). This relationship can be plotted in a so-called Volcano plot (Figure 13.2). Details on the calculation of hydrogen chemisorption per atom ΔE_H and the Gibbs free energy for hydrogen adsorption ΔG_{H*} can be found in (Norskov *et al.* 2005).

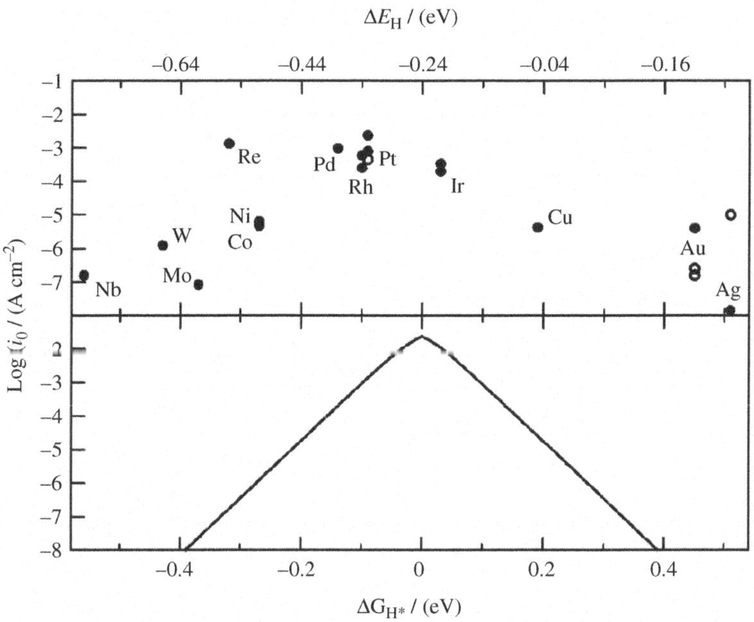

Figure 13.2. Example of a Volcano plot of the HER for various metals – from (Norskov *et al.* 2005). The figure shows the measured (top) and calculated (bottom) exchange current density i_0 for various metals as a function of the calculated hydrogen chemisorption per atom ΔE_H and the Gibbs free energy for hydrogen adsorption ΔG_{H*}. Reproduced by permission of The Electrochemical Society

The most interesting feature of the Volcano plot is that materials with high exchange current density have a Gibbs free energy for hydrogen adsorption ΔG_{H*} close to zero. The rationale for this is that: (i) the materials on the left of the maximum exchange current density in the Volcano plot form strong bonds with atomic hydrogen, which lowers the exchange current density as a result of a slow hydrogen desorption; and (ii) the materials on the right of the maximum exchange current density of the Volcano plot form weak bonds with atomic hydrogen, which lowers the exchange current density as a result of a thermodynamically unfavourable discharge step (Hinnemann *et al.* 2005).

One of the complicating factors of using Volcano plots, is the fact that the rate-determining step of the HER can be different at different pH values, which changes the shape of the Volcano plot (Conway and Jerkiewicz 2000). Therefore, it is argued that Volcano plot can be improved by turning it into a 3D plot with pH on the third axis (Conway and Jerkiewicz 2000). Nevertheless, Volcano plots have already been proven to be very useful. The fact that good HER electrocatalysts typically fulfill the requirement of a Gibbs free energy for hydrogen adsorption ΔG_{H*} close to zero has already been used to explain and/or the predict the suitability of new materials as HER electrocatalysts, such as multicomponent metal systems (Petrii and Tsirlina 1994), composite materials (e.g. MoS_2 nanoparticles (Hinnemann *et al.* 2005)), and enzymes (see below).

The exchange current density of a suitable electrode material can be further enhanced by creating a high specific surface area (e.g. Pt vs. Pt(*plat*) in Figure 13.1) (Tilak *et al.* 1986). However, high exchange current densities are not the only factor that makes a good HER electrocatalyst. Other important factors include costs, lifetime, and catalyst stability under open circuit conditions (Tilak *et al.* 1986).

13.3.2 Hydrogen evolution catalysts

13.3.2.1 Platinum

As can be seen in Table 13.2 most MEC studies so far employed platinum as the cathode electrocatalyst. However, even with the higher economic value of hydrogen (in MECs) compared to that of electricity (in MFCs), the high costs of platinum cannot be justified (Rozendal *et al.* 2008a). Moreover, even if the costs could be reduced dramatically (e.g. by reducing Pt loading), for some reason the performance of platinum catalyzed cathodes in MECs still lags far behind what would be expected from the performance of those cathodes in conventional electrolyzers (Rozendal *et al.* 2008a). As indicated by the Volcano plot (Figure 13.2) platinum and the other members of the platinum group (e.g. palladium (Pd), rhodium (Rh)) are theoretically the best known HER electrocatalysts (Hinnemann *et al.* 2005). Nevertheless, under the working conditions typical for MECs, several hundreds of millivolts of overpotential are generally encountered (Rozendal *et al.* 2006). The reasons for this low performance are not well understood at this stage, but the relatively mild conditions in MECs (low *T*, neutral bulk pH) compared to the more extreme conditions in for example PEM electrolyzers (323–373 K, $pH < 1$) are likely to be part of the explanation. Moreover, platinum cathodes in MECs are likely susceptible to performance loss due to metal poisoning (Selembo *et al.* 2009).

Table 13.2. Overview of HER electrocatalysts, catalyst loading/configurations, and electrolytes used in various MEC studies

MEC study	Catalyst	Catalyst loading/ configuration	Cathode electrolyte
Aqueous cathodes			
(Liu *et al.* 2005)	Platinum	5 g/m^2 on carbon paper	50 mmol/lPBS[a]
(Rozendal *et al.* 2006)	Platinum	50 g/m^2 on Ti mesh	10 mmol/l PBS[a]
(Ditzig *et al.* 2007)	Platinum	5 g/m^2 on carbon paper	50 mmol/l PBS[a]
(Cheng and Logan 2007)	Platinum	5 g/m^2 on carbon cloth	200 mmol/l PBS[a]
(Hu *et al.* 2008)	Platinum	5 g/m^2 on carbon cloth	320 mmol/l PBS[a]
(Liu *et al.* 2008b)	Platinum	3.5 g/m^2 on carbon cloth	200 mmol/l Tris
(Call and Logan 2008)	Platinum	5 g/m^2 on carbon cloth	50 mmol/l PBS[a]
(Rozendal *et al.* 2008c)	Platinum	50 g/m^2 on Ti mesh	10 mmol/l PBS[a]
(Chae *et al.* 2008)	Platinum	5 g/m^2 on Ti plate	50 mmol/l PBS[a]
(Call *et al.* 2009)	Nickel	Stainless steel brush	Growth medium (50 mmol/l PBS)
(Selembo *et al.* 2009)	Nickel	Various stainless steels and Ni alloys	Growth medium (50 mmol/l PBS)
(Harnisch *et al.* 2009b)[b]	Tungsten carbide	2600 g/m^2 [c] on graphite RDE[d]	100 mmol/l PBS[a]
Gas diffusion cathodes			
(Rozendal *et al.* 2007)	Platinum	10 g/m^2 on CEM[e]	–
(Rozendal *et al.* 2007)	Platinum	10 g/m^2 on AEM[e]	–
(Tartakovsky *et al.* 2008)	Palladium/ Platinum	5 g/m^2 (50% Pd/50% Pt) on carbon fibre paper	–
(Tartakovsky *et al.* 2008)	Platinum	5 g/m^2 Pt on woven carbon	–
(Tartakovsky *et al.* 2009)	Platinum	5 g/m^2 Pt on woven carbon	–

[a] PBS: Phosphate Buffer Solution

[b] Only cyclic voltammetry and chronoamperometric half-cell experiments were carried out with WC in this study; WC has not been tested in a MEC so far.

[c] The WC loading was arbitrary and not optimized yet.

[d] RDE: rotating disc electrode

[e] The cathode was supported by a titanium mesh, which was also coated with 50 g/m^2 Pt.

Therefore, in line with the developments for MFC cathodes (see above), researchers in the MEC field are also focusing more and more on the development of alternative HER electrocatalysts. In the next paragraphs we will discuss nickel, tungsten carbide, and enzymes as potentially interesting alternatives to platinum as an HER electrocatalyst.

13.3.2.2 Nickel

In the periodic table of the elements, iron, cobalt and nickel lie above the platinum group. It is not surprising, therefore, that also these metals show significant catalytic activity for the HER. Under alkaline conditions, iron is a slightly better HER electrocatalyst than cobalt and nickel (Tilak *et al.* 1986), but unfortunately iron is not stable under open circuit conditions. Nickel, on the other hand, often in conjunction with cobalt, is one of the most widely used electrocatalysts in alkaline electrolyzers (Suffredini *et al.* 2000; Tilak *et al.* 1986). The pH at the surface of a cathode in a MEC is likely to be alkaline as a result of the proton consumption (or hydroxyl ion production) (Eq. 13.4). Therefore, nickel is likely to be a suitable catalyst for MECs as well.

It has been shown that enhanced catalytic activity for the HER can be achieved by alloying nickel with other metals, such as Co, Ir, W or with non-metals such as S, P., B, Si and Sb (Tilak *et al.* 1986). Also it has been shown that stainless steels with high nickel content can form cheap, but effective electrode materials for catalyzing the HER (Olivares-Ramirez *et al.* 2007). A widely used strategy to further enhance the catalytic activity of nickel catalyzed cathodes is to increase the specific surface area of the catalytic sites. This can be done by co-depositing nickel with metals such as Zn and Al and subsequently leaching those metals out, leaving behind extremely porous surfaces with large specific surface area (Suffredini *et al.* 2000; Tilak *et al.* 1986). This highly catalytically active material is referred to as Raney-nickel (Trasatti 1995).

The suitability of nickel and stainless steel as MEC cathodes was recently confirmed in laboratory MEC studies (Call *et al.* 2009; Selembo *et al.* 2009). It was shown that nickel and stainless steel can outperform platinum as a cathode catalyst in MECs. This performance could be further enhanced by depositing a nickel oxide layer, but this layer was found to be unstable under typical MEC conditions (Selembo *et al.* 2009).

13.3.2.3 Tungsten carbide

Rosenbaum *et al.* demonstrated that tungsten carbide (WC) is an interesting anode electrocatalyst for the oxidation of microbially produced hydrogen (Rosenbaum *et al.* 2007). From previous studies, it was also known that WC is a potentially interesting low-cost HER electrocatalyst (Armstrong and Bell 1978; Burakov *et al.* 2008). Therefore, Harnisch *et al.* recently investigated the suitability of WC as a low-cost HER electrocatalyst in MECs (Harnisch *et al.* 2009b). The researchers systematically investigated the influence of the WC content and the content of W_2C, W and WO_2 contaminants. It was shown that the catalytic performance of the cathodes was directly proportional to the WC content and that W_2C, W and WO_2 content were only of minor influence. This finding demonstrated that the electrocatalytic evolution of hydrogen depends on the same surface sites as the electrocatalytic oxidation of hydrogen (Harnisch *et al.* 2009a). Further, they confirmed that WC is a promising electrocatalyst for MECs, as it operates well in pH-neutral electrolytes and shows little corrosion (particularly compared to the use as anode material), which makes it suitable for long-term operation.

13.3.2.4 Enzymes

Hydrogenases and nitrogenases have naturally evolved to become very efficient catalysts for hydrogen production, even though they are based on non-platinum group catalysts such as Fe, Ni and Mo (Evans and Pickett 2003; Hinnemann *et al.* 2005; Lee and Holm 2003; Vignais and Colbeau 2004). Interestingly, it has been shown with density functional theory (DFT) calculations that hydrogenases and nitrogenases fulfill the requirement of a Gibbs free energy for hydrogen adsorption ΔG_{H*} close to zero (see above) (Hinnemann *et al.* 2005; Kibler 2006). Hydrogenases have indeed been tested successfully as HER electrocatalysts (Morozov *et al.* 2002; Pershad *et al.* 1999), but unfortunately they have the important drawback that they are relatively unstable, especially when the electrode potential is forced outside a certain operating range, and they are not self-regenerating (Lojou and Bianco 2004). Therefore, also biomimetic approaches are being developed that are based on the structure of the inorganic catalytic sites of hydrogenases and nitrogenases (Gloaguen *et al.* 2001; Hinnemann *et al.* 2005). In addition, also microbial biocathode concepts are being developed that are based on living electrochemically active microorganisms that continuously regenerate intracellular hydrogenases and/or nitrogenases (Chapter 15) (Rozendal *et al.* 2008b).

13.3.3 MEC cathode configurations

13.3.3.1 Aqueous cathodes

Most MECs reported in the literature consist of a cathode that is surrounded by an aqueous electrolyte. As a gaseous product is produced in the HER and no gaseous reactant needs to be dissolved first (as with the ORR in MFCs), operation with aqueous cathodes does not impose any specific disadvantages compared to designs with gas diffusion electrodes (such as larger energy requirements for aeration in MFCs). In the literature so far, the cathode electrolyte has typically consisted of a well-defined phosphate buffer solution (PBS) of pH 7. Buffer concentrations reported in the literature (Table 13.2) range from 10 mmol/l (Rozendal *et al.* 2006) to as high as 320 mmol/l (Hu *et al.* 2008). In one study, 200 mmol/l Tris buffer was used as cathode electrolyte (Liu *et al.* 2008a). Hence, there is no consistency in the concentration of the used buffers, probably mainly because of a lack of knowledge of the effect of electrolyte composition on the HER at pH neutral conditions. Some buffers, such as phosphate, may not be feasible due to possible scaling of calcium or magnesium phosphate on the cathode surface, negatively affecting cathode performance (Jeremiasse *et al.*, unpublished results).

The most reported, aqueous cathode configuration is based in the use of flat cathode materials (Table 13.2), such as carbon paper and carbon cloth, coated with, for instance, 5 g/m^2 platinum (Ditzig *et al.* 2007; Liu *et al.* 2005). This type of cathode is commonly applied in lab scale MECs and is easily prepared in the laboratory (Logan *et al.* 2008). The disadvantage of flat cathodes, however, is their low specific surface area. The specific surface area can be increased by roughening the surface, by depositing catalytic microparticles on the cathode (Kinoshita 1992; Knutson *et al.* 2008), or by using carbon cloth (Call and Logan 2008; Cheng and Logan 2007; Liu *et al.* 2008b). Other possible configurations include mesh and porous materials.

Rozendal *et al.* (2006, 2008c) used a disc-shaped titanium mesh cathode (surface area: 400 cm^2, thickness: 1 mm, specific surface area: 1.7 m^2/m^2) that was coated with 50 g/m^2 platinum in a continuous flow MEC. Porous or packed bed cathodes (Brett and Brett 1993; El-Anadouli *et al.* 1991; Kinoshita 1992) have not been applied in MECs so far. Forcing the flow of electrolyte through a porous or packed bed cathode offers several advantages over a flat cathode: (i) porous and packed bed cathodes have a high specific surface area, thereby reducing the true surface specific current density, which results in a lower activation overpotential (El-Anadouli *et al.* 1991), and (ii) the porous structure promotes turbulence at the catalyst surface, decreasing mass transfer limitations. However, porous and packed bed cathodes may have a high ohmic resistance and be susceptible to clogging when used in a membrane-less MEC design that treats wastewater.

13.3.3.2 Gas diffusion cathodes

A gas diffusion cathode is a cathode that is not surrounded by an aqueous electrolyte. Gas diffusion cathodes are placed in a gas collection chamber to collect the hydrogen gas. The advantage of using a gas diffusion cathode is the reduced liquid volume and consequently, the reduced footprint of the MEC.

Rozendal *et al.* (2007) were the first to apply a gas diffusion cathode in an MEC (Table 13.2). The gas diffusion cathode was combined with a membrane in a membrane electrode assembly (MEA). The MEA was constructed by electrodeless plating of one side of a CEM or AEM with platinum, following a method developed by Millet *et al.* (1990). This resulted in an MEA with a platinum loading of about 10 g m^{-2}. A titanium mesh coated with 50 g m^{-2} platinum supported the MEA, and functioned both as part of the cathode and as the current collector. Both configurations (CEM and AEM) performed comparably.

Tartakovsky *et al.* (2008) tested two different gas diffusion cathodes in an MEC (Table 13.2): (i) carbon fibre paper that was coated with 5 g m^{-2} Pd/Pt alloy (50%/50%), and hot-pressed onto a Nafion 117 membrane, and (ii) an E-TEK gas diffusion cathode that was coated with 5 g m^{-2} Pt, and hot-pressed onto a Nafion 117 membrane. Both cathodes performed similarly. Tartakovsky *et al.* also tested the E-TEK gas diffusion cathode in a membrane-less design (Tartakovsky *et al.* 2009).

13.4 FUTURE POSSIBILITIES

The ORR and HER are just two possibilities of cathodic reactions that can be coupled to the anodic oxidation of organics in BESs and it is likely that in the future many other bioelectrochemical processes will be developed by coupling other cathodic processes. The reductive power derived from waste streams could be used for the cathodic destruction of hazardous compounds such as chromate (Frenzel *et al.* 2006), bromate (Kishimoto and Matsuda 2009), and chlorinated compounds (Cheng *et al.* 1997). Also BESs could be used for the production of valuable chemicals (Clauwaert *et al.* 2008b; Rozendal *et al.* 2008a). For instance, in the literature, several studies have shown that hydrogen peroxide can efficiently be produced at a cathode by the partial reduction of oxygen (Eq. 13.2) (Foller and Bombard 1995; Panizza and Cerisola 2008). By coupling the cathodic production of hydrogen peroxide to the anodic oxidation of organics in a BES, a method could be developed for the selective production of hydrogen peroxide from wastewater, which could have many interesting industrial applications, such as for pulp bleaching (Foller and Bombard 1995; Panizza and Cerisola 2008) and/or for electro-Fenton processes (Da Pozzo *et al.* 2005). Furthermore. in the context of reverse osmosis filtration, Clauwaert *et al.* have also suggested the possibility of using cathodic formation of hydrogen peroxide for a combined process for disinfection and pollutant removal (Clauwaert *et al.* 2008a).

In summary, electricity production with MFCs and hydrogen production with MECs are just the first two examples of what is possible in BESs. The reductive power generated in the biological oxidation of organics at the anode can be coupled to a wide range of possible cathode reactions, which is expected to result in the development of many more innovative bioelectrochemical processes in the future. Interesting bioelectrochemical processes can also be expected from the future developments in the field of biological cathodes that are catalyzed by enzymes of whole cell micro-organisms. These exciting possibilities will be discussed in Chapter 15.

REFERENCES

Andersen, T.N., Dandapani, B.S. and Berry, J.M. (1993) Hydrogen evolution studies in neutral media. *J. Electroanal. Chem.* **357**(1–2), 77–89.

Armstrong, R.D. and Bell, M.F. (1978) Tungsten carbide catalysts for hydrogen evolution. *Electrochim. Acta* **23**(11), 1111–1115.

Baranton, S., Coutanceau, C., Leger, J.M., Roux, C. and Capron, P. (2005) Alternative cathodes based on iron phthalocyanine catalysts for mini- or micro-DMFC working at room temperature. *Electrochim. Acta* **51**(3), 517–525.

Bard, A.J., Parsons, R. and Jordan, J. (eds) (1985) *Standard Potentials in Aqueous Solution*. Marcel Dekker, New York.

Bard, A.J., Inzelt, G. and Scholz, F. (eds) (2008) *Electrochemical Dictionary*. Springer, Berlin Heidelberg.

Brett, C.M.A. and Brett, A.M.O. (1993) *Electrochemistry: Principles, Methods, and Applications*. Oxford University Press, Oxford.

Burakov, V.S., Butsen, A.V., Bruser, V., Harnisch, F., Misakov, P.Y., Nevar, E.A., Rosenbaum, M., Savastenko, N.A. and Tarasenko, N.V. (2008) Synthesis of tungsten carbide nanopowder via submerged discharge method. *J. Nanopart. Res.* **10**(5), 881–886.

Call, D. and Logan, B.E. (2008) Hydrogen production in a single chamber microbial electrolysis cell lacking a membrane. *Environ. Sci. Technol.* **42**(9), 3401–3406.

Call, D.F., Merrill, M.D. and Logan, B.E. (2009) High surface area stainless steel brushes as cathodes in microbial electrolysis cells. *Environ. Sci. Technol.* **43**(6), 2179–2183.

Chae, K.J., Choi, M.J., Lee, J., Ajayi, F.F. and Kim, I.S. (2008) Biohydrogen production via biocatalyzed electrolysis in acetate-fed bioelectrochemical cells and microbial community analysis. *Int. J. Hydrogen Energy* **33**(19), 5184–5192.

Cheng, I.F., Fernando, Q. and Korte, N. (1997) Electrochemical dechlorination of 4-chlorophenol to phenol. *Environ. Sci. Technol.* **31**(4), 1074–1078.

Cheng, S., Liu, H. and Logan, B.E. (2006a) Increased performance of single-chamber microbial fuel cells using an improved cathode structure. *Electrochem. Comm.* **8**, 489–494.

Cheng, S., Liu, H. and Logan, B.E. (2006b) Power densities using different cathode catalysts (Pt and CoTMPP) and polymer binders (Nafion and PTFE) in single chamber microbial fuel cells. *Environ. Sci. Technol.* **40**(1), 364–369.

Cheng, S. and Logan, B.E. (2007) Sustainable and efficient biohydrogen production via electrohydrogenesis. *Proc. Natl. Acad. Sci. USA* **104**(47), 18871–18873.

Clauwaert, P., Aelterman, P., Pham, T.H., De Schamphelaire, L., Carballa, M., Rabaey, K. and Verstraete, W. (2008a) Minimizing losses in bio-electrochemical systems: the road to applications. *Appl. Microbiol. Biotechnol.* **79**(6), 901–913.

Clauwaert, P., Tolêdo, R., Van der Ha, D., Crab, R., Verstraete, W., Hu, H., Udert, K.M. and Rabaey, K. (2008b) Combining biocatalyzed electrolysis with anaerobic digestion. *Water Sci. Technol.* **57**(4), 575–579.

Conway, B.E. and Jerkiewicz, G. (2000) Relation of energies and coverages of underpotential and overpotential deposited H at Pt and other metals to the 'volcano curve' for cathodic H-2 evolution kinetics. *Electrochim. Acta* **45**(25–26), 4075–4083.

Da Pozzo, A., Di Palma, L., Merli, C. and Petrucci, E. (2005) An experimental comparison of a graphite electrode and a gas diffusion electrode for the cathodic production of hydrogen peroxide. *J. Appl. Electrochem.* **35**(4), 413–419.

Ditzig, J., Liu, H. and Logan, B.E. (2007) Production of hydrogen from domestic wastewater using a bioelectrochemically assisted microbial reactor (BEAMR). *Int. J. Hydrogen Energy* **32**, 2296–2304.

El-Anadouli, B.E., Khader, M.M., Saleh, M.M. and Ateya, B.G. (1991) Hydrogen evolution on metallized plastic packed-bed electrodes. *Electrochim. Acta* **36**(13), 1899–1905.

Evans, D.J. and Pickett, C.J. (2003) Chemistry and the hydrogenases. *Chemical Society Reviews* **32**(5), 268–275.

Fan, Y.Z., Hu, H.Q. and Liu, H. (2007) Enhanced Coulombic efficiency and power density of air-cathode microbial fuel cells with an improved cell configuration. *J. Power Sources* **171**, 348–354.

Foller, P.C. and Bombard, R.T. (1995) Processes for the production of mixtures of caustic soda and hydrogen peroxide via the reduction of oxygen. *J. Appl. Electrochem.* **25**(7), 613–627.

Fornero, J.J., Rosenbaum, M., Cotta, M.A. and Angenent, L.T. (2008) Microbial fuel cell performance with a pressurized cathode chamber. *Environ. Sci. Technol.* **42**(22), 8578–8584.

Freguia, S., Rabaey, K., Yuan, Z. and Keller, J. (2007) Non-catalyzed cathodic oxygen reduction at graphite granules in microbial fuel cells. *Electrochim. Acta* **53**, 598–603.

Frenzel, I., Holdik, H., Barmashenko, V., Stamatialis, D.F. and Wessling, M. (2006) Electrochemical reduction of dilute chromate solutions on carbon felt electrodes. *J. Appl. Electrochem.* **36**(3), 323–332.

Fukuzumi, S. and Imahori, H. (2001) Biomimetic electron-transfer chemistry of porphyrins and metalloporphyrins. In *Electron Transfer in Chemistry*. Balzani, V. (ed.), pp. 927–975.

Gloaguen, F., Lawrence, J.D. and Rauchfuss, T.B. (2001) Biomimetic hydrogen evolution catalyzed by an iron carbonyl thiolate. *J. Am. Chem. Soc.* **123**(38), 9476–9477.

Harnisch, F., Schröder, U., Quaas, M. and Scholz, F. (2009a) Electrocatalytic and corrosion behaviour of tungsten carbide in near-neutral pH electrolytes *Appl. Catal. B-Environ.* **87**, 63–69.

Harnisch, F., Sievers, G. and Schröder, U. (2009b) Tungsten carbide as electrocatalyst for the hydrogen evolution reaction in pH neutral solutions. (In press), doi:10.1016/j.apcatb.2009.01.003

Hinnemann, B., Moses, P.G., Bonde, J., Jorgensen, K.P., Nielsen, J.H., Horch, S., Chorkendorff, I. and Norskov, J.K. (2005) Biomimetic hydrogen evolution: MoS2 nanoparticies as catalyst for hydrogen evolution. *J. Am. Chem. Soc.* **127**(15), 5308–5309.

Hu, H.Q., Fan, Y.Z. and Liu, H. (2008) Hydrogen production using single-chamber membrane-free microbial electrolysis cells. *Water Res.* **42**(15), 4172–4178.

Jahnke, H., Schönborn, M. and Zimmermann, G. (1976) Organic dyestuffs as catalysts for fuel cells. Springer, Berlin Heidelberg, pp. 133–181.

Jasinski, R. (1964) A new fuel cell cathode catalyst. *Nature* **201**(492), 1212–1213.

Kibler, L.A. (2006) Hydrogen electrocatalysis. *Chemphyschem* **7**(5), 985–991.

Kinoshita, K. (1992) *Electrochemical Oxygen Technology*. John Wiley & Sons, Inc., New York.

Kishimoto, N. and Matsuda, N. (2009) Bromate ion removal by electrochemical reduction using an activated carbon felt electrode. *Environ. Sci. Technol.* **43**(6), 2054–2059.

Knutson, T.L., Bollinger, M.L. and Smyrl, W.H. (2008) Mesoporous microspheres of pt and Pt/Ru formed by electrodeposition: characterization of high surface area. *J. Electrochem. Soc.* **155**(2), F17–F21.

Kreysa, G., Schenck, K., Sell, D. and Vuorilehto, K. (1994) Bioelectrochemical hydrogen-production. *Int. J. Hydrogen Energy* **19**(8), 673–676.

Lee, S.C. and Holm, R.H. (2003) Speculative synthetic chemistry and the nitrogenase problem. *Proc. Natl. Acad. Sci. USA* **100**(7), 3595–3600.

Lide, R.L. (ed.) (2004) *CRC Handbook of Chemistry and Physics*. CRC Press, Boca Raton.

Liu, H., Grot, S. and Logan, B.E. (2005) Electrochemically assisted microbial production of hydrogen from acetate. *Environ. Sci. Technol.* **39**(11), 4317–4320.

Liu, H., Cheng, S., Huang, L. and Logan, B.E. (2008a) Scale-up of membrane-free single-chamber microbial fuel cells. *J. Power Sources* **179**, 274–279.

Liu, W.Z., Wang, A.J., Ren, N.Q., Zhao, X.Y., Liu, L.H., Yu, Z.G. and Lee, D.J. (2008b) Electrochemically assisted biohydrogen production from acetate. *Energy Fuels* **22**(1), 159–163.

Logan, B.E., Hamelers, B., Rozendal, R., Schröder, U., Keller, J., Freguia, S., Aelterman, P., Verstraete, W. and Rabaey, K. (2006) Microbial fuel cells: methodology and technology. *Environ. Sci. Technol.* **40**(17), 5181–5192.

Logan, B.E., Call, D., Cheng, S., Hamelers, H.V.M., Sleutels, T.H.J.A., Jeremiasse, A.W. and Rozendal, R.A. (2008) Microbial electrolysis cells for high yield hydrogen gas production from organic matter. *Environ. Sci. Technol.* **42**(23), 8630–8640.

Lojou, E. and Bianco, P. (2004) Electrocatalytic reactions at hydrogenase-modified electrodes and their applications to biosensors: From the isolated enzymes to the whole cells. *Electroanalysis* **16**(13–14), 1093–1100.

Millet, P., Durand, R. and Pineri, M. (1990) Preparation of new solid polymer electrolyte composites for water electrolysis. *Int. J. Hydrogen Energy* **15**(4), 245–253.

Miura, Y., Tsujimura, S., Kamitaka, Y., Kurose, S., Kataoka, K., Sakurai, T. and Kano, K. (2007) Bioelectrocatalytic reduction of O_2 catalyzed by CueO from Escherichia coli adsorbed on a highly oriented pyrolytic graphite electrode. *Chem. Lett.* **36**(1), 132–133.

Morozov, S.V., Karyakina, E.E., Zadvornyi, O.A., Zorin, N.A., Varfolomeev, S.D. and Karyakin, A.A. (2002) Bioelectrocatalysis by hydrogenase *Th. roseopersicina* immobilized on carbon materials. *Russ. J. Electrochem.* **38**(1), 97–102.

Morris, J.M., Jin, S., Wang, J.Q., Zhu, C.Z. and Urynowicz, M.A. (2007) Lead dioxide as an alternative catalyst to platinum in microbial fuel cells. *Electrochem. Comm.* **9**(7), 1730–1734.

Norskov, J.K., Bligaard, T., Logadottir, A., Kitchin, J.R., Chen, J.G. and Pandelov, S. (2005) Trends in the exchange current for hydrogen evolution. *J. Electrochem. Soc.* **152**(3), J23–J26.

Olivares-Ramirez, J.M., Campos-Cornelio, M.L., Godinez, J.U., Borja-Arco, E. and Castellanos, R.H. (2007) Studies on the hydrogen evolution reaction on different stainless steels. *Int. J. Hydrogen Energy* **32**(15), 3170–3173.

Panizza, M. and Cerisola, G. (2008) Electrochemical generation of H_2O_2 in low ionic strength media on gas diffusion cathode fed with air. *Electrochim. Acta* **54**, 876–878.

Pershad, H.R., Duff, J.L.C., Heering, H.A., Duin, E.C., Albracht, S.P.J. and Armstrong, F.A. (1999) Catalytic electron transport in *Chromatium vinosum* [NiFe]-hydrogenase: Application of voltammetry in detecting redox-active centers and establishing that hydrogen oxidation is very fast even at potentials close to the reversible H+/H-2 value. *Biochemistry* **38**(28), 8992–8999.

Petrii, O.A. and Tsirlina, G.A. (1994) Electrocatalytic activity prediction for hydrogen electrode-reaction: intuition, art, science. *Electrochim. Acta* **39**(11–12), 1739–1747.

Rabaey, K., Clauwaert, P., Aelterman, P. and Verstraete, W. (2005) Tubular microbial fuel cells for efficient electricity generation. *Environ. Sci. Technol.* **39**(20), 8077–8082.

Roche, I., Chainet, E., Chatenet, M. and Vondrak, J. (2007) Carbon-supported manganese oxide nanoparticles as electrocatalysts for the oxygen reduction reaction (ORR) in alkaline medium: physical characterizations and ORR mechanism. *J. Phys. Chem. C* **111**(3), 1434–1443.

Rosenbaum, M., Zhao, F., Quaas, M., Wulff, H., Schroder, U. and Scholz, F. (2007) Evaluation of catalytic properties of tungsten carbide for the anode of microbial fuel cells. *Appl. Catal. B-Environ.* **74**(3–4), 261–269.

Rozendal, R.A. and Buisman, C.J.N. (2005) Process for producing hydrogen, patent WO2005005981, Wageningen Universiteit.

Rozendal, R.A., Hamelers, H.V.M., Euverink, G.J.W., Metz, S.J. and Buisman, C.J.N. (2006) Principle and perspectives of hydrogen production through biocatalyzed electrolysis. *Int. J. Hydrogen Energy* **31**(12), 1632–1640.

Rozendal, R.A., Hamelers, H.V.M., Molenkamp, R.J. and Buisman, C.J.N. (2007) Performance of single chamber biocatalyzed electrolysis with different types of ion exchange membranes. *Water Res.* **41**, 1984–1994.

Rozendal, R.A., Hamelers, H.V.M., Rabaey, K., Keller, J. and Buisman, C.J.N. (2008a) Towards practical implementation of bioelectrochemical wastewater treatment. *Trends Biotechnol.* **26**(8), 450–459.

Rozendal, R.A., Jeremiasse, A.W., Hamelers, H.V.M. and Buisman, C.J.N. (2008b) Hydrogen production with a microbial biocathode. *Environ. Sci. Technol.* **42**, 629–634.

Rozendal, R.A., Sleutels, T.H.J.A., Hamelers, H.V.M. and Buisman, C.J.N. (2008c) Effect of the type of ion exchange membrane on performance, ion transport, and pH in biocatalyzed electrolysis of wastewater. *Water Sci. Technol.* **57**(11), 1757–1762.

Saraby-Reintjes, A. (1986) The hydrogen evolution reaction under mixed kinetic control. *J. Chem. Soc., Faraday Trans. I* **82**, 3343–3355.

Schaetzle, O., Barrière, F. and Schröder, U. (2009) An improved microbial fuel cell with laccase as the oxygen reduction catalyst. *Energy Environ. Sci.* **2**, 96–99.

Selembo, P.A., Merrill, M.D. and Logan, B.E. (2009) The use of stainless steel and nickel alloys as low-cost cathodes in microbial electrolysis cells. *Journal of Power Sources*. (In press) doi:10.1016/j.jpowsour.2008.12.144

Sell, D., Kramer, P. and Kreysa, G. (1989) Use of an oxygen gas diffusion cathode and a three-dimensional packed-bed anode in a bioelectrochemical fuel cell. *Appl. Microbiol. Biotechnol.* **31**(2), 211–213.

Suffredini, H.B., Cerne, J.L., Crnkovic, F.C., Machado, S.A.S. and Avaca, L.A. (2000) Recent developments in electrode materials for water electrolysis. *Int. J. Hydrogen Energy* **25**(5), 415–423.

Tartakovsky, B., Manuel, M.F., Neburchilov, V., Wang, H. and Guiot, S.R. (2008) Biocatalyzed hydrogen production in a continuous flow microbial fuel cell with a gas phase cathode. *J. Power Sources* **182**(1), 291–297.

Tartakovsky, B., Manuel, M.F., Wang, H. and Guiot, S.R. (2009) High rate membrane-less microbial electrolysis cell for continuous hydrogen production *Int. J. Hydrogen Energy* **34**(2), 672–677.

Tilak, B.V., Ramamurthy, A.C. and Conway, B.E. (1986) High performance electrode materials for the hydrogen evolution reaction from alkaline media. *Proc. Indian Acad. Sci. (Chem. Sci.)* **97**(3–4), 359–393.

Trasatti, S. (1972) Work function, electronegativity, and electrochemical behavior of metals: III. Electrolytic hydrogen evolution in acid solutions. *J. Electroanal. Chem.* **39**(1), 163–184.

Trasatti, S. (1995) Electrochemistry and environment: the role of electrocatalysis. *Int. J. Hydrogen Energy* **20**(10), 835–844.

Tsujimura, S., Kawaharada, M., Nakagawa, T., Kano, K. and Ikeda, T. (2003) Mediated bioelectrocatalytic O-2 reduction to water at highly positive electrode potentials near neutral pH. *Electrochem. Comm.* **5**(2), 138–141.

Vetter, K.J. (1967) *Electrochemical Kinetics: Theoretical and Experimental Aspects.* Academic Press, New York.

Vignais, P.M. and Colbeau, A. (2004) Molecular biology of microbial hydrogenases. *Curr. Issues Mol. Biol.* **6**, 159–188.

Zhao, F., Harnisch, F., Schröder, U., Scholz, F., Bogdanoff, P. and Herrmann, I. (2005) Application of pyrolysed iron(II) phthalocyanine and CoTMPP based oxygen reduction catalysts as cathode materials in microbial fuel cells. *Electrochem. Commun.* **7**(12), 1405–1410.

Zhao, F., Harnisch, F., Schröder, U., Scholz, F., Bogdanoff, P. and Herrmann, I. (2006) Challenges and constraints of using oxygen cathodes in microbial fuel cells. *Environ. Sci. Technol.* **40**(17), 5193–5199.

14

Bioelectrochemical reductions in reactor systems

Peter Clauwaert

14.1 INTRODUCTION

In BESs, at least one of the anodic or cathodic reactions is biologically catalyzed (Rabaey *et al.* 2007). Up till now, most BESs consisted of a biocatalyzed anode combined with an abiotic cathode. Typically, platinum is used as a catalyst as a way to lower cathode activation overpotentials for oxygen or proton reduction (Logan *et al.* 2007; Rozendal *et al.* 2006). The sustainability of platinum as a cathode catalyst in BESs has been questioned since the reactive sites can be poisoned easily by phosphate, carbonate, chloride and microbial intermediates like sulfides (Zhao *et al.* submitted). Moreover, the production of platinum containing cathodes is not environmentally friendly, which conflicts with the sustainable nature of BESs (Clauwaert *et al.* 2007b). Cheaper alternatives such as pyrolyzed

 Bioelectrochemical Systems: From Extracellular Electron Transfer to Biotechnological Application. Edited by Korneel Rabaey, Largus Angenent, Uwe Schröder and Jürg Keller. ISBN: 9781843392330. Published by IWA Publishing, London, UK.

iron(II) phthalocyanine and CoTMPP have been proposed as catalysts for oxygen reductions in BESs (Zhao *et al.* 2005) (see previous chapter).

Recently, there has been an increasing interest in replacing abiotic cathodes for biocathodes in which enzymes enhance the reduction catalysis (He and Angenent 2006). Two types of biocathodes can be distinguished: biocathodes in which living microorganisms consume part of the electrical energy from the cathodic electron flow that goes through their respiratory chain with concomitant ATP production (type I) and biocathodes where no electrical energy is converted into ATP (type II). Enzymatic biocathodes where enzymes are incorporated in the cathode (Cracknell *et al.* 2008) or biocathodes in which algae produce oxygen for chemical reduction (Berk and Canfield 1964) are examples of biocathodes where the cathode reduction is not resulting in ATP generation (type II biocathodes). In this chapter, only biocathodes that are believed to be type I biocathodes are discussed.

The effectiveness of the reduction catalysis of a biocathode in a BES is typically directly related to the current density, thus the cathodic conversion rate. As microorganisms in a biocathode develop strategies that lower the cathode overpotentials, higher current densities can be obtained. Overpotentials are defined as the difference between the equilibrium potential (electrode potential during OCV) and the actual electrode potential at a specific current density (Clauwaert *et al.* 2008a). In biocathodes, living microorganisms enhance the reduction catalysis and they virtually act as catalysts. However strictly spoken, the term catalyst conflicts with the energy consuming, self-renewing and self-organizing nature of the microorganisms (Clauwaert *et al.* 2008a; Schroder 2007). Typically, chemolithoautotrophic microorganisms are believed to be the main actors in these type I biocathodes since the cathode acts as the electron donor for their metabolism. As a consequence, carbon needs to be fixed from carbon dioxide or bicarbonate in the environment of these chemolithoautotrophic microorganisms. A growth period with an organic electron donor has been described to enable biocathode activity (Rozendal *et al.* 2008; Strycharz *et al.* 2008). In some cases, the effectiveness of a biocathode is measured by the maximization of the biomass production (Trash and Coates 2008 and references therein).

14.2 AEROBIC BIOCATHODES

A biofilm is readily formed on stainless steel in aerobic natural (marine) environments inducing an increase in the equilibrium potential to values between +0.440 and +0.690 V vs. SHE (Nguyen *et al.* 2007). In this microbially induced corrosion (MIC) process, electrons are provided through oxidation from the underlaying alloys for enhanced aerobic reduction catalysis at the surface (Bergel *et al.* 2005). This biologically induced increase in the potential is called *potential Ennoblement* and several mechanisms have been proposed to explain this phenomenon.

- If the electrochemical reduction of oxygen to hydrogen peroxide is coupled with an enzymatic decomposition by catalase yielding oxygen, higher cathodic currents could be obtained due to higher oxygen concentrations in the proximity of the surface (Busalmen *et al.* 2002) (Type II biocathode).
- Hydrogen peroxide production by heterotrophic microorganisms, possibly coupled with biological acidification might enhance the reduction catalysis at the stainless steel surface (Scotto and Lai 1998) (Type II biocathode).
- Microorganisms can scavenge oxidized iron with the help of iron chelators (siderophores) in order to transport them into the microbial cells, where they are assimilatorily reduced and incorporated in proteins (Scotto and Lai 1998) (Type II biocathode).
- MnO_2 and MnOOH can electrochemically be reduced to Mn^{2+} at electrode potentials in the order of +0.600 V vs. SHE, while microbial oxidation of Mn^{2+} by manganese oxidizing bacteria (MOB) (eg. species from the genera *Leptothrix* and *Siderocapsa*) with MnO_2 deposition can complete this depolarizing cycle (Dickinson and Lewandowski 1998) (Type I biocathode).

Bergel and coworkers combined a chemical hydrogen oxidizing anode with a stainless steel cathode in a saline medium. The development of a biofilm on the cathode enhanced the current production to maximum 1.89 A m^{-2} (95 A m^{-3} MFC) (maximum power production: 0.32 W m^{-2} (13 W m^{-3} MFC)) (Bergel *et al.* 2005). Orfei and co-workers combined a bio-anode buried in marine sediment with a stainless steel cathode covered with a well developed biofilm and demonstrated that cathodical protection could be established (Orfei *et al.* 2006). Also on carbon fiber cathodes in sea-water batteries, biofilms have been associated with enhanced oxygen reduction catalysis (Hasvold *et al.* 1997). It was unclear in these three types of BESs if the oxygen reduction was performed by chemolithoautotrophic microorganisms generating proton motive by the passage of current trough their electron transfer chain, or if they just enhanced the oxygen reduction properties of the cathode itself (see higher for hypotheses for *Potential Ennoblement*). The community of a marine aerobic biofilm on a stainless steel cathode was investigated and only 2 out of 24 isolates were capable of generating cathodic current on stainless steel coupons: *Winogradskyella poriferorum* and *Acinetobacter johnsonii* (Erable, unpublished results).

Rhoads and coworkers established a first complete biological microbial fuel cell with reticulated vitreous carbon electrodes, where 2-hydroxy-1,4 naphtaquinone was used as the electron shuttle in the anode for glucose oxidation by *Klebsiella pneumoniae* and biomineralized manganese was used as the electron shuttle for oxygen reduction by *Leptothrix discophora* (maximum power production of 0.13 W m^{-2} (3.3 W m^{-3} MFC)) (Rhoads *et al.* 2005). The solid nature of the manganese oxides is likely to slow down the wash-out of the cathodic electron shuttle in a continuous system.

Besides manganese, iron also can be used as the electron shuttle in a biocathode. The growth of aerobic acidophilic iron oxidizing organisms, such as *Acidothiobacillus ferrooxidans* or *Leptospirillum ferrooxidans*, could be enhanced by the electrochemical regeneration of Fe(II) by a cathode in an acid solution (Kinsel and Umbreit 1964; Thrash and Coates 2008). Typically, these biocathodes were coupled with a chemical anode where water or methanol was oxidized with the help of a power supply (Blake *et al.* 1994; Lopez-Lopez *et al.* 1999). TerHeijne and co-workers demonstrated that an acetate oxidizing bio-anode could be combined with an iron reducing abiotic cathode producing electrical energy, while iron was aerobically reoxidized in an external bioreactor (Ter Heijne *et al.* 2007). A bipolar membrane between anode and cathode lowered the need for acid supply to the cathode significantly (Ter Heijne *et al.* 2007). The maximum current and power density obtained was 4.4 A m^{-2} (103 A m^{-3} MFC) and the maximum power density was 1.2 W m^{-2} (28 W m^{-3} MFC).

Cathodically produced hydrogen can also be used to mediate electrons between the cathode and the aerobic microorganism. So called 'Knallgas' bacteria (e.g. *Ralstonia eutropha* H16) could be grown in an electrolysis cell where the anode supplied the bacteria with oxygen while the cathode provided hydrogen (Schlegel and Lafferty 1965; Thrash and Coates 2008). The *in-situ* production of these gases might be used to safely grow cells with high concentrations of biopolymers with an added value like polyhydroxyalkanoates, when operated under optimal conditions (Clauwaert *et al.* 2008a; Ishizaki *et al.* 1993).

Oxygen reducing biocathodes have also been demonstrated without the help of externally added electron mediators (Clauwaert *et al.* 2007b; Rabaey *et al.* 2008). After a start-up of approximately 20 days, the cathode equilibrium potential typically increased from +0.100 V vs. SHE to values higher than +0.400 V vs. SHE, comparable to what is observed during *Potential Ennoblement* on stainless steel. Chemically precipitated manganese oxides were not necessary as electron shuttles, however the start-up period was approximately 30% shorter compared with biocathodes not requiring manganese oxides (Clauwaert *et al.* 2007b). The maximum current density was 3.6 A m^{-2} (251 A m^{-3} MFC) and the maximum power density was 1.4 W m^{-2} (95 W m^{-3} MFC) (Clauwaert *et al.* 2007b;

Clauwaert *et al.* 2008c). It is possible that peroxide is formed on such carbon felts exposed to the air, which might prolong the start-up of the biocathode. Manganese oxides are able to catalyze the peroxide oxidation *in-situ* and in that way the conditions for biocathode development might have been favored (Hasan *et al.* 1999).

Different types of sludges and sediments were mixed as an inoculum and it was found that mainly α-Proteobacteria and γ-Proteobacteria were enriched on these biocathodes under nitrogen limited conditions (Clauwaert *et al.* 2007b). Rabaey and co-workers, on the other hand, found enrichment of Bacteroidetes, β-Proteobacteria and γ-Proteobacteria from mixed inoculums originating from activated sludge, pond sediment and samples from rusted metal poles (Rabaey *et al.* 2008). In that research, it was found that an MFC with a biocathode containing a mixed biocathode culture was able to establish higher current and power densities compared with MFCs with pure cultures isolated from this biocathode. However, the lower performance of the pure cultures could have been due to the stronger uncompensated alkalinization at the cathode (Rabaey *et al.* 2008). Freguia and co-workers used the effluents of the anodes of their MFCs as the influent for the cathodes as a way to compensate for the alkalinization in the cathode. After 5 to 10 days, they observed a higher current and power production due to biocathode activity, compared to the not catalyzed cathodes (Freguia *et al.* 2008). A continuous recirculation of a common electrolyte over anodes and cathodes was demonstrated as a way to operate 2.5 l MFCs with open air biocathodes in an overall pH neutral way since no acid nor base supply was needed (Clauwaert *et al.* 2009).

14.3 ANOXIC AND ANAEROBIC BIOCATHODES

Autotrophic denitrification is the biological process in which microorganisms respire an inorganic electron donor, typically H_2 or reduced sulfur, iron or manganese species, while using nitrate, nitrite, nitric oxide or nitrous oxide as electron acceptor (Van Rijn *et al.* 2006). Heterotrophic denitrifying microorganisms respire an organic electron donor in the denitrification process (Van Rijn *et al.* 2006).

Autotrophic denitrification is considered to produce less biomass compared to heterotrophic denitrification, while also contamination with organic carbon can be avoided in the case of autotrophic denitrification (Van Rijn *et al.* 2006). This is especially relevant in the case of drink water purification. Hydrogenotrophic denitrification rates up 0.7 kg N m^{-3} d^{-1} have been described for *Alcaligenes eutrophus* (Chang *et al.* 1999) while, to our knowledge, the maximum hydrogenotrophic denitrification rate by a mixed microbial community described in the literature is 1.4 kg N m^{-3} d^{-1} (Germonpre *et al.* 1991). A low hydrogen solubility (1.6 mg H_2 l^{-1} at 22°C) and safety issues with regard to hydrogen transportation and storage have raised interest in *in-situ* production of hydrogen by electrolysis of water. However, the denitrification rates in such electrolysis cells, consuming large amounts of electrical energy, have been limited to 0.06 kg N m^{-3} d^{-1} (Feleke *et al.* 1998; Sakakibara and Kuroda 1993). In some cases the anodically produced oxygen could be used for nitrification while the hydrogen produced in the cathode was used for the denitrification (Kuroda *et al.* 1996). In other cases, no oxygen was produced at the anode, but carbon oxidation of the anode resulted in an anoxic operation of this cell, while continuously producing a bicarbonate pH buffer (Feleke *et al.* 1998; Sakakibara *et al.* 1994). Local alkalinization, besides hydrogen overproduction, was identified as one of the main limiting factors (Trash and Coates 2008 and therein).

In 1966, the use of a nitrate reducing biocathode was proposed (while mentioning that it actually may not function) (Lewis 1966). Only in 2004 did Gregory and co-workers demonstrate for the first time the ability of species, enriched from a sediment, to retrieve electrons directly from a cathode poised at approximately −0.300 V vs. SHE for nitrate to nitrite reduction (4 A m^{-3} Nett Cathode Compartment (NCC) (Gregory *et al.* 2004). There was an enrichment of γ-Proteobacteria, Gram-positive bacteria and especially of Geobacteraceae (δ-Proteobacteria), while there was a decrease of α-Proteobacteria, β-Proteobacteria and Cytophaga-Flavobacterium-Bacteriodes (CFB) compared to a control. A pure culture of *Geobacter metallireducens* was shown to reduce nitrate to nitrite by a cathode poised at approximately −0.300 V vs. SHE. This is remarkable, since *Geobacter metallireducens* is unable to oxidize hydrogen as an electron donor, which excludes hydrogen as an electron shuttle for nitrate reduction (Gregory *et al.* 2004).

When a mixture of different types of sludge and sediment was used as an inoculum for a denitrifying biocathode, a complete denitrification from nitrate to N_2 could be established (Clauwaert *et al.* 2007a). This biocathode could be combined with an acetate oxidizing bioanode and this tubular microbial fuel cell was still able to produce electrical energy up to 8 W m^{-3} NCC and establishing a maximum denitrification rate of 0.15 kg N m^{-3} NCC d^{-1} (58 A m^{-3} NCC). Operation under pH controlled conditions resulted in an increased denitrification rate of 0.59 kg N m^{-3} NCC d^{-1} (235 A m^{-3} NCC) (Clauwaert *et al.* submitted). Virdis and co-workers demonstrated the possibility of using the anode effluent of an MFC as the influent for a denitrifying biocathode after a nitrification step (Virdis *et al.* 2008). This strategy does not only enable a sequential COD and nitrogen removal in an MFC, but it also allows an overall pH neutral operation since the alkalinization in the cathode could be compensated with the acidification of the acetate oxidation in the anode and ammonium oxidation in the external nitrification tank (Virdis *et al.* 2008). In this way, the denitrification rate could be increased to 0.33 kg N m^{-3} NCC d^{-1} (133 A m^{-3} TCC) (Virdis *et al.* 2008). In these studies, cation selective membranes were used. Since ammonium can diffuse from anode to cathode without nitrification, the use of anion selective membranes could avoid the presence of ammonium in the cathode effluent.

Besides denitrification, perchlorate reduction with the help of an anoxic biocathode was demonstrated in the presence of 0.5–5.0 mmol/l 2,6-anthraquinone disulfonate (AQDS) (Thrash *et al.* 2007). Perchlorate is widely used as a propellant in the aerospace and defense industries, and is of environmental concern due to its high mobility and inhibiting effect on thyroid function (Shea *et al.* 2008). After 70 days of operation a novel strain was isolated that could reduce perchlorate in the absence of AQDS at approximately −0.300 V vs. SHE. However, the hydrogen production rate by electrolysis in this half-cell could have accounted for the reduction of perchlorate (Thrash *et al.* 2007). Shea and co-workers adapted a mediatorless denitrifying biocathode to increasing concentrations of perchlorate while lowering the nitrate concentrations (Shea *et al.* 2008). This (per)chlorate reducing biocathode could be combined with an acetate oxidizing bioanode in a MFC configuration (Shea *et al.* 2008).

The biological reduction of other groundwater pollutants like selenate (SeO_4^{2-}), arsenate ($H_2AsO_4^{2-}$) and chromate (CrO_4^{2-}) can be achieved in the presence of hydrogen (Chung *et al.* 2007). The reduction of these compounds in a biocathode has not been demonstrated yet. On the other hand, chemical reduction of chromate in a microbial fuel has already been demonstrated without microbial catalysis (Wang *et al.* 2008).

When the cathode potential is below the theoretical equilibrium potential of −0.421 V vs. SHE (Table 14.1), hydrogen can be electrochemically produced for anaerobic oxidation. Reduction of iron(III) (Ohmura *et al.* 2002), trichloroethene (Aulenta *et al.* 2008c), 2,6-dichlorophenol (Skadberg *et al.* 1999), fumarate (Park *et al.* 1999), sulfate (Lewis 1966) and carbon dioxide (Aulenta *et al.* 2008c; Clauwaert *et al.* 2008b; Clauwaert and Verstraete 2009; Park *et al.* 1999) with hydrogen as an intermediate have been demonstrated in bioelectrochemical systems. If microorganisms effectively lower the hydrogen partial pressure from the electrode surface, the cathode overpotentials might decrease (cathodic depolarization) allowing higher turnover rates in the biocathodes (Clauwaert *et al.* 2008b; Lewis 1966). However, in one research it was recently shown that the presence of H_2 or a H_2-scavenger did not alter the current production in an MEC with a current density between 4 and 70 A m^{-3} MEC (Cheng *et al.* 2009), but it might have an effect at higher current densities.

Table 14.1. Overview of half-reactions that might be relevant in BESs. The equilibrium potentials are given between brackets in volts versus Standard Hydrogen Electrode at pH 7 ($E^{\circ\prime}$; V vs SHE). (Cochran *et al.* 1986; de Best 1999; Nerenberg and Rittmann 2004; Shantaram *et al.* 2005; Sober 1970; Thauer *et al.* 1977; Zumdahl 1998)

Half-reaction	$E^{\circ\prime}$ (V vs SHE)
$Au(III)+3\ e^- \rightarrow Au(0)$	(+1.500)
$H_2O_2+2\ e^-+2\ H^+ \rightarrow 2\ H_2O$	(+1.356)
$\frac{1}{2}\ N_2O+e^-+H^+ \rightarrow \frac{1}{2}\ N_2+\frac{1}{2}\ H_2O$	(+1.355)
$NO+e^-+H^+ \rightarrow \frac{1}{2}\ N_2O+\frac{1}{2}\ H_2O$	(+1.175)
$MnO_4^-+3\ e^-+4\ H^+ \rightarrow MnO_2+2\ H_2O$	(+1.138)
$BrO_3^-+6\ e^-+6\ H^+ \rightarrow Br^-+3\ H_2O$	(+0.996)
$O_2+4\ e^-+4\ H^+ \rightarrow 2\ H_2O$	(+0.816)
$ClO_3^-+6\ e^-+6\ H^+ \rightarrow Cl^-+3\ H_2O$	(+0.809) £
$ClO_4^-+8\ e^-+8\ H^+ \rightarrow Cl^-+4\ H_2O$	(+0.809) £
$Ag^++e^- \rightarrow Ag(0)$	(+0.800)
$ClO_4^-+2\ e^-+2\ H^+ \rightarrow ClO_3^-+H_2O$	(+0.776)
$Fe^{3+}+e^- \rightarrow Fe^{2+}$	(+0.771)
$C_2H_4Cl_2+2\ e^- \rightarrow C_2H_4+2\ Cl^-$	(+0.739) $
$MnO_{2(s)}+e^-+H^+ \rightarrow MnOOH_{(s)}$	(+0.638)
$MnOOH_{(s)}+e^-+3\ H^+ \rightarrow Mn^{2+}+2\ H_2O$	(+0.615)
$C_2Cl_4+2\ e^-+H^+ \rightarrow C_2HCl_3+Cl^-$	(+0.574) $
$C_2HCl_3+2\ e^-+H^+ \rightarrow cis\text{-}C_2H_2Cl_2+Cl^-$	(+0.550) $
$C_2H_3Cl+2\ e^-+H^+ \rightarrow C_2H_4+Cl^-$	(+0.450) $
$NO_3^-+2\ e^-+2\ H^+ \rightarrow NO_2^-+H_2O$	(+0.433)
$Cis\text{-}C_2H_2Cl_2+2\ e^-+H^+ \rightarrow C_2H_3Cl+Cl^-$	(+0.397) $
$[Fe(CN)_6]^{3-}+e^- \rightarrow [Fe(CN)_6]^{4-}$	(+0.360)
$NO_2^-+e^-+2\ H^+ \rightarrow NO+H_2O$	(+0.350)
$NO_2^-+6\ e^-+8\ H^+ \rightarrow NH_4^++2\ H_2O$	(+0.340)
$SeO_4^{2-}+6\ e^-+8\ H^+ \rightarrow Se^0+4\ H_2O$	(+0.322)
$O_{2(gas)}+2\ e^-+2\ H^+ \rightarrow H_2O_2$	(+0.295)
$HSeO_3^-+4\ e^-+5\ H^+ \rightarrow Se^0+3\ H_2O$	(+0.260)
$U(VI)+2\ e^- \rightarrow U(IV)$	(+0.246)
Ubiquinone/Ubihydroquinone	(+0.100) *
$H_2AsO_4^-+2\ e^-+3\ H^+ \rightarrow H_3AsO_3+H_2O$	(+0.052)
$Fumarate^{2-}+2\ e^-+2\ H^+ \rightarrow Succinate^{2-}$	(+0.031)
$AQDS+2\ e^-+2\ H^+ \rightarrow AQDSH_2$	(−0.184)
$2\ SO_4^{2-}+19\ H^++16\ e^- \rightarrow H_2S+HS^-+8\ H_2O$	(−0.217)
$HCO_3^-+8\ e^-+9\ H^+ \rightarrow CH_4+3\ H_2O$	(−0.244)
$2\ HCO_3^-+8\ e^-+9\ H^+ \rightarrow C_2O_2H_3^-+4\ H_2O$	(−0.278)
$NAD^++2\ e^-+H^+ \rightarrow NADH$	(−0.320)
$CrO_4^{2-}+3\ e^-+5\ H^+ \rightarrow Cr(OH)_3+H_2O$	(−0.321)
Neutral red_{ox}/neutral red_{red}	(−0.325)
$Cystine+2\ e^-+2\ H^+ \rightarrow 2\ cysteine$	(−0.340)
$2\ e^-+2\ H^+ \rightarrow H_2$	(−0.421)
Methyl $viologen_{ox}$/methyl $viologen_{red}$	(−0.440)

(£: chlorite disproportionation into chloride excluded and oxygen reduction included, $: assuming $[Cl^-] = 1$ mmol/l, *: in 95% ethanol)

Besides hydrogen, also other electron shuttles can be used for anaerobic respiration. Aulenta and co-workers demonstrated that methyl viologen, that was reduced by an electrode poised at −0.500 V vs. SHE, could be used as electron mediator for reductive dechlorination of trichloroethene, a toxic chlorinated soil and groundwater pollutant that is suspected to be a human carcinogen (Aulenta *et al.* 2007). When methyl viologen was applied in concentrations higher than 750 μmol/l in a trichloroethene reducing biocathode (−0.450 V vs. SHE), hydrogen was biologically produced (Aulenta *et al.* 2008a). These results suggest that biological hydrogen production occurs as an alternative electron sink in the absence of a more favorable electron acceptor (Aulenta *et al.* 2008a). Rozendal and co-workers demonstrated that microorganisms on a graphite electrode, pregrown as an anode for acetate and hydrogen oxidation, could also catalyze the proton reduction in mediatorless biocathodes (Rozendal *et al.* 2008).

Cathodically reduced methyl viologen has also been used for sulfate reduction in Postgate's medium C and organic sulfur reduction in crude oil (Kim *et al.* 1990). Park and Zeikus demonstrated the possibility to reduce neutral red at a cathode for biological fumarate and carbon dioxide reduction (Park *et al.* 1999).

Even in the absence of hydrogen or externally added electron mediators, reduction catalysis in anaerobic biocathodes could be demonstrated. The reduction of fumarate to succinate by *Geobacter sulfurreducens* (Gregory *et al.* 2004) and *Geobacter lovleyi* (Strycharz *et al.* 2008) on graphite electrodes could also be demonstrated on a cathode poised at −0.300 V vs. SHE. Dumas and co-workers found that the reduction rate of fumarate by *Geobacter sulfurreducens* on stainless steel electrodes (1 cm × 2.5 cm × 0.1 cm; 20.5 A m^{-2}) was significantly higher than on graphite electrodes (5 cm × 2.5 cm × 0.5 cm; 0.75 A m^{-2}) at a poised cathode potential of −0.400 V vs. SHE, although the average roughness (Ra) of the graphite (5.6 μm) was higher than for stainless steel (0.3 μm) (Dumas *et al.* 2008).

Strycharz and co-workers published preliminary results that demonstrate the use of *Geobacter lovelyi* for mediatorless reduction of tetrachlorethene to *cis*-dichloroethene after operation as a bioanode in the presence of acetate (Strycharz *et al.* 2008). Aulenta and co-workers recently demonstrated that a mixed culture containing *Dehalococcoides* spp. was able to reduce trichloroethene to more reduced products such as vinyl chloride and ethene (Aulenta *et al.* 2008b). It was remarkable that an, as yet unknown, bacterial produced electron mediator was detected (midpoint potential −0.4 V vs. SHE) that appeared to be involved in the extracellular electron transfer (Aulenta *et al.* 2008b). *Geobacter sulfurreducens* was also shown to reduce uranium(VI) to relatively insoluble uranium(IV) in an anaerobic biocathode poised at approximately −0.300 V vs. SHE (Gregory and Lovley 2005). Further research will reveal whether a wide variety of dehalogenation reactions can also be expected to take place in anaerobic biocathodes (Field and Sierra-Alvarez 2004).

Recently, a biocathode biofilm dominated by *Methanobacterium palustre* was shown to produce methane in a direct way without hydrogen as an intermediate at cathode potentials lower than −0.6 V vs. SHE (Cheng *et al.* 2009). Since biocathodes appear to be able to produce hydrogen and methane in a direct way, biocathodical acetogenesis might also be possible.

14.4 ELECTRON TRANSFER IN BIOCATHODES

In a basic aerobic respiratory electron transfer chain (NADH-Q-cytbc1-cytc-cytaa3), 5 mol protons are pumped into the periplasm per passage of 1 mol electrons. Three to four moles of protons are needed to make 1 mol ATP (ADP + $P_i \rightarrow$ ATP) in the ATP synthase (1.25 mol ATP mol^{-1} electrons) and 30.5 kJ of energy is fixed per mol ATP and thus 38 kJ of energy is fixed per mol electrons. The potential energy that can be liberated from NADH oxidation (−0.320 V vs. SHE) to oxygen reduction (+0.816 V vs. SHE) is 110 kJ mol^{-1} electrons. This means that in this aerobic electron transfer chain that only 0.40 V of the potential energy of 1.14 V can be converted into ATP (35 %). In practice however, different and more complicated electron transfer chains can simultaneously be present and the ATP yield can be lower. However, at least one proton pump needs to be present to generate ATP.

In biocathodes, the equilibrium potential of the cathode might be interpreted as a rough estimation of the equilibrium potential of the enzymatic complex that functions as the electron donor for the microbial electron transfer chains in the biocathode. In aerobic biocathodes, the equilibrium potential is around +0.5 V vs. SHE, which would suggest that electrons enter the electron transfer chain by cytochromes which divert the electrons to the oxidase enzyme. In that way, only about 1 mol of protons would be pumped to the periplasm per mol of electrons (assuming 1.25 mol ATP mol^{-1} electrons). In that way, the energy yield under the form of ATP for the biocathodic microorganisms would be 7.6 kJ mol^{-1} electrons. The 0.3 V voltage difference between the equilibrium potential and the oxygen reduction potential accounts for a potential energy of 29 kJ mol^{-1} electrons (26% efficiency). For chemically catalyzed cathodes, the difference between the equilibrium cathode potential and the theoretical equilibrium cathode potential calculated from the Nernst equation is attributed to mixed potential effects (see previous chapter). It needs further investigation whether mixed potential effects are also present in biocathodes. As the cathode potential decreases with increasing currents, it is not unlikely that alternative electron transfer mechanisms become thermodynamically feasible. In that way, extra proton pumps might become active, which can increase the ATP yield for the biocathodic microorganisms.

For denitrifying biocathodes, the equilibrium cathode potential is typically around 0 V vs. SHE. It can be hypothesized that, either directly or with the help of endogenous electron mediators, quinone (Q)-like molecules are reduced and that several enzymes retrieve electrons from a common quinone pool while generating proton motive force for ATP synthesis (Clauwaert *et al.* 2007a). The orientation (cytoplasm, periplasm) of the different denitrification enzymes and their stoichiometry (Figure 14.1) in combination with the different standard potentials of their final electron acceptors results in different theoretical energy recovery efficiencies under the form of ATP. Moreover, it is unclear whether the cation, that was 'accompanying' the electron before it entered the electron transfer chain of the biocathodic microorganisms can contribute to the buildup of proton motive force (similar to the protons that remain in the periplasm after periplasmic hydrogen oxidation). It's worth noting that besides protons, also sodium ions can be used for ATP synthesis (Stahlberg *et al.* 2001).

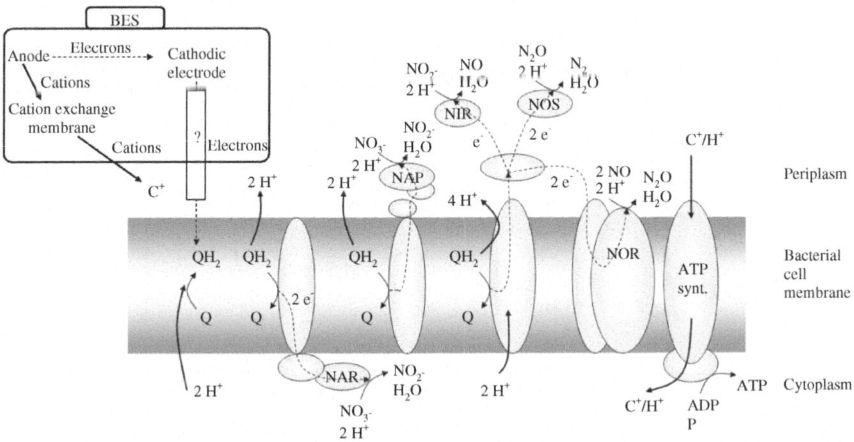

Figure 14.1. Hypothesized model for the electron transfer chain in denitrifying biocathodes (Clauwaert *et al.* 2007a; Richardson 2000)

Only recently has the possibility of methane producing biocathodes been demonstrated (Cheng *et al.* 2009), however here also the mechanisms remain unclear. If methanophenazine (MP), an electron shuttle in the cytoplasmic membrane of methanogens, can be directly reduced by an electrode, proton motive force buildup could result from the reduction of heterodisulfide by oxidation of reduced methanophenazine at the heterodisulfide reductase enzyme (Figure 14.2) (Deppenmeier 2002). However, methanogens should also be able to reduce carbon dioxide by an electrode (Figure 14.2). ATP synthesis can occur from both proton and sodium pumps (Deppenmeier *et al.* 1996). Rozendal and co-workers suggested the possibility of direct methane production by methanogens in a biocathode as well as hydrogen production (Rozendal *et al.* 2008).

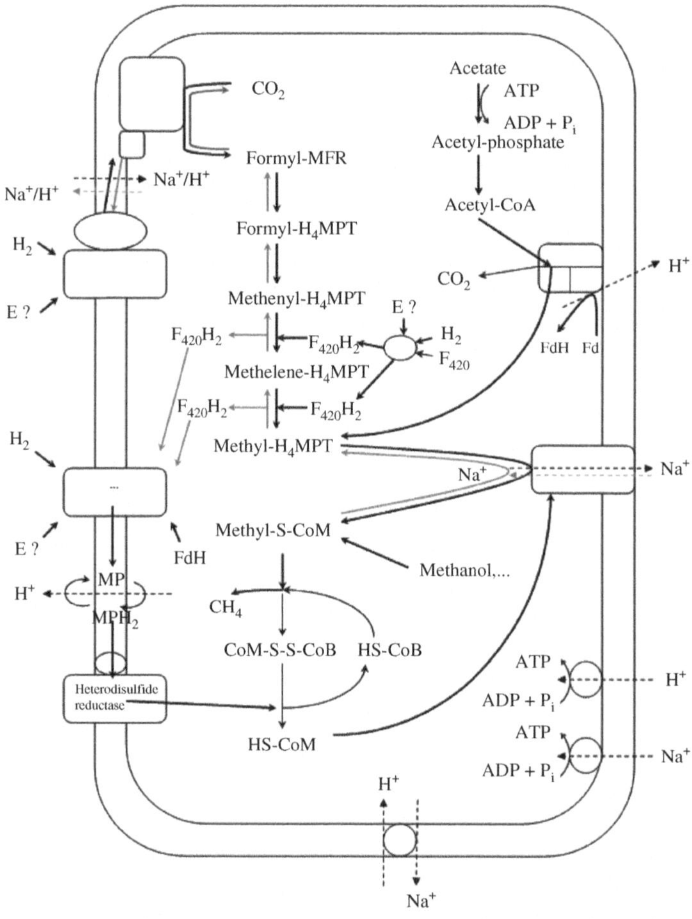

Figure 14.2. Tentative scheme of the electron transfer pathways during methanogenesis driven by acetate, methylated C1 compound, hydrogen or an electrode (E) (Deppenmeier 2002; Deppenmeier *et al.* 1996)

Hydrogen production was also hypothesized as a branched metabolism to remove excessive reducing equivalents for sulfate reducing microorganisms that were possibly retrieving electrons in a direct way from Fe^0 (Dinh *et al.* 2004) and for dehalogenating microorganisms that oxidized reduced methyl viologen (Aulenta *et al.* 2008a).

At present, it is not clear what the mechanism for biocatalyzed hydrogen production in the absence of electron acceptors other than protons in biocathodes would be (Rozendal *et al.* 2008). In the research of Rozendal and co-workers, the cathode potential was −0.7 V vs. SHE, so a cytoplasmatic ferredoxin reducing enzyme coupled with proton translocation could be a possibility.

Little is known about microbial interactions in bioelectrochemical active biofilms. Sharing of electron mediators has been described between Gram-positive and Gram-negative bacteria (Pham *et al.* 2008). Further research must elucidate whether wired electron channels between electrodes and microorganisms enable them to respire in BES but also in natural environments. Since several *Geobacter* strains are able to use an electrode both as an electron donor and acceptor (Strycharz *et al.* 2008), the presence of wired microorganisms, sharing the electrical energy that flows through their electron transfer chains, might appear more general than assumed. It can be concluded that at present bioelectrochemical active microorganisms are mainly limited by their physico-chemical environment, however as the conversion rates and biomass densities in BES increase, more and more attention will have to be paid to possible synergetic and antagonistic effects between the microorganisms in bioelectrochemical active biofilms.

14.5 LIMITING FACTORS

The highest conversion rates in biocathodes reported up to now are 20.5 A m^{-2} (or expressed in COD equivalents: 0.15 kg COD m^{-2} d^{-1}) and 1150 A m^{-3} Total Cathode Compartment (TCC) (or expressed in COD equivalents: 8.2 kg COD m^{-3} TCC d^{-1}) for small electrodes and reactor systems (Clauwaert *et al.* 2007b; Dumas *et al.* 2008). This has at present not resulted in commercial large-scale applications. Besides limitations associated with the anode, the membrane and the BES configuration (Clauwaert *et al.* 2008a), several specific limiting factors in biocathodes can be identified such as electrode properties, environmental and biological restrictions that might be interlinked with each other.

Some electrode properties that might influence the biocathode performance are: surface charge, hydrophobicity, roughness, specific surface area, porosity, ohmic electrode resistance, electrical double layer, functional groups and catalyst activity. In some cases it might be necessary to pre-treat the electrode surface in order to establish a shorter biological start-up or a higher performance, entailing, however, extra costs (Cheng and Logan 2007; Clauwaert *et al.* 2007b). It might be that the available surface area for microorganisms in a biocathode decreases because of precipitation layers (e.g. calcium and iron precipitations) or because of dead biomass.

The peculiarity of electrochemical active biofilms is their ability to keep a dynamic contact with both sides of the biofilm: the bulk solution and the electrode (Clauwaert *et al.* 2008a). If the oxidant flux by diffusion or convection from the bulk solution to the electrochemical active microorganism in the biocathode is not high enough or if the flux of products, like hydroxyl anions, away from the microorganisms to the bulk is too low, the conversion rate will be limited due to high cathodic concentration overpotentials. The biofilm structure, composition and density is affecting the flux of oxidants and products within the biofilm. The concentration of these compounds might be measured experimentally with micro-sensors in function of the biofilm depth or mathematically modeled. Also the presence or production of specific compounds in the cathode might have a toxic effect on the biocathodic community (e.g. signaling molecules, electrochemical hydrogen peroxide production in aerobic cathodes, . . .). Further research is needed to reveal how the cross-over of COD compounds or electron mediators from the anode to the cathode affects the biocathode activity, in the case in which anode and cathode share the same electrolyte (e.g. single chamber BESs, loop operated BESs).

Even if the microorganisms are not limited by their environment, there are also intrinsic limitations associated with the biological conversion processes. Besides growth rate, electron acceptor affinity, the uptake or fixation rate of carbon and nutrients, maybe the most important intrinsic property of electrochemical microorganisms in a biocathode is their ability to withdraw electrons at a high rate from the cathode to their respiratory electron transfer chain. The exact electron transfer mechanisms in biocathodes are still unknown in most cases. Most likely, similar electron transfer mechanisms, as described for biocatalyzed anodes, might be revealed in the near future. These mechanisms might either be direct (outer membrane cytochromes, nanowires or conductive extracellular DNA) or indirect (organic or inorganic electron mediators).

14.6 OUTLOOK

Microbial fuel cells with biocathodes have been proposed as a sustainable alternative for chemical cathodes (Clauwaert *et al.* 2007b). However, biocathodes hold more promise than oxygen reduction in microbial fuel cells alone. With regard to the versatility of microbial respiration pathways, a whole range of reduction processes might be catalyzed in BESs, ranging from denitrification (Clauwaert *et al.* 2007a; Gregory *et al.* 2004), bioremediation processes (Aulenta *et al.* 2007; Shea *et al.* 2008) as well as production of reduced microbial products, like hydrogen gas (Rozendal *et al.* 2008). Besides COD removal and production of electrical energy or conversion of energy carriers, the production of compounds with a high added value might be of specific interest to make biocathode processes economically viable. The synthesis of drugs (Shin *et al.* 2001), hydrogen (Rozendal *et al.* 2008), ethanol (Steinbusch *et al.* 2008), polyhydroxyalkanoates (Clauwaert *et al.* 2008a) as added value chemicals have been proposed.

Alternatively, biocathodes could be used as biosensors for monitoring toxic/inhibiting compounds, to detect biofilm growth and to detect the presence and concentration of oxidants that can be biologically reduced. Different strains of *Pseudomonas aeruginosa*, including clinical isolates, pathogens like *Escherichia coli* strains and strains of *Staphylococcocus*, as well Gram-positive bacteria changed the electrochemical properties of glassy carbon electrodes by biofilm growth, and could in this way be used to detect the presence of undesired microorganisms. These bacteria started to show an enhanced reduction catalysis after 1 to 6 hours of contact with the electrode (Cournet *et al.* unpublished results). Interestingly, when operated as microbial fuel cells, these sensors might be able to power the wireless transmission of the data obtained (Shantaram *et al.* 2005).

14.7 ACKNOWLEDGEMENTS

This research was funded by a PhD grant (IWT grant 53305) of the Institute for the Promotion of Innovation through Science and Technology in Flanders (IWT-Vlaanderen).

REFERENCES

Aulenta, F., Canosa, A., Majone, M., Panero, S., Reale, P. and Rossetti, S. (2008a) Trichloroethene dechlorination and H-2 evolution are alternative biological pathways of electric charge utilization by a dechlorinating culture in a bioelectrochemical system. *Environ. Sci. Technol.* **42**(16), 6185–6190.

Aulenta, F., Canosa, A., Reale, P., Rossetti, S., Panero, S. and Majone, M. (2008b) Microbial reductive dechlorination of trichloroethene to ethene with electrodes serving as electron donors without the external addition of redox mediators. *Biotechnol Bioeng* **103**(1), 85–91.

Aulenta, F., Catervi, A., Majone, M., Panero, S., Reale, P. and Rossetti, S. (2007) Electron transfer from a solid-state electrode assisted by methyl viologen sustains efficient microbial reductive dechlorination of TCE. *Environ. Sci. Technol.* **41**(7), 2554–2559.

Aulenta, F., Reale, P., Catervi, A., Panero, S. and Majone, M. (2008c) Kinetics of trichloroethene dechlorination and methane formation by a mixed anaerobic culture in a bio-electrochemical system. *Electrochim. Acta* **53**(16), 5300–5305.

Bergel, A., Feron, D. and Mollica, A. (2005) Catalysis of oxygen reduction in PEM fuel cell by seawater biofilm. *Electrochem. Commun.* **7**(9), 900–904.

Berk, R.S. and Canfield, J.H. (1964) Bioelectrochemical energy conversion. *Appl. Microbiology* **12**(1), 10–12.

Blake, R.C., Howard, G.T. and McGinness, S. (1994) Enhanced yields of iron-oxidizing bacteria by in-situ electrochemical reduction of soluble iron in the growth-medium. *Appl. Environ. Microbiol.* **60**(8), 2704–2710.

Busalmen, J.P., Vazquez, M. and de Sanchez, S.R. (2002) New evidences on the catalase mechanism of microbial corrosion. *Electrochim. Acta* **47**(12), 1857–1865.

Chang, C.C., Tseng, S.K. and Huang, H.K. (1999) Hydrogenotrophic denitrification with immobilized *Alcaligenes eutrophus* for drinking water treatment. *Bioresour. Technol.* **69**(1), 53–58.

Cheng, S. and Logan, B.E. (2007) Ammonia treatment of carbon cloth anodes to enhance power generation of microbial fuel cells. *Electrochem Commun.* **9**, 492–496.

Cheng, S., Xing, D., Call, D. and Logan, B.E. (2009) Direct biological conversion of electrical current into methane by electromethanogenesis. *Environ. Sci. Technol.* **43**, 3953–3958.

Chung, J., Rittmann, B.E., Wright, W.F. and Bowman, R.H. (2007) Simultaneous bio-reduction of nitrate, perchlorate, selenate, chromate, arsenate, and dibromochloropropane using a hydrogen-based membrane biofilm reactor. *Biodegradation* **18**(2), 199–209.

Clauwaert, P. and Verstraete, W. (2009) Methanogenesis in membraneless microbial electrolysis cells. *Appl. Microbiol. Biotechnol.* **82**(5), 829–836.

Clauwaert, P., Desloover, J., Shea, C., Nerenberg, R., Boon, N. and Verstraete, W. (submitted) Enhanced nitrogen removal in bio-electrochemical systems.

Clauwaert, P., Rabaey, K., Aelterman, P., DeSchamphelaire, L., Pham, T.H., Boeckx, P., Boon, N. and Verstraete, W. (2007a) Biological denitrification in microbial fuel cells. *Environ. Sci. Technol.* **41**(9), 3354–3360.

Clauwaert, P., Van der Ha, D., Boon, N., Verbeken, K., Verhaege, M., Rabaey, K. and Verstraete, W. (2007b) Open air biocathode enables effective electricity generation with microbial fuel cells. *Environ. Sci. Technol.* **41**(21), 7564–7569.

Clauwaert, P., Aelterman, P., Pham, T.H., De Schamphelaire, L., Carballa, M., Rabaey, K. and Verstraete, W. (2008a) Minimizing losses in bio-electrochemical systems: the road to applications. *Appl. Microbiol. Biotechnol.* **79**(6), 901–913.

Clauwaert, P., Toledo, R., Van der Ha, D., Crab, R., Verstraete, W., Hu, H., Udert, K.M. and Rabaey, K. (2008b) Combining biocatalyzed electrolysis with anaerobic digestion. *Water Sci. Technol.* **57**(4), 575–579.

Clauwaert, P., van der Ha, D. and Verstraete, W. (2008c) Energy recovery from energy rich vegetable products with microbial fuel cells. *Biotechnol. Lett.* **30**(11), 1947–1951.

Clauwaert, P., Mulenga, S., Aelterman, P. and Verstraete, W. (2009) Litre-scale microbial fuel cells operated in a complete loop. *Appl. Microbiol. Biotechnol.* published online.

Cochran, J.K., Carey, A.E., Sholkovitz, E.R. and Surprenant, L.D. (1986) The geochemistry of uranium and thorium in coastal marine sediments and sediment pore waters. *Geochim. Cosmochim. Acta* **50**(5), 663–680.

Cracknell, J.A., Vincent, K.A. and Armstrong, F.A. (2008) Enzymes as working or inspirational electrocatalysts for fuel cells and electrolysis. *Chem. Rev.* **108**(7), 2439–2461.

de Best, J.H. (1999) Anaerobic transformation of chlorinated hydrocarbons in a packed-bed reactor. Ph.D Thesis, University of Groningen, Groningen, The Netherlands.

Deppenmeier, U. (2002) Redox-driven proton translocation in methanogenic *Archaea. Cell. Molec. Life Sci.* **59**(9), 1513–1533.

Deppenmeier, U., Muller, V. and Gottschalk, G. (1996) Pathways of energy conservation in methanogenic archaea. *Arch. Microbiol.* **165**(3), 149–163.

Dickinson, W.H. and Lewandowski, Z. (1998) Electrochemical concepts and techniques in the study of stainless steel Ennoblement. *Biodegradation* **9**(1), 11–21.

Dinh, H.T., Kuever, J., Mussmann, M., Hassel, A.W., Stratmann, M. and Widdel, F. (2004) Iron corrosion by novel anaerobic microorganisms. *Nature* **427**(6977), 829–832.

Dumas, C., Basseguy, R. and Bergel, A. (2008) Microbial electrocatalysis with *Geobacter sulfurreducens* biofilm on stainless steel cathodes. *Electrochim. Acta* **53**(5), 2494–2500.

Feleke, Z., Araki, K., Sakakibara, Y., Watanabe, T. and Kuroda, M. (1998) Selective reduction of nitrate to nitrogen gas in a biofilm-electrode reactor. *Water Res.* **32**(9), 2728–2734.

Field, J.A. and Sierra-Alvarez, R. (2004) Biodegradability of chlorinated solvents and related chlorinated aliphatic compounds. *Rev. Env. Sci. Biotechnol.* **3**(3), 185–254.

Freguia, S., Rabaey, K., Yuan, Z.G. and Keller, J. (2008) Sequential anode-cathode configuration improves cathodic oxygen reduction and effluent quality of microbial fuel cells. *Water Res.* **42**(6–7), 1387–1396.

Germonpre, R., Liessens, J., Verstraete, W. and Beernaert, S. (1991) Methylotrophic and hydrogenotrophic denitrification at the blankaart plant. *Water Supply* **10**(3), 53–64.

Gregory, K.B. and Lovley, D.R. (2005) Remediation and recovery of uranium from contaminated subsurface environments with electrodes. *Environ. Sci. Technol.* **39**(22), 8943–8947.

Gregory, K.B., Bond, D.R. and Lovley, D.R. (2004) Graphite electrodes as electron donors for anaerobic respiration. *Environ. Microbiol.* **6**(6), 596–604.

Hasan, M.A., Zaki, M.I., Pasupulety, L. and Kumari, K. (1999) Promotion of the hydrogen peroxide decomposition activity of manganese oxide catalysts. *Appl. Catal. A-Gen.* **181**(1), 171–179.

Hasvold, O., Henriksen, H., Melvaer, E., Citi, G., Johansen, B.O., Kjonigsen, T. and Galetti, R. (1997) Sea-water battery for subsea control systems. *J. Power Sources* **65**, 253–261.

He, Z. and Angenent, L.T. (2006) Application of bacterial biocathodes in microbial fuel cells. *Electroanalysis* **18**(19–20), 2009–2015.

Ishizaki, A., Tanaka, K., Takeshita, T., Kanemaru, T., Shimoji, T. and Kawano, T. (1993) Equipment and operation for fermentative PHB production using gaseous substrate to guarantee safety from explosion. *J. Chem. Engng. Jpn.* **26**(2), 225–227.

Kim, T.S., Kim, H.Y. and Kim, B.H. (1990) Petroleum desulfurization by desulfovibrio-desulfuricans M6 using electrochemically supplied reducing equivalent. *Biotechnol. Lett.* **12**(10), 757–760.

Kinsel, N.A. and Umbreit, W.W. (1964) Method for electrolysis of culture medium to increase growth of sulfur oxidizing iron bacterium *Ferrobacillus sulfooxidans. J. Bacteriol.* **87**(5), 1243–1244.

Kuroda, M., Watanabe, T. and Umedu, Y. (1996) Simultaneous oxidation and reduction treatments of polluted water by a bio-electro reactor. *Water Sci. Technol.* **34**(9), 101–108.

Lewis, K. (1966) Symposium on Bioelectrochemistry of Microorganisms 4. Biochemical fuel cells. *Bacteriol Rev* **30**(1), 101–113.

Logan, B.E., Cheng, S., Watson, V. and Estadt, G. (2007) Graphite fiber brush anodes for increased power production in air-cathode microbial fuel cells. *Environ. Sci. Technol.* **41**(9), 3341–3346.

Lopez-Lopez, A., Exposito, E., Anton, J., Rodriguez-Valera, F. and Aldaz, A. (1999) Use of Thiobacillus ferrooxidans in a coupled microbiological-electrochemical system for wastewater detoxification. *Biotechnol. Bioengng.* **63**(1), 79–86.

Nerenberg, R. and Rittmann, B.E. (2004) Hydrogen-based, hollow-fiber membrane biofilm reactor for reduction of perchlorate and other oxidized contaminants. *Water Sci. Technol.* **49**(11–12), 223–230.

Nguyen, T.A., Lu, Y.Z., Yang, X.H. and Shi, X.M. (2007) Carbon and steel surfaces modified by *Leptothrix discophora SP-6*: Characterization and implications. *Environ. Sci. Technol.* **41**(23), 7987–7996.

Ohmura, N., Matsumoto, N., Sasaki, K. and Saiki, H. (2002) Electrochemical regeneration of Fe(III) to support growth on anaerobic iron respiration. *Appl. Environ. Microbiol.* **68**(1), 405–407.

Orfei, L.H., Simison, S. and Busalmen, J.P. (2006) Stainless steels can be cathodically protected using energy stored at the marine sediment/seawater interface. *Environ. Sci. Technol.* **40**(20), 6473–6478.

Park, D.H., Laivenieks, M., Guettler, M.V., Jain, M.K. and Zeikus, J.G. (1999) Microbial utilization of electrically reduced neutral red as the sole electron donor for growth and metabolite production. *Appl. Environ. Microbiol.* **65**(7), 2912–2917.

Pham, T.H., Boon, N., Aelterman, P., Clauwaert, P., De Schamphelaire, L., Vanhaecke, L., De Maeyer, K., Hofte, M., Verstraete, W. and Rabaey, K. (2008) Metabolites produced by *Pseudomonas* sp enable a Gram-positive bacterium to achieve extracellular electron transfer. *Appl. Microbiol. Biotechnol.* **77**(5), 1119–1129.

Rabaey, K., Rodríguez, J., Blackall, L.L., Keller, J., Gross, P., Batstone, D., Verstraete, W. and Nealson, K.H. (2007) Microbial ecology meets electrochemistry: Electricity-driven and driving communities. *Isme J.* **1**(1), 9–18.

Rabaey, K., Read, S.T., Clauwaert, P., Freguia, S., Bond, P.L., Blackall, L.L. and Keller, J. (2008) Cathodic oxygen reduction catalyzed by bacteria in microbial fuel cells. *Isme J.* **2**(5), 519–527.

Rhoads, A., Beyenal, H. and Lewandowski, Z. (2005) Microbial fuel cell using anaerobic respiration as an anodic reaction and biomineralized manganese as a cathodic reactant. *Environ. Sci. Technol.* **39**(12), 4666–4671.

Richardson, D.J. (2000) Bacterial respiration: A flexible process for a changing environment. *Microbiology-Sgm* **146**(3), 551–571.

Rozendal, R.A., Hamelers, H.V.M., Euverink, G.J.W., Metz, S.J. and Buisman, C.J.N. (2006) Principle and perspectives of hydrogen production through biocatalyzed electrolysis. *Int. J. Hydrogen Energy* **31**(12), 1632–1640.

Rozendal, R.A., Jeremiasse, A.W., Hamelers, H.V.M. and Buisman, C.J.N. (2008) Hydrogen production with a microbial biocathode. *Environ. Sci. Technol.* **42**(2), 629–634.

Sakakibara, Y. and Kuroda, M. (1993) Electric prompting and control of denitrification. *Biotechnol. Bioeng.* **42**(4), 535–537.

Sakakibara, Y., Araki, K., Tanaka, T., Watanabe, T. and Kuroda, M. (1994) Denitrification and neutralization with an electrochemical and biological reactor. *Water Sci. Technol.* **30**(6), 151–155.

Schlegel, H.G. and Lafferty, R. (1965) Growth of knallgas bacteria (*Hydrogenomonas*) using direct electrolysis of culture medium. *Nature* **205**(4968), 308–309.

Schroder, U. (2007) Anodic electron transfer mechanisms in microbial fuel cells and their energy efficiency. *Phys. Chem. Chem. Phys.* **9**(21), 2619–2629.

Scotto, V. and Lai, M.E. (1998) The ennoblement of stainless steels in seawater: A likely explanation coming from the field. *Corros. Sci.* **40**(6), 1007–1018.

Shantaram, A., Beyenal, H., Raajan, R., Veluchamy, A. and Lewandowski, Z. (2005) Wireless sensors powered by microbial fuel cells. *Environ. Sci. Technol.* **39**(13), 5037–5042.

Shea, C., Clauwaert, P., Verstraete, W. and Nerenberg, R. (2008) Adapting a denitrifying biocathode for perchlorate reduction. *Water Sci. Technol.* **58**(10), 1941–1946.

Shin, H.S., Jain, M.K., Chartrain, M. and Zeikus, J.G. (2001) Evaluation of an electrochemical bioreactor system in the biotransformation of 6-bromo-2-tetralone to 6-bromo-2-tetralol. *Appl. Microbiol. Biotechnol.* **57**(4), 506–510.

Skadberg, B., Geoly-Horn, S.L., Sangamalli, V. and Flora, J.R.V. (1999) Influence of pH, current and copper on the biological dechlorination of 2,6-dichlorophenol in an electrochemical cell. *Water Res.* **33**(9), 1997–2010.

Sober, H.A. (1970) *Handbook of Biochemistry, Selected Data for Molecular Biology*, The Chemical Rubber Co., Ohio, USA.

Stahlberg, H., Muller, D.J., Suda, K., Fotiadis, D., Engel, A., Meier, T., Matthey, U. and Dimroth, P. (2001) Bacterial Na+-ATP synthase has an undecameric rotor. *EMBO Rep.* **2**(3), 229–233.

Steinbusch, K.J.J., Hamelers, H.V.M. and Buisman, C.J.N. (2008) Alcohol production through volatile fatty acids reduction with hydrogen as electron donor by mixed cultures. *Water Res.* **42**(15), 4059–4066.

Strycharz, S.M., Woodard, T.L., Johnson, J.P., Nevin, K.P., Sanford, R.A., Loffler, F.E. and Lovley, D.R. (2008) Graphite electrode as a sole electron donor for reductive dechlorination of tetrachlorethene by *Geobacter lovleyi*. *Appl. Environ. Microbiol.* **74**(19), 5943–5947.

Ter Heijne, A., Hamelers, H.V.M. and Buisman, C.J.N. (2007) Microbial fuel cell operation with continuous biological ferrous iron oxidation of the catholyte. *Environ. Sci. Technol.* **41**(11), 4130–4134.

Thauer, R.K., Jungermann, K. and Decker, K. (1977) Energy conservation in chemothropic anaerobic bacteria. *Bacteriol. Rev.* **41**(1), 100–180.

Thrash, J.C. and Coates, J.D. (2008) Review: Direct and indirect electrical stimulation of microbial metabolism. *Environ. Sci. Technol.* **42**(11), 3921–3931.

Thrash, J.C., Van Trump, J.I., Weber, K.A., Miller, E., Achenbach, L.A. and Coates, J.D. (2007) Electrochemical stimulation of microbial perchlorate reduction. *Environ. Sci. Technol.* **41**(5), 1740–1746.

Van Rijn, J., Tal, Y. and Schreier, H.J. (2006) Denitrification in recirculating systems: Theory and applications, *Aquacultural Engineering* **34**, 364–376.

Virdis, B., Rabaey, K., Yuan, Z. and Keller, J. (2008) Microbial fuel cells for simultaneous carbon and nitrogen removal. *Water Res.* **42**(12), 3013–3024.

Wang, G., Huang, L.P. and Zhang, Y.F. (2008) Cathodic reduction of hexavalent chromium [Cr(VI)] coupled with electricity generation in microbial fuel cells. *Biotechnol. Lett.* **30**(11), 1959–1966.

Zhao, F., Harnisch, F., Schroder, U., Scholz, F., Bogdanoff, P. and Herrmann, I. (2005) Application of pyrolysed iron(II) phthalocyanine and CoTMPP based oxygen reduction catalysts as cathode materials in microbial fuel cells. *Electrochem. Commun.* **7**(12), 1405–1410.

Zhao, F., Slade, R.C.T. and Varcoe, J.R. (submitted) Techniques for the study and development of microbial fuel cells: an electrochemical perspective.

Zumdahl, S.S. (1998) *Chemical Principles.* Houghton Mifflin Company, New York, USA.

15

Bioelectrochemical systems (BES) for subsurface remediation

Federico Aulenta and Mauro Majone

15.1 BIOREMEDIATION OF CONTAMINATED SOILS AND AQUIFERS

Soil and aquifer contamination has grown as a societal problem over the past two decades and has become a key issue for the political sphere, the business and scientific communities, and is catching the public interest. Contaminated sites generally result from the manufacturing, storage, use, and disposal of hazardous materials. It is now widely recognized that polluted sites pose threats to human health and the environment.

Bioremediation, i.e. the destruction, detoxification, or immobilization of harmful compounds by living microorganisms, has gained wide acceptance as a viable, environmentally friendly, and cost-effective alternative to physico-chemical cleanup options (Stenuit *et al.* 2008). The remarkable metabolic

 Bioelectrochemical Systems: From Extracellular Electron Transfer to Biotechnological Application. Edited by Korneel Rabaey, Largus Angenent, Uwe Schröder and Jürg Keller. ISBN: 9781843392330. Published by IWA Publishing, London, UK.

versatility of microorganisms makes bioremediation applicable to treat a large and ever-increasing number of environmental pollutants such as pesticides, industrial chemicals, jet fuel and gasoline, and metals. Even compounds that were once believed to be recalcitrant, such as chlorinated solvents, polychlorobphenyls (PCBs), methyl tert-butyl ether (MTBE), and other stable synthetic organics have been shown to be degraded by microorganisms.

Bioremediation techniques can be implemented either *in situ*, i.e. the contaminated sample is treated in its natural setting, or *ex situ*, i.e. the contaminated sample is taken out and treated in above-ground facilities. Generally speaking, because of the need to excavate or pump out the contaminated soil or groundwater, *ex situ* bioremediation treatments are often less economical than *in situ* approaches. On the other hand, they typically permit a greater control over treatment conditions and are often more reliable in terms of process outcome.

The underlying principle for bioremediation is to enhance pollutant degradation/transformation/ immobilization kinetics through stimulation of the growth and metabolism of microorganisms that are involved in the bioremediation processes. Some of the successful approaches have been based on addition of nutrients or electron donors/acceptors, on manipulation of environmental parameters such as pH, redox state, or temperature, or by increasing contaminant availability. In some other cases, pollutant degradation was increased through exogenous introduction of microorganisms harbouring the desired metabolic capabilities (bioaugmentation).

This review illustrates the fundamental aspects of some recently developed bioelectrochemical systems (BES) for subsurface remediation. The main attention is placed on describing the biological treatment of reducible contaminants (chlorinated hydrocarbons, nitrate, perchlorate, and U(VI)) with electrodes serving as electron donors. Some outlooks and perspectives on the potential application of such methods for treatment of reducible or oxidisable pollutants or for monitoring purposes are also presented.

15.2 CHEMICAL VS. ELECTROCHEMICAL STRATEGIES OF ELECTRON DELIVERY

A promising bioremediation approach for reducible contaminants consists in their microbiological reduction to end-products which are either non-toxic, less mobile, or both (Loffler and Edwards 2006). This is the case of chlorinated aliphatic hydrocarbons, nitrate, perchlorate, and metals, as discussed in the following paragraphs.

To reduce contaminants, microorganisms require suitable electron donors. Therefore, engineered approaches for bioremediation of these contaminants are typically based on the continual addition of chemicals, as electron donors, to sustain microbial degradative activity. These include hydrogen gas, acetate, lactate, alcohols, as well as long-lasting, slowly biodegradable (H_2- and acetate-releasing) organic substrates. The main perceived limitation of this approach is that adding the appropriate amounts of electron donors without stimulating unwanted metabolic reactions (e.g. methane generation) and/or accumulating large amounts of (side) fermentation products, as well as biomass in the subsurface, is extremely challenging. Moreover, the lack of tools for controlling and manipulating the microbial communities carrying out biodegradative reactions, and the difficulty to homogenously distribute chemicals in highly heterogeneous porous media, often limit process reliability.

To possibly overcome some of these drawbacks, bioelectrochemical approaches have been proposed. A bioelectrochemical remediation method employs electrochemical measures (e.g. application of a current or a voltage) to enhance biodegradative processes (Lohner *et al.* 2008a; Lohner *et al.* 2008b). The basic idea behind the use of electrodes and electrochemical systems to stimulate biological reductions in the subsurface is that the delivery of electrons can, in principle, be continuously monitored/controlled in terms of current and potential; no chemicals need to be injected, which eliminates the need for transport, storage, dosing and post-treatment (Rabaey and Keller 2008). Moreover, since the microorganisms responsible of the biodegradation can be grown directly at the surface of the electrodes or in their proximity, an effective usage of the supplied electrons is expected (Rabaey and Keller 2008).

At present, bioelectrochemical systems for the reduction of oxidized contaminants can be broadly grouped in four categories, based on the mechanism(s) through which the electrons are ultimately transferred to the microorganisms (Figure 15.1).

The first mechanism (Figure 15.1, panel a), is based on the generation of hydrogen through water electrolysis and its use as electron donor for microbial reductions. In theory, in aqueous systems, at neutral pH, hydrogen production requires an electrode (cathode) potential more negative than –414 mV vs. standard hydrogen electrode (SHE) (or a potential difference between anode and cathode of 1230 mV). In practice, when noble metal free cathodes are employed (i.e. carbon-based electrodes), more negative cathode potential values (< -700 mV vs. SHE) are required to overcome electrochemical losses. In principle this approach can be adopted to stimulate virtually any hydrogenotrophic metabolism. However, due to its very limited solubility in water, hydrogen tends to rapidly escape from the biologically reactive zone without being utilized for the desired reaction. Consequently, the process is often highly inefficient, i.e. only a marginal fraction of the consumed electric charge is ultimately used for the reduction of oxidized contaminants. Moreover, hydrogen gas bubbling may cause reduction in aquifer permeability and changes in groundwater flow path.

The second mechanism (Figure 15.1, panel b), is based on the presence of a redox shuttling molecule which is reduced at the cathode surface and reoxidized by microorganisms. Even though the redox mediator is the ultimate electron carrier, it is, in principle, never consumed during the reaction, being continuously recycled within the system.

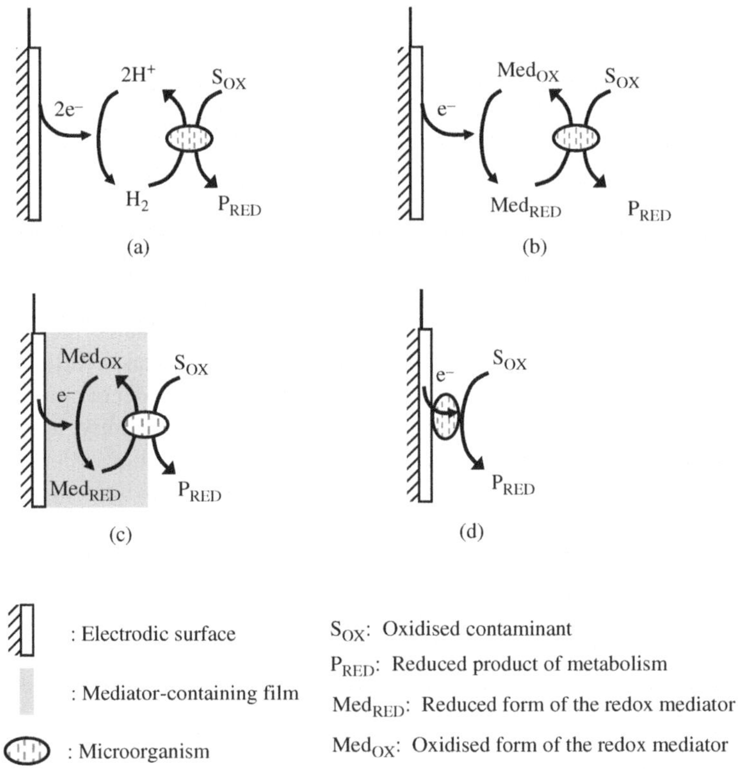

Figure 15.1. Bioelectrochemical reduction of oxidised contaminants: possible electron transfer strategies

Redox mediators facilitating extracellular electron transfer processes can be exogenous (i.e. externally added) (Aulenta *et al.* 2007) or endogenous (i.e. self-produced by microorganisms) (Aulenta *et al.* 2009; Biffinger *et al.* 2008;

Marsili *et al.* 2008; Rabaey *et al.* 2004). Redox mediators have recently been shown to have a remarkable potential for increasing process selectivity. As an example, 2,6-anthrahydroquinone disulfonate (AH_2DS) was recently shown to selectively stimulate the biological reduction of perchlorate over competitive sulfate reduction and methanogenesis (Van Trump and Coates 2009). Typically, mediators with a redox potential less negative than that required for H_2 evolution are commonly used. In this way, the process is sustained without intermediate H_2 formation.

The main drawbacks of the redox-mediated approach are related to difficulties in retaining the mediator in a continuous-flow system (e.g. contaminated aquifers). The anchoring of redox mediators in thin films at the surface of electrodes has been proposed as a strategy to overcome this limitation (Figure 15.1, panel c) (Aulenta *et al.* 2007). In this case, microorganisms need to establish a close interaction with the modified electrode in order to use it as an electron donor. Depending on the mediator involved, several physical (e.g. chemisorption, adsorption, polymer film coating) or chemical (e.g. covalent bonding) strategies to develop such modified electrodes have been described in the literature (Bard *et al.* 2007).

The last mechanism depicted in Figure 15.1 (panel d) is based on the newly discovered ability of certain microorganisms to directly accept electrons from the surface of polarized solid-state cathodes and use them to reduce physiological electron acceptors (Gregory *et al.* 2004; Strycharz *et al.* 2008). Also in this case, the process is driven exclusively by those microorganisms which establish a close physical association with the electrodic surface. Additionally, since mediators (which often are expensive and/or environmentally unsustainable compounds) are not required to drive the extracellular electron transfer process, the process is very attractive for *in situ* applications.

15.2.1 Chlorinated hydrocarbons

Perchloroethylene (PCE), trichloroethylene (TCE) and other highly chlorinated aliphatic hydrocarbons (CAH) such as 1,1,2-trichloroethane (TCA), 1,2-dichloroethane (DCA) and chloroform (CF) are widely used in many different industrial processes, mainly as solvents in dry-cleaning operations and semiconductor manufacture. Careless storage, handling, and disposal, as well as their high chemical stability, have contributed to their status of most frequently encountered subsurface contaminants. CAH are highly toxic, as well as being known or suspected carcinogens, therefore their presence in the environment poses important health risks and has prompted investigations concerning their fate in the subsurface. CAH are transformed in the subsurface through a range of biotic and abiotic reactions. Most abiotic transformations are slow, although in some cases, still significant within the time scales commonly associated with groundwater remediation. Biotic reactions are typically much faster, provided that suitable environmental conditions are present such as sufficient substrate, nutrients, and suitable microbial populations (De Wildeman and Verstraete 2003; Vogel *et al.* 1987).

Highly chlorinated aliphatic hydrocarbons, such as PCE, TCE, TCA, or CF do not support microbial growth in aerobic environment. However, many of these CAH are anaerobically transformed to lesser or non-chlorinated compounds through sequential reductive dechlorination (RD) reactions. Over the last years, fundamental and applied research on RD has demonstrated that the transformation of CAH into harmless non-chlorinated end products can be practically achieved by enhancing of bacterial dechlorination reactions in the field and, accordingly, exploited to remediate CAH contamination plumes and source zones (Aulenta *et al.* 2006; Morse *et al.* 1997; Sleep *et al.* 2006).

Anaerobic reductive dechlorination is the most important metabolic pathway for highly chlorinated CAH. The reductive dechlorination is a form of anaerobic respiration in which the chlorinated compound is used, by certain microorganisms, as terminal electron acceptor (Holliger and Schumacher 1994). Typically, the reductive dechlorination involves sequential replacements of chlorine substituents with hydrogen atoms, resulting in the formation of less- or eventually non-chlorinated end-products (Figure 15.2).

Hydrogen is the most important electron donor for many dechlorinating microorganisms. *Dehalococcoides* spp. uses only hydrogen to dechlorinate PCE to ethene (Maymo-Gatell *et al.* 1997; Maymo-Gatell *et al.* 1995). Accordingly, *in situ* enhancement of microbial reductive dechlorination is typically accomplished by providing indigenous dechlorinating populations with organic substrates which release H_2 through their fermentation. Common electron donors for CAH dechlorination include acetate, lactate, formate, and pyruvate, which slowly release H_2 upon fermentation within the aquifer. Notably, as for chloroethenes, complete dechlorination to harmless ethene has never been reported with electron donors and microorganisms other than H_2 and *Dehalococcoides* spp., respectively.

Recently, some researchers have proposed the use of bioelectrochemical systems (BES) for the reduction of chlorinated compounds.

Skadberg and colleagues (Skadberg *et al.* 1999) performed batch studies with a mixed hydrogen-consuming culture and showed that 2,6-dichlorophenol was reductively dechlorinated when an electric current was passed in the circuit. Electrochemically generated H_2, through electrolysis of water at the cathode, served as the ultimate electron donor in the biological reaction (Table 15.1). Zhang and colleagues (Zhang *et al.* 2001) used polarized iron wires to generate H_2 and stimulate the microbial dechlorination of PCE. A similar approach was also adopted by our group (Aulenta *et al.* 2008b) to stimulate the reductive dechlorination of TCE to ethene in a *Dehalococcoides*-containing mixed culture. In that study we showed that a graphite cathode (unpolished graphite rods) polarized in the range –600 mV ÷ −800 mV vs. SHE could be employed to generate H_2 at controlled rates and, in so doing, enhance the hydrogenotrophic dechlorination of TCE (Table 15.1). Notably, the rate and efficiency (i.e. electrons channelled to dechlorination vs. total electrons consumed) of the process was mostly influenced by the value of the applied cathode potential. Specifically, in the narrow range –650 mV to –675 mV vs. SHE, both high dechlorination rates and current efficiencies were achieved.

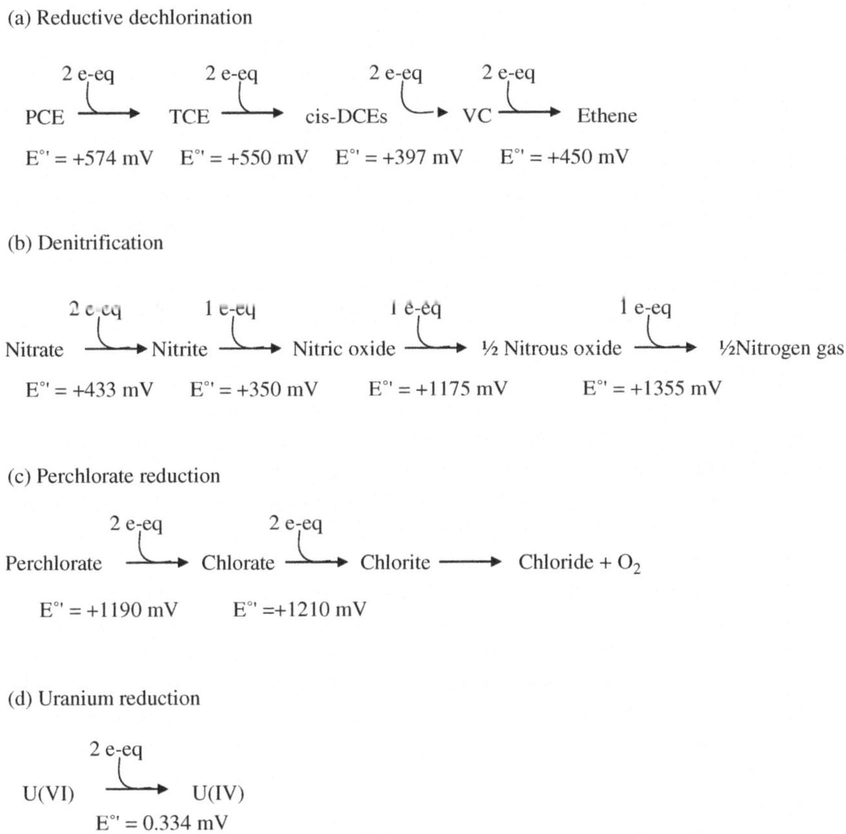

Figure 15.2. Reduction sequence and electron equivalent requirements for the microbial reduction of perchloroethene, nitrate, perchlorate, and uranium (VI) (Dolfing and Janssen 1994; Thauer *et al.* 1977; Vogel *et al.* 1987; Weast 1982)

Table 15.1. Overview of conditions and performance of bioelectrochemical remediation systems

Target contaminant	Microorganism	Cathode material	Cathode potential (vs. SHE)	Redox mediator	Degradation rate	Reference
2,4-dichlorophenol	Mixed-culture	Stainless steel	n.a.[1]	H_2[2]	40.9 μeq/l d	(Skadberg *et al.* 1999)
TCE	Mixed-culture	Graphite	−800 mV	H_2[2]	657 μeq/l d	(Aulenta *et al.* 2008b)
TCE	Mixed-culture	Glassy carbon	−500 mV	Methyl Viologen (500 μmol/l)	12.1 μeq/l d	(Aulenta *et al.* 2007)
TCE	Mixed-culture	Glassy carbon	−450 mV	Methyl Viologen (5000 μmol/l)	653 μeq/l d	(Aulenta *et al.* 2008a)
TCE	Mixed-culture	Glassy carbon	−500 mV	Methyl Viologen[4]	5.50 μeq/l d (0.26 μeq/cm^2 d)[3]	(Aulenta *et al.* 2007)
TCE	Mixed-culture	Carbon paper	−450 mV	Self-produced (unidentified)	2.7 μeq/l d (0.03 μeq/ cm^2 d)[3]	(Aulenta *et al.* 2009)
PCE	*Geobacter lovleyi*	Graphite	−300 mV	No	200 μeq/l d	(Strycharz *et al.* 2008)
Perchlorate	*Azospira spp.*	Graphite	−250 mV	AQDS (500 μmol/l)	21700 μeq/l d	(Thrash *et al.* 2007)
Perchlorate	*Desulforomonas* sp.	Graphite	−250 mV	AQDS (5000 μmol/l)	4024 μeq/l d	(Thrash *et al.* 2007)
Perchlorate	Strain VDY	Graphite	−300 mV	H_2[2]	4830 μeq/l d	(Thrash *et al.* 2007)

Perchlorate	Mixed-culture	Graphite	n.a.[1]	No	965 μeq/l d (0.29 μeq/cm^2 d)[3]	(Shea *et al.* 2008)
Nitrate	Mixed-culture	Stainless steel	n.a.[1]	H_2[2]	1891 μeq/l d	(Hayes *et al.* 1998)
Nitrate	*Geobacter metallireducens*	Graphite	−300 mV	No	667 μeq/ld	(Gregory *et al.* 2004)
Nitrate	Mixed culture	Graphite	∼−100 mV	No	28571 μeq/l d (3.5 μeq/cm^2 d)[3]	(Clauwaert *et al.* 2007)
U(VI)	*Geobacter sulfurreducens*	Graphite	−300 mV	No	26.6 μeq/l d	(Gregory and Lovley 2005)

[1]n.a.: not available;
[2]The H2 generated from water electrolysis can also be regarded as a soluble redox mediator;
[3]Rates have been normalized with respect to the projected surface area of the cathode;
[4]Methyl viologen was immobilized in a film at the electrode surface.

A remarkable advancement in the application of electrolytic systems for the biotreatment of chlorinated aliphatics was the development of the electrolytic methanogenic/methanotrophic coupled (eMaMoC) process (Guiot *et al.* 2008). This is an integrated anaerobic/aerobic biofilm process which takes advantage of both the hydrogen and the oxygen generated from water electrolysis. Sequential anaerobic and aerobic biodegradation had been proposed as a treatment strategy for CAH. Higher chlorinated compounds such as PCE, are anaerobically transformed to lower chlorinated compounds, such as *cis*-DCE, which are subsequently oxidized, through metabolic or co-metabolic processes, under aerobic condition. With the eMaMoC process, Guiot and colleagues (Guiot *et al.* 2008) demonstrated that anaerobic and aerobic populations may coexist within a single biofilm, consisting of granular biomass. Hydrogen serves as the electron donor for the anaerobic reductive dechlorination of PCE and methanogenesis, with both metabolic processes taking place in the inner core of the granule. Oxygen is used as electron acceptor by methanotrophic bacteria, growing in the outer layers of the granule, to oxidize indigenous methane while cometabolically degrading the lower chlorinated intermediates. According to the authors, the eMaMoC concept could be implemented into a variety of remediation schemes either *in situ* or *ex situ*. For instance, the eMaMoC design can retrofit an above-ground reactor for *ex situ* treatment; it can be used in a permeable biobarrier, or be used to create an *in situ* underground reactive zone with combined anaerobic/aerobic conditions (Guiot *et al.* 2008).

In the attempt to characterize the dehalogenase activity of *Dehalococcoides ethenogenes* strain 195, researchers at Cornell University (Nijenhuis and Zinder 2005) found that several chemically-reduced redox-active molecules acted as artificial electron donors for PCE dechlorination by whole cells of this microorganism. Among them, low-potential viologen dyes, and particularly methyl viologen ($E^{\circ\prime} = -446$ mV vs. SHE) yielded the fastest dechlorination. Conversely, negligible activity was observed in the presence of higher potential compounds, such as anthrahydroquinone disulfonate (AQDS), phenosafranine and phenazine methosulfate. Instead these compounds appeared to inhibit reductive dechlorination. Similar results were previously found for *S. multivorans* in which only viologen dyes supported dechlorination of PCE and TCE. It is worth noting that chloroethenes are quite energetic electron acceptors having a redox potential ($E^{\circ\prime}$) of +260 to +480 mV vs. SHE (Dolfing and Janssen 1994; Vogel *et al.* 1987). Accordingly, it is somewhat surprising that only highly reducing, low-potential electron donors seem to sustain their dechlorination. Inspired by the above mentioned findings, our research group investigated the possibility of using a polarized (–500 mV vs. SHE) glassy carbon electrode, in combination with methyl viologen, as electron donor for the microbial reductive dechlorination of TCE (Aulenta *et al.* 2008a; Aulenta *et al.* 2007) (Table 15.1). This bioelectrochemically assisted reductive dechlorination process (BEARD), is based on the extracellular electron transfer mechanism depicted in Figure 15.1, panel b: the mediator is reduced at the electrode surface and is re-oxidised by microorganisms. A key aspect of the process is the fact that H_2 generation is prevented by maintaining the cathode potential at a value that is only slightly more reducing (–500 mV vs. SHE) than that theoretically needed for H_2 derived from electrolysis of water, i.e. $E^{\circ\prime} = -442$ mV vs. SHE, at pH = 7.5, and not sufficient to prevail the overpotential at the glassy carbon electrode. By eliminating H_2 production, competitive metabolic reactions could be minimized and the process turned out to be highly efficient in terms of current consumption. Notably, with the above described process, TCE dechlorination was stimulated also when the redox mediator was physically anchored within a thin conductive Nafion™ film, deposited at the surface of the electrode (Table 15.1). This provided the first evidence of heterogeneous-phase, extracellular electron transfer to dechlorinating microorganisms. A further step was the recent discovery that *Geobacter lovleyi* is capable of partially dechlorinating PCE to *cis*-DCE using a polarized (−300 mV vs. SHE) graphite cathode as the sole electron donor, in the absence of any externally added redox mediator. In addition to dechlorinating PCE to *cis*-DCE, *G. lovleyi* also reduced fumarate to succinate with the polarized electrode serving as the sole electron donor or generate electricity from the oxidation of acetate using the electrode as electron acceptor (Strycharz *et al.* 2008) (Table 15.1). Our research group recently demonstrated that complete dechlorination of TCE to ethene is also achievable using graphite electrodes polarized at lower potentials (−450 mV vs. SHE) and a microbial culture containing *Dehalococcoides* spp. (Aulenta *et al.* 2009). In spite of the formation of a biofilm at the surface of the carbon electrodes, it was suggested that a soluble, self-produced redox mediator was also involved in the extracellular electron transfer (Aulenta *et al.* 2009) (Table 15.1).

15.2.2 Inorganic pollutants

Also for inorganic pollutants, bioelectrochemical processes pose an attractive alternative to the traditional biological treatment methods of supplementing microorganisms with organic electron donors. Indeed, several groundwater inorganic pollutants, such as nitrate, U(VI), perchlorate are reduced to harmless end products through microbially catalysed, dissimilatory reductions (Figure 15.2b–d). An overview of bioelectrochemical systems for remediation of inorganic pollutants is reported in Table 15.1.

Nitrate is estimated to be the most common agrochemical contaminant of the world's aquifers, especially in humid regions where drain effluent, carrying residual fertilizer nitrate, contributes significantly to pollution. The increased awareness of the detrimental effects of nitrate pollution has led to the development of water quality management programs for nitrate pollution control (Ugwuegbu *et al.* 2001).

Bioelectrochemical systems based on the electrolytic H_2 generation have been extensively applied for nitrogen removal from civil or industrial wastewater (Sakakibara and Kuroda 1993). On the other hand, fewer studies have investigated the application of this process to address nitrate-contaminated groundwater. Hayes and colleagues (Hayes *et al.* 1998) demonstrated that, by inserting stainless steel cathodes within laboratory-scale sand columns inoculated with a mixed culture of denitrifying bacteria, denitrification was stimulated when electric current (as high as 20 mA) was passed in the circuit. Even though the nitrogen removal efficiency increased with an increase in current, the Faradic current efficiency decreased at higher currents due to an inefficient utilization of the generated hydrogen (Table 15.1).

Recently, some researchers (Komori and Sakakibara 2008) equipped a fluidized-bed biofilm reactor with a solid-polymer-electrolyte membrane electrode (SPEME) cell in order to enhance hydrogenotrophic denitrification of groundwater. Experimental results showed that efficient production and dissolution of hydrogen were achieved by application of electric current as well as high-rate denitrification. Moreover, denitrification rates of nitrite increased with the increase of electric current.

A different approach, based on the use of methyl viologen to shuttle the electrons from a polarized electrode and the denitrifying bacteria was employed by Oh and colleagues (Oh *et al.* 2006). According to the authors, the use of a mediator decreased the overpotential required to transfer the electrons to denitrifying microorganisms.

The discovery of the direct, mediator-less, electron transfer from a polarized (−500 mV vs. Ag/AgCl) graphite electrode to a nitrate-reducing microorganism was due to Gregory and colleagues (Gregory *et al.* 2004) (Table 15.1). The authors showed that a pure culture of *Geobacter metallireducens* was able to reduce nitrate to nitrite with the electrode as the sole electron donor, with the expected stoichiometry of electron consumption (Figure 15.2). According to the authors, cells attached onto the electrode were responsible for the nitrate reduction. Nitrate reduction to nitrite is, however, only the first out of four reduction steps toward harmless nitrogen gas (Figure 15.2).

Researchers at Ghent University (Clauwaert *et al.* 2007) were the first to demonstrate, in mixed cultures, the possibility to achieve complete denitrification to nitrogen gas by exploiting direct electron transfer from a granular graphite cathode. Interestingly, significant nitrate reduction rates were obtained only when the cathode potential was lower than 0 mV vs. SHE (Table 15.1). Similar results were subsequently obtained by other research groups (Jia *et al.* 2008; Virdis *et al.* 2008).

It is worth noting that, in spite of its great potential, such a denitrification process based on direct extracellular electron transfer has been almost exclusively studied in the context of civil and industrial wastewater treatment, and never applied to the case of groundwater contamination.

Perchlorate (ClO_4^-) salts have been largely used as ingredients in solid rocket fuels, highway safety flares, airbag inflators, fireworks, and matches. Due to improper storage, handling, and disposal practices, perchlorate has become a widespread environmental contaminant worldwide (Engelhaupt 2007; Urbansky 2002), often detected in soil, groundwater, and drinking water supplies. Perchlorate contamination is of high concern due to the inhibiting effects of this compound on the human thyroid hormone production (Urbansky 2002). Perchlorate is a highly soluble anion, which does not significantly sorb to soils or sediments (Tipton *et al.* 2003). Perchlorate is used as terminal electron acceptor by pure and mixed cultures of microorganisms during the oxidation of many different organic and inorganic electron donors (Coates and Achenbach 2004; Ju *et al.* 2007; Logan 2001; Logan and LaPoint 2002; Logan *et al.* 2001; Shrout and Parkin 2006; Son *et al.* 2006). Dissimilatory perchlorate-reducing bacteria are mainly facultative anaerobes and denitrifiers, ubiquitous in the environment. The reduction sequence is depicted in Figure 15.2.

So far, very little information is available on the application of electrochemical methods to stimulate the metabolism of dissimilatory perchlorate-reducing bacteria. Recently, Thrash and colleagues (Thrash *et al.* 2007) investigated the reduction of perchlorate in the cathodic chamber of a bioelectrochemical reactor. The results showed that washed cells of *Dechloromonas* and *Azospira* species reduced 90 mg/l perchlorate in the presence of a graphite cathode polarized at −450 mV vs. Ag/AgCl and 500 μmol/l of AQDS at rates comparable to those obtained in the presence of acetate as electron donor (Table 15.1). Similar results were also obtained when a natural microbial community was inoculated into the bioelectrochemical reactor, thereby indicating that perchlorate-reducing bacteria are ubiquitous in the environment. Following this preliminary investigation, the authors developed a continuous up-flow bioelectrochemical reactor packed with powdered graphite which completely reduced 60 mg perchlorate/l d. Some hydrogen was produced at the cathode from water electrolysis that may have supported bacterial perchlorate reduction.

More recently, Shea and colleagues (Shea *et al.* 2008) investigated a functioning microbial fuel cell (MFC) with a denitrifying biocathode for perchlorate reduction. In contrast to the previous work, the cathode of the MFC reactor was not maintained at a controlled potential and no exogenous shuttles were added. The maximum attained perchlorate removal rate was 12 mg/l d (Table 15.1). Since no soluble electron donors were detected in the medium the authors suggested that perchlorate reducing bacteria were directly using the cathode (consisting of graphite granules) as electron donor via an extracellular electron transfer mechanism.

Uranium has been discharged into the environment for more than 60 years as a result of nuclear energy and weapons production and incorrect mine management. The remediation of uranium-contaminated soils and groundwater is an environmental challenge worldwide. Groundwater contamination is of particular concern because oxidised uranium is toxic, generally soluble in groundwater, and therefore highly mobile within the subsurface at many of these sites.

The substantial decrease in solubility accompanying the reduction of U(VI) to U(IV), producing the insoluble mineral uraninite, has been viewed as a potential mechanism for containment of uranium contamination (Luo *et al.* 2007; Wu *et al.* 2007; Yabusaki *et al.* 2007). Over the last few years, it has been firmly established that a wide variety of bacteria exhibit this reductive capacity as part of their energy metabolism (Cardenas *et al.* 2008; Gao and Francis 2008; Hwang *et al.* 2008; Lovley and Phillips 1992; N'Guessan *et al.* 2008; Shelobolina *et al.* 2008; Sung *et al.* 2006; Vrionis *et al.* 2005; Wall and Krumholz 2006).

Accordingly, the stimulation of such bacteria through addition of organic electron donors or hydrogen has been extensively considered for subsurface remediation of U-contaminated aquifers and sediments (Anderson *et al.* 2003; Lovley 1995). However, similarly to perchlorate, electrochemical methods have only recently been considered for the stimulation of uranium reducing bacteria.

Previous evidence that *Geobacter* species can reduce uranium with organic compounds serving as electron donors (Holmes *et al.* 2002) and also directly accept electrons from negatively poised electrons (Gregory *et al.* 2004), suggested the possibility of using insoluble electrodes as the sole electron donors for uranium reduction by a pure culture of *Geobacter sulfurreducens* and microorganisms in uranium-contaminated sediments (Gregory and Lovley 2005). Electrodes poised at –500 mV vs. Ag/AgCl rapidly removed U(VI) from solution in the absence of cells. However, when poise to the electrode was removed, all of the U(VI) returned to solution, indicating that no uranium reduction occurred. In contrast, if *G. sulfurreducens* was present on the graphite electrode, U(VI) did not return to solution unless the electrode was exposed to oxygen. This finding suggested that *G. sulfurreducens* reduced U(VI) to U(IV) (using the electrode as the sole electron donor), which was stably precipitated until reoxidized in the presence of oxygen (Table 15.1). Similar behaviour was observed when an electrode was placed in uranium-contaminated subsurface sediments. Notably, by this approach, uranium precipitates could even be recovered by pulling the electrode out of the contaminated subsurface.

15.3 OUTLOOKS, PERSPECTIVES, AND CHALLENGES TOWARDS FIELD APPLICATIONS

This review has mostly focused on electrically stimulated microbial reductions. However, bioelectrochemical processes could find important applications also in the treatment of contaminants undergoing microbial oxidation, such as petroleum hydrocarbons or lower chlorinated solvents. Among them, the enhancement of oxidative biodegradation processes through *in situ* electrolytic oxygen generation is probably the most straightforward (Franz *et al.* 2002).

On the other hand, the remarkable advancements made in the last few years in the understanding of anodic extracellular electron transfer processes in microbial fuel cells (Schroder 2007) may contribute to the development of new remediation methods in which insoluble electrodes dug in the subsurface serve as direct electron acceptors during the microbial oxidation of contaminants. This approach would provide a non-exhaustible source of electron acceptors (i.e. a solid-state anode) that is thermodynamically similar to or as favourable as oxygen but without the need for subsurface aeration. To our best knowledge, the feasibility of this approach has been, so far, proved only once (Morris and Jin 2008), using a laboratory-scale microbial fuel cell and a mixture of petroleum hydrocarbons as the substrate.

In spite of the extensive research currently carried out worldwide, *bioelectrochemical* systems for subsurface remediation are, however, still in their infancy, being investigated almost exclusively at the laboratory scale under well-controlled reaction conditions. Indeed, detailed information concerning full-scale or field-scale application of BES is lacking. On the other hand, *electrochemical* remediation systems represent a mature technology for *in situ* treatment of a variety of inorganic and organic pollutants (Trombly 1994). Although different mechanisms (e.g. electromigration, electroosmosis, electrophoresis) guide the process, electrochemical remediation occurs when (relatively) low-voltage direct current electric fields are applied to contaminated sites by means of electrode arrays. Electric current flows between pairs of anodes and cathodes placed in the ground (either in the vadose zone or in the aquifer). Current density is generally on the order of milliamperes per square centimetre (mA/cm^2) or an electric potential difference on the order of a few Volts per centimetre across electrodes. Whether situated horizontally in trenches or vertically in drilled tunnels around the marked site, the electrodes serve as goal posts to migrating heavy metals, soluble organics and, in some cases insoluble organics. Once the charged ionic contaminants or solubilized pollutants have accumulated at the electrode, they are more easily extracted for treatment by conventional methods (Trombly 1994). In principle, similar configurations could also be adopted for the bioelectrochemical remediation systems. The schematic of a potential bioelectrochemical system for *in situ* treatment of perchloroethene (PCE) contaminated groundwater is illustrated in Figure 15.3. The electrodes, possibly made of conductive and non-corrosive carbon materials (e.g. graphite granules), are placed within the aquifer so as to form a permeable and reactive barrier which intercepts (and treats) the contamination plume. Bacteria reductively dechlorinate PCE to *cis*-DCE using the cathode as electron donor. Produced *cis*-DCE migrates downgradient where, in principle, it could be oxidized by microorganisms using the anode as electron acceptor. Other oxidized contaminants (e.g. nitrate, perchlorate, U(VI)) could be reduced at the cathode with the same approach. Analogously, besides *cis*-DCE or lower chlorinated hydrocarbons, other oxidable pollutants (e.g. petroleum hydrocarbons) could also be oxidized at the anode. If needed, the potential generated by the reaction can be augmented using an external power supply.

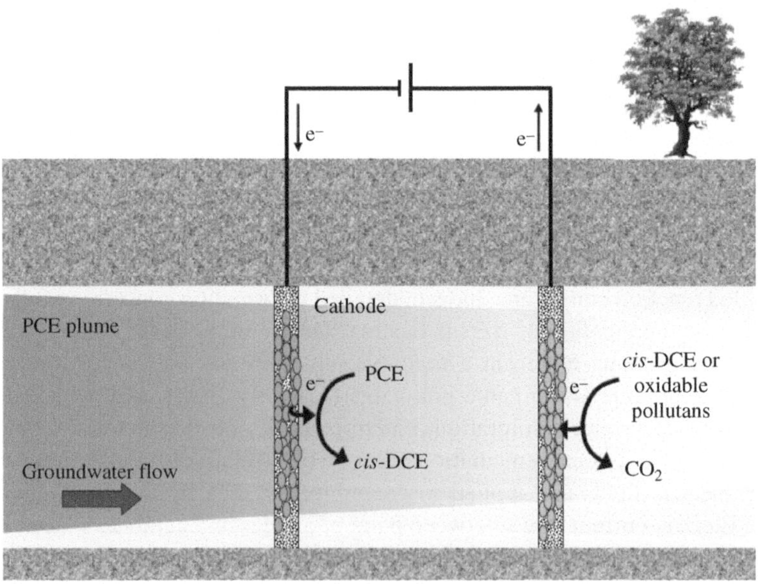

Figure 15.3. Schematic of a potential bioelectrochemical system for *in situ* treatment of groundwater contaminated by perchloroethene (PCE) and/or oxidable pollutants

Clearly, in order to make such processes competitive with existing remediation technologies, extensive testing at the field scale is needed. Major engineering challenges towards field application include the identification of suitable and sustainable electrode materials, strategies to control individual electrode potentials, site-specific design, operation and realization criteria.

Moreover, the subsoil chemical and biological factors that may limit the application of bioelectrochemical remediation will also have to be adequately quantified, as well as the impact that bioelectrochemical remediation may have on the soil's physical, chemical, and biological characteristics. Finally, as for most *in situ* remediation technologies, it is anticipated that process performance will be dramatically dependent upon subsurface characteristics such as soil type and grain size, contaminant type, concentration, and distribution, total ionic concentration, etc.

Besides subsurface remediation, direct or mediated extracellular electron transfer processes could find noticeable applications also in the field of subsurface sensing. Indeed, in microbial fuel cells (or in BES) current generation is directly linked to substrate concentration/availability and microbial respiratory activity; therefore, upon miniaturization these devices could be used in *in situ* monitoring of contaminants concentration as well as in evaluating bioremediation potential. Recently, Tront and colleagues (Tront *et al.* 2008a; Tront *et al.* 2008b) demonstrated that, in a microbial fuel cell inoculated with *Geobacter sulfurreducens*, current generation was mirrored by bulk phase acetate concentration, and an excellent correlation was developed between current values and acetate concentrations. The microbial fuel cell-based sensor also responded to pulses of oxygen with a reversible drop of current generation, thereby demonstrating that the microbial fuel cell-sensor provided real-time data for electron donor availability and biological activity.

In conclusion, this review has illustrated the exciting potential of bioelectrochemical systems (BES) for subsurface remediation. Most of the above described systems take advantage of solid-state electrodes to deliver electrons instead of chemicals to microorganisms carrying out biodegradative processes. BESs have several potential advantages over traditional approaches based on the subsurface injection of chemicals and hold great promise as sustainable and highly versatile bioremediation methods.

The 'proof-of-concep' of BES has been shown only for a few contaminants, biodegradative processes, microorganisms, and almost exclusively at the laboratory scale. Additionally, the cathodic extracellular electron transfer mechanisms are still poorly understood. Further research efforts and extensive field testing are therefore needed in order to fully establish the potential of these new remediation methods. Most probably, all these issues will be the subject of investigations over the next years.

REFERENCES

Anderson, R.T., Vrionis, H.A., Ortiz-Bernad, I., Resch, C.T., Long, P.E., Dayvault, R., Karp, K., Marutzky, S., Metzler, D.R., Peacock, A. *et al.* (2003) Stimulating the *in situ* activity of *Geobacter* species to remove uranium from the groundwater of a uranium-contaminated aquifer. *Appl. Environ. Microbiol.* **69**(10), 5884–5891.

Aulenta, F., Canosa, A., Majone, M., Panero, S., Reale, P. and Rossetti, S. (2008a) Trichloroethene dechlorination and H_2 evolution are alternative biological pathways of electric charge utilization by a dechlorinating culture in a bioelectrochemical system. *Environ. Sci. Technol.* **42**(16), 6185–6190.

Aulenta, F., Canosa, A., Reale, P., Rossetti, S., Panero, S. and Majone, M. (2009) Microbial reductive dechlorination of trichloroethene to ethene with electrodes serving as electron donors without the external addition of redox mediators. *Biotechnol. Bioeng.* **103**(1), 85–91.

Aulenta, F., Catervi, A., Majone, M., Panero, S., Reale, P. and Rossetti, S. (2007) Electron transfer from a solid-state electrode assisted by methyl viologen sustains efficient microbial reductive dechlorination of TCE. *Environ. Sci. Technol.* **41**(7), 2554–2559.

Aulenta, F., Majone, M. and Tandoi, V. (2006) Enhanced anaerobic bioremediation of chlorinated solvents: environmental factors influencing microbial activity and their relevance under field conditions. *J. Chem. Technol. Biotechnol.* **81**(9), 1463–1474.

Aulenta, F., Reale, P., Catervi, A., Panero, S. and Majone, M. (2008b) Kinetics of trichloroethene dechlorination and methane formation by a mixed anaerobic culture in a bio-electrochemical system. *Electrochim. Acta* **53**(16), 5300–5305.

Bard, A., Stratmann, M., Rubinstein, I., Fujihira, M. and Rusling, J. (2007) *Encyclopedia of Electrochemistry*, Vol. 10. Modified Electrodes: Wiley.

Biffinger, J.C., Pietron, J., Bretschger, O., Nadeau, L.J., Johnson, G.R., Williams, C.C., Nealson, K.H. and Ringeisen, B.R. (2008) The influence of acidity on microbial fuel cells containing *Shewanella oneidensis*. *Biosens. Bioelectron.* **24**(4), 906–911.

Cardenas, E., Wu, W.M., Leigh, M.B., Carley, J., Carroll, S., Gentry, T., Luo, J., Watson, D., Gu, B., Ginder-Vogel, M. *et al.* (2008) Microbial communities in contaminated sediments, associated with bioremediation of uranium to submicromolar levels. *Appl. Environ. Microbiol.* **74**(12), 3718–3729.

Clauwaert, P., Rabaey, K., Aelterman, P., de Schamphelaire, L., Pham, T.H., Boeckx, P., Boon, N. and Verstraete, W. (2007) Biological denitrification in microbial fuel cells. *Environ. Sci. Technol.* **41**(9), 3354–3360.

Coates, J.D. and Achenbach, L.A. (2004) Microbial perchlorate reduction: rocket-fueled metabolism. *Nature Rev. Microbiol.* **2**(7), 569–580.

De Wildeman, S. and Verstraete, W. (2003) The quest for microbial reductive dechlorination of C_2 to C_4 chloroalkanes is warranted. *Appl. Microbiol. Biotechnol.* **61**(2), 94–102.

Dolfing, J. and Janssen, D.B. (1994) Estimates of Gibbs free energies of formation of chlorinated aliphatic compounds. *Biodegradation* **5**(1), 21–28.

Engelhaupt, E. (2007) Fingerprinting perchlorate sources. *Environ. Sci. Technol.* **41**(8), 2654.

Franz, J.A., Williams, R.J., Flora, J.R.V., Meadows, M.E. and Irwin, W.G. (2002) Electrolytic oxygen generation for subsurface delivery: effects of precipitation at the cathode and an assessment of side reactions. *Water Res.* **36**(9), 2243–2254.

Gao, W. and Francis, A.J. (2008) Reduction of uranium(VI) to uranium(IV) by clostridia. *Appl. Environ. Microbiol.* **74**(14), 4580–4.

Gregory, K.B., Bond, D.R. and Lovley, D.R. (2004) Graphite electrodes as electron donors for anaerobic respiration. *Environ. Microbiol.* **6**(6), 596–604.

Gregory, K.B. and Lovley, D.R. (2005) Remediation and recovery of uranium from contaminated subsurface environments with electrodes. *Environ. Sci. Technol.* **39**(22), 8943–8947.

Guiot, S.R., Cimpoia, R., Kuhn, R. and Alaplantive, A. (2008) Electrolytic methanogenic–methanotrophic coupling for tetrachloroethylene bioremediation: proof of concept. *Environ. Sci. Technol.* **42**(8), 3011–3017.

Hayes, A.M., Flora, J.R. and Khan, J. (1998) Electrolytic stimulation of denitrification in sand columns. *Water Res.* **32**(9), 2830–2834.

Holliger, C. and Schumacher, W. (1994) Reductive dehalogenation as a respiratory process. *Antonie Van Leeuwenhoek* **66**(1–3), 239–246.

Holmes, D.E., Finneran, K.T., O'Neil, R.A. and Lovley, D.R. (2002) Enrichment of members of the family Geobacteraceae associated with stimulation of dissimilatory metal reduction in uranium-contaminated aquifer sediments. *Appl. Environ. Microbiol.* **68**(5), 2300–2306.

Hwang, C., Wu, W., Gentry, T.J., Carley, J., Corbin, G.A., Carroll, S.L., Watson, D.B., Jardine, P.M, Zhou, J., Criddle, C.S. *et al.* (2008) Bacterial community succession during in situ uranium bioremediation: spatial similarities along controlled flow paths. *Isme J.* (In press).

Jia, Y.H., Tran, H.T., Kim, D.H., Oh, S.J., Park, D.H., Zhang, R.H. and Ahn, D.H. (2008) Simultaneous organics removal and bio-electrochemical denitrification in microbial fuel cells. *Bioprocess Biosyst. Engng.* **31**(4), 315–321.

Ju, X., Field, J.A., Sierra-Alvarez, R., Salazar, M., Bentley, H. and Bentley, R. (2007) Chemolithotrophic perchlorate reduction linked to the oxidation of elemental sulfur. *Biotechnol. Bioeng.* **96**(6), 1073–1082.

Komori, M. and Sakakibara, Y. (2008) High-rate hydrogenotrophic denitrification in a fluidized-bed biofilm reactor using solid-polymer-electrolyte membrane electrode (SPEME). *Water Sci. Technol.* **58**(7), 1441–1446.

Loffler, F.E. and Edwards, E.A. (2006) Harnessing microbial activities for environmental cleanup. *Curr. Opin. Biotechnol.* **17**(3), 274–284.

Logan, B.E. (2001) Assessing the outlook for perchlorate remediation. *Environ. Sci. Technol.* **35**(23), 482A–487A.

Logan, B.E. and LaPoint, D. (2002) Treatment of perchlorate- and nitrate-contaminated groundwater in an autotrophic, gas phase, packed-bed bioreactor. *Water Res.* **36**(14), 3647–3653.

Logan, B.E., Wu, J. and Unz, R.F. (2001) Biological perchlorate reduction in high-salinity solutions. *Water Res.* **35**(12), 3034–3038.

Lohner, S.T., Katzoreck, D. and Tiehm, A. (2008a) Electromigration of microbial electron acceptors and nutrients: (I) transport in synthetic media. *J. Environ. Sci. Health A Tox. Hazard Subst. Environ. Engng.* **43**(8), 913–921.

Lohner, ST, Katzoreck, D, Tiehm, A. 2008b. Electromigration of microbial electron acceptors and nutrients: (II) transport in groundwater. *J. Environ. Sci. Health A Tox. Hazard Subst. Environ. Engng.* **43**(8), 922–925.

Lovley, D.R. (1995) Bioremediation of organic and metal contaminants with dissimilatory metal reduction. *J. Ind. Microbiol.* **14**(2), 85–93.

Lovley, D.R. and Phillips, E.J. (1992) Reduction of uranium by *Desulfovibrio desulfuricans*. *Appl. Environ. Microbiol.* **58**(3), 850–856.

Luo, J., Weber, F.A., Cirpka, O.A., Wu, W.M., Nyman, J.L., Carley, J., Jardine, P.M., Criddle, C.S. and Kitanidis, P.K. (2007) Modeling *in-situ* uranium(VI) bioreduction by sulfate-reducing bacteria. *J. Contam. Hydrol.* **92**(1–2), 129–148.

Marsili, E., Baron, D.B., Shikhare, I.D., Coursolle, D., Gralnick, J.A. and Bond, D.R. (2008) Shewanella secretes flavins that mediate extracellular electron transfer. *Proc. Natl. Acad. Sci. USA* **105**(10), 3968–3973.

Maymo-Gatell, X., Chien, Y., Gossett, J.M. and Zinder, S.H. (1997) Isolation of a bacterium that reductively dechlorinates tetrachloroethene to ethene. *Science* **276**(5318), 1568–1571.

Maymo-Gatell, X., Tandoi, V., Gossett, J.M. and Zinder, S.H. (1995) Characterization of an H_2-utilizing enrichment culture that reductively dechlorinates tetrachloroethene to vinyl chloride and ethene in the absence of methanogenesis and acetogenesis. *Appl. Environ. Microbiol.* **61**(11), 3928–3933.

Morris, J.M. and Jin, S. (2008) Feasibility of using microbial fuel cell technology for bioremediation of hydrocarbons in groundwater. *J. Environ. Sci. Health, Part A* **43**(1), 18–23.

Morse, J.J., Alleman, B.C., Gossett, J.M., Zinder, S.H., Fennell, D.E., Sewell, G.W. and Vogel, C.M. (1997) A treatability test for evaluating the potential applicability of the reductive anaerobic biological *in situ* treatment technology (RABITT) to remediate chloroethenes. A draft technical protocol developed for the Environmental Security Technology Certification Program, Department of Defence; US Government Printing Office, Washington DC.

N'Guessan, A.L., Vrionis, H.A., Resch, C.T., Long, P.E. and Lovley, D.R. (2008) Sustained removal of uranium from contaminated groundwater following stimulation of dissimilatory metal reduction. *Environ. Sci. Technol.* **42**(8), 2999–3004.

Nijenhuis, I. and Zinder, S.H. (2005) Characterization of hydrogenase and reductive dehalogenase activities of Dehalococcoides ethenogenes strain 195. *Appl. Environ. Microbiol.* **71**(3), 1664–1667.

Oh, C.K., Song, S.H., Kim, Y.H., Park, D.H. and Yoo, Y.J. (2006) Bioelectrochemical denitrification using permeabilized *Ochrobactrum antropi* SY509. *J. Microbiol. Biotechnol.* **16**(5), 678–682.

Rabaey, K., Boon, N., Siciliano, S.D., Verhaege, M. and Verstraete, W. (2004) Biofuel cells select for microbial consortia that self-mediate electron transfer. *Appl. Environ. Microbiol.* **70**(9), 5373–5382.

Rabaey, K. and Keller, J. (2008) Microbial fuel cell cathodes: from bottleneck to prime opportunity? *Water Sci. Technol.* **57**(5), 655–659.

Sakakibara, Y. and Kuroda, M. (1993) Electric prompting and control of denitrification. *Biotechnol. Bioeng.* **42**(4), 535–537.

Schroder, U. (2007) Anodic electron transfer mechanisms in microbial fuel cells and their energy efficiency. *Phys. Chem. Chem. Phys.* **9**(21), 2619–2629.

Shea, C., Clauwaert, P., Verstraete, W. and Nerenberg, R. (2008) Adapting a denitrifying biocathode for perchlorate reduction. *Water Sci. Technol.* **58**(10), 1941–1946.

Shelobolina, E.S., Vrionis, H.A., Findlay, R.H. and Lovley, D.R. (2008) *Geobacter uraniireducens* sp. nov., isolated from subsurface sediment undergoing uranium bioremediation. *Int. J. Syst. Evol. Microbiol.* **58**(Pt 5), 1075–1078.

Shrout, J.D. and Parkin, G.F. (2006) Influence of electron donor, oxygen, and redox potential on bacterial perchlorate degradation. *Water Res.* **40**(6), 1191–1199.

Skadberg, B., Geoly-Horn, S.L., Sangamalli, V., Flora, J.R.V. (1999) Influence of pH, current and copper on the biological dechlorination of 2,6-dichlorophenol in an electrochemical cell. *Water Res.* **33**, 1997–2010.

Sleep, B.E., Seepersad, D.J., Kaiguo, M.O., Heidorn, C.M., Hrapovic, L., Morrill, P.L., McMaster, M.L., Hood, E.D., Lebron, C., Lollar, B.S. *et al.* (2006) Biological enhancement of tetrachloroethene dissolution and associated microbial community changes. *Environ. Sci. Technol.* **40**(11), 3623–3633.

Son, A., Lee, J., Chiu, P.C., Kim, B.J. and Cha, D.K. (2006) Microbial reduction of perchlorate with zero-valent iron. *Water Res.* **40**(10), 2027–2032.

Stenuit, B., Eyers, L., Schuler, L., Agathos, S.N. and George, I. (2008) Emerging high-throughput approaches to analyze bioremediation of sites contaminated with hazardous and/or recalcitrant wastes. *Biotechnol. Adv.* **26**(6), 561–575.

Strycharz, S.M., Woodard, T.L., Johnson, J.P., Nevin, K.P., Sanford, R.A., Loffler, F.E., Lovley, D.R. (2008) Graphite electrode as a sole electron donor for reductive dechlorination of tetrachlorethene by *Geobacter lovleyi*. *Appl. Environ. Microbiol.* **74**(19), 5943–5947.

Sung, Y., Fletcher, K.E., Ritalahti, K.M., Apkarian, R.P., Ramos-Hernandez, N., Sanford, R.A., Mesbah, N.M. and Loffler, F.E. (2006) *Geobacter lovleyi* sp. nov. strain SZ, a novel metal-reducing and tetrachloroethene-dechlorinating bacterium. *Appl. Environ. Microbiol.* **72**(4), 2775–2782.

Thauer, R.K., Jungermann, K. and Decker, K. (1977) Energy conservation in chemotrophic anaerobic bacteria. *Bacteriol. Rev.* **41**(1), 100–180.

Thrash, J.C., Van Trump, J.I., Weber, K.A., Miller, E., Achenbach, L.A. and Coates, J.D. (2007) Electrochemical stimulation of microbial perchlorate reduction. *Environ. Sci. Technol.* **41**(5), 1740–1746.

Tipton, D.K., Rolston, D.E. and Scow, K.M. (2003) Transport and biodegradation of perchlorate in soils. *J. Environ. Qual.* **32**(1), 40–46.

Trombly, J. (1994) Electrochemical remediation takes to the field. *Environ. Sci. Technol.* **28**(6), 289A–291A.

Tront, J., Fortner, J., Plötze, M., Hughes, J. and Puzrin, A. (2008a) Microbial fuel cell technology for measurement of microbial respiration of lactate as an example of bioremediation amendment. *Biotechnol. Lett.* **30**(8), 1385–1390.

Tront, J.M., Fortner, J.D., Plötze, M., Hughes, J.B. and Puzrin, A.M. (2008b) Microbial fuel cell biosensor for *in situ* assessment of microbial activity. *Biosensors Bioelectron.* **24**(4), 586–590.

Ugwuegbu, B.U., Prasher, S.O. and Ahmad, D. (2001) Bioremediation of residual fertilizer nitrate: I. Laboratory demonstration of an on-farm *in situ* pollution control system. *J. Environ. Qual.* **30**(1), 1–10.

Urbansky, E.T. (2002) Perchlorate as an environmental contaminant. *Environ. Sci. Pollut. Res. Int.* **9**(3), 187–192.

Van Trump, J.I. and Coates, J.D. (2009) Thermodynamic targeting of microbial perchlorate reduction by selective electron donors. *ISME J.* **3**, 466–476.

Virdis, B., Rabaey, K., Yuan, Z. and Keller, J. (2008) Microbial fuel cells for simultaneous carbon and nitrogen removal. *Water Res.* **42**(12), 3013–3024.

Vogel, T.M., Criddle, C.S. and McCarty, P.L. (1987) Transformations of halogenated aliphatic compounds. *Environ. Sci. Technol.* **21**(8), 722–736.

Vrionis, H.A., Anderson, R.T., Ortiz-Bernad, I., O'Neill, K.R., Resch, C.T., Peacock, A.D., Dayvault, R., White, D.C., Long, P.E. and Lovley, D.R. (2005) Microbiological and geochemical heterogeneity in an in situ uranium bioremediation field site. *Appl. Environ. Microbiol.* **71**(10), 6308–6318.

Wall, J.D. and Krumholz, L.R. (2006) Uranium reduction. *Ann. Rev. Microbiol.* **60**, 149–166.

Weast, R. (1982) *Handbook of Chemistry and Physics.* Weast, R. (ed.) Boca Raton, FL: CRC Press.

Wu, W.M., Carley, J., Luo, J., Ginder-Vogel, M.A., Cardenas, E., Leigh, M.B., Hwang, C., Kelly, S.D., Ruan, C., Wu, L. *et al.* (2007) *In situ* bioreduction of uranium (VI) to submicromolar levels and reoxidation by dissolved oxygen. *Environ. Sci. Technol.* **41**(16), 5716–5723.

Yabusaki, S.B., Fang, Y., Long, P.E., Resch, C.T., Peacock, A.D., Komlos, J., Jaffe, P.R., Morrison, S.J., Dayvault, R.D. and White, D.C. *et al.* (2007) Uranium removal from groundwater via *in situ* biostimulation: field-scale modeling of transport and biological processes. *J. Contam. Hydrol.* **93**(1–4), 216–235.

Zhang, X.H., Sewell, G.W. and Cui, S.Y. (2001) An improved method of hydrogen production as electron donor for anaerobic bioremediation. *J. Environ. Sci. Health A Tox. Hazard Subst. Environ. Engng.* **36**(9), 1661–1670.

16

Fundamentals of benthic microbial fuel cells: theory, development and application

Peter R. Girguis, Mark E. Nielsen and Clare E. Reimers

16.1 INTRODUCTION

Recent years have been a watershed for fuel cell research, in particular for microbial fuel cells (see for example: Bullen *et al.* 2006; Du *et al.* 2007; He *et al.* 2005; Logan *et al.* 2006; Logan 2008; Lovley 2006; Rabaey *et al.* 2007; Schröder 2007). While the majority of microbial fuel cell (MFC) research has focused on the performance and characteristics of laboratory reactors, benthic microbial fuel cells (BMFCs), or those that rely on a potential gradient at a sediment–water interface, have played a significant – sometimes

 Bioelectrochemical Systems: From Extracellular Electron Transfer to Biotechnological Application. Edited by Korneel Rabaey, Largus Angenent, Uwe Schröder and Jürg Keller. ISBN: 9781843392330. Published by IWA Publishing, London, UK.

understated – role in the evolution of MFC technology. Whereas laboratory MFCs offer a unique opportunity to examine fundamental attributes of MFC performance and microbial physiology (e.g. ohmic resistance, current density, and mechanisms of electron transfer by laboratory cultivated microbes), BMFCs offer a unique opportunity to examine how energy moves through microbial communities, the efficiency of harvesting electricity from natural systems, and the potential role(s) that BMFCs may play in power generation or bioremediation in the natural world. Accordingly, BMFCs that have been deployed in the ocean have also furthered our understanding of marine biogeochemical processes. A comprehensive review of BMFCs would entail, by necessity, a detailed discussion of all aspects of MFC technology, and is beyond the scope of this chapter. Rather, here we provide a review of the underlying principles of BMFCs, including theoretical and geochemical considerations, aspects of material selection and system design, the trends in microbial ecology among BMFCs, and the potential that BMFCs have as practical – and commercially viable – power systems. We examine these fundamentals of BMFC development and application with the goal of illustrating how BMFCs have evolved in design, performance and application over the last decade.

16.2 FUNDAMENTAL PRINCIPLES OF SEDIMENT REDUCTION-OXIDATION CHEMISTRY

In nature, reduction–oxidation gradients are typically found across sediment–water interfaces and within aquatic and terrestrial sediments (Alberte *et al.* 2005; Reimers *et al.* 2001; Tender *et al.* 2002; Whitfield 1972). These gradients are the result of spatially separated microbially mediated reduction oxidation reactions, linked to the microbial degradation of organic matter (Froelich et al. 1979). At the sediment–seawater interface, oxygen is present and microbial aerobic respiration is the dominant process. In deeper sediments where oxygen is depleted (termed suboxic sediments), other microbially mediated processes – such as nitrate, iron, and manganese reduction – are used as oxidants to support organic matter degradation. In anoxic sediments, microbial sulfate reduction, fermentation and methanogenesis dominate, as they are the most thermodynamically favored reactions in these environments. The biogeochemical zonation arises because each of the processes listed above occur in sequence according to the free energy (ΔG) yielded by the reaction, with the most energetically favorable processes occurring first. Theoretically, if all reactions were known, the difference in redox potentials between the electron-accepting and electron-donating couple (ΔE) could be characterized and related to ΔG according to:

$$\Delta E = -nF/\Delta G$$

where n is the number of electrons transferred and F is the faraday constant (96,485 coulomb mol^{-1}). From this, the theoretical redox potential difference (ΔE) between oxic seawater and anoxic sulfide-rich sediments is approximately 0.86 to 0.91 V at the conditions found *in situ* (Stumm and Morgan 1996). In practice, observed redox potential differences in BMFCs are 0.7 to 0.8 V as measured by redox electrodes (Schulz 2006), a consequence of environmental heterogeneity such as variations in the concentrations of redox species, *in situ* mixing of porewaters and bioturbation (mixing caused by biological processes) (Stumm and Morgan 1996).

Based on a combination of the redox characteristics described above, the physical properties of sediments and the dynamics of natural environments, there are a special set of factors that affect the performance of BMFCs in ways that do not apply to traditional laboratory or wastewater MFCs. First, mass transport through sediments is generally slow, thus, limiting the availability of fuel transport to the anode (and the transport of reaction products away from the anode) (Nielsen *et al.* 2007; Tender *et al.* 2002). Electrodes may also become passivated by electrochemical deposition (of which sulfur is one example) or by biofouling (such as biofilm formation or sedimentation on the cathode that reduces the availability of oxygen)(Reimers *et al.* 2006). By necessity, anodes and cathodes are spatially separated due to the scale of natural redox gradients and this separation may lead to high internal resistance with increasing sediment depth. Finally, working in underwater field settings presents a unique set of challenges that require extremely robust instrument designs that can be deployed with available resources. We will return to these considerations – and how this may influence commercialization – after describing the fundamental aspects of BMFCs below.

16.3 PRINCIPLES OF DESIGN AND APPROACHES TO TESTING BENTHIC MICROBIAL FUEL CELLS (BMFCs)

BMFCs are electrochemical devices that generate electrical current from the natural redox gradients that commonly occur across the sediment–water interface. In general, a BMFC consists of a circuit in which inert but electrically conductive electrodes are placed in an anoxic zone (the anode) and an oxic zone (cathode).

The electrodes are connected through a load, such as an external resistor or potentiostat (in research applications) or a sensor system (in field demonstrations or applications). Electrons resulting from microbial metabolic activity are transferred from electron donors to the anode and flow through the load to the cathode, where they lead to the reduction of dissolved oxygen to water.

Monitoring the performance of BMFCs (with respect to power production) entails making several measurements. Typically, a voltmeter may be used to monitor cell voltage (E_{cell}) – which is the potential between anode and cathode – and a combination of a voltmeter and reference electrode (typically a Ag/AgCl electrode in the oxic media) may be used to measure individual electrode potentials (E_{anode} and $E_{cathode}$). A potentiostat may be used to poise the cell potential at a fixed value (or, if desired, may be used to fix a single electrode potential against the reference electrode). Current (I) may be calculated according to Ohm's law if a known resistor is used, whereas the use of a potentiostat requires the investigator to directly measure current if they are interested in calculating power density. There is some debate about the merits of fixed resistance BMFCs (which results in varying electrode potentials) versus poised potential BMFCs (which results in varying current). The relative effects of varying current and potential on microbial community composition and activity have yet to be thoroughly addressed and it remains to be determined which of these approaches – if either – is better suited for commercial applications. Nevertheless, future studies should better quantify the effect of these factors on microbial processes.

BMFC investigators are often interested in determining the *efficiency* of energy transfer from endogenous fuels to the BMFC. One measure of efficiency is the current efficiency (also known as Coulombic efficiency), which is the percentage of electrons coming from oxidation of the organic matter that are harnessed by the electrode (measured as current, Bard and Faulkner, 2001). This metric provides investigators with a better understanding of how much energy in organic matter is available to – and ultimately harnessed by – the BMFC. Through the application of laboratory and *in situ* measurement techniques, investigators set out to determine the influence of electrode materials and design, as well as overall system design and deployment, to BMFC performance (discussed in detail below).

16.4 ANODE MATERIAL AND DESIGN

In all BMFCs, the anode serves to accept electrons from biological and abiotic processes. Primarily, the anode is a solid-phase terminal electron acceptor for microbial metabolism. To ensure consistent, long-term performance, the ideal BMFC anode should be impervious to biological or abiotic corrosion, passivation or degradation. To that end, different materials that have been investigated including platinum mesh (Reimers *et al.* 2001; Whitfield 1972); graphite discs, plates or rods (Donovan *et al.* 2008; Reimers *et al.* 2006; Ryckelynck *et al.* 2005; Tender *et al.* 2002; Tender *et al.* 2008); corrosion-resistant stainless steel (Dumas *et al.* 2007); carbon fiber cloth (Rezaei *et al.* 2007; Scott *et al.* 2008b); glassy carbon and modified graphite (Lowy *et al.* 2006); as well as carbon sponge and reticulated vitreous carbon (Scott *et al.* 2008a).

Platinum is instinctively appealing in light of its catalytic efficiency. However, in one of the earliest studies, there was no observed increase in power production when platinum was used as an anode and cathode (relative to graphite plate; Reimers *et al.* 2001). While this may be due to limitations in other aspects of BMFC power production, such as mass transport limitations, it may also result from chemical fouling by hydrogen sulfide (which is common in organic-rich marine sediments; Reimers *et al.* 2006).

A more recent study has tested the use of stainless steel as both an anode and cathode in BMFCs, in laboratory and field settings (Dumas *et al.* 2007). In these deployments, BMFCs with stainless steel electrodes produced power at densities of 4 $mW \cdot m^{-2}$, which is lower than those reported previously with marine BMFCs using graphite electrodes (e.g. Bond *et al.* 2002; Bond *et al.* 2003; Holmes *et al.* 2004; Reimers *et al.* 2001; Reimers *et al.* 2006; Tender *et al.* 2002). Nevertheless, it is apparent that stainless steel is not suitable as an anode. Despite its name, stainless steel will readily corrode when kept in anoxic environments (the chromium oxide film that prevents iron oxidation is not stable in the absence of oxygen).

Conductive allotropes of carbon, including graphite, amorphous, and glassy carbon, are typically impervious to biological degradation and are effective as electrodes in BMFCs. Indeed, the number of BMFC studies that employed graphite or other conductive carbon electrodes are many (for review, see Lovley *et al.* 2006). In general, graphite sheet, plates and rods have been used as anodes because of the material's strength, inertness, and commercial availability (Bond *et al.* 2003; Reimers *et al.* 2001; Tender *et al.* 2002). While carbon allotropes bear a superficial similarity, they have different physical and electrical properties, including resistivity. To understand the effect of carbon allotrope composition, as well as form factor, on power production in BMFCs, Scott *et al.* (2008a) tested the performance of carbon sponge, carbon cloth, carbon fiber, and reticulated vitreous carbon (RVC) as anodes (with graphite cloth as a cathode). Similar open circuit voltages (up to 0.7 V) were observed for most cells, though carbon sponge produced the highest power density at 55 $mW \cdot m^{-2}$, nearly twice that of carbon cloth (Scott *et al.* 2008a). Notably, mass transport at both the anode and cathode was found to be the limiting factor in all treatments.

In a further effort to improve upon BMFC performance by alleviating limitations in surface reactions, investigators have modified graphite to examine the effects of surface treatments, such as doping with manganese, nickel, iron and compounds previously shown to be effective as electron shuttles in laboratory MFCs (Lowy *et al.* 2006). They typically observed a five-fold increases in power density in treated anodes over the course of the incubation, though it remains to be determined whether this improvement in performance is sustained during long-term incubations.

As mentioned, seafloor BMFC electrodes have shown signs of passivation by elemental sulfur deposits on the anode (Reimers *et al.* 2006). This is largely due to the abiotic oxidation of hydrogen sulfide (which is the byproduct of microbial sulfate reduction in marine sediments) at the anode surface. It has been suggested that anodes can be electrically poised at potentials that would 'strip'sulfur from the anodes, though it has yet to be determined if this is feasible in long-term deployments. Regardless of anode material, the issue of passivation is one that must be addressed prior to long-term use or commercialization.

16.5 CATHODE MATERIALS AND DESIGN

The function of a BMFC cathode is to donate electrons to available acceptors, thus, enabling the continuous flow of current from the anode. In most cases, the terminal electron acceptor at the cathode is dissolved oxygen in the surrounding seawater (though other electron acceptors are possible, and are discussed elsewhere in this volume). Indeed, cathodes are as important as anodes in terms of BMFC design, as they can limit MFC power production (Fornero *et al.* 2008). In general, BMFCs have typically employed the same carbon-based electrode materials that are used at the anode. At the cathode, platinum is effective in catalyzing the formation of water from oxygen and protons (proton equivalents), which could alleviate cathode limitation. However, its high cost has limited its utility in BMFCs, though platinum-doped graphite is an effective cathode that costs a fraction of platinum mesh or sheet and may prove useful in long-term deployments (Logan *et al.* 2006). Furthermore, in a recent study, Scott *et al.* (2008b) conducted a systematic comparison of 10 different cathode materials, with and without catalysts, such as Fe–Co tetramethoxyphenyl porphyrin (FeCoTMPP) and platinum. Surprisingly, the investigators found that the FeCoTMPP cathode supported power densities twice that of untreated carbon cloth. As mentioned, platinum is poisoned by hydrogen sulfide, so its utility in sulfur-rich systems may be limited. In practice, it is easier to reduce cathode limitation in BMFCs as cathodes are typically not subject to the spatial constraints of anodes, such as being buried or placed within a chamber. Thus, cathodes can be of much greater size. However, research on cathode performance in BMFCs is particularly sparse, and future research should aim to better characterize the microbial ecology, catalytic properties and overall performance of BMFC cathodes. Several research groups have designed BMFCs with stainless steel cathodes (Donovan *et al.* 2008; Dumas *et al.* 2007). In the work by Dumas *et al.*, the investigators decoupled the anode and cathode in laboratory BMFC experiments, and showed that the cathode was able to sustain driven current densities up to 140 mA m^{-2} (at +0.05 V versus Ag/AgCl), suggesting that stainless steel cathodes are potentially useful as BMFC applications (though their observed current densities are likely to be an order of magnitude greater than what might be observed in field deployments).

16.6 PERFORMANCE AND PRACTICAL CONSIDERATIONS OF BMFC DESIGNS

BMFCs, like all MFCs, are influenced by a variety of factors, including electrode properties, internal resistivity, environmental geochemistry, and microbial ecology. BMFC research has, in general, approached these matters via field experiments and laboratory simulations of field deployments, wherein one or more variables were manipulated to assess the influence on system performance over time. In the sections above, we have briefly reviewed the effect of anode and cathode materials and treatments specifically in the context of BMFCs. Here we present a brief summary of key field deployments, and key changes in design and implementation that yielded improvements in performance.

In the earliest BMFC studies, investigators deployed solid graphite plates as anodes and cathodes, burying the anodes at arbitrary depths within sediments, and suspending the cathodes in the overlying seawater on non-conductive frames. In these early deployments, investigators were largely focused on examining differences in power production among sites, or in relation to the bulk sediment properties (Reimers *et al.* 2001; Tender *et al.* 2002). For example, in Tender *et al.* (2002), the investigators deployed 48 cm-diameter graphite plates as anodes and cathodes, which yielded average power densities of 25 to 28 mW · m^{-2} of seafloor over a 3-month deployment (Tender *et al.* 2002). While the graphite plates provided good surface area, their aspect often made them difficult to deploy in marine sediments. In particular, anodes had to be well buried to avoid shorting by wave action or bioturbation (where marine animals irrigate the sediments with oxygen-rich seawater, thereby short-circuiting the electrode and ceasing power production). Tender and colleagues, however, have recently built a larger scale system called the Benthic Unattended Generator, or B.U.G., which uses large graphite plates to generate sufficient power for submarine sensors (Lovley 2006; Tender *et al.* 2008). To date, these systems have been less susceptible to the problems associated with earlier deployments perhaps because of an advantage of size. However, the size of these electrode arrays can require significant infrastructure for deployment, to insure that the arrays are properly anchored in the sediments.

To test a different anode and cathode configuration, Reimers *et al.* (2006) deployed a 1 m × 9 cm solid graphite rod in hydrocarbon seep sediments, coupled to a titanium-wound graphite brush cathode that was suspended approximately 1 m above the sediment-seawater interface. This BMFC yielded power densities of 34 $mW \cdot m^{-2}$ of anode, roughly three times higher than power densities obtained with similar electrode materials in coastal sediments (Ryckelynck *et al.* 2005). This particular anode and cathode configuration made for easier deployment by the remotely operated submersible in the deep ocean sediments. The improved performance, however, was largely attributable to the differences in porewater chemistry found at hydrocarbon seeps, including higher organic carbon and sulfide concentrations.

While the graphite rod facilitated deployment in sediments, the primary disadvantage came from its relatively limited surface area. Inspired by sediment benthic chambers (e.g. Smith and Teal 1973). Nielsen *et al.* (2007) developed a chambered BMFC design in which the anode is contained in a semi-enclosed chamber that is either pushed into or set atop the sediments. The chamber design has several advantages over previous BMFCs, including the ability to use high surface area electrodes (e.g. carbon-fiber brushes) and the ability to enhance mass transport either mechanically or through natural advection (Nielsen *et al.* 2007, 2008). These advantages yielded greater power densities and – to date – longer-term performance. In Nielsen *et al.* (2007), a chambered BMFC produced power densities more than an order of magnitude greater (with peak power densities of 380 mW/m^{-2}) than those achieved by previous BMFCs. All chambered BMFCs were found to generate current continuously for over 200 days. It should be noted that the highest power densities were achieved by occasionally pumping on the chamber. The cost of pumping a chambered BMFC system using BMFC-derived power remains to be determined. Chambered BMFCs have also been subject to failure due to scouring of sediments by strong bottom currents, or bioturbation, that results in oxygenation of the anode chamber.

Another effort to improve power production by BMFCs involved alleviating cathode limitation by rotating the cathode (He *et al.* 2007). This had the effect of increasing oxygen availability to the cathode, and therefore improving the cathode reaction rate (this method also incurs an energetic cost). This resulted in a higher power density (49mW · m^{-2}), nearly 70% greater than the non-rotating BMFC control. Interestingly, the data suggest that the increased oxygenation also had an adverse effect on the anode potential (possibly due to irrigation of the underlying sediment). This in turn constrained maximum power density. The authors suggest that future experiments should aim to address this issue, and determine an optimum level of rotation for maximal power production.

There has also been an effort to examine the efficacy of substrate enhancement in BMFCs. Razaei and Logan (2007) examined the influence of supplementing a BMFC with chitin and cellulose. Chitin in particular is common in marine sediments and, along with keratin and cellulose, is among the more common of recalcitrant carbon compounds in both marine and terrestrial settings. The authors observed maximal power densities from 64 to 83 mW · m^{-2} for chitin and cellulose, respectively (though cellulose-stimulated power production rapidly declined). Given the availability of chitin and cellulose from industrial processing, the authors consider the benefits of supplementing BMFCs with sufficient chitin or cellulose to sustain and stimulate prolonged power production. While this is a compelling idea, it remains to be determined how much chitin or cellulose would be needed to sustain power production for longer-term deployments, as both of these compounds may be catabolized by free-living microbial phylotypes that do not contribute power to the BMFC. Further studies should examine the utility of chitin or cellulose supplementation *in situ*.

16.7 MICROBIAL ECOLOGY OF BMFCs

In contrast to wastewater or other laboratory MFCs, there is no reliable or repeatable procedure for inoculating field deployed BMFCs; they are populated by the indigenous microbial species (or phylotypes) found in the environment. Even if laboratory microbial phylotypes could be inoculated onto field-deployed BMFCs, it remains unclear if they would remain a dominant member of the electrode community, or if they would be subject to competition and predation along with the indigenous microbes. In either case, microorganisms play two distinct roles in the generation of power by BMFCs. First, the redox gradients that are naturally found in marine sediments are established and maintained by microbially mediated oxidation (or remineralization) of organic matter (for review see Froelich *et al.* 1979; Jørgensen 2006; Nealson 1997). In absence of these microbial processes, there would be no electrochemical potential, no voltage differential between anode and cathode, and as such BMFCs would not function.

Second, electrode-hosted microorganisms play the primary role in facilitating the transfer of electrons between the electrodes (anode or cathode) and the indigenous oxidants or reductants. There are, in principal, two modes of electron transfer from microbes to solid-state oxidants: direct electron transfer (DET) and mediated electron transfer (MET). As these have been extensively reviewed herein and elsewhere (e.g. Schröder 2007; Logan 2008), we will simply note that both of these mechanisms have been observed in BMFC-hosted microbial communities. However, the degree to which these mechanisms are differentially responsible for power production in BMFCs, as well as the extent to which these mechanisms are found within microbial communities, remains unknown.

Since the earliest experiments, most BMFC studies have examined, to varying degrees, the diversity of microbes associated with BMFC anodes. In general, BMFC studies have implicated two major phylogenetic groups in power production; the delta- and gamma-proteobacteria (e.g. Bond *et al.* 2002; Holmes *et al.* 2004; Nielsen *et al.* 2008; Reimers *et al.* 2006; Tender *et al.* 2002). Delta-proteobacteria are ubiquitous in marine sediments, many of which are involved in sulfate reduction via the oxidation of organic matter or hydrogen (Lovley *et al.* 2006 and references therein). Bond and colleagues (2002) observed that over two-thirds of the clones recovered in a 16S rRNA gene library (constructed from a BMFC anode) were allied to the genus *Desulfuromonas*. Indeed, power production in an MFC inoculated with pure cultures of *Desulfuromonas acetoxidans* as well as *Geobacter metallireducens* yielded power densities comparable to those observed *in situ*. *D. acetoxidans* is a sulfur-reducing organism and can use elemental sulfur or L-malate and fumarate as electron acceptors. This organism expresses cytochromes used to reduce metals such as Fe(III) and Mn(IV), which may be involved in direct electron transfer to MFC anodes. *Geobacter sulfurreducens* may also employ direct electron transfer in power production, though the precise mechanism remains the subject of discussion (Reguera *et al.* 2006). In general, *Geobacter* species are known for their ability to oxidize acetate to CO_2 with solid-state oxidants – including MFC anodes – serving as the sole electron acceptor (Bond *et al.* 2002; Bond and Lovley 2003). Comparing power production in laboratory MFCs to field deployed MFCs, however, may be problematic. Lab cultures are typically supplemented with refined organic acids and are not subject to the variations in conditions that are found *in situ*. Nevertheless, the dominance of the delta-proteobacteria, and the independent observation that cultivated strains are capable of generating electricity, clearly establish their role in power production.

A later study by Ryckelynck *et al.* (2005), examining the geochemical impacts of BMFCs, found evidence for a mediated electron transfer in which dissolved and mineral forms of reduced sulfur were oxidized to S^0 at the anode.

Ryckelynck and colleagues suggested that sulfur cycling plays a significant role in power production in marine BMFCs (analogous to an engineered system described by Habermann *et al.* 1991 in which microbiologically generated sulfide was oxidized abiotically at an anode). Also, the isolation of sulfur disproportionating organisms allied to *Desulfobulbus* and *Desulfocapsa* spp. from buried anodes suggests that S^0 deposited on the anode serves as a substrate for regeneration of sulfate and sulfide (Holmes *et al.* 2004).

These studies clearly illustrate the role of the delta-proteobacteria and sulfur cycling in power production. However, in Reimers *et al.* (2006), the investigators found tremendous microbial diversity along the length of a 1-m sediment-hosted anode. When the electrodes were under load, nearly 65 distinct microbial phylotypes were enriched on the anode. This appears to confound prior studies in which delta-proteobacteria were dominant (Bond *et al.* 2003; Holmes *et al.* 2004). Notably, there were key differences in this study that likely resulted in the increased microbial diversity. In Reimers *et al.* the BMFC anode was deployed down to a depth of 1 m within a hydrocarbon seep, thus, transecting distinct geochemical milieus that were not found in the previous deployments. Furthermore, the investigators examined nearly 2000 sequences – compared to 60 sequences per deployment analyzed by Bond *et al.* (2002) – which provided significantly more depth in phylogenetic diversity. Finally, the investigators analyzed the distribution of these microbial phylotypes along the length of the anode, and found that delta-proteobacteria were clearly dominant in the upper 20 cm of the anode (which is consistent with previous studies by Bond *et al.* 2002, 2003; and Holmes *et al.* 2004). At deeper horizons, however, where sulfate is depleted, no delta-proteobacteria were observed, and the investigators recovered microbial phylotypes that have not been seen in previous libraries. In some cases, dominant phylotypes (such as an epsilon-proteobacteria allied to *Arcobacter nitrofigilis*) were found at densities 3 to 4 orders of magnitude greater on the anode than in the surrounding sediment (as determined via quantitative PCR; unpublished data). Their enrichment on the MFC suggests they are capable of extracellular electron transfer. However, these observations are circumstantial, as the direct role of these phylotypes in power production was not established.

Several earlier BMFC studies set out to examine the effect of biofilm formation on cathode performance. These investigators observed that electron transfer in a BMFC appears to be enhanced by cathode-hosted microbial biofilms, which may contain bacteria that effectively use the cathode as an electron donor (Bergel *et al.* 2005; Gregory *et al.* 2004; Rhoads *et al.* 2005). While this suggests that microbial phylotypes on the cathode also have an influential role on power production, there is limited information on the microbial diversity associated with BMFC cathodes. Reimers *et al.* (2006)

found high representation of gamma-proteobacteria allied to *Pseudomonas flourescens* – nearly 50% – in libraries constructed from cathode biofilms. In light of previous studies showing that Pseudomonads are capable of electron transfer via quinones (Newman and Kolter 2000; Hernandez *et al.* 2004), it is likely that *P. fluorescens* was involved in electron transfer, though again the details of their role in power production are unknown.

It is apparent that power production in BMFCs is governed by a complex series of ecological, geochemical and electrochemical interactions. The data on BMFC microbial ecology has informed us of the major groups that appear to be involved in power production. However, our understanding of which groups are primarily responsible for contributing to – or deterring from – power production is limited. The molecular microbiological approaches used in all these previous studies are primarily descriptive. These studies underscore the need for further quantitative investigations of microbial ecology in laboratory and field-deployed BMFCs (as in White *et al.* 2009), and future studies should aim to use quantitative molecular approaches (e.g. qPCR or fluorescent *in-situ* hybridization) to advance our understanding of how quantitative changes in microbiological factors influence system performance.

16.8 FACTORS GOVERNING POWER OUTPUT

Ultimately, the electron source for BMFCs is organic matter stored in sediments. Energy is liberated through spatially separated redox half-reactions that occur at the anode and cathode. The net reaction is, therefore, the oxidation of organic carbon by oxygen. While there is a wide range of potential electron donors for BMFCs (that is not adequately described by the bulk label of 'sediment organic matter'), the general model is that complex organic matter undergoes microbially-mediated hydrolysis and is separated into fractions including long-chain fatty acids, aromatic compounds, fermentable sugars and amino acids (Lovley 2006). The latter two fractions are further processed through microbial fermentation and produce organic acids such as acetate, lactate, propionate, butyrate, succinate and formate. These products, in turn, are known to support the growth of dissimilatory metal reducing bacteria capable of extracellular electron transfer (e.g. *Geobacter* spp.). Fermentation products may also fuel sulfate reduction resulting in dissolved sulfide and/or solid phase iron sulfide minerals, which are electrochemically active and can donate electrons to an anode (Ryckelynck *et al.* 2005).

As previously mentioned, it is likely that current produced by a given BMFC results from a combination of electron donors (which in turn are found in different geochemical milieus). While factors that limit power production (P) in prototype laboratory MFCs are well described (Larminie and Dicks 2000; Logan *et al.* 2006; Rabaey and Verstraete 2005), they are not necessarily the same in BMFCs. Per Logan (2008), $P = E_{cell} \times I$, and limitations are described in terms of loss factors that affect E_{cell}, which are related to overpotentials. Current dependent overpotentials are defined as the difference between electrode potential when current is passing and the potential at zero current. Overpotentials occur at both the anode and cathode and occur as consequences of ohmic losses, activation losses, and mass transport losses. A key difference between laboratory MFCs and BMFCs is the importance and scale of mass transport losses. As mentioned, mass transport is often the main limiting factor in BMFCs due to the relatively low concentration of electron donors and the slow rate of mass transport in sediments (Scott *et al.* 2008a; Tender *et al.* 2002). However, from the data presented in these investigations, it is difficult to ascertain if power density is indeed limited by bulk transport of substrates, or by micro-scale chemical diffusion, such as occurs within microbial biofilms. Nielsen *et al.* (2007) calculate the internal resistance of a chambered BMFC under conditions of natural and artificially enhanced mass transport. Under the natural condition the apparent internal resistance of the BMFC was far greater than when pumping eliminated mass transport limitation. In contrast, mass transport limitation is observed at different scales in laboratory and wastewater MFCs. In such systems there is little evidence that investigators have achieved maximum power based on substrate flux to the electrodes (Logan 2008). However, limitations arise due to transport limitations of protons through biofilms, which result in pH changes that could adversely affect bacterial physiology. Laboratory MFCs tend to be run either as batch reactor studies to investigate a particular aspect of the system or with a flow-through system. In either case, the concentration of electron donors can be easily manipulated so as not to be a limiting factor. Furthermore, in laboratory MFCs ohmic losses are addressed primarily by minimizing the spacing between electrodes and by careful design of the membranes. BMFCs do not have a membrane and there is a limit to how close the electrodes can be located since the oxic and anoxic zones are spatially separated (Rezaei *et al.* 2007). On the other hand, the electrolyte in most BMFCs has been seawater, which has a conductivity of approximately 25–40 mS cm^{-2} (varying as a function of temperature). Owing to the high conductivity of seawater, BMFC performance is not – in most circumstances – influenced by electrode spacing.

Calculating the Coulombic efficiency (defined above) in a laboratory MFC fed with refined fuels (e.g. acetate), is a relatively straightforward exercise, as the number of electrons available per mole of fuel via respiration to CO_2 (with the anode as a terminal electron acceptor) can be compared to the measured current to determine these efficiencies. However, in practice, the Coulombic efficiencies of MFCs varies widely from 0.7% to greater than 89% (He *et al.* 2005; Liu and Logan 2004; Liu *et al.* 2004; Liu *et al.* 2005; Rabaey *et al.* 2003; Reimers *et al.* 2007). This is primarily due to differences in fuel quality, including the presence of recalcitrant carbon that likely elevates the estimate of total organic carbon while contributing very little to power production. In addition, it has been shown that other factors affect Coulombic efficiencies, such as ionic strength, temperature and electrode spacing (Liu and Logan 2004); multiple, and possibly competing, anodic reactions in seawater-based MFCs (Ryckelynck *et al.* 2003); and the degree to which the particular microbial community expends energy on respiration versus growth and biosynthesis (Finkelstein *et al.* 2006). For example, Finkelstein *et al.* (2006) suggested as much as 95% of energy available from organic carbon oxidation may not contribute to power production, as it is used by the anode-hosted microbial biofilm for growth and biosynthesis (though it should be noted that that other studies, e.g. Aelterman *et al.* (2008) have found that a greater percentage of the energy available in organic matter may be harnessed by the electrode). Similar to laboratory MFCs, BMFCs have shown a wide range of Coulombic efficiencies with an upper bound of about 65% (Nielsen 2008). The observed differences may be due to environmental differences, or energy partitioning by the indigenous microbial community. In a BMFC, available 'fuel' is particularly subject to consumption by the free-living microbial community, and as such it is difficult to determine which fraction of the available fuel is being used by the anode-hosted community. In general, the amount of potential energy stored in sediment that is available for electrical work appears to have been overestimated by previous calculations, and requires further consideration (Reimers *et al.* 2001; Tender *et al.* 2002).

16.9 SCALING AND ENVIRONMENTAL VARIABILITY IN BMFCs

The relationship between performance and device size is currently being investigated for both laboratory MFCs and BMFCs. Dewan *et al.* (2008) recently compiled results from 20 studies, including laboratory MFCs and field-deployed BMFCs, that clearly showed that MFCs with the largest total anode surface area had the lowest power densities. They also conducted systematic experiments and observed that power density is – to a degree – inversely proportional to the logarithm of the surface area of the anode (though we and others have observed that power density is dependent on a great many factors, such as mass transport, quality of fuel, and density and distribution of the microbial community). In summary, this draws into question the validity of simply extrapolating power densities from small-scale experiments to estimate how much power might be delivered by scaling up various MFC designs. This question is central to BMFC experiments since the underlying motivation for this research is to develop useful power supplies.

It should also be noted that the natural environment is extremely dynamic, and not subject to control like a typical laboratory or industrial setting. Accordingly, BMFC studies should make a concerted effort to understand the degree to which natural fluctuations influence BMFC performance and spatial or temporal variability. Natural fluctuations, including temperature changes, tidal cycles, pressure variations, changes in water flow, salinity and conductivity changes, and dissolved oxygen changes, can readily influence microbial physiology and ultimately BMFC performance. For example, Reimers *et al.* (2006) observed a relationship between current (produced by the abiotic oxidation of sulfide) and temperature on the order of 1.3 $mA \cdot m^{-2}$ for every °C. In estuarine environments, Nielsen *et al.* (2007) observed current oscillations, the frequency of which were consistent with tidal pumping. In the latter example, Nielsen *et al.* determined that the total resistance in a BMFC decreases by 100 Ω from high tide to low tide due to changes in seawater conductivity. Even larger relative power fluctuations were observed in the power records from chambered BMFCs deployed at cold seeps in Monterey Canyon (Nielsen *et al.* 2008). Curiously, these chambered BMFCs were deployed at a water depth of 960 m and the frequency of the power oscillations did not correlate with tidal changes or any measured environmental factor. The most likely explanation for the variability was variable sediment porewater seepage and mixing with the chambers that would control the delivery of electroactive species to the anode.

In summary, significant power fluctuations appear to be a common feature of all chambered BMFC deployments regardless of where they are deployed. Regardless of the underlying cause, variable power generation affects the requirement for power management devices in that they must be able to store fluctuating levels of power for subsequent use in powering a sensor with steady or variable demands.

16.10 COMMERCIAL VIABILITY OF BMFCs

As of late, there has been a surge in the development of low power sensor and monitoring systems for remote and rural applications. Such systems are widely used in both science and industry, to increase the footprint of data collection or to monitor materials and infrastructure. Examples of parameters that are of interest to both science and industry include temperature, salinity, water level, irradiance, pressure, pH, fluid flow including tidal patterns, migration patterns of fish and other animals, and pollution monitoring. Long-term monitoring requires instrumentation that is powered to collect, store and transmit data, and to date most sensor systems have relied on solar power or battery packs for extended deployments. However, solar systems are of limited long-term use above certain latitudes, and batteries, while improving in energy density and duration, are typically unable to provide power for any deployment longer than two or three years (depending upon the sensor system). With respect to oceanographic applications, batteries can typically power a sensor for about one year. For logistical and financial reasons, this limits the ability of the end user to deploy extensive sensor systems in nature. BMFCs should, in principal, be able to meet the power requirements of a number of sensors systems that are also robust enough to operate for many years.

For the reasons described above, the practical use of electricity from BMFCs requires power management systems that satisfy three requirements (Nielsen 2008). First, current is produced at a low voltage and must be converted to higher voltages to satisfy sensor requirements, second, modest and steady power production requires storage to satisfy variable load cycles, and third, cell voltage should be regulated to operate at the most efficient and maximum sustainable level. Recently, the development and deployment of the B.U.G. (Tender *et al.* 2008), has shown that small sensor systems on buoys may be powered by BMFCs. Shantaram *et al.* (2005) were able to power a wireless sensor and telemetry system, though their system employed a sacrificial anode that provided higher voltage, which in turn enabled the use of commercial electronics for power management (the low voltages generated by most BMFCs limits the use of commercial electronics for power management). More recently, Donovan *et al.* (2008) developed a power management system that enabled sufficient power to be harnessed from a BMFC without a sacrificial anode, by using a capacitor to store charge, and a voltage comparator and DC–DC converter to boost the voltage to 3.3 V (sufficient to power most solid-state devices). Here too, the duty cycle of the sensor was decreased to meet the performance of the BMFC. In sum, the potential for BMFCs to power sensor systems in the environment has been largely established. With respect to potential commercialization, the success of BMFCs as power systems will depend on their ability to compete – in terms of performance, cost and reliability – with conventional and alternative systems. Ultimately, this will require a concerted and thorough investigation of the aforementioned factors, from electrode material to microbial ecology, to further optimize BMFCs for powering future sensor infrastructure.

On the whole, BMFC studies comprise a small fraction of the total body of research on microbial fuel cells. Nevertheless, BMFC research has provided critical insights into the geochemical and microbiological factors that contribute to – and may ultimately limit – power production in complex systems. Whether BMFC research grows in proportion to the field remains to be determined. However, it is likely that BMFCs will continue to be used as a testbed for those interested in understanding the intricacies of geochemical and microbial interactions in natural systems, as well as those in pursuit of developing BMFCs as reliable and cost-effective means of remote power generation.

REFERENCES

Aelterman, P., Freguia, S., Keller, J., Verstraete, W. and Rabaey, K. (2008) The anode potential regulates bacterial activity in microbial fuel cells. *Appl. Microbiol. Biotechnol.* **78**, 409–418.

Alberte, R., Bright, H., Reimers, C. and Tender, L. (2005) United States.

Bard, A.J. and Faulkner, L.R. (2001) *Electrochemical Methods*. New York: John Wiley & Sons, Inc.

Bergel, A., Feron, D. and Mollica, A. (2005) Catalysis of oxygen reduction in PEM fuel cell by seawater biofilm. *Electrochem. Comm.* **7**, 900–904

Bond, D.R. and Lovley, D.R. (2003) Electricity production by *Geobacter sulfurreducens* attached to electrodes. *Appl. Environ. Microbiol.* **69**, 1548–1555.

Bond, D.R., Holmes, D.E., Tender, L.M. and Lovley, D.R. (2002) Electrode-reducing microorganisms that harvest energy from marine sediments. *Science* **295**, 483–485.

Bullen, R.A., Arnot, T.C., Lakeman, J.B. and Walsh, F.C. (2006) Biofuel cells and their development. *Biosensors Bioelectron.* **21**, 2015–2045.

Dewan, A., Beyenal, H. and Lewandowski, Z. (2008) Scaling up microbial fuel cells. *Environ. Sci. Technol.* **42**, 7643–7648.

Donovan, C., Dewan, A., Heo, D. and Beyenal, H. (2008) Batteryless, wireless sensor powered by a sediment microbial fuel cell. *Environ. Sci. Technol.* **42**, 8591–8596.

Du, Z.W., Li, H.R. and Gu, T.Y. (2007) A state of the art review on microbial fuel cells: a promising technology for wastewater treatment and bioenergy. *Biotechnol. Adv.* **25**, 464–482.

Dumas, C., Mollica, A., Feron, D., Basseguy, R., Etcheverry, L. and Bergel, A. (2007) Marine microbial fuel cell: use of stainless steel electrodes as anode and cathode materials. *Electrochim. Acta* **53**, 468–473.

Finkelstein, D.A., Tender, L.M. and Zeikus, J.G. (2006) Effect of electrode potential on electrode-reducing microbiota. *Environ. Sci. Technol.* **40**, 6990–6995.

Fornero, J.J., Rosenbaum, M., Cotta, M.A. and Angenent, L.T. (2008) Microbial fuel cell performance with a pressurized cathode chamber. *Environ. Sci. Technol.* **42**, 8578–8584.

Froelich, P.N., Klinkhammer, G.P., Bender, M.L., Luedtke, N.A., Heath, G.R. *et al.* (1979) Early oxidation of organic matter in pelagic sediments of the eastern equatorial Atlantic: suboxic diagenesis. *Geochim. Cosmochim. Acta* **43**, 1075–1090.

Gregory, K.B., Bond, D.R. and Lovley, D.R. (2004) Graphite electrodes as electron donors for anaerobic respiration. *Environ. Microbiol.* **6**, 596–604.

Habermann, W. and Pommer, E.H. (1991) Biological fuel-cells with sulfide storage capacity. *Appl. Microbiol. Biotechnol.* **35**, 128–133.

He, Z., Minteer, S.D. and Angenent, L.T. (2005) Electricity generation from artificial wastewater using an upflow microbial fuel cell. *Environ. Sci. Technol.* **39**, 5262–5267.

He, Z., Shao, H. and Angenent, L.T. (2007) Increased power production from a sediment microbial fuel cell with a rotating cathode. *Biosensors Bioelectron.* **22**, 3252–3255.

Hernandez, M.E., Kappler, A. and Newman, D.K. (2004) Phenazines and other redox-active antibiotics promote microbial mineral reduction. *Appl. Environ. Microbiol.* **70**, 921–928.

Holmes, D.E., Bond, D.R., O'Neill, R.A., Reimers, C.E., Tender, L.R. and Lovley, D.R. (2004) Microbial communities associated with electrodes harvesting electricity from a variety of aquatic sediments. *Microbial Ecol.* **48**, 178–190.

Jørgensen, B.B. (2006) Bacteria and Marine Biogeochemistry. In *Marine Geochemistry*. Schulz, H.D. and Zabel, M. (eds). Berlin: Springer.

Larminie, J. and Dicks, A. (2000) *Fuel Cell Systems Explained*. Chichester: John Wiley & Sons Ltd.

Liu, H. and Logan, B. (2004) Electricity generation using an air-cathode single chamber microbial fuel cell (MFC) in the absence of a proton exchange membrane. *Abstracts of Papers of the American Chemical Society* **228**, U622–U622.

Liu, H., Cheng, S.A. and Logan, B.E. (2005) Production of electricity from acetate or butyrate using a single-chamber microbial fuel cell. *Environ. Sci. Technol.* **39**, 658–662.

Liu, H., Ramnarayanan, R. and Logan, B.E. (2004) Production of electricity during wastewater treatment using a single chamber microbial fuel cell. *Environ. Sci. Technol.* **38**, 2281–2285.

Logan, B.E. (2008) *Microbial Fuel Cells*. Hoboken, N.J., U.S.A.: John Wiley & Sons, Inc.

Logan, B.E., Hamelers, B., Rozendal, R., Schröder, U., Keller, J. *et al.* (2006) Microbial fuel cells: methodology and technology. *Environ. Sci. Technol.* **40**, 5181–5192.

Lovley, D.R. (2006) Bug juice: harvesting electricity with microorganisms. *Nature Rev. Microbiol.* **4**, 497–508.

Lowy, D.A., Tender, L.M., Zeikus, J.G., Park, D.H. and Lovley, D.R. (2006) Harvesting energy from the marine sediment-water interface II – Kinetic activity of anode materials. *Biosensors Bioelectron.* **21**, 2058–2063.

Nealson, K.H. (1997) Sediment bacteria: who's there, what are they doing, and what's new? *Ann. Rev. Earth Planet. Sci.* **25**, 403–434.

Newman, D.K. and Kolter, R. (2000) A role for excreted quinones in extracellular electron transfer. *Nature* **405**, 94–97.

Nielsen, M.E. (2008) *Utilization of Natural and Supplemental Biofuels for Harvesting Energy from Marine Sediments*. Oregon State University, Corvallis, OR, p. 137.

Nielsen, M.E., Reimers, C.E. and Stecher, H.A. (2007) Enhanced power from chambered benthic microbial fuel cells. *Environ. Sci. Technol.* **41**, 7895–7900.

Nielsen, M.E., Reimers, C.E., White, H.K., Sharma, S. and Girguis, P.R. (2008) Sustainable energy from deep ocean cold seeps. *Energy Environ. Sci.* **1**, 584–593.

Rabaey, K. and Verstraete, W. (2005) Microbial fuel cells: novel biotechnology for energy generation. *Trends Biotechnol.* **23**, 291–298.

Rabaey, K., Lissens, G., Siciliano, S.D. and Verstraete, W. (2003) A microbial fuel cell capable of converting glucose to electricity at high rate and efficiency. *Biotechnol. Lett.* **25**, 1531–1535.

Rabaey, K., Rodriguez, J., Blackall, L.L., Keller, J., Gross, P. *et al.* (2007) Microbial ecology meets electrochemistry: electricity-driven and driving communities. *ISME J.* **1**, 9–18.

Reguera, G., Nevin, K.P., Nicoll, J.S., Covalla, S.F., Woodard, T.L. and Lovley, D.R. (2006) Biofilm and nanowire production leads to increased current in *Geobacter sulfurreducens* fuel cells. *Appl. Environ. Microbiol.* **72**, 7345–7348.

Reimers, C.E., Tender, L.M., Fertig, S. and Wang, W. (2001) Harvesting energy from the marine sediment-water interface. *Environ. Sci. Technol.* **35**, 192–195.

Reimers, C.E., Girguis, P., Stecher, H.A., Tender, L.M. and Ryckelynck, N. (2006) Microbial fuel cell energy from an ocean cold seep. *Geobiology* **4**, 123–136.

Reimers, C.E., Stecher, H.A., Westall, J.C., Alleau, Y., Howell, K.A. *et al.* (2007) Substrate degradation kinetics, microbial diversity and current efficiency of microbial fuel cells supplied with marine plankton. *Appl. Environ. Microbiol.* **73**, 7029–7040.

Rezaei, F., Richard, T.L., Brennan, R.A. and Logan, B.E. (2007) Substrate-enhanced microbial fuel cells for improved remote power generation from sediment-based systems. *Environ. Sci. Technol.* **41**, 4053–4058.

Rhoads, A., Beyenal, H. and Lewandowski, Z. (2005) Microbial fuel cell using anaerobic respiration as an anodic reaction and biomineralized manganese as a cathodic reactant. *Environ. Sci. Technol.* **39**, 4666–4671.

Ryckelynck, N., Stecher, H.A. and Reimers, C.E. (2005) Understanding the anodic mechanism of a seafloor fuel cell: Interactions between geochemistry and microbial activity. *Biogeochemistry* **76**, 113–139.

Schröder, U. (2007) Anodic electron transfer mechanisms in microbial fuel cells and their energy efficiency. *Phys. Chem. Chem. Phys.* **9**, 2619–2629.

Schulz, H.D. (2006) Quantification of early diagenesis: dissolved constituents in pore water and signals in the solid phase. In *Marine Geochemistry*. Schulz, H.D. and Zabel, M. (eds). Berlin: Springer.

Scott, K., Cotlarciuc, I., Hall, D., Lakeman, J.B. and Browning, D. (2008a) Power from marine sediment fuel cells: the influence of anode material. *J. Appl. Electrochem.* **38**, 1313–1319.

Scott, K., Cotlarciuc, I., Head, I., Katuri, K.P., Hall, D. *et al.* (2008b) Fuel cell power generation from marine sediments: investigation of cathode materials. *J. Chem. Technol. Biotechnol.* **83**, 1244–1254.

Shantaram, A., Beyenal, H., Raajan, R., Veluchamy, A. and Lewandowski, Z. (2005) Wireless sensors powered by microbial fuel cells. *Environ. Sci. Technol.* **39**, 5037–5042.

Smith, K.L., Jr. and Teal, J.M. (1973) Deep-sea benthic community respiration: an *in situ* study at 1850 meters. *Science* **179**, 282–283.

Stumm, W. and Morgan, J.J. (1996) *Aquatic Chemistry: Chemical Equilibria and Rates in Natural Waters*. Wiley-Interscience.

Tender, L.M., Reimers, C.E., Stecher, H.A., Holmes, D.E., Bond, D.R. *et al.* (2002) Harnessing microbially generated power on the seafloor. *Nature Biotechnol.* **20**, 821–825.

Tender, L.M., Gray, S.A., Groveman, E., Lowy, D.A., Kauffman, P. *et al.* (2008) The first demonstration of a microbial fuel cell as a viable power supply: powering a meteorological buoy. *J. Power Sources* **179**, 571–575.

White, H.K., Reimers, C.E., Cordes, E.E., Dilly, G.F. and Girguis, P.R. (2009) Quantitative population dynamics of microbial communities in plankton-fed microbial fuel cells. *ISME J.* (1–12).

Whitfield, M. (1972) The electrochemical characteristics of natural redox cells. *Limnol. Oceanogr.* **17**, 383–393.

17

Microbial fuel cells as biochemical oxygen demand (BOD) and toxicity sensors

Byung Hong Kim, I.N. Seop Chang and Geoffrey M. Gadd

17.1 INTRODUCTION

Biological oxygen demand (BOD) is the most widely-used parameter to measure the organic content of wastewater (Eaton *et al.* 1994). The standard method to determine BOD involves measurement of the dissolved oxygen consumed by microbes in the biochemical oxidation of organic matter. This method is time-consuming (5 days of incubation) and usually requires experience and skill to achieve reproducible results (Bourgeois *et al.* 2001). Rapid and reliable on-line methods to measure BOD values are a prerequisite for real-time control in the water and wastewater industry and BOD sensors of various designs and theoretical backgrounds have been studied and commercialized. These are

 Bioelectrochemical Systems: From Extracellular Electron Transfer to Biotechnological Application. Edited by Korneel Rabaey, Largus Angenent, Uwe Schröder and Jürg Keller. ISBN: 9781843392330. Published by IWA Publishing, London, UK.

mainly microbial sensors employing various transducers and include dissolved oxygen probes, manometers, photometers, and microbial fuel cells among others.

17.1.1 Dissolved oxygen probe-based BOD sensors

A number of microbial BOD sensors have been reported, mostly based on dissolved O_2 (DO) monitoring systems, and this topic has been well reviewed by Liu and Mattiasson (2002). These DO-based BOD sensors consist of a synthetic membrane with immobilized microorganisms as the biological recognition element, and an oxygen electrode as the physical transducer. The BOD is then determined by monitoring any decrease in the DO concentration due to oxidation of biodegradable compounds by the microorganisms used. A variety of microorganisms have been tested for this purpose, including bacteria and yeasts, where the selection was based on the ability of the microbial strain to assimilate a suitable spectrum of substances (Kim and Kwon 1999; Li and Chu 1991; Sangeetha *et al.* 1996).

Although BOD sensors based on a single strain have relatively good stability and a long lifetime, they have a limited capacity to metabolize a wide range of contaminants which results in inaccurate BOD values. Thus, to alleviate this problem, mixed cultures (Tan *et al.* 1993) or activated sludge (Liu and Mattiasson 2002; Liu *et al.* 2000; Liu *et al.* 2004) have been used.

BOD sensors based on DO monitoring have a limited stability due to the nature of the DO probe used. A typical Clark-type DO probe is a two-electrode system consisting of a silver anode and a gold or platinum cathode (Liu and Mattiasson 2002). The anode metal is oxidized during use and therefore the electrolyte needs to be changed and the anode surface of the DO probe cleaned regularly. Membrane fouling is another serious problem in biosensors based on a DO electrode and frequent maintenance is required to maintain a high sensitivity. Furthermore, oxygen-based BOD sensors have a narrow response range due to the limited oxygen solubility in water of less than 9 mg/l (Trosok *et al.* 2001). Fluctuating oxygen levels in a sample can also mean poor reproducibility and reliability unless a sufficient oxygen supply is incorporated into the sensor (Kumlanghan *et al.* 2008; Riedel 1991).

The changes in pressure due to microbial respiration can be monitored using barometry to avoid problems associated with an oxygen electrode (Tzoris *et al.* 2005). A BOD sensor based on pressure changes due to microbial respiration is commercially available (http://www.envitech.co.uk/Prod_BioxBOD.html).

17.1.2 Photometric BOD sensors

Photometric techniques have been applied for rapid estimation of the BOD based on the UV absorbance (at 254 nm) of dissolved organic matter in water, including wastewater (Bari and Farooq 1985; Edwards and Cresser 1987; Nataraja *et al.* 2006), and the fluorescent properties of humic substances and organic matter in natural surface water (Vodacek and Philpot 1987).

Reynolds and Ahmad (1995a, b) reported a linear relationship between the BOD values of wastewater and the fluorescence intensities at 340 nm for organisms growing in wastewater. The reduced forms of pyridine nucleotides (NADH and NADPH) are fluorescent. Thus, in the presence of an electron donor, pyridine nucleotides are reduced and the ratio NAD(P)H/NAD(P)+ is related to the concentration of the electron donor.

The luminous bacterium Photobacterium phosphoreum can also be used for a similar purpose where the intensity of the luminescence, which is proportional to the amount of assimilable organic compounds in wastewater, is measured using a photodiode (Hyun *et al.* 1993; Sakakuchi *et al.* 2007).

However, these techniques are not suitable for accurate BOD determination due to the effects of interfering substances and the fact that some organic compounds absorb more UV (or are more fluorescent) than others.

Bacterial catabolism has been monitored by replacing oxygen with redox dyes that capture electrons from a redox molecule in the electron transport chain. An oxidized dye is reduced during its interaction with a reduced electron transport chain molecule and the reduced dye is then quantified spectrophotometrically or electrochemically. Various redox dyes have been tested including 2,6-dichlorophenolindophenol (DCPIP) (Nakamura *et al.* 2007a), and ferricyanide (Nakamura *et al.* 2007b; Pasco *et al.* 2004). Since redox dyes are highly soluble in water, the sensors employing them can determine a wide range of BOD values, but they are not suitable for long-term operation because the redox dyes interfere with bacterial metabolism.

17.1.3 Titration and respirometric sensors

Microbial activity during the assimilation of organic contaminants can be measured through determination of nitrate consumption or carbon dioxide evolution. Good correlation was found between nitrate consumption measured by an anoxic titration method and BOD values (Onnis *et al.* 2006). Carbon dioxide content was measured in the off gas from a bioreactor containing samples by infrared spectrophotometry, and this strongly correlated with off-line BOD measurements (Vaiopoulou *et al.* 2005). However, these methods are not suitable for field applications because expensive analytical equipment is required.

17.1.4 Electrochemical BOD sensors with mediators

As mentioned above, bacterial catabolism has been monitored by replacing oxygen with redox dyes that capture electrons from redox molecules in the electron transport chain, and these reduced dyes can be quantified electrochemically. When a dye is present in a culture under anoxic conditions, it acts as an electron acceptor and is reduced during the metabolic oxidation of organic substances. The reduced form of the dye is then reoxidized at a working electrode (anode) which is maintained at a sufficiently high electric potential. This system has been studied not only as a BOD sensor but also as a microbial fuel cell (MFC), a device that can produce electricity from waste substances. Various redox dyes have been used for this purpose including thionine, methyl viologen, 2-hydroxy-1, 4-naphthoquinone, neutral red, and other hydrophobic compounds (Table 17.1).

Table 17.1. Redox dyes used to monitor bacterial metabolism as alternative electron acceptors

Redox mediators	Redox potential ($E^{o'}$, mV)	References
ABB, alizarin brilliant blue	−173	(Delaney *et al.* 1984; Roller *et al.* 1984; Thurston *et al.* 1985)
BV, benzyl viologen	−359	(Delaney *et al.* 1984)
BCB, brilliant cresyl blue	−99	(Delaney *et al.* 1984)
DCPIP, 2, 6-dichlorophenolindophenol	+217	(Delaney *et al.* 1984)
DMST, *N, N*-dimethyl disulfonated thionine	+220	(Delaney *et al.* 1984)
Fe(III)/Fe(II) chelate complexes*	+200**	(Vega and Fernández 1987)
GC, gallocyanine	+21	(Delaney *et al.* 1984)
HNQ, 2-hydroxy-1,4-naphthoquinone	−139	(Ardeleanu *et al.* 1983; Roller *et al.* 1984)
MB, methylene blue	+5	(Ardeleanu *et al.* 1983; Roller *et al.* 1984)
MV, methyl viologen	−440	(Akiba *et al.* 1987)
NR, neutral red	−325	(Park and Zeikus 2000)
PES, phenazine ethosulphate	+65	(Delaney *et al.* 1984; Thurston *et al.* 1985)
PTZ, phenothiazinone	+130	(Roller *et al.* 1984)
RS, resorufin	−51	(Delaney *et al.* 1984; Roller *et al.* 1984; Thurston *et al.* 1985)
TH, thionoine	+64	(Bennetto *et al.* 1985; Kim *et al.* 2000)

*Ferric(III) chelate complexes: Fe(III)CyDTA, ferric 1,2-cyclohexanediamine tetraacetic acid; Fe(III)EDADPA, ferric ethylenediamine diacetic acid dipropionic acid; Fe(III)EDTA, ferric ethylenediaminetetracetic acid; Fe(III)TTHA, ferric triethylenetetraamine hexaacetic acid.
**Iron chelate complexes were assumed the redox potential of Fe(III)/Fe(II) at pH 7.0.

An MFC system can be used for various purposes, including biosensors, bioelectrochemical synthetic processes, and electricity generation (Tayhas *et al.* 1994). In particular, mediated MFCs have been studied as BOD sensors (Pasco *et al.* 2004; Trosok *et al.* 2001; Yoshida *et al.* 2000). The sensor responses are linear with BOD values from 2 to 200 mg/l and have a response time of 5 to 15 min. However, mediated MFCs are not suitable for long-term operation as the dyes used in the device are toxic to the organisms used, which means the microorganisms need to be replaced regularly. Typically, a mediated MFC cannot be operated for more than 7 days.

As an alternative method for on-line BOD determination, Karube *et al.* (1977a, b) investigated a hydrogen fuel cell coupled to anaerobic fermentation by *Clostridium butyricum* immobilized onto the electrode. A linear relationship was obtained between the steady state current and the BOD of a standard solution up to 300 mg/l. A steady state current was attained within 30–40 min, and the standard deviation was 2 mg/l. The BOD of an industrial wastewater was estimated using this device, and the current output of the MFC was almost constant for 30 days. This BOD sensor cannot be used to measure BOD values of samples containing non-fermentable contaminants and not all fermentable contaminants are fermentable by the bacterium.

17.2 THE MEDIATOR-LESS MICROBIAL FUEL CELL

17.2.1 Electrochemically-active bacteria

A number of bacteria have been isolated based on their ability to use Fe(III) as a terminal electron acceptor (Lovley 1981; Nealson and Saffarini 1994). Although there is some evidence that soluble electron mediators are involved in the reduction of a water-insoluble electron acceptor (Newman and Kolter 2000; Turick *et al.* 2002), direct contact between the bacterial cells and the electron acceptor is required for dissimilatory Fe(III) reduction (Nevin and Lovley 2000). Among the Fe(III) reducers, *Shewanella oneidensis* (Bretschger *et al.* 2007; Myers and Myers 1992) and *Geobacter sulfurreducens* (Lloyd *et al.* 2000) localize the majority of membrane-bound cytochromes on the outer membrane, are electrochemically-active (Bond and Lovley 2003; Kim *et al.* 1999a,b,c) and can be used as the anode catalyst in an MFC. In addition, *S. putrefaciens* was found to grow on lactate in the absence of electron acceptors in an electrochemical fuel cell, yet did not metabolize lactate when the anode was disconnected from the cathode (Kim *et al.* 1999b, c). An MFC using *S. putrefaciens* can therefore also be used as a lactate sensor (Kim *et al.* 1999c), because lactate is one of the few electron donors used by this bacterium. Electrochemical activities have also been found in other

Fe(III)-reducers, including *Clostridium butyricum* (Park *et al.* 2001), *Aeromonas hydrophila* (Pham *et al.* 2003), *Rhodoferax ferrireducens* (Chaudhuri and Lovley 2003), *Desulfobulbus propionicus* (Holmes *et al.* 2004), and *Enterococcus gallinarum* (Kim *et al.* 2005).

MFCs using a single organism have an intrinsic disadvantage due to their limited range of fuel utilization. As such, electrochemically-active microbial communities with different nutritional characteristics have been successfully enriched using fuel cell-type electrochemical cells. Anode mediators were not used in these cases, and the coulombic efficiency was over 90% in some of the MFCs examined (Chang *et al.* 2004).

17.2.2 Enrichment of an electrochemically-active bacterial community

The anode compartment of a fuel cell-type electrochemical device was inoculated with sludge collected from a wastewater treatment plant of a corn-processing factory and fed with wastewater from the same source with continuous aeration to the cathode compartment (Kim *et al.* 2004). It was found that an open circuit potential of around 0.5–0.7 V was observed immediately after inoculation, while a stable current of around 2.0 mA was generated within 3 weeks after the circuit was closed by connecting the electrodes through a resistance of 10 Ω (Chang *et al.* 2004; Lee *et al.* 2003), These results showed that the inoculum contained electrochemically-active microbes (EAM), and that these organisms propagated in the fuel cell used.

The anode was removed from the MFC, enriched, and operated for about a year for microscopic observations. A thick biofilm was observed on the surface of the electrode (Kim *et al.* 2004) indicating that the electrochemically-active bacteria had colonized the electrode surface, using the electrode as an electron acceptor/sink in the absence of electron acceptors, with other bacteria developing and being associated with them. Current generation was the result of the electrons being transferred to the electrode by the electrochemically-active microbes after they had metabolized electron donors in the wastewater. The current was generated stably for over 6 years with consumption of the chemical oxygen demand. It is well documented that the microbial community in a biofilm is more stable than a free-living community which would explain why the biofilm that developed on the electrode performed better than artificially immobilized microbial cells (Kim *et al.* 2004).

Enrichment cultures were made with various nutritional characteristics, including copiotrophic cultures enriched with artificial wastewater containing acetate (Lee *et al.* 2003) or artificial wastewater containing glucose and glutamate (Chang *et al.* 2004), and oligotrophic cultures using artificial wastewater or river water (Kang *et al.* 2003; Kang *et al.* 2004; Phung *et al.* 2004). The current generated from the MFC was directly proportional to the organic matter concentration ('strength') of the wastewater. Therefore, based on these observations, MFCs have been developed as BOD sensors (Kim *et al.* 2003a; Moon *et al.* 2004).

17.2.3 Microbiology of a mediator-less MFC

DNA was extracted from the anodes of MFCs that were enriched and operated under various nutritional regimes for molecular analyses. Denaturing gradient gel electrophoresis (DGGE) revealed that the bacterial population in the MFC was different from that in the inoculum, and that the microbial populations that developed in the MFCs were dependent on the fuel used [65, 66, 69]. Analyses of 16S rDNA sequences also showed that dominant bacterial classes differed, thereby substantiating DGGE results. In all cases, Gram-negative bacteria were dominant. Table 17.2 compares the bacterial populations in the MFCs enriched with different fuels.

Studies have been conducted to determine the effects of alternative electron acceptors on the performance of an MFC. When MFCs were fed with fuel containing electron acceptors with a high redox potential such as oxygen and nitrate, the current as well as coulombic efficiency were reduced. However, sulfate showed less deteriorative effects (Kim *et al.* 2004). These results showed that the microbes in the anode compartment used oxygen and nitrate preferentially over the electrode reduction. Since nitrate is commonly present in the effluent of wastewater treatment processes and oxygen is ubiquitous, their effects should be eliminated for accurate BOD measurement using MFCs (Chang *et al.* 2005).

17.2.4 Optimization of MFC performance

It has been shown that current generation from a MFC is determined by several physical and biochemical factors, including (1) the ability and activity of the microbes to oxidize the fuel, (2) electron transfer rate to the electrode from the microbes, (3) circuit resistance, (4) proton transfer from the anode compartment to the cathode compartment, (5) oxygen supply and reduction at the cathode, and (6) oxygen diffusion into the anode compartment through the membrane (Gil *et al.* 2003). Among these, the most serious problem in using an MFC as a BOD sensor is oxygen diffusion into the anode compartment.

Table 17.2. Comparison of bacterial communities in MFCs enriched with different fuels

Fuel (value as COD)[a]	Class (%)						Reference
	α-Proteobacteria	*β-Proteobacteria*	*γ-Proteobacteria*	*δ-Proteobacteria*	*Firmicutes*	Others	
Glucose/glutamate (copiotrophic, 200)	1.4	6.8	**36.5**	14.9	27.0	13.4	(Choo *et al.* 2006)
Glucose/glutamate (oligotrophic, 10)	**64.4**	21.1	3.3	0	0	11.1	(Kang *et al.* 2004)
SPW[b] (400)	27.2	**40.9**	0	0	4.5	27.1	(Tender *et al.* 2002)
River water (≈ 5)	10.8	**46.2**	12.9	12.9	0	17.2	(Kang *et al.* 2003)
Acetate (300)	7.0	1.7	17.3	**68.8**	1.0	3.8	(Lee *et al.* 2003)
Propionate (100)	0	19.4	22.4	10.2	0	**41.8**	In preparation

[a]COD unit was measured in mg l^{-1}.
[b]SPW: starch processing wastewater.

When graphite was used as the cathode, the critical oxygen concentration was around 6 mg/l. This high critical oxygen concentration was apparently due to the poor oxygen-reducing activity of the graphite electrode. The incorporation of a platinum-coated graphite cathode generated a maximum current that was 3–4 times higher than that occurring with the graphite cathode with a critical oxygen concentration of 2.0 mg/l (Pham *et al.* 2004).

A considerable amount of oxygen diffuses from the cathode compartment to the anode compartment, thereby reducing the coulombic efficiency (Pham *et al.* 2004). However, a coulombic efficiency of over 90% was achieved in an MFC fed with a fuel concentration of over 100 mg/l, whereas an oligotrophic MFC fed with a fuel concentration of 10 mg/l showed a coulombic efficiency of less than 30%. Thus, to reduce oxygen diffusion through the membrane, an oligotrophic MFC was designed with a reduced membrane size (Kang *et al.* 2003).

17.3 DESIGN AND PERFORMANCE OF AN MFC USED AS BOD SENSOR

Mediator-less MFCs can be used to determine BOD by recording the current under continuous conditions or by measuring the coulomb generation in batch mode. Since the form and concentration of the fuel determines the microbial population in an MFC, MFCs should be enriched using actual wastewater or appropriate substances determined from analysis of the wastewater. For this reason, MFCs appropriate for low BOD values are discussed separately from those to be used for high BOD samples.

Various types of reactors have been tested to improve MFC performance (Du *et al.* 2007). Those reactors designed for power generation are also suitable as BOD sensors. Reactors have been designed specifically for BOD sensors and are termed C-type, S-type and O-type. C-type reactors are used for continuous determination of BOD values by recording the produced current (Moon *et al.* 2004). The S-type reactor is a down-sized C-type reactor designed to improve flow characteristics in the anode compartment (Moon *et al.* 2006). The O-type MFC was designed as a low value BOD sensor to enhance the cathode reaction and lower the O_2 diffusion into the anode compartment by reducing the size of the membrane (Kang *et al.* 2003; Moon *et al.* 2005). The performance of this new MFC was greatly improved in terms of the maximum current and coulombic efficiency. Figure 17.1 shows several MFC type sensors.

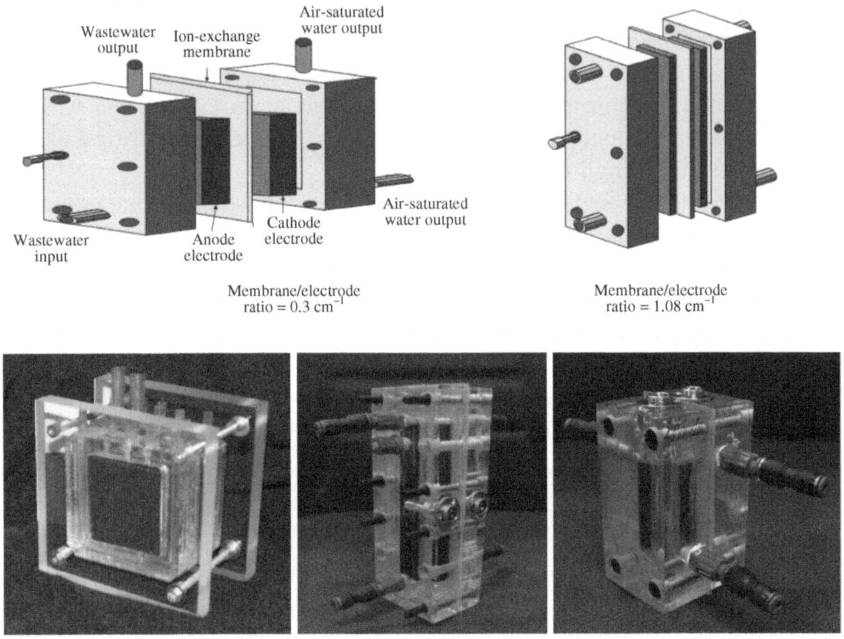

Figure 17.1. Several MFC-type BOD sensors and pictures. Top: O-type (left), C-type (right). Bottom: Batch type, C-type and S- type (from left)

17.3.1 MFC to measure BOD values higher than 10 mg/L

17.3.1.1 MFC design

Polyacrylic plastic is used to construct the frame of a C-type MFC (Moon *et al.* 2004), consisting of anode and cathode compartments separated by a cation exchange membrane sealed with silicon rubber gaskets. Graphite felt is used as the electrode after being cleaned with 0.1 mol/l HCl and thoroughly rinsed with deionized water. Platinum wire is then used to connect the electrodes and the electronic component, and the MFC is installed in a temperature-controlled chamber. A cross-section of each compartment reveals a rectangular shape with a narrow width through which fuel or air-saturated tap water flows as the catholyte flows upward.

As such, this design minimizes the fuel channeling and flow short-circuiting problems encountered in continuous operation. Each compartment is installed with two ports at the top and bottom, respectively. The electrodes are packed to occupy the entire anode and cathode compartments in order to ensure direct contact between the cation-specific membrane and the electrodes. The anode and cathode are connected through a voltameter with a resistance of 10 Ω to monitor the current under closed-circuit conditions. The current is calculated as $I = V/R_{load}$, where V is the potential drop across R_{load}. Figure 17.2 shows a schematic diagram of a typical BOD monitoring system, including an MFC as the sensor.

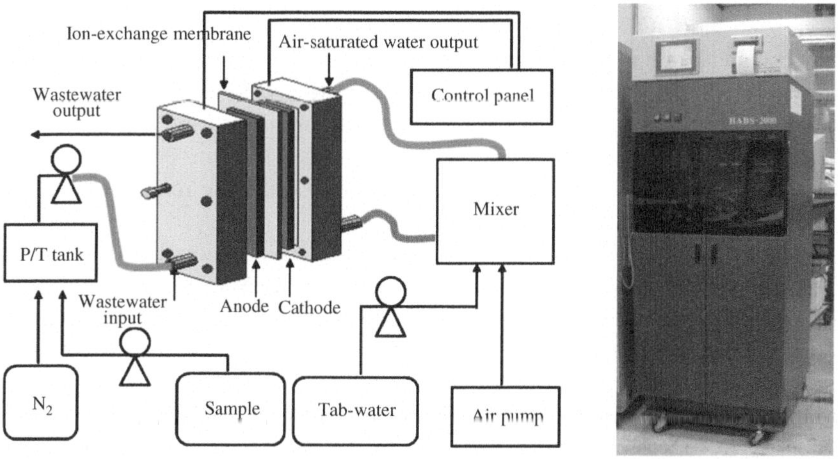

Figure 17.2. Schematic diagram (left) and picture (right) of MFC type BOD monitoring system

17.3.1.2 Enrichment and operation

The anode compartment of the MFC is loaded with a freshly collected sludge suspension, while the cathode compartment is fed with air-saturated tap water. The anode compartment should be kept as anoxic as possible. The sample is purged with oxygen-free nitrogen gas in a pretreatment tank before being fed to the anode compartment at a hydraulic retention time of about 14 min. Normally, within 3 weeks of feeding the fuel, a stable current is generated, suggesting that electrochemically-active bacteria have been successfully enriched. The MFC can be used to measure the BOD. The peak current or coulombic efficiency is monitored for BOD estimation in a batch operation, while the current is monitored in a continuous operation: a precise sample amount needs to be supplied for batch operation, whereas a constant feeding speed is required for continuous measurement.

17.3.1.3 Performance

The performance of an MFC as a BOD sensor is described here according to the criteria suggested by Liu and Mattiasson (2002). As the current from a copiotrophic MFC has a linear relationship with a fuel concentration of 20 to 400 mg/l, a BOD sample over 400 mg/l needs to be diluted to this range.

A short response time is important for on-line BOD monitoring and this can vary from 5 min to 10 h depending on the microbial fuel cell type, operational conditions, and BOD concentration. A non-ideal plug flow in the anode compartment results in an increased response time and therefore, the feeding rate needs to be determined to reduce a non-ideal flow (Moon *et al.* 2004). The smaller the MFC, the shorter the response time. The measuring time is shorter with continuous operation than for batch operation. Thus, at optimal operating conditions using an MFC with an anode volume of 5 ml, the response time can be reduced to 5 min.

The American Public Health Association (APHA) standard method for a 5-day BOD test allows a standard deviation of 30.5 mg/l using artificial wastewater containing glucose and glutamate with an average BOD_5 value of 198 mg/l, which is equivalent to a ±15.4% variation. As shown in Table 17.3, the repeatability of MFCs ranges from ±3 to ±13%, which is within the acceptable variation set by APHA. The long-term stability of a mediator-less MFC-type BOD sensor is remarkable when compared with a DO electrode-based BOD sensor and mediator-aided MFC-based BOD sensor. Without any maintenance, a mediator-less MFC-type BOD sensor can be operated for over 6 years with good accuracy. An MFC-type BOD sensor system was tested under field conditions at a sewage plant (Kim *et al.* 2003b), where the sensor was successfully operated in batch mode with high accuracy and good stability for over 60 days with a sample measurement time of 45 min. Recently, an MFC enriched with glucose was described as a BOD sensor with a measuring range from 25 mg/l up to 25 g/l (Kumlanghan *et al.* 2007).

Table 17.3. Comparison of MFC-based BOD sensor performances

	Copiotrophic	Copiotrophic	Oligotrophic
Mode	Batch	Continuous	Continuous
Range (mg/l)	20–400	20–200	2–10
Response time (min)	30–600	5–15	60
Repeatability (%)	3–12	3–4	–
Operational stability	>5 years	>1 year	>1 year
Reference	(Phung *et al.* 2004; Kim *et al.* 2003a)	(Kim *et al.* 2005)	(Kang *et al.* 2003; Moon *et al.* 2004)

17.3.2 MFC to measure BOD values lower than 10 mg/l

17.3.2.1 Background

The BOD sensors reported have almost all been developed to determine high BOD values in industrial wastewater. Yet, surface water and secondary effluents usually contain less biodegradable organic compounds, such as humic acid, lignin, tannic acid, gum arabic, and surfactants at low concentrations (Chee *et al.* 1999, 2000). Some microorganisms are known to grow optimally in environments with low levels of nutrients. These microorganisms are known as 'oligotrophs' in contrast to 'copiotrophs', which are common in environments with greater nutritional opportunities. The ability of oligotrophic microorganisms to grow in this way has a number of important biotechnological, medical, and environmental implications, and the authors have been involved in developing an MFC using oligotrophs obtained from surface water for the rapid determination of low BOD values.

The coulombic efficiency is very low, less than 10%, in oligotrophic MFCs enriched using C- or S-type reactors, as used in MFCs for measuring BOD values higher than 10 mg/l. This low yield is apparently due to fuel consumption through aerobic respiration in the anode compartment using O_2 that has diffused through the membrane (Pham *et al.* 2004). Thus, a new oligotrophic MFC (O type MFC) was designed as a low value BOD sensor to enhance the cathode reaction and lower O_2 diffusion into the anode compartment by reducing the size of the membrane (Kang *et al.* 2003; Moon *et al.* 2005). The performance of the new MFC was greatly improved in terms of the maximum current and coulombic efficiency. In this regard it is interesting to note that the anode was a preferred electron acceptor over O_2 in a MFC inoculated with an acidophilic bacterium (Malki *et al.* 2008).

17.3.2.2 Oligotrophic sensor design and performance

O-type MFCs were inoculated with sediment collected from a local river and fed with river water collected from the same source to enrich electrochemically active oligotrophic bacteria. Alternatively, the MFCs were fed with artificial wastewater containing 10 mg/l glucose and glutamate as the BOD. The cathode was fed continuously with air-saturated tap water. Within 8 weeks of continuous fuel feeding, the MFCs generated a stable current of over 0.01 mA with a resistance of 10 Ω, showing that electrochemically active oligotrophs had been enriched under these conditions.

When the MFCs enriched with surface water were run in batch mode, the maximum current generated was 0.02 mA with a coulombic efficiency of 20%, which was much higher than for an oligotrophic MFC enriched in a C-type MFC. In continuous operation (Moon *et al.* 2005), the dynamic linear range of the calibration curve was 2.0–10.0 mg BOD/l. The response time to a change of 2 mg BOD/l was about 60 min.

17.3.3 BOD determination of samples containing oxygen and nitrate

17.3.3.1 Oxygen and nitrate reduce current and coulombic efficiency

As discussed above (Section 17.2.4), oxygen and nitrate are reduced preferentially over electron transfer to the anode, thereby reducing the coulombic efficiency. Effluents from sewage treatment plants and certain types of industrial wastewaters contain high concentrations of nitrate, and oxygen can be easily dissolved into such samples. For accurate BOD measurement, these electron acceptors with a high redox potential should be removed from the samples. Oxygen can be purged from the samples using inert gases, such as nitrogen, before the samples are fed to the MFCs for BOD measurement (Chang *et al.* 2004; Kim *et al.* 2004; Moon *et al.* 2004). However, a considerable amount of oxygen still diffuses into the anode compartment through the ion-exchange membrane, which separates the anode compartment from the cathode compartment (Chang *et al.* 2005). It is not easy to eliminate nitrate from the samples.

17.3.3.2 Use of respiratory inhibitors

Azide (Carlson *et al.* 1982) and cyanide (Richardson *et al.* 1990) are known to inhibit nitrate reductase, the key enzyme of dissimilatory nitrate reduction (Stewart 1988), and cytochrome oxidase, the terminal oxidase of aerobic respiration is inhibited by azide and cyanide (Bayoumi Hamuda *et al.* 1996). Since azide and cyanide do not inhibit current generation in MFCs (Tender *et al.* 2002), they have been used to improve the performance of an MFC as a BOD sensor. The signal from MFCs decreases in the presence of electron acceptors with a higher redox potential such as nitrate and oxygen. The respiratory inhibitors eliminated the effects of oxygen and nitrate on current generation from MFCs. The addition of azide or cyanide did not change the signal in the absence of the electron acceptors (Chang *et al.* 2005). Similar results were also obtained using oligotrophic MFCs fed with an environmental sample that contained nitrate. Accordingly, the use of respiratory inhibitors is recommended for accurate BOD measurements of environmental samples containing nitrate and/or oxygen with a MFC-type BOD sensor. Nonetheless, aerobic respiration that is resistant to azide and cyanide is not uncommon in prokaryotes, and studies are therefore needed to see if inhibitor-resistant bacteria are enriched in MFCs during long-term operation with the inhibitors.

17.4 MFC AS A TOXICITY SENSOR

There are increasing needs for rapid, simple and sensitive methods for monitoring toxic substances in the water industry. Analytical instrumental methods, including HPLC, GC-MS and LC-MS, are widely used for such purposes. However, they are not suitable for real-time monitoring on-site because they require complex pretreatment and analytical protocols. Biomonitoring methods using fish, invertebrates, algae, plants and bioluminescent bacteria have been reported but these systems do not directly generate electric signals. A microbial fuel cell-type BOD sensor was successfully used to monitor toxic substances including an organophosphorus compound, PCBs and heavy metals such as lead and mercury (Kim *et al.* 2007). Since the signal is concentration dependent, an MFC-type sensor can be a useful tool in the water industry for monitoring toxic compounds in water.

17.5 CONCLUSIONS

A microbial fuel cell can be used not only as a waste to electricity-generating process but also as a biochemical oxygen demand sensor because the electrical output from the device is directly proportional to the amount of biodegradable organic matter used. A copiotrophic microbial fuel cell can be used to monitor BOD values up to 400 mg/l. BOD values lower than 10 mg/l can be measured using an oligotrophic microbial fuel cell. Since oxygen and nitrate are preferred electron acceptors over the anode in an MFC, respiratory inhibitors may be used for accurate measurement of BOD values of samples containing such electron acceptors with a high redox potential. These BOD sensors have been successfully used in a field test (Kim *et al.* 2003b) and commercialized (http://www.korbi.com/eng/index.htm).

17.6 ACKNOWLEDGEMENTS

Parts of this chapter have been adapted from a chapter in the ENCYCLOPEDIA of SENSORS [Kim B. H., I. S. Chang and H. Moon (2006), Microbial fuel cell type biochemical oxygen demand sensor In: Encyclopedia of sensors. Eds.

Grimes C. A., Dickey E. C., Pishko M. V., American Scientific Publishers, Washington, DC, USA]. This work was partly supported by a grant from the Korea Science and Engineering Foundation (KOSEF) NRL Program (No. R0A-2008-000-20088-0).

REFERENCES

Akiba, T., Bennetto, H.P., Stirling, J.L. and Tanaka, K. (1987) Electricity production from alkalophilic organisms. *Biotechnol. Lett.* **9**, 611–616.

Ardeleanu, I., Márgineanu, D.G. and Vais, H. (1983) Electrochemical conversion in biofuel cells using *Clostridium butyricum* or *Staphylococcus aureus* Oxford. *Bioelectrochem. Bioenerg.* **11**, 273–277.

Bari, A. and Farooq, S. (1985) Measurement of wastewater treatment efficiency by fluorescence and UV absorbance. *Environ. Monit. Assess.* **5**, 423–434.

Bayoumi Hamuda, H.E.A.F., Kiss, Z., Varady, G., Kucsma, N., Balazsy, S. and Kecskes, M. (1996) The influence of nitrapyrin and sodium azide on the growth and respiration of some *Rhizobium strains. Microbiol. Res.* **151**, 247–251.

Bennetto Delaney, H.P.G.M. and Mason, J.R. (1985) The sucrose fuel cell: Efficient biomass conversion using a microbial catalyst. *Biotechnol. Lett.* **7**, 699–704.

Bond, D.R. and Lovley, D.R. (2003) Electricity production by Geobacter sulfurreducens attached to electrodes. *Appl. Environ. Microbiol.* **69**, 1548–1555.

Bourgeois, W., Burgess, J.E. and Stuetz, R.M. (2001) On-line monitoring of wastewater quality: a review. *J. Chem. Technol. Biotechnol.* **76**, 337–348.

Bretschger, O., Obraztsova, A., Sturm, C.A., Chang, I.S., Gorby, Y.A., Reed, S.A., Culley, D.E., Reardon, C.L., Barua, S., Romine, M.F., Zhou, J., Beliaev, A.S., Bouhenni, R., Saffarini, D., Mansfeld, F., Kim, B.H., Fredrickson, J.K. and Nealson, K.H. (2007) Current production and metal oxide reduction by Shewanella oneidensis MR-1 wild type and mutants. *Appl. Environ. Microbiol.* **73**, 7003–7012.

Carlson, C.A., Ferguson, L.P. and Ingraham, J.L. (1982) Properties of dissimilatory nitrate reductase purified from the denitrifier Pseudomonas aeruginosa. *J. Bacteriol.* **151**, 162–171.

Chang, I.S., Jang, J.K., Gil, G.C., Kim, M., Kim, H.J., Cho, B.W. and Kim, B.H. (2004) Continuous determination of biochemical oxygen demand using microbial fuel cell type biosensor. *Biosens. Bioelectron.* **19**, 607–613.

Chang, I.S., Moon, H., Jang, J.K. and Kim, B.H. (2005) Improvement of a microbial fuel cell performance as a BOD sensor using respiratory inhibitors. *Biosens. Bioelectron.* **20**, 1865–1859.

Chaudhuri, S.K. and Lovley, D.R. (2003) Electricity generation by direct oxidation of glucose in mediatorless microbial fuel cells. *Nature Biotechnol.* **21**, 1229–1232.

Chee, G.J., Nomura, Y. and Karube, I. (1999) Biosensor for the estimation of low biochemical oxygen demand. *Anal. Chim. Acta* **379**, 185–191.

Chee, G.J., Nomura, Y., Ikebukuro, K. and Karube, I. (2000) Optical fiber biosensor for the determination of low biochemical oxygen demand. *Biosens. Bioelectron.* **15**, 371–376.

Choo, Y.P., Lee, J., Chang, I.S. and Kim, B.H. (2006) Bacterial communities in microbial fuel cells with high concentrations of glucose and glutamate. *J. Microbiol. Biotechnol.* **16**, 1481–1484.

Delaney, G.M., Bennetto, H.P., Mason, J.R., Roller, S.D., Stirling, J.L. and Thurston, C.F. (1984) Electron-transfer coupling in microbial fuel cells. 2. Performance of fuel cells containing selected microorganism-mediator-substrate combinations. *J. Chem. Tech. Biotechnol.* **34B**, 13–27.

Du, Z.W., Li, H.R. and Gu, T.Y. (2007) A state of the art review on microbial fuel cells: a promising technology for wastewater treatment and bioenergy. *Biotechnol. Adv.* **25**, 464–482.

Eaton, A.D., Clesceri, L.S. and Greenberg, A.E. (1994) Standard methods for the examination of water and wastewater, Washington, D. C., American Public Health Association. Edwards, A.C. and Cresser, M.S. (1987) Relationships between ultraviolet absorbance and total organic carbon in two upland catchments. *Water Res.* **21**, 49–56.

Gil, G.C., Chang, I.S., Kim, B.H., Kim, M., Jang, J.K., Park, H.S. and Kim, H.J. (2003) Operational parameters affecting the performance of a mediator-less microbial fuel cell. *Biosens. Bioelectron.* **18**, 327–334.

Holmes, D.E., Bond, D.R. and Lovley, D.R. (2004) Electron transfer by Desulfobulbus propionicus to Fe(III) and graphite electrodes. *Appl. Environ. Microbiol.* **70,** 1234–1237.

Hyun, C.K., Tamiya, E., Takeuchi, T. and Karube, I. (1993) A novel BOD sensor based on bacterial luminescence. *Biotechnol. Bioeng.* **41**, 1107–1111.

Kang, K.H., Jang, J.K., Pham, T.H., Moon, H., Chang, I.S. and Kim, B.H. (2003) A microbial fuel cell with improved cathode reaction as a low biochemical oxygen demand sensor. *Biotechnol. Lett.* **25**, 1357–1361.

Kang, K.H., Jang, J.K., Lee, J.Y., Moon, H., Chang, I.S., Kim, J.M. and Kim, B.H. (2004) A low BOD sensor using a microbial fuel cell. *J. KSEE* **26**, 58–63.

Karube, I., Matsunaga, T., Mitsuda, S. and Suzuki, S. (1977a) Microbial electrode BOD sensors. *Biotechnol. Bioeng.* **19**, 1535–1547.

Karube, I., Matsunaga, T., Tsuru, S. and Suzuki, S. (1977b) Biochemical fuel cell utilizing immobilized cells of Clostridium butyricum. *Biotechnol. Bioeng.* **19**, 1727–1733.

Kim, B.H., Ikeda, T., Park, H.S., Kim, H.J., Hyun, M.S., Kano, K., Takagi, K. and Tatsumi, H. (1999a) Electrochemical activity of an Fe(III)-reducing bacterium, Shewanella putrefaciens IR-1, in the presence of alternative electron acceptors. *Biotechnol. Tech.* **13**, 475–478.

Kim, B.H., Chang, I.S., Gil, G.C., Park, H.S. and Kim, H.J. (2003a) Novel BOD (biological oxygen demand) sensor using mediator-less microbial fuel cell. *Biotechnol. Lett.* **25,** 541–545.

Kim, B.H., Kim, H.J., Hyun, M.S. and Park, D.H. (1999b) Direct electrode reaction of Fe(III)-reducing bacterium, Shewanella putrefaciens. *J. Microbiol. Biotechnol.* **9**, 127–131.

Kim, B.H., Park, H.S., Kim, H.J., Kim, G.T., Chang, I.S., Lee, J. and Phung, N.T. (2004) Enrichment of microbial community generating electricity using a fuel-cell-type electrochemical cell. *Appl. Microbiol. Biotechnol.* **63**, 672–681.

Kim, G.T., Hyun, M.S., Chang, I.S., Kim, H.J., Park, H.S., Kim, B.H., Kim, S.D., Wimpenny, J.W.T. and Weightman, A.J. (2005) Dissimilatory Fe(III) reduction by an electrochemically active lactic acid bacterium phylogenetically related to Enterococcus gallinarum isolated from submerged soil. *J. Appl. Microbiol.* **99**, 978–987.

Kim, H.J., Hyun, M.S., Chang, I.S. and Kim, B.H. (1999c) A microbial fuel cell type lactate biosensor using a metal-reducing bacterium, *Shewanella putrefaciens*. *J. Microbiol. Biotechnol.* **9**, 365–367.

Kim, M., Youn, S.M., Shin, S.H., Jang, J.G., Han, S.H., Hyun, M.S., Gadd, G.M. and Kim, H.J. (2003b) Practical field application of a novel BOD monitoring system. J. *Environ. Monit.* **5**, 640–643.

Kim, M., Hyun, M.S., Gadd, G.M. and Kim, H.J. (2007) A novel biomonitoring system using microbial fuel cells. *J. Environ. Monit.* **9**, 1323–1328.

Kim, M.N. and Kwon, H.S. (1999) Biochemical oxygen demand sensor using serratia marcescens LSY 4. *Biosens. Bioelectron.* **14**, 1–7.

Kumlanghan, A., Kanatharana, P., Asawatreratanakul, P., Mattiasson, B. and Thavarungkul, P. (2008) Microbial BOD sensor for monitoring treatment of wastewater from a rubber latex industry. *Enzyme Microbial Technol.* **42**, 483–491.

Kumlanghan, A., Liu, J., Thavarungkul, P., Kanatharana, P. and Mattiasson, B. (2007) Microbial fuel cell-based biosensor for fast analysis of biodegradable organic matter. *Biosens. Bioelectron.* **22**, 2939–2944.

Lee, J., Phung, N.T., Chang, I.S., Kim, B.H. and Sung, H.C. (2003) Use of acetate for enrichment of electrochemically active microorganisms and their 16S rDNA analyses. *FEMS Microbiol. Lett.* **223**, 185–191.

Li, Y.R. and Chu, J. (1991) Study of BOD microbial sensors for waste water treatment control. *Appl. Biochem. Biotechnol.* **28–29**, 855–863.

Liu, J. and Mattiasson, B. (2002) Microbial BOD sensors for wastewater analysis. *Water Res.* **36**, 3786–3802.

Liu, J., Bjornsson, L. and Mattiasson, B. (2000) Immobilised activated sludge based biosensor for biochemical oxygen demand measurement. *Biosens. Bioelectron.* **14**, 883–893.

Liu, J., Olsson, G. and Mattiasson, B. (2004a) Short-term BOD (BODst) as a parameter for on-line monitoring of biological treatment process: Part I. A novel design of BOD biosensor for easy renewal of bio-receptor. *Biosens. Bioelectron.* **20**, 562–570.

Lloyd, J.R., Sole, V.A., Van Praagh, C.V.G. and Lovley, D.R. (2000) Direct and Fe(II)-mediated reduction of technetium by Fe(III)-reducing bacteria. *Appl. Environ. Microbiol.* **66**, 3743–3749.

Lovley, D.R. (1981) Dissimilatory Fe(III) and Mn(IV) reduction. *Microbiol. Rev.* **55**, 259–287.

Malki, M., De Lacey, A.L., Rodriguez, N., Amils, R. and Fernandez, V.M. (2008) Preferential use of an anode as an electron acceptor by an acidophilic bacterium in the presence of oxygen. *Appl. Environ. Microbiol.* **74**, 4472–4476.

Moon, H., Chang, I.S., Kang, K.H., Jang, J.K. and Kim, B.H. (2004) Improving the dynamic response of a mediator-less microbial fuel cell as a biochemical oxygen demand (BOD) sensor. *Biotechnol. Lett.* **26**, 1717–1721.

Moon, H., Chang, I.S., Jang, J.K., Kim, K.S., Lee, J., Lovitt, R.W. and Kim, B.H. (2005) On-line monitoring of low biochemical oxygen demand through continuous operation of a mediator-less microbial fuel cell. *J. Microbiol. Biotechnol.* **15**, 192–196

Moon, H., Chang, I.S. and Kim, B.H. (2006) Continuous electricity production from artificial wastewater using a mediator-less microbial fuel cell. *Bioresour. Technol.* **97**, 621–627.

Myers, C.R. and Myers, J.M. (1992) Localization of cytochromes to the outer membrane of anaerobically grown Shewanella putrefaciens MR-1. *J. Bacteriol.* **174**, 3429–3438.

Nakamura, H., Kobayashi, S., Hirata, Y., Suzuki, K., Mogi, Y. and Karube, I. (2007a) A spectroscopic biochemical oxygen demand determination method using 2,6-dichlorophenolindophenol as the redox color indicator and the eukaryote Saccharomyces cerevisiae. *Anal. Biochem.* **369**, 168–174.

Nakamura, H., Abe, Y., Kozumi, R., Suzuki, K., Mogi, Y., Hiroyama, T. and Karube, I. (2007b) A chemiluminescense biochemical oxygen demand measuring method. *Anal. Chim. Acta* **602**, 94–100.

Nataraja, M., Qin, Y. and Seagren, E.A. (2006) Ultraviolet spetrophotometry as an index parameter for estimating the biochemical oxygen demand of domestic wastewater. *Environ. Technol.* **27**, 789–800.

Nealson, K.H. and Saffarini, D. (1994) Iron and manganese in anaerobic respiration: Environmental significance, physiology, and regulation. *Ann. Rev.* **48**, 311–343.

Nevin, K.P. and Lovley, D.R. (2000) Lack of production of electron-shuttling compounds or solubilization of Fe(III) during reduction of insoluble Fe(III) oxide by *Geobacter metallireducens*. *Appl. Environ. Microbiol.* **66**, 2248–2251.

Newman, D.K. and Kolter, R. (2000) A role for excreted quinones in extracellular electron transfer. *Nature* **405**, 94–97.

Onnis, A., Carucci, A. and Cappai, G. (2006) Titration biosensors for the estimation of the biochemical nitrate demand of municipal and industrial waste. *J. Ind. Microbiol. Biotechnol.* **33**, 243–246.

Park, D.H. and Zeikus, J.G. (2000) Electricity generation in microbial fuel cells using neutral red as an electronophore. *Appl. Environ. Microbiol.* **66**, 1292–1297

Park, H.S., Kim, B.H., Kim, H.S., Kim, H.J., Kim, G.T., Kim, M., Chang, I.S., Park, Y.K. and Chang, H.I. (2001) A novel electrochemically active and Fe(III)-reducing bacterium phylogenetically related to Clostridium butyricum isolated from a microbial fuel cell. *Anaerobe* **7**, 297–306.

Pasco, N., Baronian, K., Jeffries, C., Webber, J. and Hay, J. (2004) MICREDOX® – Development of a ferricyanide-mediated rapid biochemical oxygen demand method using an immobilised *Proteus vulgaris* biocomponent. *Biosens. Bioelectron.* **20**, 524–532.

Pham, C.A., Jung, S.J., Phung, N.T., Lee, J., Chang, I.S., Kim, B.H., Yi, H. and Chun, J. (2003) A novel electrochemically active and Fe(III)-reducing bacterium phylogenetically related to *Aeromonas hydrophila*, isolated from a microbial fuel cell. *FEMS Microbiol. Lett.* **223**, 129–134.

Pham, T.H., Jang, J.K., Chang, I.S. and Kim, B.H. (2004) Improvement of cathode reaction of a mediatorless microbial fuel cell. *J. Microbiol. Biotechnol.* **14**, 324–329.

Phung, N.T., Lee, J., Kang, K.H., Chang, I.S., Gadd, G.M. and Kim, B.H. (2004) Analysis of microbial diversity in oligotrophic microbial fuel cells using 16S rDNA sequences. *FEMS Microbiol. Lett.* **233**, 77–82.

Reynolds, D.M. and Ahmad, S.R. (1995a) The effect of metal ions on the fluorescence of sewage wastewater. *Water Res.* **29**, 2214–2216.

Reynolds, D.M. and Ahmad, S.R. (1995b) Rapid and direct determination of wastewater BOD values using a fluorescence technique. *Water Res.* **31**, 2012–2018.

Richardson, D.J., McEwan, A.G., Page, M.D., Jackson, J.B. and Ferguson, S.J. (1990) The identification of cytochromes involved in the transfer of electrons to the periplasmic NO3 – reductase of Rhodobacter capsulatus and resolution of a soluble NO3-reductase – cytochrome-c552 redox complex. *Eur. J. Biochem.* **194**, 263–270.

Riedel, K. (1991) Biochemical fundamentals and improvement of the selectivity of microbial sensors – a minireview. *Bioelectrochem. Bioenerg.* **25**, 19–30.

Roller, S.D., Bennetto, H.P., Delaney, G.M., Mason, J.R., Stirling, J.L. and Thurston, C.F. (1984) Electron-transfer coupling in microbial fuel cells: 1. Comparison of redox-mediator reduction rates and respiratory rates of bacteria. *J. Chem. Tech. Biotechnol.* **34B**, 3–12.

Sakakuchi, T., Morioka, Y., Yamasaki, M., Iwanaga, J., Beppu, K., Maeda, H., Morita, Y. and Tamiya, F. (2007) *Biosens. Bioelectron.* **22**, 1345–1350.

Sangeetha, S., Sugandhi, G., Murugesan, M., Murali Madhav, V., Berchmans, S., Rajasekar, R., Rajasekar, S., Jeyakumar, D. and Prabhakara Rao, G. (1996) Torulopsis candida based sensor for the estimation of biochemical oxygen demand and its evaluation. *Electroanalysis* **8**, 698–701.

Stewart, V. (1988) Nitrate respiration in relation to facultative metabolism in enterobacteria. *Microbiol. Rev.* **52**, 190–232.

Tan, T.C., Li, F. and Neoh, K.G. (1993) Measurement of BOD by initial rate of response of a microbial sensor. *Sensors and Actuators: B. Chemical* **10**, 137–142.

Tayhas, G., Palmore, R. and Whitesides, M. (1994) Microbial and enzyme biofuel cells. In *Enzymatic Conversion of Biomass for Fuels Production.* Washington DC, American Chemical Society.

Tender, L.M., Reimers, C.E., Stecher Iii, H.A., Holmes, D.E., Bond, D.R., Lowy, D.A., Pilobello, K., Fertig, S.J. and Lovley, D.R. (2002) Harnessing microbially generated power on the seafloor. *Nature Biotechnol.* **20**, 821–825.

Thurston, C.F., Bennetto, H.P. and Delaney, G.M. (1985) Glucose metabolism in a microbial fuel cell. Stoichiometry of product formation in a thionine-mediated Proteus vulgaris fuel cell and its relation to coulombic yields. *J. Gen. Microbiol.* **131**, 1393–1401.

Trosok, S.P., Driscoll, B.T. and Luong, J.H.T. (2001) Mediated microbial biosensor using a novel yeast strain for wastewater BOD measurement. *Appl. Microbiol. Biotechnol.* **56**, 550–554.

Turick, C.E., Tisa, L.S. and Caccavo Jr., F. (2002) Melanin production and use as a soluble electron shuttle for Fe(III) oxide reduction and as a terminal electron acceptor by *Shewanella algae* BrY. *Appl. Environ. Microbiol.* **68**, 2436–2444.

Tzoris, A., Fernandez-Perez, V. and Hall, E.A.H. (2005) Direct toxicity assessment with a mini portable respirometer. *Sensors and Actuators B: Chemical* **105**, 39–49.

Vaiopoulou, E., Melidis, P., Campragou, E. and Aivasidis, A. (2005) *Biosens. Bioelectron.* **21**, 365–371.

Vega, C.A. and Fernández, I. (1987) Mediating effect of ferric chelate compounds in microbial fuel cells with Lactobacillus plantarum, Streptococcus lactis, and Erwinia dissolvens. *Bioelectrochem. Bioenerg.* **17**, 217–222.

Vodacek, A. and Philpot, W.D. (1987) Environmental effects on laser-induced fluorescence spectra of natural waters. *Remote Sens. Environ.* **21**, 83–95.

Wilkinson, S. (2000) "Gastrobots" – Benefits and challenges of microbial fuel cells in food powered robot applications. *Auton. Robot.* **9**, 99–111.

Yagishita, T., Sawayama, S., Tsukahara, K.I. and Ogi, T. (1997) Behavior of glucose degradation in *Synechocystis* sp. M-203 in bioelectrochemical fuel cells. *Bioelectrochem. Bioenerg.* **43**, 177–180.

Yoshida, N., Yano, K., Morita, T., McNiven, S.J., Nakamura, H. and Karube, I. (2000) A mediator-type biosensor as a new approach to biochemical oxygen demand estimation. *Analyst* **125**, 2280–2284.

18

Feedstocks for BES conversions

Freda R. Hawkes, Jung Rae Kim, Godfrey Kyazze and Giuliano C. Premier

18.1 INTRODUCTION

Any material that can provide substances oxidizable by electrochemically-active bacteria can potentially be used as a feedstock for bioelectrochemical systems (BES). However, organic wastes, wastewaters and agri-food co-products are likely to be the most sustainable and cost-effective feedstocks. BESs could be incorporated into the series of physico-chemical and biological treatment processes common to wastewater treatment plants. In the future, the treatment industry may cease to depend on energy-intensive aerobic processes such as activated sludge treatment and move towards anaerobically-based treatment technologies which include dark fermentative hydrogen production/acidogenesis, anaerobic digestion, MFCs and MECs. Anaerobic treatment processes generate less microbial biomass for disposal, avoid energy expenditure in oxygenation and generate the energy carriers H_2 (dark fermentation and

© 2010 IWA Publishing. *Bioelectrochemical Systems: From Extracellular Electron Transfer to Biotechnological Application*. Edited by Korneel Rabaey, Largus Angenent, Uwe Schröder and Jürg Keller. ISBN: 9781843392330. Published by IWA Publishing, London, UK.

microbial electrolysis in MEC), CH_4 (anaerobic digestion and as an undesired by-product of MFCs and MEC) and electricity (MFCs). For BES technologies to be considered for inclusion in the process train, it is important to obtain information on operational parameters such as:

- organic loading rates (OLRs) (reactor sizing relative to feedstock)
- the alkalinity required at given OLRs and the need for additional chemicals e.g. nutrients, ions to increase conductivity (dosing cost determination)
- the ability of BES to reduce COD/BOD in the chosen feedstock (so discharge limits can be met)
- ability of BES to degrade refractory/xenobiotic components (so the range of possible feedstocks can be defined)

The feedstocks for MFCs and MECs (agro-industrial and chemical industry wastewaters, domestic wastewater, etc.) may either be used directly or after passing through processes such as physico-chemical hydrolysis, flocculation, gravity settling, acidogenesis or dark fermentative H_2 production and anaerobic digestion. The last three of these break down complex substrates to end-products including volatile fatty acids (VFA). Higher concentrations of VFA will arise from acidogenesis or dark fermentative H_2 production and lower concentrations from anaerobic digestion. It is important to determine what range of substrate concentrations can be efficiently utilized and treated by BES. Generally for acidogenesis/fermentative H_2 production/anaerobic digestion a higher substrate concentration gives more cost-effective treatment, as heat required to warm the influent in temperate/cold climates will be less and the reactor size will be smaller. However, if MFCs can remove acetate and other sources of BOD from wastewaters at low concentrations, they may replace the aerobic polishing step which is required after anaerobic digestion for industrial wastewaters before discharge to the consent levels of around 20 mg/l BOD.

Municipal sewage has a relatively low COD level. The role of MFCs in domestic wastewater treatment works receiving influent at around 200–300 mg/l COD has been discussed by Logan (2008). Agro-industrial wastewaters, however commonly have CODs in the range 1.5–80 gCOD/l (Angenent and Wrenn 2008). The sequence in which BES can receive these effluents in the treatment process could be determined by the initial degradability of the substrates, the feedstock concentrations/organic loading rates which the BESs are designed to handle, the effluent quality required and the mix of methane, H_2 and electricity desired.

The gross feedstock energy conversion efficiency, measured as coulombic efficiency (CE) in MFCs or H_2 yield in MECs, and the removal efficiency, whether measured as COD, BOD or total solids removal, are related not only to the design of the BES but also to the nature of the substrate and the volumetric organic loading rate (OLR or B_v). The OLR is an important concept in feedstock utilization. For high rate anaerobic digesters, OLRs of 10–20 gCOD/l/d may be expected (Speece 1996), while for MFCs, reported OLRs are generally considerably lower. MECs could compete with traditional wastewater treatment plants if the OLR reached 6.5 gCOD/l/day and hydrogen recovery from the substrate was 100%, giving 10 m^3 H_2/m^3/d, which has not yet been reported (Logan *et al.* 2008).

The composition of the wastes and wastewaters which may be used as feedstocks is often not well characterized, unlike the many laboratory studies of BES where the substrate is defined. Aspects of the feedstocks which are important for electricity and H_2 production include the nature of the organic components, the nutrient composition, buffering capacity (alkalinity) and conductivity. Other physically significant properties include fat content and particulate matter, so a prior settling or disintegration/hydrolysis stage may be needed. BES used with non-sterile feedstocks will contain a mixed bacterial community, with the advantage of greater substrate flexibility but the disadvantage of side-reactions e.g. fermentations and aerobic metabolism. Methanogenesis is an often-reported side reaction in BES (Kim *et al.* 2005).

In summary, organic substrates for BES should:

- be cheap,
- be readily degradable by microbial consortia from commonly available natural sources,
- contain suitable macro- and micro-nutrients, buffering capacity and conductivity,
- be able to be dosed by commercially available feeding devices during continuous operation without mechanical problems such as blockage by suspended solids or coating with fats.

This chapter considers the range of defined chemical compounds and complex undefined wastewaters so far reported in the literature as feedstocks for BES. The concentration range and OLRs at which these feedstocks have been applied to BES and the resulting treatment efficiency is also examined. The use of agro-industrial wastewaters after passage through acidogenesis/dark fermentative H_2 production/anaerobic digestion processes is considered.

18.2 DEFINED SUBSTRATES UTILIZED BY BES

Substrates for bacteria in the anode compartment studied at laboratory scale have been divided into those which are non-fermentable, such as acetate and other fermentation end products, and those which are fermentable, such as carbohydrates. In batch-fed MFCs there is evidence that acetate and lactate out-perform glucose (all at 200 mgCOD/l) in terms of power and current density (Jung and Regan 2007). Glucose conversion to current and CO_2 in an MFC is proposed to depend on a syntrophic association of fermenters producing acetate and H_2 and exoelectrogens consuming these products and providing conditions in which fermentation proceeds more efficiently (Freguia *et al.* 2008).

18.2.1 Volatile fatty acids and other fermentation end products

A major category of exoelectrogenic bacteria, *Geobacter* species, grow preferentially on acetate, oxidizing it to CO_2, and some species use monoaromatic compounds and ethanol (Lovley and Nevin 2008). The well-characterized species *Geobacter sulfurreducens* readily grows on acetate at low oxygen concentrations, apparently with O_2 as the sole electron acceptor (Lin *et al.* 2004). Lee *et al.* (2008) give the K_s for acetate for *G. sulfurreducens* in continuous culture (not in an MFC) as 0.64 mg COD/l and for acetoclastic methanogens as 177–427 mgCOD/l. Thus *G. sulfurreducens* might out-compete methanogens for acetate if the acetate concentration remains sufficiently low in the continuously-fed reactor.

Bacteria from the genus *Shewanella* are well-studied exoelectrogens and most do not use acetate as electron donor. *Shewanella* species are facultative anaerobes, have multiple pathways for electron transfer to soluble metal oxides and to solid electrodes and can produce their own soluble mediators (quinones) (Nealson and Scott 2006). The carbon source of many strains is quite restricted, mainly lactate, pyruvate, some amino acids, formate and H_2; thus in nature they live syntrophically with fermentative communities, converting lactate to acetate and removing H_2 (Nealson and Scott 2006). However, Biffinger *et al.* (2008) have shown that oxygen exposure in a chamber outside a mini-MFC enabled a *Shewanella oneidensis* strain to use glucose or fructose as the sole electron donor in the anaerobic anode compartment, presumably as enzymes of the citric acid cycle are induced. For *S. oneidensis* strain MR-1 acetate was also an exoelectrogenic substrate (Biffinger *et al.* 2009).

To examine the dependence of power output on substrate availability, it is necessary to use MFCs where power output is not limited by other factors such as high internal resistance (Kim *et al.* 2007b). A direct relation of current to acetate concentration was observed for MFCs with *G. sulfurreducens* up to 2.3 mmol/l (138 mg/l) (Tront *et al.* 2008). Above 2.3 mmol/l no further increase in current was seen, attributed to system design parameter limitations. In MFCs with mixed microflora continuously operated on acetate, a low external resistance was needed before an increase in OLR was accompanied by an increase in current generation and CE. A lower external resistance allows a higher electron flux, lower anode potential and higher energy gain for the bacteria (Aelterman *et al.* 2008).

Acetate and butyrate are major end-products of fermentative H_2 production and anaerobic digestion, where BES are candidates for a follow-on treatment step. Acetate and butyrate utilization using mixed cultures were studied by Liu *et al.* (2005) in a single-chamber batch-fed MFC without a PEM. The concentrations used and maximum power densities are shown in Table 18.1, which compares power densities from feedstocks tested in this same reactor type. When influent acetate concentration was increased in the range 80–800 mg/l, CE decreased, possibly due to ingress of O_2 at the longer batch cycle times occurring, and to higher acetate concentrations favouring methanogenesis. In the same MFC operated batch-wise, ethanol (tested at 3–25 mmol/l) also acted as an electron donor (Table 18.1). Acetate appears to be an intermediate in butyrate, ethanol and lactate utilization in mixed culture MFCs (Liu *et al.* 2005; Jung and Regan 2007; Kim *et al.* 2007a; Richter *et al.* 2007).

To study acetate conversion to H_2 in MECs, a number of researchers have used batch and fed batch operational modes with mixed cultures. Results are summarized in Table 18.2. In a batch two chamber reactor Rozendal *et al.* (2006) used acetate (2 mmol/l) and obtained a H_2 yield of 2.12 mol/mol acetate consumed. Similar results were obtained in fed batch operation (Chae *et al.* 2008). A lower H_2 yield was obtained by Liu *et al.* (2008a) using a fed batch H-type reactor. A typical batch cycle lasted 240–260 hours with an acetate consumption of 35%. So far the highest H_2 yield from acetate known to the authors, whether in batch or continuous operation, was obtained by Call and Logan (2008) using a membrane-free single chamber batch MEC with a brush anode. In this study the H_2 production rate reached 3.12 l/l reactor/day, still only one-third of the target set by Logan *et al.* (2008). For experiments with MECs using continuous operation, acetate was fed at 13.75 h HRT, 1.12 gCOD/l/day OLR, in a single chamber reactor configuration where the cathode served as a gas collection chamber (Rozendal *et al.* 2007). With a two chamber MEC where the cathode served as a gas collection chamber, Tartakovsky *et al.* (2008) using acetate at HRT 10h, OLR 1.78 gCOD/l/day obtained a higher H_2 yield (Table 18.2) and a production rate of 0.48 l H_2/l anode/d. There is need to increase the organic loadings to reach 6.5 gCOD/l/day (Logan *et al.* 2008).

Table 18.1. Comparison of power outputs from various substrates in batch-fed studies in a single chamber membrane-free air cathode MFC

Substrate	Conc. (mgCOD/l)	Max. power density mW/m^2 (W/m^3)	% removal as mg/l (COD/BOD where stated)	Reference
Acetate	853	506 (12.7)	>98	Liu *et al.* (2005)
Butyrate	1,818	305 (7.6)	>98	Liu *et al.* (2005)
Ethanol	960	488	>99	Kim *et al.* (2007a)
Glucose	640	494 (12.4)	>98	Liu and Logan (2004)
BSA,	1,100	354 (8.9)	90	Heilmann and Logan (2006)
Peptone	300	269	86 (BOD)	Heilmann and Logan (2006)
Domestic wastewater	200–300	146 (3.7)	75	Liu and Logan (2004)
Swine wastewater	8,320∗	261	86	Min *et al.* (2005)
Meat packing wastewater	1,420	139	93 (BOD) 87 (COD)	Heilmann and Logan (2006)
Beer brewery wastewater	2,250	205 (5.1)	87 (COD)	Feng *et al.* (2008)
Corn stover hydrolysates	250–1,000	371	93 (BOD) 60 (COD)	Zuo *et al.* (2006)

∗ Soluble COD only

Table 18.2. Comparison of hydrogen production from various substrates using MECs with mixed microflora

Substrate (mg COD/l)	Mode of operation	H_2 yield mol/mol added unless otherwise stated	Voltage applied (mV)	COD removal (%)	Reference
Acetate (128)	batch	2.12	500	NG	Rozendal *et al.* (2006)
Acetate (781)	batch	3.7	800	89	Call and Logan (2008)
Acetate (NG)	fed batch	2.9	850	95	Liu *et al.* (2005)
Acetate (128)	fed batch	2.1	800	NG	Chae *et al.* (2008)
Acetate (640)	fed batch	0.46 per mol consumed	540	35	Liu *et al.* (2008b)

(continued)

Table 18.2. *Continued*

Substrate (mg COD/l)	Mode of operation	H_2 yield mol/mol added unless otherwise stated	Voltage applied (mV)	COD removal (%)	Reference
Acetate (1,280)	fed batch	2.5 1.0	600 400	NG	Hu (2008)
Acetate (640)	continuous	0.92	1000	NG	Rozendal *et al.* (2007)
Acetate (725)	continuous	2.0	1160	94	Tartakovsky *et al.* (2008)
Butyrate (1,000)		8.01			
Lactate (1,000)	batch	5.45	600	NG	Cheng and Logan (2007)
Propionate (1,000)		6.25			
Valerate (1,000)		8.77			
Glucose (1,000)	batch	8.55	600	NG	Cheng and Logan (2007)
Glucose (725)	continuous	1.63	700	92	Tartakovsky *et al.* (2008)
Cellulose (1,000)	batch	8.20*	600	NG	Cheng and Logan (2007)
Domestic waste-water (204–481)	batch	154 ml/gCOD consumed	500	90	Ditzig *et al.* (2007)
Glycerol waste (6,160)	batch	126 ml/gCOD added	200 Vs Ag/AgCl at anode	85	Sakai and Yagishita (2007)

NG = not given; * per mol hexose equivalent

Few studies have used butyrate as a substrate for H_2 production in MECs. The rate of consumption of butyrate by anode respiring bacteria was shown to be low compared to acetate (Torres *et al.* 2007), suggesting that in continuous operation butyrate would require longer retention times. Cheng and Logan (2007) compared H_2 yields from butyric, lactic, propionic and valeric acids in batch MECs, the highest yields being obtained for butyric and valeric acids (see Table 18.2). Minor fermentation end products e.g. formate, caproate, ethanol, butanol, succinate should also be suitable feedstocks for MECs and more information about the usability of these substrates is required.

18.2.2 Soluble carbohydrates, amino acids and xenobiotics

Table 18.1 shows the maximum power density obtained with glucose was close to that for acetate and ethanol in the same MFC. However, glucose has not been clearly shown to generate electricity directly and methanogenesis is often reported to occur during operation on carbohydrates, lowering coulombic efficiency. Both acetate and H_2 produced from 0.5 gCOD/l glucose (OLR 2.4 kgCOD/m^3/d) were used to generate current (Freguia *et al.* 2008). A syntrophic interaction between the fermenters and anodophiles was proposed where the latter make fermentation more favourable. For hydrogen this is analogous to interspecies hydrogen transfer in anaerobic digestion. From this viewpoint, fermentation is not detrimental to electricity generation. Results indicated that methanogens competed for H_2 but did not consume acetate, probably related to their high K_s for acetate (around 138 mg/l) (ADM1 2002). The K_s for H_2 for methanogens is known to be low, while the K_s for H_2 for exoelectrogens is not known, so H_2 may be the main substrate for CH_4 production in MFCs and MECs operating on fermentable substrates.

A comparative study of acetate and glucose as substrates (Lee *et al.* 2008) showed 71% of acetate (as COD applied) was accounted for as current, and that exoelectrogens appeared to out-compete acetoclastic methanogens at the low acetate concentration used (360 mg/l) in this batch study, as no methane was detected. For glucose, in contrast, only 49% (as COD) gave rise to current, and methane and more biomass were produced. Biomass yields and biomass density on the anode were greater when glucose rather than acetate was substrate. They conclude that in MFCs fed fermentable substrates, a lower power output with the production of methane and extra biomass may be unavoidable. In continuous operation, the CEs for fermentable substrates (glucose, dextran and starch) were only 14–21% compared to those for acetate and butyrate at 65 and 50% (Min and Logan 2004).

Soluble carbohydrates e.g. glucose have also been used for H_2 production in MECs. Using 1 gCOD/l as substrate in a batch MEC, a H_2 yield of 8.55 mol/mol glucose was obtained (Cheng and Logan 2007) (Table 18.2). In continuous culture at 10 h HRT in the anode compartment, OLR 1.78 gCOD/l/d, a lower H_2 yield and H_2 production rate of 0.34 l/l/day from glucose were reported (Tartakovsky *et al.* 2008). When the OLR was doubled at the same HRT to 3.56 gCOD/l/day (above half way to the target value of 6.5 gCOD/l/d), there was no increase in the H_2 production rate. It appears that glucose gets converted into acetate and propionate before being utilized for H_2 production; for continuous systems involving high organic loadings, long retention times may be required, as from batch studies, cycle times of 2 to 10 days are common.

MFCs have been operated continuously using the standard BOD mixture of glucose and glutamate and used as an online BOD monitor (Chang *et al.* 2004). The linear response range was up to 100 mgBOD/l at a HRT of 1 h. Using the same substrate in a continuously-fed MFC, Moon *et al.* (2006) obtained an effluent containing below 10 mg/l COD from an influent of 300 mg/l COD at an OLR of 9.2 kg/m^3/d, comparable with OLRs for anaerobic digesters. Another amino acid, cysteine, has been shown to support power generation. Cysteine was used as the only available oxidizable substrate (concentration range up to 800 mg/l) in a two-chambered batch fed system (Logan *et al.* 2005). *Shewanella* species were dominant. It appeared that cysteine was not just partially oxidized to cystine but more likely completely oxidized to CO_2 and no other organic end-products were detectable. Heilmann and Logan (2006) studied electricity production from the well characterized standard purified protein bovine serum albumen (BSA) and from peptone as used in microbial growth media. Their results are compared with those from other substrates in Table 18.1, showing power densities of the order of that for butyrate. The CE for BSA was higher than that for peptone (11–21 *cf* 2–6%).

Nitrilotriacetic acid (NTA) is a chelating agent used in detergents and other industrial processes and is known to have implications for human health. It is 70–90% eliminated by activated sludge treatment, and is of environmental concern as it solubilizes heavy metals. NTA is known to be a substrate for *Shewanella* species and its degradation in MFCs was examined by Jang *et al.* (2006) operating continuously at HRT 2.2 h with defined substrates. Degradation of NTA (22 and 49 mg/l) was studied with 10 mg/l glucose/glutamate or 5 mmol/l acetate as co-substrates, giving 85–91% removal. At the higher co-substrate concentrations, NTA removal was around 20%. When NTA at 14, 30 and 49 mg/l was used as the sole organic fuel, current was generated in the MFC proportional to the NTA concentration. Thus under suitable conditions of low organic substrates (for example at the end of the treatment train), NTA may be removed by MFC operating at a short HRT with efficiencies comparing well with those for aerobic systems.

18.3 COMPLEX SUBSTRATES AND WASTEWATERS UTILIZED BY BES

Municipal, agricultural and industrial wastewaters contain organic matter that can be used as feedstocks for BES. Compared to synthetic substrates e.g. acetate, real wastewaters are reported to show decreased electrochemical performance in BESs (Rozendal *et al.* 2008) attributable to the low degradation rates, possibility of side reactions, e.g. methanogenesis and low conductivity. Table 18.1 in which wastewaters and pure substrates are compared in the same MFC system shows this to be true for meat packing, brewery and swine wastewaters but not for corn stover hydrolysates.

18.3.1 Cellulosic feedstocks

So far, to the authors' knowledge no microorganism is known to degrade cellulose and be exoelectrogenic. However, syntrophic microbial communities could accomplish this. Ren *et al.* (2007) used *Clostridium cellulolyticum* and *Geobacter sulfurreducens* to degrade soluble carboxymethyl cellulose with a power density comparable to acetate, and a particulate cellulose preparation containing amorphous and microcrystalline cellulose showing less than half this power density. Paper recycling wastewater was studied in batch mode by Huang and Logan (2008). In 350 hours with feedstock containing 1.2 g/l cellulose, 85% of the cellulose was removed if 50 mmol/l phosphate buffer was added compared to 16% with no addition. TCOD removal was 60% with buffer and 29% without. This difference was attributed mainly to the increase in conductivity, although without phosphate buffer addition the pH decreased to 5.9. The maximum power production increased with TCOD in the range tested of 181–1464 mg/l. In batch mode the longer batch cycle time at higher substrate concentrations is proposed to allow O_2 ingress and less efficient power generation. Information from continuous studies would be very useful.

Studies on H_2 production from cellulose using batch MECs gave a H_2 yield of 8.2 mol/mol hexose equivalent (68% of theoretical) at an applied voltage of 600 mV (Cheng and Logan 2007). However the H_2 production rate was very low (0.11 l/l/day) suggesting long cycle times reflective of the slow hydrolysis rate.

Much work worldwide is concentrated on lignocellulose degradation to a fermentable feedstock (Hendriks and Zeeman 2009), with the focus at present being on hexose production for ethanol fermentation. Pentoses are significant products of such pretreatment. In batch-fed studies, xylose at 0.5–9.7 mmol/l has been shown to be a suitable substrate for electricity generation by MFCs with COD removals around 24% (Huang *et al.* 2008). Hydrolysates arising from acid and neutral steam exploded pre-treatment of corn stover were examined by Zuo *et al.* (2006) with power densities shown in Table 18.1. CEs from 20–30% were reported. At 1000 mgCOD/l influent, around 20–30% of the colour was also removed by the MFC. An understanding of the inhibitory influences in BES of compounds such as furfural and hydroxymethylfurfural (HMF) is important.

18.3.2 Chitin

Chitin is the second most abundant material in the world derived from living matter after lignocellulose and widely available as a seafood industry co-product. Sediment microbial fuel cells, reliant on such detritus, could be useful to power remote sensors, but the organic matter content of many sediments is low. Rezaei *et al.* (2007) added crushed crab shells (20% chitin, 25% protein, 40% calcium carbonate, 15% water) or treated crab shells (95% chitin) to the sediment used in sediment MFCs. Power generation was significantly increased but there was variability in reactor performance. For autonomous robots which would eventually be built to obtain their own food from the environment, chitin, prawns and flies were successfully tested as substrates for the MFC power source (Melhuish *et al.* 2006).

18.3.3 Domestic wastewater

In comparison to pure substrates and other wastewaters tested, batch-fed studies using primary clarifier effluent from a domestic wastewater treatment plant (Liu and Logan 2004) showed almost the lowest power density (see Table 18.1). CE was 20% and BOD removal was 78%, giving a final BOD of 48 mg/l. In a continuously fed reactor, Liu *et al.* (2004) used primary clarifier effluent from a domestic wastewater treatment plant (see Table 18.4). A maximum power density of 26 mW/m^2 was achieved and BOD removal reached 78%. COD removal increased from 40% at HRT = 3 h to 80% at HRT = 33 h. However, a large amount of organic matter was removed by processes which did not directly contribute to electricity generation. CE ranged from 3 to 12% also indicating COD removal not associated with power generation. In another type of application, Kim *et al.* (2007b) operated an MFC as a BOD monitor continuously on a wastewater treatment facility (average BOD 35 mg/l), detecting diurnal changes in BOD content and responding to spiked toxicants (Pb, Hg, PCBs, organophosphate).

For H_2 production in MECs, primary clarifier effluent from a domestic wastewater treatment plant was used in batch 2-chambered MECs (see Table 18.2) (Ditzig *et al.* 2007), although the group report the design used was not as efficient as more recent systems (Logan *et al.* 2008). COD destruction was 90% which theoretically could give 1414 ml H_2/g COD consumed, but only 154 ml H_2/gCOD consumed was obtained.

18.3.4 Simulated and actual industrial wastewaters

Meat packing wastewater, arising at 6,010 mgCOD/l (approx 54% of COD as BOD), used as feedstock in an MFC after dilution, achieved a final BOD of

62 mg/l (Heilmann and Logan 2006). Swine wastewater at 8,320 mgSCOD/l (diluted 1:2 from initial wastewater) showed soluble COD removal of 88–92% while ammonia decreased from 198 to 34 mg/l (83% removal) (Min *et al.* 2005). A comparison of maximum power densities obtained is given in Table 18.1. Kim *et al.* (2008b) reported that mal odour mainly resulted from volatile organic acids is efficiently removed ($>99\%$) from swine wastewater in an MFC with electricity generation of 228 mW/m^2. The acetate, butyrate and propionate removal rate was increased in a closed circuit MFC as compared with a slow decrease in an open circuit MFC and consequently a slow removal of odourous chemicals. This work indicates that MFCs can remove a number of organic compounds which contribute to odour. Feng *et al.* (2008) used full strength beer brewery wastewater (COD 2,240 mg/l, BOD 1,340 mg/l, TSS 480 mg/l, conductivity 3.23 mS/cm) in the MFC type reported in Table 18.1. Adding 50 and 200 mmol/l phosphate buffer raised conductivity and increased power densities from 205 to 483 and 528 mW/m^2 and CE from 10% to 16% and 20% respectively. In the same reactor, CE for glucose was 9–20%, 10% for swine wastewater and 20% for domestic wastewater. For full strength beer brewery wastewater, the batch cycle time was 94 h, achieving 87% COD removal. HRTs of days rather than hours may be needed if an MFC was operated on undiluted wastewater. Prior fermentation of the wastewater to yield H_2 or methane may be preferable. Pilot scale reactors have been in operation in Australia on brewery wastewater.

Other wastewaters have been treated batch-wise in MFCs to meet discharge standards, but with long batch cycle times. Shimoyama *et al.* (2008) propose a cassette design of MFC such that the electrodes could be directly inserted into an anaerobic digester. A simulated wastewater containing starch, peptone and fish extract (289 gCOD/l) was used in fed-batch mode at an OLR of up to 5.8 gCOD/l/d, at which COD removal was 93% with an effluent COD of 21 mg/l. However the equivalent HRT was 50 days. Cereal processing plant wastewater diluted 10 times to 892 mgCOD/l gave a power density close to that obtained for acetate in the same MFC. The BOD was <30 mg/l BOD after approx. 450 h (Oh and Logan 2005).

Landfill leachate is a high COD effluent (5,000–35,000 mgCOD/l) rich in VFA and ammonium-N depending on the phase of degradation and with high conductivity. Landfills may be equipped for energy generation, based at present on methane recovery, thus locating MFCs at landfill sites may be acceptable to operators. Leachate diluted in the range 10–4,900 mg/l COD with the addition of phosphate was used in fed-batch operation in an MFC. CE decreased with increasing substrate concentration, again possibly as a result of O_2 ingress during the increased cycle times in these batch experiments (You *et al.* 2006b).

Continuously fed MFCs were studied by Greenman *et al.* (2009) using leachate (12,900–35,300 mg/l COD, BOD:COD between 0.34 and 0.57) in a range of dilutions to 1:8 and adjusted to pH 7 with phosphate buffer. BOD removal efficiency was up to 57%, with a maximal removal efficiency at OLR 0.81 kg BOD/m^3/d and HRT of 18.75 h. These results were similar to those from a biological aerated filter operated simultaneously but the MFC did not require power for aeration. HRTs in the range 5–38 h were examined and found to have a big influence on performance, data which is significant for process design. The amount of phosphate added was not discussed, but the need to adjust to a suitable COD:N:P is a feature of all biological treatment methods.

For H_2 production in MECs, glycerol, a waste product of biodiesel manufacture, was used in batch operation with 2-chamber MECs using 2 mmol/l thionine as mediator (Sakai and Yagishita 2007) (see Table 18.2). With increasing output of biodiesel, glycerol supply is expected to increase. Production of H_2 from glycerol-containing wastes in MECs may be an attractive treatment option. For the range of MEC substrates in Table 18.2, where information on substrate removal is given, with one exception removal was between 89 and 95%. Substrate concentrations used in MECs were generally around 130–1,300 mgCOD/l with one exception of 6,160 mgCOD/l (Table 18.2). Waste streams that contain high levels of ammonium e.g. animal waste water and urine could be another useful source of H_2 production using MECs. Such streams have been simulated using acetate as carbon source and variable amounts of ammonium phosphate in continuous two chambered MECs (Clauwaert *et al.* 2008). At an OLR of 2.56 gCOD/l anode compartment/day (2.2 hr HRT), applied voltage of 600 mV and 5 g/l ammonium nitrogen, a H_2 yield of 1.85 mol/mol acetate was obtained. Hydrogen production was severely hampered by ammonium nitrogen concentrations of 5.5 g/l. Waste streams containing high levels of ammonium nitrogen would thus have to be diluted down for effective MEC operation.

18.4 OTHER ASPECTS OF FEEDSTOCK COMPOSITION

Ionic strength can have a more important effect than wastewater strength on power output, as described for beer brewery wastewater in Section 18.3.4. It would aid research if, as suggested by Huang and Logan (2008), data on the conductivity of wastewaters was available. For example, the conductivity of settled paper recycling wastewater was 0.8 mS/cm, increased to 10.2 mS by addition of 100 mmol/l phosphate buffer which increased power density up to 245% (Huang and Logan 2008). For practical purposes this addition is expensive, there is no simple recovery method, and wastewater treatment aims to remove phosphate, e.g. to prevent eutrophication and to conserve this limited global resource. Fan *et al.* (2007) suggested substituting phosphate with the carbonate–bicarbonate buffer system based on CO_2 produced in the system, as commonly used in anaerobic digestion, and obtained good performance. Torres *et al.* (2008a) obtained a 45% increase in power density by providing acid to the cathode compartment by mixing CO_2 in the air supplied to an air cathode. Carbonate species then migrate to the anode compartment through an anion exchange membrane. The importance of the buffer to overall performance of a MEC has been demonstrated since current density was shown to be limited by proton transport out of the anode biofilm, that is, by transport of the associated form of the buffer (Torres *et al.* 2008b).

The theoretical minimum COD:N:P for highly loaded anaerobic digesters is 350:7:1 (Speece 1996). For BES receiving effluent as part of the process train N and P will have already been supplied, and for those operating on domestic wastewater and animal manures N and P will be plentiful and buffering capacity (alkalinity) may be sufficient. However, industrial wastes which are low in nutrients need to be enhanced e.g. through co-digestion. The addition of phosphate buffer to such a wastewater (paper recycling) much improved performance but the beneficial effect of P as a growth requirement was unquantified (Huang and Logan 2008). Other wastes likely to be deficient in P and N while rich in organic carbon include distillery effluents.

To meet consent levels for discharge to watercourse, nitrogen needs to be removed during treatment of wastewaters. Kim *et al.* (2008c) investigated ammonium–nitrogen loss in two chamber and single chamber MFCs using swine wastewater. Ammonium ions were transported from anode to cathode in a two-chamber MFC and this transport was increased with electricity generation. Oxidation and reduction reactions in anode and cathode chambers resulted in ion species transport including ammonium ($pK_a = 9.3$). No ammonia-oxidizing bacteria were found at the anode, but they were detected at the cathode. Therefore nitrogen loss driven by physico-chemical reactions should be considered in MFCs, along with biological ammonia oxidation at the cathode. Clauwaert *et al.* (2007) demonstrated the removal of nitrate (KNO_3 added as sole electron acceptor at 0.16 $kgN/m^3/d$ to the cathode compartment) by microorganisms at the cathode of an MFC operated on acetate. In the conventional anaerobic treatment process such as AD, ammonia-N cannot be removed efficiently. Therefore a prior aerobic nitrification step followed by denitrification (oxic/anoxic process) would be necessary, using an energy intensive aeration process. However recent results have shown that nitrogen removal can be combined with MFC technologies with simultaneous energy generation (Virdis *et al.* 2008).

18.5 FEEDSTOCKS AND BES INTEGRATION IN WASTEWATER TREATMENT PROCESSES

The usefulness of the integration of MFC/MEC with fermentative biohydrogen and conventional anaerobic digestion processes has been widely recognized (Pham *et al.* 2006; Rozendal *et al.* 2008). Since non-fermentable substrates appear to outperform those which are readily fermentable, more energy might be recovered if fermentable feedstocks went through acidogenic/H_2 producing/ methanogenic stages to produce H_2/CH_4 and VFA such as acetate before the effluent passed to a BES. The energy balance and cost implications of this should be evaluated. Information on the concentrations of VFA etc. arising in the effluents from these up-stream processes is important. Schröder *et al.* (2003) reported the operation of an MFC on effluent from a dark hydrogen-producing reactor where *E. coli* fermented 55 mmol/l (9.9g/l) glucose. The concentration of fermentation end products was not reported but from the stoichiometry for glucose fermentation for *E. coli* given by Niessen *et al.* (2004) there could have been up to 1.5 g/l ethanol and acetate and 3.1 g/l lactate.

If the effluent of a H_2-producing stage, usually with several grams of organic acids per litre, is fed to an anaerobic digester, lower levels of these acids (typically several hundred milligrams per litre) are obtained. Some examples are given in Table 18.3. Kyazze *et al.* (2007) showed acidogenesis of sucrose optimized for H_2 production (12 h HRT) followed by anaerobic digestion (48 h HRT) could operate continuously stably. A simulated potato wastewater when passed through a fermentative H_2-producing reactor (6 h HRT) and a methanogenic stage (90 h HRT) showed overall TCOD removal of 64% (Zhu *et al.* 2008). For food waste from a cafeteria, bench and field pilot-scale acidification/methanogenesis reactors showed 90% TCOD removal (Kim *et al.* 2008a).

Essential considerations for proper system design would be the characteristics of the wastewater feedstock, the OLR which can be applied and the discharge requirements for waste treatment. Table 18.4 summarizes OLRs which have been applied to BES at laboratory scale with various feedstocks. It is important to obtain information on what COD/BOD reductions can be achieved with such OLRs in order to site the BES in a suitable position in the wastewater treatment train.

It has recently been shown possible at laboratory scale to electrically couple an MFC to an MEC with no requirement for an external electricity supply, as the MFC provided the voltage for electrolysis (Sun *et al.* 2008). Using sodium acetate (100 mg/l) as substrate in a coupled batch MFC–MEC system, Sun *et al.* (2008) reported that a voltage of 348 mV can be applied to the MEC, giving a H_2 yield of 3.24 mol/mol acetate in the MEC and a yield of 1.60 mol/mol acetate for the whole coupled system. However the batch cycle time until complete exhaustion of acetate was long – of the order of 11.5 days – and therefore the system needs further optimization.

Table 18.3. Major organic components of some feedstocks after dark fermentative hydrogen production followed by anaerobic digestion

Feedstock (conc.)	Conc. after dark fermentative H2 reactor	Conc. after anaerobic digester	Reference
Sucrose (20 g/l)	acetate 2.29 g/l butyrate 4.48 g/l	acetate 0.52 g/l butyrate 1.37 g/l	Kyazze *et al.* (2007)
Potato waste (simulated) (12.6 g/l TCOD 2.22 g/l SCOD carbohydrates 573 mg/l)	acetate 1.87 g/l butyrate 1.37 g/l	acetate 0.44 g/l propionate 0.53 g/l	Zhu *et al.* (2008)
Cafeteria waste (71 g/l TCOD)	acetate 4.8 g/l lactate 13.4 g/l	acetate 0.55 g/l lactate up to 0.4 g/l	Kim *et al.* (2008a)

The K_s of a system, the half-saturation constant, indicates the substrate concentration at which the biomass is working at half of its maximum capacity. Although K_s values are often quoted for MFCs and MECs, it is important to note that these only represent true K_s values if it is substrate concentration which is rate limiting and not other parameters such as electrode potential (see Chapter 21 Section 21.6.1 Bioelectrode kinetics). The apparent K_s for organic substrates so far reported are generally low (see Table 18.5) and at low concentrations BES can consume and respond sensitively to substrates and out-compete methanogenesis from acetate. Hence BES systems could treat low levels of organic contaminants successfully, although methanogenesis as a side reaction may still occur due to interspecies hydrogen transfer.

Figure 18.1 suggests configurations which could supply substrates at suitable OLRs to BESs. Industrial wastewaters (typically >1000 mg/l COD) from agriculture and food processing plants which contain fermentable substrates could enter a dark H_2 fermentation stage followed by anaerobic digestion. Effluent from the anaerobic digester should contain total VFA in concentrations (<1000 mgCOD/l) that bioelectrochemical systems can handle. MFC/MEC technologies might be usefully applied as post-treatment processes to the AD effluent to meet discharge consents and simultaneously generate electricity, even at low COD concentration. Methane from AD could be converted into electricity

Table 18.4. Operational parameters reported for continuous BES

Substrate and reactor	Feed conc. mgCOD/l or BOD where stated	OLR gCOD/*/d or BOD where stated	HRT h or d where stated	% removal as COD unless otherwise stated	Reference
Acetate MFC/*G. sulfurreducens*	147	0.45	8	NG**	Tront *et al.* (2008)
Acetate MFC	770	3.3	5.6	NG	Aelterman *et al.* (2008)
Acetate MEC	640	1.12	13.75	NG	Rozendal *et al.* (2007)
Acetate MEC	723	1.78	10	87	Tartakovsky et al. (2008)
Glucose MEC	2170	3.52	10	NG	
Acetate MFC	NG	1	NG	NG	Freguia *et al.* (2007)
Glucose MFC	NG	1	NG		
Glucose/glutamate MFC	300	9.2	0.8	97	Moon *et al.* (2006)
Domestic wastewater MFC	210–220	3	3	40	Liu *et al.* (2004)
		0.75	12	70	
Domestic wastewater MFC	246	3	2	58	Min and Logan (2004)
	280	1.7	4	79	
Starch, peptone, fish extract fed-batch MFC	289g/l	2.9	96d	95	Shimoyama *et al.* (2008)
		5.8	50d	93	
Landfill leachate	630 mg/l BOD	0.81 BOD	18.75	57 BOD	Greenman *et al.* (2009)

*per litre of anode compartment **NG not given

Table 18.5. Apparent K_s values* reported for various substrates in BES with mixed cultures unless stated

Feedstock	Reported apparent K_s* (mg/l)	Reference
Acetate	141	Liu *et al.* (2005)
Butyrate	93	Liu *et al.* (2005)
Ethanol	225	Kim *et al.* (2007a)
Glucose	103	Liu and Logan (2004)
Glucose	204	Hu (2008)
Cysteine	460	Logan *et al.* (2005)
Xylose	44	Huang *et al.* (2008)
Corn stover hydrolysate	347 COD	Zuo *et al.* (2006)
neutral acid	170 COD	
Swine wastewater	1,510 TCOD	Min *et al.* (2005)
Domestic wastewater	461–719 COD	Min and Logan (2004)
Landfill leachate	252 COD	You *et al.* (2006a)
Beer brewery wastewater	115–144 COD	Feng *et al.* (2008)

*Note: Apparent K_s values are system-specific and so cannot be compared across systems or used to represent Monod kinetics.

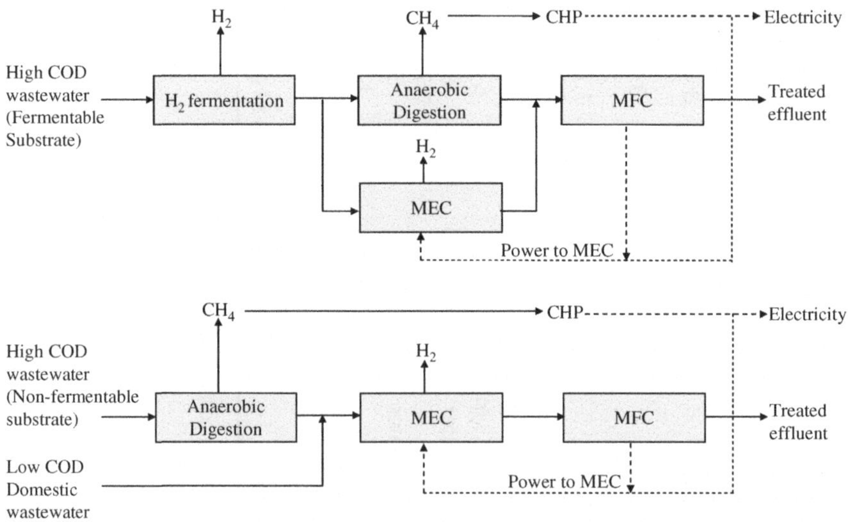

Figure 18.1. Use of dark hydrogen fermentation/anaerobic digestion to provide suitable substrate concentrations to bioelectrochemical systems (MFC and MEC) to optimize the biological conversion of feedstock to energy (H_2, methane, electricity) and give treated effluent. (Dashed lines – electrical power)

(e.g. using a combined heat and power (CHP) unit) and used in MECs for further H_2 generation. Low voltage electricity production by MFCs can be used to operate MECs. Potato wastewater (Table 18.3) could be suitably treated by this process train. Low strength feedstocks e.g. domestic wastewater (typically 200–300 mg/l COD) (Logan 2008) could be fed directly to an MFC or MEC with the objective of minimizing energy consumption and producing electricity/H_2. High COD feedstocks which are less readily available for fermentation or non-fermentable, e.g. lignocellulosic-containing animal manures and landfill leachate already high in VFA, could be passed directly to the anaerobic digester. The objective would be to extract useful energy vectors while meeting discharge standards. The integration of BES into wastewater treatment trains is a useful subject for further research and the following chapter expands on these principles.

18.6 CONCLUSIONS

A wide range of feedstocks has been shown capable of supporting microbial electricity generation by mixed cultures, including some industrial wastewaters without pretreatment. In comparative studies with the same reactor, pure substrates gave higher power densities than wastewaters. The CEs for fermentable substrates are considerably lower than that for acetate, probably due to increased methane and biomass production. In principle, various industrial, municipal and agricultural wastes and wastewaters should be suitable for H_2 production using MECs, allowing for distributed H_2 production. The yield of H_2 from natural wastewaters is low compared to simulated wastewaters. Each wastewater would have to be examined independently to assess its usability for H_2 production through MECs.

BES so far appear to have low apparent system K_s values and therefore may be better for treating low concentrations of organic contaminants than conventional anaerobic digestion processes. Feedstock concentrations applied to MFCs in the literature reviewed are generally in the range 0.1–1 g/l. The majority of these studies have been performed in batch or batch-fed reactors. There is a pressing need to extend these studies to continuous operation and to provide design data for scale-up and economic assessment. For example it appears that HRT may have a big influence on MFC performance.

The need for buffering capacity addition may be overcome by using the alkalinity source common to anaerobic digestion, CO_2–bicarbonate–carbonate. The influence of conductivity on performance may determine the wastewaters most suitable for BEC applications, since increasing conductivity is not possible in a treatment process without addition of a contaminant such as NaCl.

The feedstock concentration and OLRs which can be supported by BESs with good removal efficiency will determine the position of BESs in the treatment train. Since data so far suggests good substrate removal at the lower COD/BOD ranges, a positioning of BESs after an anaerobic digestion process seems suitable for agro- and food-industry wastewaters.

18.7 ACKNOWLEDGEMENTS

This work has been made possible by the UK Engineering and Physical Sciences Research Council's SUPERGEN programme funding the Biological Fuel Cells Consortium (grant 68-3A75-3-150, EP/E040071/1) and Sustainable Hydrogen Energy Consortium project (EP/D047943/1).

REFERENCES

ADM1 (2002) Anaerobic Digestion Model No. 1. IWA Task Group for mathematical modelling of anerobic digestion processes. IWA London.

Aelterman, P., Versichele, M., Marzorati, M., Boon, N. and Verstraete, W. (2008) Loading rate and external resistance control the electricity generation of microbial fuel cells with different three-dimensional anodes. *Bioresour. Technol.* **99**(18), 8895–8902.

Angenent, L.T. and Wrenn, B.A. (2008) Optimising mixed-culture bioprocessing to convert wastes into bioenergy. In *Bioenergy*. Wall, J.D., Harwood, C.S. and Demain, A., (eds). ASM Press, Washington, DC, pp. 179–193.

Biffinger, J.C., Byrd, J.N., Dudley, B.L. and Ringeisen, B.R. (2008) Oxygen exposure promotes fuel diversity for *Shewanella oneidensis* microbial fuel cells. *Biosens. Bioelectron.* **23**(6), 820–826.

Biffinger, J., Ribbens, M., Ringeisen, B., Pietron, J., Finkel, S. and Nealson, K. (2009) Characterization of electrochemically active bacteria utilizing a high-throughput voltage-based screening assay. *Biotechnol. Bioeng.* **102**(2), 436–444.

Call, D. and Logan, B.E. (2008) Hydrogen production in a single chamber microbial electrolysis cell lacking a membrane. *Environ. Sci. Tech.* **42**(9), 3401–3406.

Chae, K.J., Choi, M.J., Lee, J., Ajayi, F.F. and Kim, I.S. (2008) Biohydrogen production via biocatalyzed electrolysis in acetate-fed bioelectrochemical cells and microbial community analysis. *Int. J. Hydrogen Energy* **33**(19), 5184–5192.

Chang, I.S., Jang, J.K., Gil, G.C., Kim, M., Kim, H.J., Cho, B.W. and Kim, B.H. (2004) Continuous determination of biochemical oxygen demand using microbial fuel cell type biosensor. *Biosens. Bioelectron.* **19**(6), 607–613.

Cheng, S. and Logan, B.E. (2007) Sustainable and efficient biohydrogen production via electrohydrogenesis. *Proc. Natl. Acad. Sci. USA* **104**(47), 18871–18873.

Clauwaert, P., Rabaey, K., Aelterman, P., Deschamphelaire, L., Pham, T.H., Boeckx, P., Boon, N. and Verstraete, W. (2007) Biological denitrification in microbial fuel cells. *Environ. Sci. Technol.* **41**(9), 3354–3360.

Clauwaert, P., Toledo, R., Van Der Ha, D., Crab, R., Hu, H., Udert, K.M., Rabaey, K. and Verstraete, W. (2008). Combining biocatalysed electrolysis with anaerobic digestion. *Wat. Sci. Technol.* **57**(4) 574–579.

Ditzig, J., Liu, H. and Logan, B.E. (2007) Production of hydrogen from domestic wastewater using a bioelectrochemically assisted microbial reactor (BEAMR). *Int. J. Hydrogen Energy* **32**(13), 2296–2304.

Fan, Y., Hu, H. and Liu, H. (2007) Sustainable Power Generation in microbial fuel cells using bicarbonate buffer and proton transfer mechanisms. *Environ. Sci. Technol.* **41**(23), 8154–8158.

Feng, Y., Wang, X., Logan, B. and Lee, H. (2008) Brewery wastewater treatment using air-cathode microbial fuel cells. *Appl. Microbiol. Biotechnol.* **78**(5), 873–880.

Freguia, S., Rabaey, K., Yuan, Z. and Keller, J. (2007) Electron and carbon balances in microbial fuel cells reveal temporary bacterial storage behavior during electricity generation. *Environ. Sci. Technol.* **41**(8), 2915–2921.

Freguia, S., Rabaey, K., Yuan, Z. and Keller, J.R. (2008) Syntrophic processes drive the conversion of glucose in microbial fuel cell anodes. *Environ. Sci. Tech.* **42**(21), 7937–7943.

Greenman, J., Galvez, A., Giusti, L. and Ieropoulos, I. (2009) Electricity from landfill leachate using microbial fuel cells: comparison with a biological aerated filter. *Enzyme Microb. Technol.* **44**(2), 112–119.

Heilmann, J. and Logan, B.E. (2006) Production of electricity from proteins using a single chamber microbial fuel cell. *Water Environ. Res.* **78(5)**, 531–537.

Hendriks, A.T.W.M. and Zeeman, G. (2009) Pretreatments to enhance the digestibility of lignocellulosic biomass. *Bioresour. Technol.* **100**(1), 10–18.

Hu, Z. (2008) Electrochemical determination of anaerobic microbial decay coefficients. *Chemosphere* **72**(2), 312–318.

Huang, L. and Logan, B. (2008) Electricity generation and treatment of paper recycling wastewater using a microbial fuel cell. *Appl. Microbiol. Biotechnol.* **80**(2), 349–355.

Huang, L.P., Zeng, R.J. and Angelidaki, I. (2008) Electricity production from xylose using a mediator-less microbial fuel cell. *Bioresour. Technol.* **99**(10), 4178–4184.

Jang, J.K., Chang, I.S., Moon, H., Kang, K.H. and Kim, B.H. (2006) Nitrilotriacetic acid degradation under microbial fuel cell environment. *Biotechnol. Bioengng* **95**(4), 772–774.

Jung, S. and Regan, J. (2007) Comparison of anode bacterial communities and performance in microbial fuel cells with different electron donors. *Appl. Microbiol. Biotechnol.* **77**(2), 393–402.

Kim, J.K., Han, G.H., Oh, B.R., Chun, Y.N., Eom, C.Y. and Kim, S.W. (2008a) Volumetric scale-up of a three stage fermentation system for food waste treatment. *Bioresour. Technol.* **99**(10), 4394–4399.

Kim, J.R., Dec, J., Bruns, M.A. and Logan, B.E. (2008b) Removal of odors from swine wastewater by using microbial fuel cells. *Appl. Environ. Microbiol.* **74**(8), 2540–2543.

Kim, J.R., Jung, S.H., Regan, J.M. and Logan, B.E. (2007a) Electricity generation and microbial community analysis of alcohol powered microbial fuel cells. *Bioresour. Technol.* **98**(13), 2568–2577.

Kim, J.R., Min, B. and Logan, B.E. (2005) Evaluation of procedures to acclimate a microbial fuel cell for electricity production. *Appl. Microbiol. Biotechnol.* **68**(1), 23–30.

Kim, J.R., Zou, Y., Regan, J.M. and Logan, B.E. (2008c) Analysis of ammonia loss mechanisms in microbial fuel cells treating animal wastewater. *Biotechnol. Bioengng.* **99**(5), 1120–1127.

Kim, M., Hyun, M.S., Gadd, G.M. and Kim, H.J. (2007b) A novel biomonitoring system using microbial fuel cells. *J. Environ. Monitor.* **9**(12), 1323–1328.

Kyazze, G., Dinsdale, R., Guwy, A.I., Hawkes, F.R., Premier, G.C. and Hawkes, D.L. (2007) Performance characteristics of a two-stage dark fermentative system producing hydrogen and methane continuously. *Biotechnol. Bioengng.* **97**(4), 759–770.

Lee, H.-S., Parameswaran, P., Kato-Marcus, A., Torres, C.A.I. and Rittmann, B.E. (2008) Evaluation of energy-conversion efficiencies in microbial fuel cells (MFCs) utilizing fermentable and non-fermentable substrates. *Water Res.* **42**(6–7), 1501–1510.

Lin, W.C., Coppi, M.V. and Lovley, D.R. (2004) *Geobacter sulfurreducens* can grow with oxygen as a terminal electron acceptor. *Appl. Environ. Microbiol.* **70**(4), 2525–2528.

Liu, H., Cheng, S. and Logan, B.E. (2005) Production of electricity from acetate or butyrate using a single-chamber microbial fuel cell. *Environ. Sci. Technol.* **39**(2), 658–662.

Liu, H. and Logan, B.E. (2004) Electricity generation using an air-cathode single chamber microbial fuel cell in the presence and absence of a proton exchange membrane. *Environ. Sci. Technol.* **38**(14), 4040–4046.

Liu, H., Ramnarayanan, R. and Logan, B.E. (2004) Production of electricity during wastewater treatment using a single chamber microbial fuel cell. *Environ. Sci. Technol.* **38**(7), 2281–2285.

Liu, W.-Z., Wang, A.-J., Ren, N.-Q., Zhao, X.-Y., Liu, L.-H., Yu, Z.-G. and Lee, D.-J. (2008a) Electrochemically assisted biohydrogen production from acetate. *Energy and Fuels* **22**(1), 159–163.

Liu, W.Z., Wang, A.J., Ren, N.Q., Zhao, X.Y., Liu, L.H., Yu, Z.G. and Lee, D.J. (2008b) Electrochemically assisted biohydrogen production from acetate. *Energy Fuels* **22**(1), 159–163.

Logan, B.E. (2008) *Microbial Fuel Cells.* Wiley-Interscience, Hoboken, NJ.

Logan, B.E., Call, D., Cheng, S., Hamelers, H.V.M., Sleutels, T.H.J.A., Jeremiasse, A.W. and Rozendal, R.A. (2008) Microbial electrolysis cells for high yield hydrogen gas production from organic matter. *Environ. Sci. Technol.* **42**(23), 8630–8640.

Logan, B.E., Murano, C., Scott, K., Gray, N.D. and Head, I.M. (2005) Electricity generation from cysteine in a microbial fuel cell. *Water Res.* **39**(5), 942–952.

Lovley, D.R. and Nevin, K.P. (2008) Electricity production with electricigens. In *Bioenergy*. Wall, J.D., Harwood, C.S. and Demain, A., (eds). ASM Press, Washington, DC, pp. 295–306

Melhuish, C., Ieropoulos, I., Greenman, J. and Horsfield, I. (2006) Energetically autonomous robots: food for thought. *Auton.s Robots* **21**(3), 187–198.

Min, B., Kim, J.R., Oh, S., Regan, J.M. and Logan, B.E. (2005) Electricity generation from swine wastewater using microbial fuel cells. *Water Res.* **39**(20), 4961–4968.

Min, B. and Logan, B.E. (2004) Continuous electricity generation from domestic wastewater and organic substrates in a flat plate microbial fuel cell. *Environ. Sci. Technol.* **38**(21), 5809–5814.

Moon, H., Chang, I.S. and Kim, B.H. (2006) Continuous electricity production from artificial wastewater using a mediator-less microbial fuel cell. *Biores. Technol.* **97**(4), 621–627.

Nealson, K.H. and Scott, J. (2006) Ecophysiology of the genus Shewanella. In *The Prokaryotes: a Handbook on the Biology of Bacteria*. Dworkin, M. and Falkow, S., (eds). Springer, New York.

Niessen, J., Schröder, U., Rosenbaum, M. and Scholz, F. (2004) Fluorinated polyanilines as superior materials for electrocatalytic anodes in bacterial fuel cells. *Electrochem. Comm.* **6**(6), 571–575.

Oh, S. and Logan, B.E. (2005) Hydrogen and electricity production from a food processing wastewater using fermentation and microbial fuel cell technologies. *Water Res.* **39**(19), 4673–4682.

Pham, T.H., Rabaey, P., Aelterman, P., Clauwaert, L., Schamphelaire, D., Boon, N. and Verstraete, W. (2006) Microbial fuel cells in relation to conventional anaerobic digestion technology. *Engng Life Sci.* **6**(3), 285–292.

Ren, Z., Ward, T.E. and Regan, J.M. (2007) Electricity production from cellulose in a microbial fuel cell using a defined binary culture. *Environ. Sci. Technol.* **41**(13), 4781–4786.

Rezaei, F., Richard, T.L., Brennan, R.A. and Logan, B.E. (2007) Substrate-enhanced microbial fuel cells for improved remote power generation from sediment-based systems. *Environ. Sci. Technol.* **41**(11), 4053–4058.

Richter, H., Lanthier, M., Nevin, K.P. and Lovley, D.R. (2007) Lack of electricity production by *Pelobacter carbinolicus* indicates that the capacity for Fe(III) oxide reduction does not necessarily confer electron transfer ability to fuel cell anodes. *Appl. Environ. Microbiol.* **73**(16), 5347–5353.

Rozendal, R.A., Hamelers, H.V.M., Euverink, G.J.W., Metz, S.J. and Buisman, C.J.N. (2006) Principle and perspectives of hydrogen production through biocatalyzed electrolysis. *Int. J. Hydrogen Energy* **31**(12), 1632–1640.

Rozendal, R.A., Hamelers, H.V.M., Molenkmp, R.J. and Buisman, J.N. (2007) Performance of single chamber biocatalyzed electrolysis with different types of ion exchange membranes. *Water Res.* **41**(9), 1984–1994.

Rozendal, R.A., Hamelers, H.V.M., Rabaey, K., Keller, J. and Buisman, C.J.N. (2008) Towards practical implementation of bioelectrochemical wastewater treatment. *Trends Biotechnol.* **26**(8), 450–459.

Sakai, S. and Yagishita, T. (2007) Microbial production of hydrogen and ethanol from glycerol-containing wastes discharged from a biodiesel fuel production plant in a bioelectrochemical reactor with thionine. *Biotechnol. Bioengng* **98**(2), 340–348.

Schröder, U., Niessen, J. and Scholz, F. (2003) A generation of microbial fuel cells with current outputs boosted by more than one order of magnitude. *Angewandte Chemie-Int. Edn.* **42**(25), 2880–2883.

Shimoyama, T., Komukai, S., Yamazawa, A., Ueno, Y., Logan, B.E. and Watanabe, K. (2008) Electricity generation from model organic wastewater in a cassette-electrode microbial fuel cell. *Appl. Microbiol. Biotechnol.* **80**(2), 325–330.

Speece, R.E. (1996) *Anaerobic Biotechnology for Industrial Wastewaters.* Archae Press, Nashville.

Sun, M., Sheng, G.P., Zhang, L., Xia, C.R., Mu, Z.X., Liu, X.W., Wang, H.L., Yu, H.Q., Qi, R., Yu, T. and Yang, M. (2008) An MEC-MR-coupled system for biohydrogen production from acetate. *Environ. Sci. Technol.* **42**(21), 8095–8100.

Tartakovsky, B., Manuel, M.F., Neburchilov, V., Wang, H. and Guiot, S.R. (2008) Biocatalyzed hydrogen production in a continuous flow microbial fuel cell with a gas phase cathode. *J. Power Sources* **182**(1), 291–297.

Torres, C., Kato Marcus, A. and Rittmann, B. (2007) Kinetics of consumption of fermentation products by anode-respiring bacteria. *Appl. Microbiol. Biotechnol.* **77**(3), 689–697.

Torres, C., Lee, H.-S. and Rittmenn, B. (2008a) Carbonate species as OH carriers for decreasing the pH gradient between cathode and anode in biological fuel cells. *Environ. Sci. Technol.* **42**, 8773–8777.

Torres, C.I., Marcus, A.K. and Rittmann, B.E. (2008b) Proton transport inside the biofilm limits electrical current generation by anode-respiring bacteria. *Biotechnol. Bioengng* **100**(5), 872–881.

Tront, J.M., Fortner, J.D., Plotze, M., Hughes, J.B. and Puzrin, A.M. (2008) Microbial fuel cell biosensor for in situ assessment of microbial activity. *Biosen. Bioelectron.* **24**(4), 586–590.

Virdis, B., Rabaey, K., Yaun, Z. and Keller, J. (2008) Microbial fuel cells for simultaneous carbon and nitrogen removal. . *Water Res.* **32**, 3031–3024

You, S., Zhao, Q., Zhang, J., Jiang, J. and Zhao, S. (2006a) A microbial fuel cell using permanganate as the cathodic electron acceptor. *J. Power Sources* **162**(2), 1409–1415.

You, S.J., Zhao, Q.L., Jiang, J.Q., Zhang, J.N. and Zhao, S.Q. (2006b) Sustainable approach for leachate treatment: Electricity generation in microbial fuel cell. *Journal of Environmental Science and Health Part a-Toxic/Hazardous Substances & Environmental Engineering* **41**(12), 2721–2734.

Zhu, H., Stadnyk, A., Beland, M. and Seto, P. (2008) Co-production of hydrogen and methane from potato waste using a two-stage anaerobic digestion process. *Bioresour. Technol.* **99**(11), 5078–5084.

Zuo, Y., Maness, P.C. and Logan, B.E. (2006) Electricity production from steam-exploded corn stover biomass. *Energy Fuels* **20**(4), 1716–1721.

19

Integrating BES in the wastewater and sludge treatment line

Miriam Rosenbaum, Matthew T. Agler, Jeffrey J. Fornero, Arvind Venkataraman and Largus T. Angenent

19.1 INTRODUCTION

Traditionally, wastewater treatment has been a combination of physical, biological, and chemical treatment methods. Physical methods of treatment include settling and sand filtration, biological methods include aerobic activated sludge, denitrification, and anaerobic digester treatment systems, while chemical methods include phosphate removal, dye and endocrine disruptor removal, and disinfection systems. All of these methods have been adopted and optimized to

© 2010 IWA Publishing. *Bioelectrochemical Systems: From Extracellular Electron Transfer to Biotechnological Application*. Edited by Korneel Rabaey, Largus Angenent, Uwe Schröder and Jürg Keller. ISBN: 9781843392330. Published by IWA Publishing, London, UK.

raise the quality and clarity of the final effluent stream and to minimize the quantity and severity of pollutants and harmful pathogens to be released into the environment. Biological oxygen demand (BOD) and chemical oxygen demand (COD) levels in untreated municipal wastewater are typically low and the use of aerobic activated sludge systems has become a conventional treatment method in developed nations to reduce these levels even further to meet minimum quality standards (Tchobanoglous *et al.* 2003). Activated sludge systems rely on oxygen transfer from air into the mixed liquor, which conventionally is accomplished by large blowers that are energy intensive. Burton (1996) has estimated that aeration in activated sludge systems may account for 1% of all electricity consumed in the US. Thus, conventional treatment systems are not sustainable.

To make matters worse, activated sludge systems mainly convert soluble constituents to suspended biomass (i.e. waste activated sludge [WAS]), which can then be removed from the water by physical methods, such as settling. As a consequence, vast quantities of sludge accumulate in the treatment line and must be further treated or removed. Biomass in sludges can bio-accumulate toxic substances from the influent, making sludge treatment or biosolids reuse complicated (Rulkens 2008). Various methods have been proposed and utilized for circumventing the disposal of WAS, including treatment of influent or the accumulated sludge with anaerobic digesters (Rulkens 2008; Seghezzo *et al.* 1998). This is a more sustainable treatment method because methane in biogas can partly offset the energy consumption of conventional wastewater treatment. With the advent of BES we can now potentially transform chemical energy stored in organic substrates directly to electrical energy through bacterial metabolism even at conditions outside the optimum range for anaerobic digestion (Angenent *et al.* 2004; Logan 2005).

Here, we will discuss how BES may be integrated in the wastewater and sludge treatment line with the overarching goal to increase the sustainability of waste treatment. Each section represents a description of one or two plant flow diagram(s) in which a BES is utilized, which is also visualized in Figure 19.1 A–G. Each section heading indicates which flow diagram will be discussed (A–G). We realize that wastewater must be prepared before entering BES by, for example, screens, dissolved air flotation, or settling basins to remove solids. In addition, a buffer tank may be necessary to control pH (e.g. by recirculating cathode effluent) and to equalize fluctuations in wastewater flow and strength. We have omitted these pre-treatment steps in the flow diagrams (except for when the buffer tank is used as a preacidification tank). Many more variations of flow diagrams that include BES are possible, and our goal is merely to stimulate the reader's imagination.

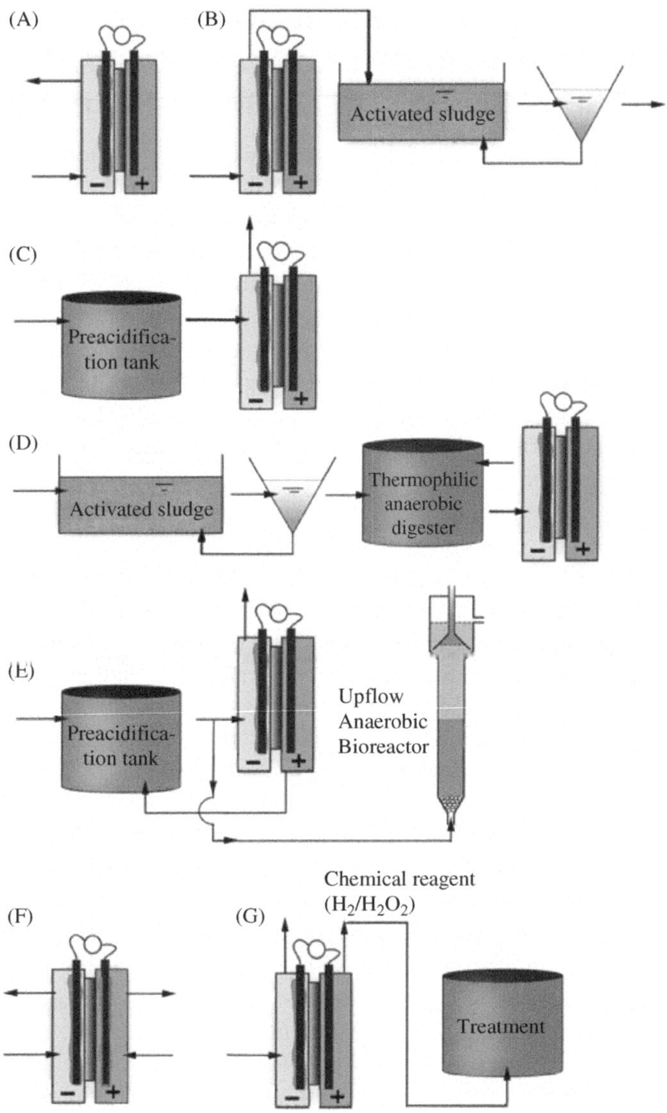

Figure 19.1. Flow diagrams of wastewater and sludge treatment line that incorporate BES. Each section in this chapter describes one or two of these alternatives (A–G)

19.2 BES AS THE SINGLE BIOLOGICAL TREATMENT UNIT (A) OR FOLLOWED BY AN ACTIVATED SLUDGE SYTEM AS A POLISHING STEP (B)

The question of where to apply BES systems within the wastewater treatment line can be addressed through consideration of the anodic and cathodic bacterial metabolisms. Most research has been performed on bacterial metabolism in the anode compartment of BES where organic constituents can be oxidized with an electrode as the electron accepter. Undefined mixed cultures that include electrode-respiring bacteria can metabolize a wide variety of organic substrates, including simple sugars, lipids, proteins, polysaccharides, and fatty acids (He *et al.* 2005; Heilmann and Logan 2006b; Huang and Logan 2008; Huang *et al.* 2008; Liu *et al.* 2005a; Min *et al.* 2005; Rabaey *et al.* 2003). Because of the metabolic diversity of undefined mixed cultures, BES systems have the flexibility to contribute to wastewater COD reduction in a variety of locations in the treatment line. Although oxygen may be the terminal electron acceptor at the cathode of BES, which is typical for microbial fuel cells (MFCs), most biochemical processes in the anode chamber proceed close to thermodynamic equilibrium, leaving little excess energy for biomass growth. The advantages of producing lower quantities of excess sludge compared to activated sludge systems and a potential for electricity generation (rather than energy consumption) make the technology attractive for secondary wastewater treatment (MFC instead of activated sludge) or as a pretreatment step to remove the bulk of organic constituents before activated sludge systems (Figure 19.1A and B).

The amount of electric power that can be generated from wastewater using BES depends on the number of electrons harvested from the oxidation of organic substrates within the wastewater. This number of electrons harvested is directly proportional to the COD removed and expressed as coulombic efficiency (CE) (Logan *et al.* 2006). In other words, the CE calculation evaluates the number of electrons captured versus the total number of electron equivalents released during the oxidation of wastewater constituents. At constant CEs, a higher COD concentration of the wastewater would result in a greater current density. Since the COD concentration of the wastewater flow decreases through the treatment line (not taking the sludge line into consideration), the introduction of BES at the start would be logical. However, CE levels for real wastewater treatment are generally low (Table 19.1). The relatively high CE levels for MFCs have been achieved with synthetic wastewater that was rich in carboxylic acid substrates (Table 19.2), such as acetate, which are directly usable by bacteria (Brock 1966).

Table 19.1. The effect of wastewater type and concentration on the coulombic efficiency for MFCs. The table is ranked based on the coulombic efficiency

Wastewater substrate	COD (mg/l)	Coulombic efficiency (%)	Reference
Paper Recycling	723	22	(Huang and Logan 2008)
Municipal	429	20	(Rabaey *et al.* 2005)
Beer Brewery	2,240	10	(Feng *et al.* 2008)
Swine Waste	8,320	8	(Min *et al.* 2005)
Municipal	379	6	(Min and Logan 2004)
Slaughterhouse	1,420	5	(Heilmann and Logan 2006)

Table 19.2. The effect of carboxylic acid substrate and concentration on the coulombic efficiency for MFCs. The Table is ranked based on the coulombic efficiency

Carboxylic acid substrate	COD (mg/l)	Coulombic efficiency (%)	Reference
Acetate	640	100	(Nevin *et al.* 2008)
Acetate	458	98	(Rabaey *et al.* 2005)
Acetate	64	94	(Bond and Lovley 2003)
Acetate	1,000	65	(Min and Logan 2004)
Butyrate	1,100	50	(Min and Logan 2004)
Acetate	800	29	(Liu *et al.* 2005a)

Real wastewater consists of many constituents (either soluble or suspended) that are not directly usable by bacteria and that must be first hydrolyzed and digested in a food web. Examples include particulate and colloidal substrates, and hydrolytic processes to solubilize these constituents to easy-to-metabolize substrates are often rate limiting (Dimock and Morgenroth 2006; Ginestet *et al.* 2002). This results in a complex food web in the anodic compartment that will expend significant amounts of energy and time – energy that can never be recovered as electric power and time that results in the requirement of large reactor volumes – for microbial breakdown. Thus, significantly lower CEs for complex wastewaters that involve complex microbial food webs are anticipated (Marcus *et al.* 2007). Many bacteria in such a food web are, thus, not directly involved with electrode reduction (Liu *et al.* 2005a; Min and Logan 2004; Rabaey *et al.* 2005), which may lead to crowding of electrode-respiring bacteria that often require direct interaction with the anode. We have observed the formation of a thick biofilm on the electrodes of a 6-L MFC treating real brewery wastewater with vast quantities of suspended solids (data not shown).

This resulted in methane formation in the outer biofilm layers due to insulation of the electrode and the accumulation of particulate matter from the wastewater inside the anode chamber. Rabaey *et al.* (2005) also observed this latter phenomenon. A solution to this problem may be to screen solids more efficiently and to place a preacidification step before BES (Section 19.3), which is a conventional treatment step before high-rate anaerobic treatment of industrial wastewaters, such as brewery wastewater.

It is highly unlikely that oxidation of organic wastewater constituents in the anode chamber of BES will ever be a stand-alone technology (Figure 19.1A). First, the effluent BOD concentration may still be higher than the local regulations allow. Second, the anaerobic conditions of the anode chamber do not fulfill the requirements to support biological nutrient removal. In the case of nitrification/denitrification, a separate aerated reactor or zone is necessary to convert ammonia-N into nitrate-N (nitrification). However, denitrification may take place in BES as a cathodic process (Clauwaert *et al.* 2007) – see Section 19.6. In the case of phosphorus removal, no report has evaluated if this process occurs in BES. Some phosphate may diffuse from the anode chamber to the cathode chamber when an anion exchange membrane is present, but this is against the ionic forces and unlikely to be sufficient to meet discharge requirements. Thus, an aerobic/anoxic polishing step after BES may be required similar to conventional anaerobic digester implementation in the treatment line. Here, this is envisioned as an activated sludge process in the flow diagram after BES (Figure 19.1B).

19.3 PREACIDIFICATION OF ORGANIC WASTEWATER BEFORE BES (C)

Very high CEs have been achieved with MFCs for simple, easy-to-metabolize, organic substrates, such as acetate (Table 19.2). These higher CEs were reported because the anodic food web does not need to support hydrolysis and digestion of high-molecular-weight constituents compared to anodic food webs in MFCs treating complex wastewater. Rather, electrode-respiring bacteria can readily ingest and oxidize the simple organic substrates, regenerate reducing equivalents with anaerobic respiration processes, and deposit electrons to the anode electrode (Gregory *et al.* 2004; Lovley 2006, 2008). Thus, aside from meeting cell maintenance and growth energy requirements, the bacteria can efficiently transform the simple organic substrates into electricity. This will result in a less-complex food web, lower biomass concentrations, and a higher concentration of electrode-respiring bacteria in the anode. We can enforce such an efficient BES by outsourcing the rate-limiting hydrolysis and digestion steps into a preacidification tank (Figure 19.1C). We anticipate that feeding mostly carboxylic acids rather than complex wastewater will increase CEs and kinetic rates, resulting in much smaller anode volumes.

Wastewater preacidification is a process commonly used to accelerate soluble organic constituent conversion and some hydrolysis. It is a required pretreatment step before upflow anaerobic digesters to prevent excessive growth of filamentous bacteria in the upflow bioreactor at high food-to-microorganism ratios, which can cause bulking of granular sludge (Alphenaar 1994). However, it must be realized that solids in the preacidification tank effluent can be high, which necessitates the use of a solids separation unit between the preacidification unit and BES. The residence time in preacidification tanks are typically short enough to prevent methanogenesis and to only promote carboxylic acid formation (i.e. acetate, butyrate, and propionate) from anaerobic microbial conversion of soluble organic constituents, such as sugars and proteins. Hydrogen and carbon dioxide gases are by-products of preacidification, and carbon dioxide increases the alkalinity of the wastewater by solution-gas equilibrium, limiting (to some extent) the pH drop associated with the formation of the carboxylic acids (often the pH level in the preacidification tank still must be equalized by, for example, circulating cathodic effluent; see Section 19.5). Finally, preacidification does slightly decrease the COD concentration of the wastewater, but this disadvantage will be greatly compensated for by the ease of carboxylic acid to electric current conversion (Figure 19.1C).

19.4 ANAEROBIC DIGESTERS FOR SLUDGE STABILIZATION FOLLOWED BY BES (D)

We have introduced WAS above, which is the biological product of activated sludge systems. Primary sludge is separated from wastewater up front of the treatment line by mechanical filtration and/or settling (Tchobanoglous *et al.* 2003). Often these two sludge streams are combined for further treatment in anaerobic digesters to lower the amount of biosolids and to recover energy (Kalyuzhnyi 2008; Rulkens 2008). For the purposes of this chapter, we will just focus on WAS treatment by a two-step process that consists of a thermophilic anaerobic digester system followed by BES (a combination of primary sludge and WAS could also be an excellent substrate for such a process). WAS is fed into the first phase in which hydrolysis removes solids and methane is formed. Next, separated carboxylic acids from the digestion effluent are introduced into the anode of the BES (e.g. MFC) to be converted into current (Figure 19.1D).

A lab-scale system that included anaerobic digester effluent flow into the anode of a MFC was reported by De Schamphelaire and Verstraete (2009).

Because much of WAS is particulate cellular material, thermophilic temperatures (55°), which increase the rate of solubilization, have been reported to double methane production at an 8-day HRT compared to mesophilic temperatures (35–37°), and achieve volatile solids (VS) removal efficiencies up to 40% (Lafitte-Trouque and Forster 2002). Thermophilic pretreatment of WAS with anaerobic digesters would also allow for sufficient pathogen removal to obtain biosolids that can be more easily dispersed into the environment (class A biosolids) (Han and Dague 1997; Welper *et al.* 1997). However, high temperatures contribute to increased accumulation of carboxylic acids (also often referred to as volatile fatty acids or short-chain fatty acids) due to higher concentrations of free ammonia and propionate, which can inhibit methanogenesis (Bocher *et al.* 2008; Lier 1996; Lier *et al.* 1993). In the past, this problem has been tackled by placing additional thermophilic methanogenic reactors/compartments in series or by placing a mesophilic anaerobic digester after the thermophilic unit to remove propionate. With the advent of BES, an ideal polishing step can convert carboxylic acids efficiently into current without limitations, such as hydrolysis and digestion, which were described in Section 19.3 (Pham *et al.* 2006; Rulkens 2008) (Figure 19.1D).

We performed a simple estimation of the potential for energy production and sludge reduction with thermophilic anaerobic digestion of WAS followed by polishing with BES (Figure 19.2). A typical WAS stream contains 90% particulate COD and 10% soluble COD. Here, we assume 40% hydrolysis of particulate COD from WAS in the anaerobic digester (Mahmoud *et al.* 2004), which is a conservative estimate under thermophilic conditions. Assuming a 90% conversion of hydrolyzed soluble COD and influent soluble COD to methane (for a thermophilic reactor), 4.6% of the original influent COD (particulate and soluble COD) is left as soluble COD in the digester effluent. A typical WAS stream contains about 30,000 mg total COD/l, and therefore a digester effluent stream with 1,380 mg soluble COD/l (mainly as acetate and propionate) could be separated from total solids and fed to BES. With an average 65% CE for carboxylic acids (Table 19.2), we predict a total of ~ 900 mg SCOD/l going to electric current in BES (3% of the COD fed). Under these assumptions, the digester would convert 41.4% of the total COD present into methane. Then, the total energy recovery from WAS may be improved by 7.2% by integrating BES in the sludge line when compared to thermophilic anaerobic digestion alone.

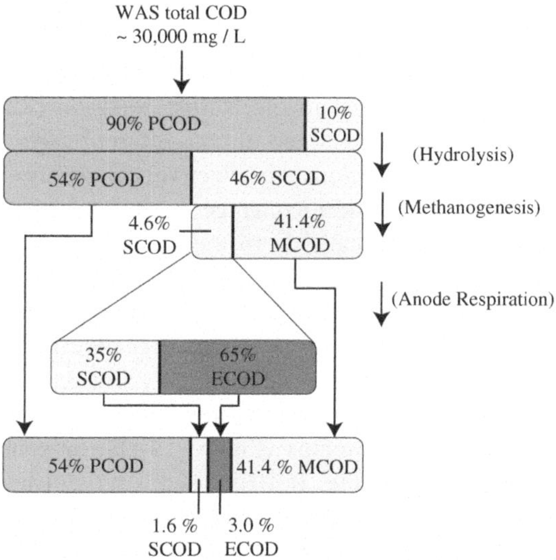

Figure 19.2. Estimation of the COD balance for a two-step process of WAS treatment. The top bar represents the relative COD constituents in WAS. The outgoing relative COD streams (lower bar) are methane (MCOD) generated in the anaerobic digestion process, electric current (ECOD) generated in BES, and left-over particulate (PCOD) and soluble (SCOD) material

19.5 GENERATING CAUSTIC IN THE CATHODE OF BES TO CONTROL ANAEROBIC DIGESTER pH (E)

BES systems balance the electron flow from anode to cathode by transport of ions across the ion exchange membrane to maintain electroneutrality. Unlike proton exchange membrane fuel cells, BESs operate in a near neutral pH range, and therefore electroneutrality is maintained with ions other than protons or hydroxide ions (Zhao *et al.* 2006). Because protons or hydroxide ions are not involved in BES electroneutrality, cathode pH increases are experienced (Fornero *et al.* 2008; Rozendal *et al.* 2006a; Zhao *et al.* 2006). The cathode pH increase results from the cathode oxygen reduction reaction, which can be represented either as consuming protons or generating hydroxide ions (Larminie and Dicks 2003).

$$O_2 + 4\,e^- + 4\,H^+ \rightarrow 2\,H_2O \text{ (Proton consumption)}$$

$$O_2 + 4\,e^- + 2\,H_2O \rightarrow 4\,OH^- \text{ (Hydroxide generation)}$$

Regardless of how the cathode oxygen reduction stoichiometry is represented, the net effect is an increase in the cathode pH, resulting in a pH gradient across the ion exchange membrane, which negatively impacts the BES potential by 0.059 V/pH according to the Nernst equation (Zhao *et al.* 2006). Also, electroneutrality can be maintained in two different ways: 1. with a cation exchange membrane for which cations are selectively transferred from the anode to the cathode; and 2. with an anion exchange membrane for which anions are selectively transferred from the cathode to the anode. It is now clear that the cathode pH increase occurs whether a cation or anion exchange membrane is used (Fornero *et al.* 2008). Thus, the catholyte pH requires attention independent of the ion exchange membrane selection (Fornero *et al.* 2008).

While an alkaline BES catholyte can negatively impact electric power, the accumulation of caustic in the catholyte can be beneficial to buffer preacidification of wastewater streams (Figure 19.1E) or some other process in the treatment line. For example, if BES uses a replenished catholyte, the alkaline catholyte produced by the oxygen reduction reactions can be discharged into the preacidification tank to partly compensate the pH decline from the production of carboxylic acids (see Section 19.3 for details on preacidification). Thus, a BES system used in parallel with high-rate anaerobic digestion can treat part of the preacidified wastewater stream and control the pH without the need of external alkalinity (Figure 19.1E). This may present a nice cost saving for the industry that is treating its wastewater by anaerobic biotechnology.

19.6 DENITRIFICATON IN THE CATHODE OF BES TO REMOVE NUTRIENTS FROM WATER (F)

Wastewater treatment systems typically not only employ biological treatment units to decrease the amount of organic carbon (activated sludge treatment and sludge conversion processes), but also to remove nutrients, such as soluble nitrogen. Conventionally, nitrogen is removed as molecular nitrogen in a two-step process: 1. reduced nitrogen species (i.e. ammonium) are oxidized by autotrophic nitrifying bacteria; and 2. generated nitrate is reduced to molecular nitrogen by heterotrophic denitrifying bacteria. Since denitrification is a reductive process, it can be coupled to reduction reactions with the cathode as an electron donor and microorganisms as biocatalysts (Figure 19.1F). For this cathodic process, the microbes perform autotrophic denitrification rather than the more conventional heterotrophic denitrification. The main advantage of using a biocathode to reduce nitrate is the absence of noble metal catalysts on the cathode (Clauwaert *et al.* 2007; He and Angenent 2006; Jia *et al.* 2008;

Virdis *et al.* 2008). In the wastewater treatment line, nitrate can originate from different sources. Examples include nitrate-rich industrial wastewater or ground water (Clauwaert *et al.* 2007); nitrate generation in activated sludge systems or other separate operation units from ammonium-rich waste streams; and nitrate generation in aerated compartments from ammonium-rich anode effluent from BES (Virdis *et al.* 2008). In the latter alternative, the anolyte becomes the catholyte after an additional treatment step. Then, the aeration and cathodic processes can act as additional COD polishing steps after anodic wastewater treatment. This may nearly eliminate the residual COD concentration of the anode effluent (Virdis *et al.* 2008; Fornero *et al.* 2008).

Using a denitrifying cathode system is advantageous for both the wastewater treatment process and BES beyond the circumvention of noble metal catalysts – anaerobic denitrifying cathodes omit the aeration of the cathode and/or the catholyte and, as a result, can prevent oxygen crossover from the cathode to the anode with negative effects on CE. Even though the theoretical potential difference for the organic substrate/nitrate couple is lower than the organic substrate/O_2 couple (Clauwaert *et al.* 2007), solubility and catalytic advantages of nitrate compared to oxygen may guarantee a comparable measured potential difference.

In this section we have discussed that an autotrophic denitrification process can occur in the cathode with the electrode as the electron donor (Figure 19.1F). As part of an additional flow diagram, we have envisioned molecular hydrogen produced in BES cathodes as the electron donor for autotrophic denitrification in a separate process unit (Section 19.7). This may be the necessary flow diagram when the anolyte cannot become the catholyte due to incompatibilities in regard to, for example, conductivity. Regardless of the most advantageous configuration, the resulting autotrophic denitrification requires a much lower carbon/nitrogen ratio than conventional heterotrophic denitrification (step two for nitrogen removal mentioned above), which is a tangible advantage because alternative carbon sources, such as methanol, do not have to be procured. Additional advantages of autotrophic versus heterotrophic denitrification are: 1. prevention of organic carbon carryover to the final effluent, and 2. decrease in biomass yield, resulting in lower biosolids production (Rezania *et al.* 2007).

19.7 GENERATING CHEMICAL REAGENTS AT CATHODES FOR TREATMENT PURPOSES (G)

A recent development in the field of BES research has been the transition from power to chemical producing systems in which energy released from anodes is stored as chemicals in the cathode. It is possible to integrate the production of some of these specific chemicals in the wastewater treatment line (Figure 19.1G). Here, the production of molecular hydrogen and hydrogen peroxide species and their application for the treatment process will be discussed:

Molecular hydrogen can be produced at the cathode of BESs by addition of a small voltage ($>$ 0.2 V in practice) and maintaining an anaerobic cathode (Liu *et al.* 2005b; Rozendal *et al.* 2006b), according to the following reaction:

$2\,e^- + 2\,H^+ \rightarrow H_2$ (Hydrogen generation)

This invention, which has also been called bio-catalyzed electrolysis, can produce hydrogen gas that can be utilized for wastewater denitrification purposes. With BES, the hydrogen required for denitrification can, thus, be produced *in situ*, which eliminates acquisition, shipment, and storage of hydrogen. Two process configurations can be used: 1. a single reactor configuration in which hydrogen generation and denitrification take place simultaneously in the cathodic chamber of BES (note the difference with the biocathode system – here hydrogen rather than the electrode is the electron donor); and 2. a two-step system where hydrogen is produced in a BES cathode and fed to an anoxic denitrification tank (Kiss *et al.* 2000) (Figure 19.1G). It is likely that the latter (two-step) process would be the preferred configuration due to microbial poisoning of the metal catalyst on the cathode.

Hydrogen peroxide can also be electrogenerated by incomplete cathodic reduction of dissolved oxygen (Brillas and Casado 2002; Duran Moreno *et al.* 2004; Figueroa *et al.* 2009). The H_2O_2 produced *in situ* can be activated by soluble Fe^{2+} to produce a strong oxidant (Fenton's reagent). This can be used for removal of hard-to-degrade wastewater components, such as dyes in textile wastewater (Figueroa *et al.* 2009) and environmentally toxic anilines (Brillas and Casado 2002). The choice of electrode material will be a key factor for the electro-production of hydrogen peroxide due to the sluggish kinetics, however, this process is considered a highly promising technique to deal with challenging contaminants. Thus far, only abiotic electrochemical cells have been employed to generate hydrogen peroxide, but a bioelectrochemical system that would already be integrated in other wastewater treatment processes (i.e. anodic processes), would be a perfect *in situ* generator of hydrogen peroxide at the cathode (Figure 19.1G).

19.8 OUTLOOK

BES technology integrated in the wastewater and sludge treatment line did not exist at full scale when this chapter was written. Even though, at the time, these systems may have had potential to accomplish a more sustainable wastewater treatment system, they may not (have) come to fruition due to scale-up problems and a nonlinear power (rate) drop with increasing reactor volumes (Dewan *et al.* 2008). On the other hand, problems are there to be solved and BES may find a niche beyond our present-day imagination. Generation of high-value products from the electrons stored in the organic constituents of wastewater may make BES sustainable, economical, and beneficial in wastewater and sludge treatment lines. In addition, lowering the carboxylic acids concentrations in thermophilic digester effluent with BESs, such as MFCs, may present a logical and feasible alternative due to the presence of high concentrations of easy-to-metabolize substrate. Gaining all the advantages of BESs by combining, for example, flow diagrams D and G may be necessary to guarantee success.

19.9 ACKNOWLEDGEMENTS

This work was supported by a US NSF CAREER grant to LTA with number 0645021.

REFERENCES

Alphenaar, A. (1994) Anaerobic granular sludge: characterization, and factors affecting its functioning. Ph. D. thesis, Wageningen Agricultural University, Wageningen, The Netherlands.

Angenent, L.T., Karim, K., Al-Dahhan, M.H., Wrenn, B.A. and Domínguez-Espinosa, R. (2004) Production of bioenergy and biochemicals from industrial and agricultural wastewater. *Trends Biotechnol.* **22**(9), 477–485.

Bocher, B.T., Agler, M.T., Garcia, M.L., Beers, A.R. and Angenent, L.T. (2008) Anaerobic digestion of secondary residuals from an anaerobic bioreactor at a brewery to enhance bioenergy generation. *J. Ind. Microbiol. Biotechnol.* **35**(5), 321–329.

Bond, D.R. and Lovley, D.R. (2003) Electricity production by *Geobacter sulfurreducens* attached to electrodes. *Appl. Environ. Microbiol.* **69**(3), 1548–1555.

Brillas, E. and Casado, J. (2002) Aniline degradation by Electro-Fenton and peroxi-coagulation processes using a flow reactor for wastewater treatment. *Chemosphere* **47**(3), 241–248.

Brock, T.D. (1966) Microbial ecology and applied microbiology. *Adv. Appl. Microbiol.* **8**, 61–75.

Burton, F.L. (1996) *Water and Wastewater Industries: Characteristics and Energy Management Opportunities*, Electric Power Research Institute, Inc. (EPRI), Community Environmental Center, St. Louis, MO.

Clauwaert, P., Rabaey, K., Aelterman, P., De Schamphelaire, L., Pham, H.T., Boeckx, P., Boon, N. and Verstraete, W. (2007) Biological denitrification in microbial fuel cells. *Environ. Sci. Technol.* **41**(9), 3354–3360.

De Schamphelaire, L. and Verstraete, W. (2009) Revival of the biological sunlight-to-biogas energy conversion system. *Biotechnol. Bioeng.* **103**(2), 296–304.

Dewan, A., Beyenal, H. and Lewandowski, Z. (2008) Scaling up microbial fuel cells. *Environ. Sci. Technol.* **42**(20), 7643–7648.

Dimock, R. and Morgenroth, E. (2006) The influence of particle size on microbial hydrolysis of protein particles in activated sludge. *Water Res.* **40**(10), 2064–2074.

Duran Moreno, A., Frontana-Uribe, B.A. and Ramirez Zamora, R.M. (2004) Electro-Fenton as a feasible advanced treatment process to produce reclaimed water. *Water Sci. Technol.* **50**(2), 83-90.

Feng, Y., Wang, X., Logan, B.E. and Lee, H. (2008) Brewery wastewater treatment using air-cathode microbial fuel cells. *Appl. Microbiol. Biotechnol.* **78**(5), 873–880.

Figueroa, S., Vazquez, L. and Alvarez-Gallegos, A. (2009) Decolorizing textile wastewater with Fenton's reagent electrogenerated with a solar photovoltaic cell. *Water Res.* **43**(2), 283–294.

Fornero, J.J., Rosenbaum, M., Cotta, M.A. and Angenent, L.T. (2008) Microbial fuel cell performance with a pressurized cathode chamber. *Environ. Sci. Technol.* **42**(22), 8578–8584.

Ginestet, P., Maisonnier, A. and Sperandio, M. (2002) Wastewater COD characterization: biodegradability of physico-chemical fractions. *Water Sci. Technol.* **45**(6), 89–97.

Gregory, K.B., Bond, D.R. and Lovley, D.R. (2004) Graphite electrode as electron donors for anaerobic respiration. *Environ. Microbiol.* **6**(6), 596–604.

Han, Y. and Dague, R.R. (1997) Laboratory studies on the temperature phased anaerobic digestion of domestic primary sludge. *Water Environ.Res.* **69**(6), 1139–1143.

He, Z. and Angenent, L.T. (2006) Application of bacterial biocathodes in microbial fuel cells. *Electroanalysis* **18**(19–20), 2009–2015.

He, Z., Minteer, S.D. and Angenent, L.T. (2005) Electricity generation from artificial wastewater using an upflow microbial fuel cell. *Environ. Sci. Technol.* **39**(14), 5262–5267.

Heilmann, J. and Logan, B.E. (2006) Production of electricity from proteins using a microbial fuel cell. *Water Environ. Res.* **78**, 531–537.

Huang, L. and Logan, B.E. (2008) Electricity generation and treatment of paper recycling wastewater using a microbial fuel cell. *Appl. Microbiol. Biotechnol.* **80**(2), 349–355.

Huang, L., Zeng, R.J. and Angelidaki, I. (2008) Electricity production from xylose using a mediator-less microbial fuel cell. *Bioresour. Technol.* **99**(10), 4178–4184.

Jia, Y.H., Tran, H.T., Kim, D.H., Oh, S.J., Park, D.H., Zhang, R.H. and Ahn, D.H. (2008) Simultaneous organics removal and bio-electrochemical denitrification in microbial fuel cells. *Bioprocess. Biosyst. Eng.* **31**(4), 315–321.

Kalyuzhnyi, S. (2008) Energy potential of anaerobic digestion of wastes produced in Russia via biogas and microbial fuel cell technologies. *Pure Appl. Chem.* **80**(10), 2115–2124.

Kiss, I., Szekeres, S., Bejerano, T.T. and Soares, M.I.M. (2000) Hydrogen-dependent denitrification: preliminary assessment of two bio-electrochemical systems. *Water Sci. Technol.* **42**(1–2), 373–379.

Lafitte-Trouque, S. and Forster, C.F. (2002) The use of ultrasound and gamma-irradiation as pre-treatments for the anaerobic digestion of waste activated sludge at mesophilic and thermophilic temperatures. *Bioresour. Technol.* **84**(2), 113–118.

Larminie, J. and Dicks, A. (2003) *Fuel Cell Systems Explained.* John Wiley & Sons, Ltd, Chisester, West Sussex, England.

Lier, J.B., van (1996) Limitations of thermophilic anaerobic wastewater treatment and the consequences for process design. *Antonie Van Leeuwenhoek* **69**, 1–14.

Lier, J.B., van, Grolle, K.C., Frijters, C.T., Stams, A.J. and Lettinga, G. (1993) Effects of acetate, propionate, and butyrate on the thermophilic anaerobic degradation of propionate by methanogenic sludge and defined cultures. *Appl. Environ. Microbiol.* **59**(4), 1003–1011.

Liu, H., Cheng, S. and Logan, B.E. (2005a) Production of electricity from acetate or butyrate using a single-chamber microbial fuel cell. *Environ. Sci. Technol.* **39**(2), 658–662.

Liu, H., Grot, S. and Logan, B.E. (2005b) Electrochemically assisted microbial production of hydrogen from acetate. *Environ. Sci. Technol.* **39**(11), 4317–4320.

Logan, B.E. (2005) Editorial – generating electricity from wastewater treatment. *Water Environ. Res.* **77**(3), 209.

Logan, B.E., Hamelers, B., Rozendal, R., Schröder, U., Keller, J., Freguia, S., Aelterman, P., Verstraete, W. and Rabaey, K. (2006) Microbial fuel cells: methodology and technology. *Environ. Sci. Technol.* **40**(17), 5181–5192.

Lovley, D.R. (2006) Microbial fuel cells: novel microbial physiologies and engineering approaches. *Curr. Opin. Biotechnol.* **17**, 1–6.

Lovley, D.R. (2008) The microbe electric: conversion of organic matter to electricity. *Curr. Opin. Biotechnol.* **19**(6), 564–571.

Mahmoud, N., Zeeman, G., Gijzen, H. and Lettinga, G. (2004) Anaerobic stabilisation and conversion of biopolymers in primary sludge – effect of temperature and sludge retention time. *Water Res.* **38**(4), 983–991.

Marcus, A.K., Torres, C.I. and Rittmann, B.E. (2007) Conduction-based modeling of the biofilm anode of a microbial fuel cell. *Biotechnol. Bioeng.* **98**(6), 1171–1182.

Min, B., Kim, J., Oh, S., Regan, J.M. and Logan, B.E. (2005) Electricity generation from swine wastewater using microbial fuel cells. *Water Res.* **39**(20), 4961–4968.

Min, B. and Logan, B.E. (2004) Continous electricity generation from domestic wastewater and organic substrates in a flat plate microbial fuel cell. *Environ. Sci. Technol.* **38**, 5809–5814.

Nevin, K.P., Richter, H., Covalla, S.F., Johnson, J.P., Woodard, T.L., Orloff, A.L., Jia, H., Zhang, M. and Lovley, D.R. (2008) Power output and columbic efficiencies from biofilms of *Geobacter sulfurreducens* comparable to mixed community microbial fuel cells. *Environ. Microbiol.* **10**(10), 2505–2514.

Pham, T.H., Rabaey, K., Aelterman, P., Clauwaert, P., De Schamphelaire, L., Boon, N. and Verstraete, W. (2006) Microbial fuel cells in relation to conventional anaerobic digestion technology. *Eng. Life Sci.* **6**(3), 285–292.

Rabaey, K., Clauwaert, P., Aelterman, P. and Verstraete, W. (2005) Tubular microbial fuel cells for efficient electricity generation. *Environ. Sci. Technol.* **39**(20), 8077–8082.

Rabaey, K., Lissens, G., Siciliano, S.D. and Verstraete, W. (2003) A microbial fuel cell capable of converting glucose to electricity at high rate and efficiency. *Biotechnol. Lett.* **25**(18), 1531–1535.

Rezania, B., Oleszkiewicz, J.A. and Cicek, N. (2007) Hydrogen-dependent denitrification of water in an anaerobic submerged membrane bioreactor coupled with a novel hydrogen delivery system. *Water Res.* **41**(5), 1074–1080.

Rozendal, R.A., Hamelers, H.V.M. and Buisman, C.J.N. (2006a) Effects of membrane cation transport on pH and microbial fuel cell performance. *Environ. Sci. Technol.* **40**, 5206–5211.

Rozendal, R.A., Hamelers, H.V.M., Euverink, G.J.W., Metz, S.J. and Buisman, C.J.N. (2006b) Principle and perspectives of hydrogen production through biocatalyzed electrolysis. *Int. J. Hydrogen Energ.* **31**(12), 1632–1640.

Rulkens, W. (2008) Sewage sludge as a biomass resource for the production of energy: overview and assessment of the various options. *Energy & Fuels* **22**(1), 9–15.

Seghezzo, L., Zeeman, G., Lier, J.B., van, Hamelers, H.V.M. and Lettinga, G. (1998) A review: the anaerobic treatment of sewage in UASB and EGSB reactors. *Bioresour. Technol.* **65**(3), 175–190.

Tchobanoglous, G., Burton, F.L. and Stensel, H.D. (2003) *Wastewater Engineering, Treatment and Reuse*. Metcalf & Eddy, McGraw Hill, New York, NY.

Virdis, B., Rabaey, K., Yuan, Z. and Keller, J. (2008) Microbial fuel cells for simultaneous carbon and nitrogen removal. *Water Res.* **42**(12), 3013–3024.

Welper, L.L., Sung, S.W. and Dague, R.R. (1997) Laboratory studies on the temperature-phased ASBR system. *Water Sci. Technol.* **36**(2–3), 295–302.

Zhao, F., Harnisch, F., Schröder, U., Scholz, F., Bogdanoff, P. and Herrmann, I. (2006) Challenges and constraints of using oxygen cathodes in microbial fuel cells. *Environ. Sci. Technol.* **40**(17), 5191–5199.

20

Peripherals of BES – small scale yet feasible (demonstrated) applications

Ioannis Ieropoulos, John Greenman, Chris Melhuish and Ian Horsfield

20.1 INTRODUCTION

This chapter describes the definition of peripherals in the context of BES and proposes different scenarios for maintaining a Bioelectrochemical System through the interaction between this and the peripherals in its environment. According to the *Concise Oxford Dictionary* a peripheral is a device which is '*…used with a computer etc. but not an integral part of it…*' The definition is obviously written with the computer environment in mind and it basically describes devices or units that connect to the main/central processing platform

 Bioelectrochemical Systems: From Extracellular Electron Transfer to Biotechnological Application. Edited by Korneel Rabaey, Largus Angenent, Uwe Schröder and Jürg Keller. ISBN: 9781843392330. Published by IWA Publishing, London, UK.

without forming a part of it – for example the keyboard[1] mouse, monitor, printer, scanner, modem, webcam, etc. In this respect, a Bioelectrochemical System's (BES) peripheral can be defined in exactly the same way – only here (at least for the time being) the BES is not the central processing unit but the central powering unit instead. However, we argue that specifically because of the 'bio' element in the BES term, a peripheral can in fact form an integral part of the BES and more importantly contribute to its operation, longevity, maintenance and viability. Most of this chapter is concerned with BESs that are designed to produce electricity (i.e. MFC). The types of peripherals that might be needed for electrolytic devices that consume energy (e.g. BES producing hydrogen) will be covered in the last section.

20.2 ARTIFICIAL SYMBIOSIS

Associations between two species of life-forms have often been described with the term symbiosis. Although true in its original context, symbiosis is now used to describe an obligate association between two species in which the two life-forms benefit from, but also depend on, each other, i.e. a form of interdependence. This type of interaction is different from proto-cooperation which is a term used to describe a non-obligate mutually beneficial association. There are many other forms of interaction between two species and the whole spectrum of associations may be divided into two groups; one, in which neither of the two parties is harmed, and one in which either or both species are harmed. The envisaged system is one that could potentially mimic symbiosis, which is obligatory and mutually beneficial. Due to the artificiality of the peripherals, an appropriate term for such a system is '*artificial symbiosis*'. One such example is the series of EcoBots, which demonstrate the above in a primitive manner through the integration between a mechatronic unit (robot) and the microbes in the Microbial Fuel Cells (MFCs/BES). The exemplar of EcoBots shows that in order for the MFC to remain viable for prolonged periods, there has to be interdependence with the artificial surroundings, i.e. the peripherals, and this is what this chapter is going to talk about.

---

[1] Although a keyboard is necessary as an input means and is required during the boot process, it is nevertheless a peripheral because the main processing unit along with everything else on the motherboard will still function whether a keyboard is connected or not.

20.3 MICROBIAL FUEL CELLS AND THEIR CONFIGURATIONS

Microbial fuel cells are an exciting technology which is receiving increased attention from the scientific community. It is a 'green' energy source employing microorganisms that break down organic biomass, which would have been broken down in nature by any other means to produce the same quantity of carbon dioxide that was fixed by photosynthesis, in the first place, (by plants or microbes). Thus, MFCs (or BES producing fuels such as methane or hydrogen) are not part of the fossil fuel carbon cycle.

The rate of reactions inside a BES is dictated by the microbial speciation and the availability of substrates that they are feeding on, and is relatively slow (compared to e.g. combustion). This is a major drawback of the technology because it implies that the energy output from individual devices remains low and thus (at the time of writing) insufficient to energize any substantive actuator continuously or for prolonged periods. In order for BES to become a viable alternative to conventional methods of energy generation, their general maintenance including fluid exchange, pH and temperature control, hydration and waste disposal need to become part of the BES entity, must not consume more power than can be generated and moreover be automated. This is the basis for the argument that a peripheral in this sense plays a vital role in the operation of a BES.

20.3.1 Definition of peripherals

A peripheral can be either 'active' or 'passive' with respect to the normal operation of a MFC-type BES, which in order to operate autonomously or at least automatically, must be provided with the necessary elements for its survival by the peripheral units. In return, the BES provides the energy that keeps the peripheral(s) running as a system. This is what we call a '*symbot*' (Melhuish *et al.* 2006). Depending on the role of a device in a network, whether being active or passive, peripherals can be defined in one of two ways in the context of BES. The first is identical to the dictionary definition and describes a 'passive' unit i.e. a device that does not form an integral part of the BES and contributes nothing to its operation or maintenance. This might include almost anything that could be driven by BESs (or stacks of BESs), which in their present state of development will be restricted to low power devices (e.g. sensors, low current LEDs, etc.).

The second type of peripheral is one that forms an inherent part of – and hence plays a vital role in – the continuous BES operation and its control.

A definition for this second type of peripheral could be: 'a device powered by the BES to provide long term viability for the BES' i.e. the two types of system are interdependent for longevity. Based on the two aforementioned definitions, the BES peripherals can thus be classified into two main categories; active peripherals (which contribute to the system's maintenance) and passive peripherals (which are 'loads' that the BES system was designed to drive). Figure 20.1 shows a block diagram of the active and passive types of peripherals.

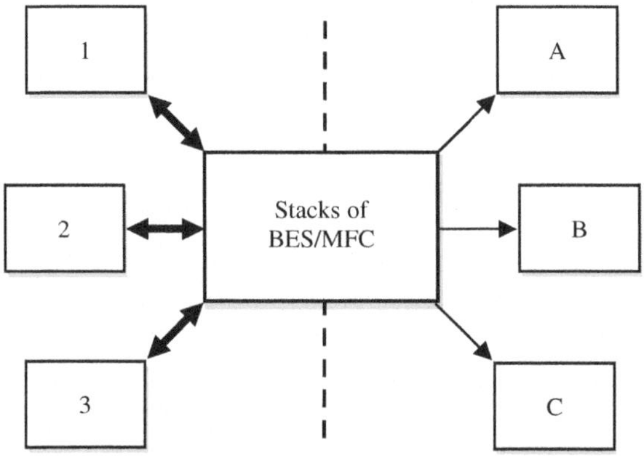

Figure 20.1. A block diagram representation of the two types of peripherals. Shown on the left are the active peripherals, 1, 2, 3 etc. (that contribute to the running of the MFC stack) and on the right (A, B, C etc.) the passive peripherals that use whatever additional net power, which may be generated or stored

In Figure 20.1, the work- and information-flow is larger and bi-directional for the active and smaller and unidirectional for the passive peripherals. This symbolism represents the complexity of the actual hardware that will have to be installed in order to operate a self-sustainable system as opposed to the relatively simple straight connection between a power supply and a consuming load. Tables 20.1 and 20.2 below give some examples of active and passive peripherals with function descriptions that may help to distinguish between these two types.

20.3.2 Bridging the power divide

In order to explain the likely role of peripherals in a BES, a crude historical approach may be taken. Since the early 1900's technological evolution helped decrease the power requirements of various devices or machines. A good historic example of this is the replacement of thermionic filament valves with first, transistors (1960's) and in their turn, silicon chips (1980's). Currently, inefficient filament lamps are being replaced by low power fluorescent light bulbs or LEDs. Similarly, energy intensive actuators [pre-rare earth magnets] are being replaced by more efficient types, and in the future, may be replaced by electroactive polymers (EAP) that require much less power to carry out their designated functions (motion).

Table 20.1. Examples of active peripherals

Example	Function
1. Feedstock medium pump or control valve if gravity fed	Keeps the system continuously or periodically fed
2. Power controller	To scavenge electrons and charge capacitor banks or accumulators and distribute power in pulsed behaviour mode
3. Accumulators, re-chargeable battery or banks of capacitors	To store energy in order to provide high power for short intervals of time
4. Internal pH sensor/controller	Optimization of pH of input fuel
5. Internal temperature sensor/controller	Optimization of temperature
6. Sensor/actuator that measures quality of feedstock and adjusts stack output (by fluidic or electrical connections).	Optimization of electron abstraction against various qualities of feedstock
7. Low power radio transceiver	To both transmit data regarding the systems state; to receive new programming

Table 20.2. Examples of passive peripherals

Example	Function
1. Sensors and low power silicon circuitry	To sense [temperature, pH, redox, pO_2, BOD, specific chemicals] and store or transmit time resolved data
2. Light emitting diodes or low power lamps	To energise indicator lights or local sources of illumination
3. Radio transmitter	Data communication and monitoring by supervisor (central computer)
4. Electric motors (small scale?)	Fans, pumps, wheels, propellers, refrigerators
5. Higher power electronic devices: calculators, laptops, mobile phones, radio-TV, battery powering devices	Various
6. Local heating (of short duration)	Sufficient to run mass-flow meters?

During the time of technological evolution of electrical devices there have also been scientific and technological developments in MFCs and more recently, BES in general. From the system first described by Potter at the beginning of the 20th century (Potter 1912) producing microWatts per MFC, to systems more recently described in the literature, which are more powerful by a factor of 2-log fold or greater (Cheng *et al.* 2006; Rabaey *et al.* 2003). An illustration of how the MFC power output has evolved from the first report and also how it is anticipated to evolve in the years to come is shown in Figure 20.2.

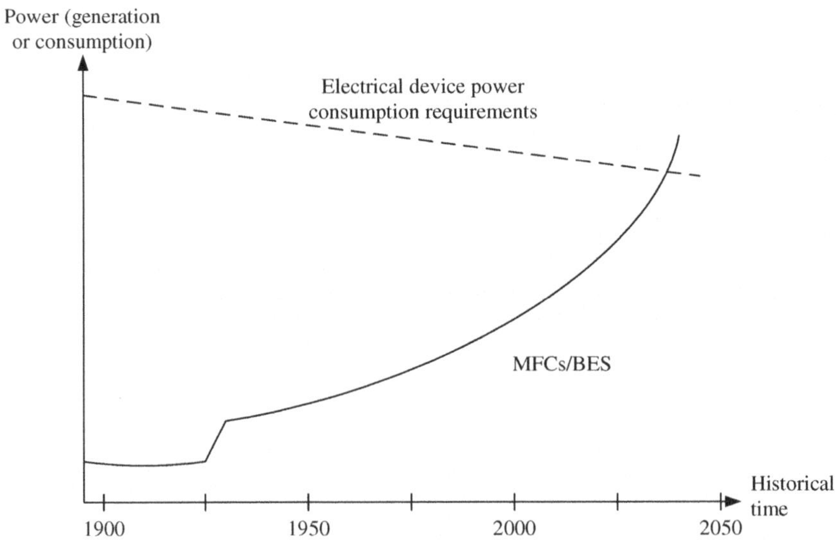

Figure 20.2. Generalization and simplification of MFC and peripherals technological development in terms of power generation and device power requirements or consumption for specific functions (produced from the main published articles in the early years: Potter 1912; Cohen 1931; DelDuca *et al.* 1963; Karube *et al.* 1976, 1977; Bennetto *et al.* 1987). The sudden increase in the MFC/BES power generation curve is a reflection of the work from Cohen (1931) who for the first time demonstrated the electrical connection of multiple MFCs as a stack which resulted in a voltage output of 35 V at a current of 2m A (70 mW)

A more recent development stems from the realization that a single MFC can never produce more than the theoretical maximum of 1.14 V (open circuit voltage – $E_{O/C}$). Typical $E_{O/C}$ values are of the order of 0.7 V – even though values of 1V have been reported (Jong *et al.* 2006; Ieropoulos *et al.* 2007) – and which decreases to about half (typically 0.35) when connected to a load resistor at maximum power transfer. This is generally insufficient to power electronic devices or charge up accumulators that might be used to store sufficient power to allow for pulsed behaviour. Thus, pluralities of MFCs (i.e. stacks) are required, which can be electrically configured to step up the output (Ieropoulos *et al.* 2008). It then becomes possible to charge up accumulators through which – in a pulsated manner – conventional devices can be energized.

Although stacks of MFCs are more powerful than individual units, they may carry their own attendant problems (discussed later) and are at present only suitable for powering the lowest power requiring devices. However, such a device can include an efficient 'electron-harvester' that can extract electrons by adjusting the load and current output to persistently charge a capacitor (accumulator). Such a system was devised, built and used to operate EcoBots I and II to control the output from 8 MFCs in order to operate the robot in what is called 'pulsed behaviour mode' (Ieropoulos *et al.* 2003; Melhuish *et al.* 2006).

20.3.3 Minimal peripheral requirements for continuous and autonomous operation

Early generations of BES (MFC) described in the literature were typically based on batch (non-continuous flow) systems (see review article from Palmore *et al.* 1994). Batch mode operation has severe limitations in the sense that the system will run out of organic fuel unless the rate limiting substrate(s) (organic fuel feedstock) are continuously or periodically replenished and end products are removed. Although replenishment could be done manually, for a stack of many BESs this is impractical, especially if located in remote areas. This brings us to the first 'minimum requirement' peripheral; one that will allow some form of longevity and automation regarding optimal operation. This can be in the form of a low power pump that allows fresh feedstock to be continually added when required. It should be noted that continuous or periodic culture may be possible without a pump, if the system is gravity fed and a control valve is used.

A second peripheral that would be ideally required is a silicon circuit-based energy harvester that is allowing for a more efficient charging of the third peripheral, an accumulator. This charge/discharge cycle can be uneven, with charging times (low power) of the order of 23 hours in the day, allowing a high energy discharge, e.g. for 1 hour/day. This automated charge/discharge system was used onboard EcoBots I, II (Ieropoulos *et al.* 2003; Melhuish *et al.* 2006). We call this behaviour (with low but long resting charge, fast and furious discharge) 'pulsed' behaviour and it requires a microcontroller to control the charge–discharge cycles and distribute the power between the peripherals on a sharing or sequential basis (known as hard wired "action selection").

If MFCs are ever exploited to give outputs at power grid scale (kW-MW), then all discussion regarding the types of peripherals becomes trivial, since (as global peripherals) this could include just about anything electrical. Since no one has yet produced an MFC that by itself is powerful enough to drive a pump directly (peripheral 1) it implies that (a) stacks must be used, and (b) pulsed behaviour mode will be employed, requiring a silicon circuit processor (peripheral 2) and bank of accumulators/capacitors (peripheral 3). These are probably the minimum requirements for energy autonomy, with some energy to spare for passive peripherals.

20.3.4 Complexity in stacks

Individual MFCs within a stack of say, $n = 8$ can be electrically connected in series, parallel or series-parallel configurations. The fluidic input can also be routed through the MFCs in 2 different ways: either all are fed equally from the same feedstock vessel directly, or the feedstock goes through each MFC in turn, with the output of the first-stage MFC acting as the input to the next MFC in the matrix. The final treated-utilised feedstock only exits the whole stack after all of the fluid has passed through all eight of the MFCs.

The first study on MFC stacks comes from Cohen (1931), which reported an output of 2 mA at 35 V from a stack of unspecified number of units. Based on the output of individual units (quoted in the article) one can estimate a probable total of 70 MFCs connected together. However, the several significant experimental details are not given such as nature of the electrodes, catholyte, external load and open circuit voltage. It is assumed that these fuel cells were operated in batch-culture mode and the produced output was sustained for only a few minutes. Further developments of MFC stacks in practical applications, were reported by Wilkinson (2000), Ieropoulos *et al.* (2003) and Ieropoulos *et al.* (2005). An interesting report of a stack configuration with fuel cells in continuous fluid flow was given by Aelterman *et al.* (2006). The units were connected together both fluidically and electrically, although details of fluidic isolation, which is critical for the operation of such stacks (Ieropoulos *et al.* 2008), were not presented. Furthermore, their described method for measuring the total internal resistance (R_{int}) is in fact a method to calculate only the ohmic part of the total internal losses as the authors state. This study was the first one to report cell reversal in a series-connected stack.

One important feature of stacks of MFC connected electrically and fluidically together is that electrons travel along the fluidic lines of interlinked MFCs, thus causing fluidic conductance, which opposes the voltage increases that would be expected from series connected stacks, (shunt losses) (Ieropoulos *et al.* 2008). This difficulty can be overcome by utilising a method of pulse feeding each individual MFC in turn using a sequence feed-line disrupter, which distributes 'a dose' of input feedstock to each of the MFCs in turn, ensuring fluidic-electrical isolation between units. Another method to provide electrical isolation or to reduce electron leakage through the fluid is to employ fluidic links of sufficient length. Pulses of feed can be of short duration in comparison to intervals between pulses, which also improves isolation. This engineering answer to the problem of fluidic shunt conductance can be automated by use of microprocessor control, a motor and a geared distributor. This system we would describe as an *ad hoc* 'specialist peripheral' designed to get round a particular local problem and may not be required for other types or designs of MFC and their stack arrangements.

It is envisaged that the majority of MFC stack power units would operate under general ambient conditions of neutral pH (e.g. pH 6.5–7.5), at mesophilic temperatures (15–25°C) and with a rich source of feedstock of fairly constant composition. However, there may be specialised applications where these normal working conditions are exceeded with deleterious effects on power output. Adjustments of these parameters may be required using special peripherals, as described herein:

(a) pH: A peripheral to measure pH of incoming feedstock (i.e. a pH electrode plus microprocessor system) may be used as a passive peripheral to merely measure and monitor the pH and transmit the data to a central controller (to record or interpret data). However, to be a useful active peripheral, it must have the means to control the pH, for homeostasis. A more sophisticated approach is to allow negative feedback (see Figure 20.3) from monitoring the system to automatically adjusting the pH; in essence to convert the system into an acid-base auto-titrator. If the pH is too high the electrode reading goes above threshold and a pump is switched ON to add acid. If the pH is too low, a different pump is activated to add base in order to maintain the pH within acceptable limits. Such a system will naturally require stocks of acid and base. Electrode, microcontroller, pumps and stock chemicals combined as a module, would comprise the peripheral for pH control.

(b) Temperature: As with pH, a passive peripheral may be used to measure, monitor and transmit data, and a demonstration that this can be driven by power from MFCs has been previously demonstrated (Melhuish *et al.* 2006). However, temperature control of a stack of MFCs (making the peripheral into an active type) is much more difficult to achieve. In large laboratory fermenters, it is common for temperature to be controlled by simply coupling a temperature sensor with a controller that can switch various heating elements on or off as required (i.e. a simple thermostat). However, heating elements are very energy intensive and if employed are likely to drain away all the power from the accumulators. For small mobile robots, an array of temperature sensors can allow the robot to detect and traverse temperature gradients; for example, *if* too hot *then* seek out shade and *if* too cold *then* go into shelter or seek sunlight but this requires more sophisticated action selection mechanisms. This can also be applied to any type of mobile application as a method for thermostatic control. For stationary applications, control of the stack temperature by water (using evaporative or circulatory cooling) is not particularly energy intensive, especially in a gravity fed system. However, heating up a large number of stacks in a cold environment is very energy expensive. Although small amounts of heat are produced by the biochemical reactions occurring in the microbial cells, additional heat will be needed unless the stack is dense and very well thermally insulated.

(c) Quality of feedstock: In cases where an MFC stack may have to operate using feedstock of highly variable composition, a peripheral could be included that can control the flow rate (and thus hydraulic retention time – HRT) of the stack in response to the quality of feedstock provided. The quality of the feedstock is measured (simply) from the total MFC-stack power output and this value can be used to influence the pump controlling the feedstock flow rate. These methods, along with electronic control of overpotentials may be combined with automation to change both electrical and fluidic configurations of the stack which may also add another useful way to optimise the stack performance across a wide range of differing nutrient conditions and environments. Peripherals would include fluidic valves, pumps and electrical connection switches (employing transistors).

For example, an MFC stack can be designed to select nutrient rich over nutrient deficient food sources by exploiting a miniature MFC (called a 'taster-MFC') in line with a nutrient suction tube. If the output from the 'taster-MFC' is low, suggesting a low nutrient source or one containing inhibitory compounds (e.g. from heavy metal pollution), the microcontroller peripheral rejects the source. If the output from the taster MFC is high (suggesting high quality feedstock) then the nutrient suction continues and the whole MFC-stack is fed. This example demonstrates the potential role of MFCs as sensing agents; not just as systems to produce energy.

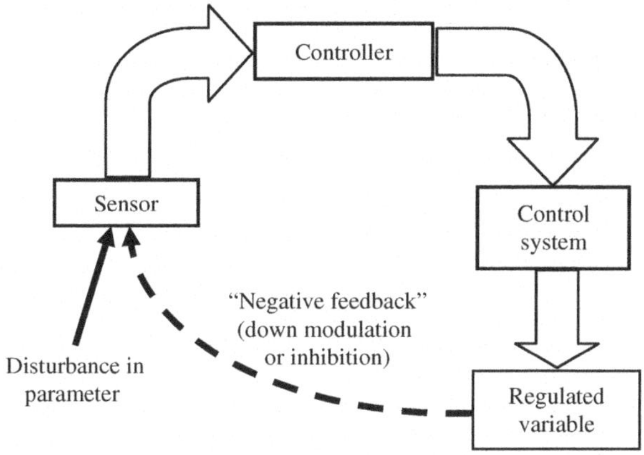

Figure 20.3. Block diagram representation of a negative feedback loop

(d) Novel types of processing units: Miniature BES in a stack can be connected to each other in such a way (using either electrical or fluidic connections) that the activity of one can modulate the activities of the others. This allows for non-conventional computing (Greenman *et al.* 2006a,b, 2008) using the BES as multi logic processing units. By integrating microprocessor control with these types of BES, it may be possible to build an integrated "processing engine" that may be used to carry out higher domain computing on board a robot. This area of research remains relatively unexplored.

20.3.5 Microbial Electrolysis Cells (MECs) that transform organic feedstocks into other types of energy (hydrogen or methane) but require input of electrical power in the process

The production of hydrogen by BES has recently been described at laboratory level (Cheng and Logan 2007; Logan 2004). These studies have not accounted for the full energy balance, which should include the energy costs of BES replenishment, temperature control and pressurisation of produced hydrogen. Furthermore, in order to poise a cathode at sufficient positive overpotential for H_2 production requires an external power supply (Logan 2004). This necessitates the use of MFC driver units (unless the extra electrical power is served from wind farms or photovoltaics. This implies a plurality of MFC as well as the BES.

20.3.6 Microbial Electrolysis Cells (MECs) that consume electrical power to drive useful reactions (e.g. denitrification)

A number of BESs have been described that utilize externally supplied electrons in order to carry out some useful transformation. One example includes denitrification. However, to the authors knowledge there have been no reports on scaled up systems that 'practically' treat high nitrate wastewaters although small-scale laboratory systems have been described (Clauwaert *et al.* 2007; Jia *et al.* 2007). In these systems, electrons supplied by microorganisms oxidizing acetate at the anode are used by microbial species in the cathode that perform complete denitrification. Thus these types of BES may have the potential for removing carbon as well as nitrogen compounds from wastewater. The denitrification rate and the power production was limited by the cathodic microorganisms, which only denitrified significantly at a cathodic electrode potential below 0 V vs. standard hydrogen electrode.

Since these types of specialized transformations have value in themselves, they can be driven from any local supply of electricity. Combination of a plant with MFC or other renewable source of electricity would again be sensible for large-scale systems. Peripherals will still be required to (a) move large bodies of liquid through the system (using gravity feed wherever possible) and (b) some form of power controller.

REFERENCES

Bennetto, H.P. (1987) Microbes come to power. *New Scientist*, April 16th, 36–39.

Cheng, S., Liu, H. and Logan, B.E. (2006) Increased power generation in a continuous flow MFC with advective flow through the porous anode and reduced electrode spacing. *Environ. Sci. Technol.* **40**, 2426–2432.

Cohen, B. (1931) The bacterial culture as an electrical half-cell. *J. Bacteriol.* **21**, 18–19.

DelDuca, M.G., Friscoe, J.M. and Zurilla, R.W. (1963) Developments in industrial microbiology. *American Institute of Biological Sciences* **4**, 81–84.

Greenman, J., Ieropoulos, I. and Melhuish, C. (2006a) Perfusion anodophile biofilm electrodes and their potential for computing. *From Utopian to Genuine Unconventional Computers*. Adamatzky, A. and Teuscher, C. (eds), pp. 67–78, ISBN: 0-9551170-9-7.

Greenman, J., Ieropoulos, I., McKenzie, C. and Melhuish, C. (2006b) Microbial computing using Geobacter biofilm electrodes: output stability and consistency. *Int. J. Unconv. Comp.* **2**(3), 2491–265.

Greenman, J., Ieropoulos, I. and Melhuish, C. (2008) Biological computing using perfusion anodophile biofilm electrodes (PABE). *Int. J. Unconv. Comp.* **4**, 23–32.

Ieropoulos, I., Melhuish, C. and Greenman, J. (2003) Artificial metabolism: towards true energetic autonomy in artificial life. In *Proc. of the 7th European Conf. in Artificial Life*. Dortmund, 14–17 September.

Ieropoulos, I., Melhuish, C. and Greenman, J. (2007) Artificial gills for robots: MFC behaviour in water. *Bioinsp. Biomim.* **2**(3), S83–S93. DOI:10.1088/1748-3182/2/3/S02.

Ieropoulos, I., Greenman, J. and Melhuish, C. (2008) Microbial fuel cells based on carbon veil electrodes: stack configuration and scalability. *Int. J. Energy Res.* **32**(13), 1228–1240. DOI: 10.1002/er.1419

Jong, B.C., Kim, B.H., Chang, I.S., Liew, P.W.Y., Choo, Y.F. and Kang, G.S. (2006) Enrichment, performance and microbial diversity of a thermophilic mediatorless microbial fuel cell. *Environ. Sci. Technol.* **40**, 6449–6454. DOI: 10.1021/es0613512.

Karube, I., Matasunga, T., Suzuki, S. and Tsuru, S. (1976) Continuous hydrogen production by immobilized whole cells of Clostridium butyricum. *Biochim. Biophys. Acta* **24**(2), 338–343.

Karube, I., Matasunga, T., Suzuki, S. and Tsuru, S. (1977) Biochemical cells utilizing immobilized cells of *Clostridium butyricum. Biotechnol. Bioeng.* **19**, 1727–1733.

Melhuish, C., Ieropoulos, I., Greenman, J. and Horsfield, I. (2006) Energetically autonomous robots: food for thought. *Auton. Robot* **21**(3), 187–198.

Palmore, G.T.R. and Whitesides, G.M. (1994) Microbial and enzymatic biofuel cells. In *Enzymatic Conversion of Biomass for Fuels Production*. Himmel, M.E., Baker, J.O. and Overend, R.P. (eds), Am. Chem. Soc. Symp. Series No. 566, 271–290, Oxford University Press, ISBN: 0841229562.

Potter, M.C. (1912) Electrical effects accompanying the decomposition of organic compounds. In *Proc. Royal Soc.* **84**(Part B), 260–276.

Rabaey, K. *et al.* (2003) A microbial fuel cell capable of converting glucose to electricity at high rate and efficiency. *Biotechnol. Lett.* **25**, 1531–1535.

21

Towards a mathematical description of bioelectrochemical systems

Jorge Rodriguez and Giuliano C. Premier

21.1 INTRODUCTION

Mathematical modelling has an important, integrating and consolidating role to play in bioelectrochemical systems (BESs) research and can be helpful in expediting the contribution of such systems to a more sustainable society. The BES-based technologies that have attracted most interest so far are microbial fuel cells and microbial electrolysis cells (MFCs and MECs respectively). Both are based on the same principle of biocatalysed electron transfer to/from solid electrodes. Researchers are fortunate to have the accumulated developmental experience of antecedent technologies such as anaerobic digestion and conventional electrolysis from which to draw lessons and these technologies

 Bioelectrochemical Systems: From Extracellular Electron Transfer to Biotechnological Application. Edited by Korneel Rabaey, Largus Angenent, Uwe Schröder and Jürg Keller. ISBN: 9781843392330. Published by IWA Publishing, London, UK.

amply illustrate the utility of modelling in developing understanding. BESs offer not only the prospect of simultaneous wastewater treatment and either electricity generation in the case of MFCs or cathodic hydrogen production in the case of MECs, but also a portfolio of very promising biocatalysed chemical conversions, both anodic and cathodic. All of these processes would be well served by conceptual direction from modelling.

The viability of larger scale BES-based technologies is currently restricted by a limited understanding of specific elements in the bioelectrochemical process (e.g. the biological electron transfer to/from the electrodes). Furthermore there is a lack of appropriate and validated mathematical descriptions of BESs, which properly account for the most important interactions between system elements and processes. Mathematical models, appropriate to differing specific objectives and BESs applications, are necessary to achieve an integrated understanding of the various coupled phenomena in order to progress BESs research and development.

In this chapter, the state of the art of BESs modelling will be summarized addressing existing models (both published and under development) and their main characteristics. Current issues and challenges in mathematical modelling of BESs will also be discussed, as well as the authors' views on modelling selected specific elements and processes. The chapter aims to provide an overview of BESs mathematical modelling to the whole BESs research community and to serve as a reference in identifying the type of model required for a specific purpose.

21.2 MATHEMATICAL MODELLING

Models, in a broad sense, are representations of real (or envisaged) systems which make conceptual ideas of the system and, as necessary, its components and their functional relationships apparent. These ideas can be represented visually (e.g. drawings), linguistically, mentally, physically (e.g. scale replicas) and also mathematically. The mathematical formulation of conceptual ideas related to a real system produces a mathematical model of that system, which can be used for example as a quantitative description of its response to external inputs.

In essence, a mathematical model can be conceived as a machine transforming inputs (u) into outputs (y) through functional relationships. The characteristics of these relationships between inputs and outputs define the basic structure of a model. The outputs are variables of interest to the model user (e.g. current, concentration values, potential) while the inputs are manipulated variables and/or perturbations that affect the system (e.g. resistance, flows, influent concentrations, temperature).

Input–output type mathematical models consist simply of transfer functions which transform inputs directly to outputs. State-space mathematical models however introduce the so-called state variables (x) as intermediates between inputs and outputs. These state variables define the state of the system at a certain instant in time (e.g. species concentrations in a reaction compartment) as determined from their recent past values and the inputs (u) to the system (e.g. a reactor influent composition and flow) through a function known as the state-transition equation (e.g. a mass balance). The defined outputs (y) (e.g. a gas flow rate, electric current, a concentration itself, electrode potential) are generated from the state variables (x) through the so-called observation equations. General expressions for both equations are presented in Eq. (21.1) where f and h are functions representing the state-transition and the observation equations respectively and θ is the parameter set of the model (Dochain and Vanrolleghem 2001).

$$\begin{aligned} \frac{dx}{dt} &= f(x, u, t, \theta); \qquad x(t=0) = x_0 \\ y &= h(x, u, t, \theta) \end{aligned} \tag{21.1}$$

All the constituents of the expressions in a mathematical model, irrespective of the model structure, are variables, constants or parameters. Variables include inputs, outputs and states; constants are those constituents that never change their value for any model application (e.g. physical or geometric constants) and parameters are those constituents for which values can change depending on circumstances and application. Parameter values (e.g. kinetic, stoichiometric or operational) must therefore be determined for each particular model application.

21.2.1 Model characteristics

It is conceivable that BESs will require models for several purposes and would therefore benefit from models derived by different approaches. There are a number of characteristics of mathematical models worth considering explicitly in order to decide on the type of model or modelling approach best suited to each particular purpose.

21.2.1.1 Mechanistics vs. empirism

Mechanistic (also *physical* or *white-box*) model structures are based on physical, chemical or biological laws and principles. The sole source of information required for developing these models would be prior knowledge of natural laws. They result from a deductive modelling strategy.

Empirical (also *phenomenological* or *black-box*) model structures are based on empiricism rather than on natural laws and first principles. The source of information would be experimental observation. They result from an inductive modelling strategy.

Grey-box (or *semi-physical*) models are those resulting from the combination of mechanistic and empirical approaches. This type of model presents the advantage of better identifiability and interpretability of their parameters when compared to purely empirical methods. In general, empirical models tend to be formulated with less complexity than mechanistic models.

21.2.1.2 Dynamic vs. stationary models

Stationary models describe the steady-state condition, with given, fixed inputs, and therefore incorporate equations to calculate outputs from inputs, constants and parameters. The solution of these models normally consists of the solution of systems of algebraic equations.

Dynamic models describe time-dependent system behaviour and therefore incorporate differential equations for their state variables. The solution of these models normally requires the numerical integration of these equations.

Dynamic models may contain elements and processes that have fast time-constants compared to other relevant and dynamically dominant (slower) state variables. These fast elements can be simplified, assuming they are instantaneous and therefore calculated using algebraic (steady-state) equations. This is often the case with acid/base equilibria or ionic speciation in the context of bioprocess modelling (Batstone *et al.* 2002).

Although numerical integration of dynamic models has become simpler using the currently available computing capabilities, they may still require long computing times, depending on their characteristics. These models may also be complex requiring, for example, many iterations to converge on a solution and hence long computing times.

21.2.1.3 Level of segregation/aggregation

The degree of segregation/aggregation greatly affects the level of complexity/simplicity of the model. *Segregated* or distributed parameter models normally provide a more detailed description of the system components separately but they involve higher complexity and larger numbers of parameters. A model is considered *aggregated* or lumped parameter compared to a reference model when some of the model variables and equations are united in a simplified form. Lumping results inevitably in a less detailed description of the system and in model errors, thus a proper compromise between complexity and accuracy must the achieved, always in view of the model objectives.

A specific level of segregation/aggregation is related to the space dependency of a model's state variables. In *space-independent* models, the state variables usually vary only as a function of time (i.e. space variations are somehow lumped, e.g. assuming CSTR) while in *space-dependent* models, variations also occur in the spatial coordinates. Space-dependent dynamic models require the use of differential equations in partial derivatives. This greatly increases the time required for numerical solutions and normally discretization methods or finite elements/difference techniques are typically used.

21.2.2 How do model characteristics affect the model user?

The model characteristics described above relate to properties relevant to the model user following a few general (not necessarily strict) rules. These properties include accuracy, interpretability of results, range of applicability of the model, its complexity and number of parameters or time required for model simulation.

In general, *mechanistic* and highly *segregated* models are *more accurate*, their results *more interpretable* and they have a *wider range of applicability* On the other hand they will in general be *more complex* and include a *larger number of parameters* that will need identification. They may require *long simulation times* particularly in the cases of dynamic, highly segregated and/or multidimensional space dependent models.

Highly *empirical* and very *aggregated* models are in general *less accurate*, their results *less interpretable* and have a *narrower range of applicability*. They are normally formulated with *less complexity and fewer parameters*. They tend to require *shorter simulation times*, particularly lumped parameter models.

Consideration of the model characteristics explained in this section is important for selecting a modelling approach and structure which is best suited to a specific application and the rules indicated above can be of great help.

21.3 BESs MODELLING OBJECTIVES

BESs are very complex dynamic systems in which numerous phenomena occur over a wide range of time scales. Biological, (electro)chemical and mass transport processes take place with strong interaction(s) with each other. These interactions largely affect the BESs behaviour as a whole and therefore their investigation is as important as that of the individual processes themselves.

The task of building adequate models for BESs offers however significant challenges, some arising from the complexity of BESs and the multiple mechanisms required to understand them. BESs modelling must also cope with an incomplete state of knowledge which still exists regarding some specific processes such as the mechanisms by which microorganisms transfer electrons to electrodes or the multi-phase microbial interactions within the biofilm ecosystem created around the electrode.

Many disciplines, including materials science, electrochemistry, microbiology, ecology and process engineering are required to reach a comprehensive understanding of BESs. Each of these disciplines is able to contribute to the mathematical description of specific elements and processes relevant to BESs at different levels of detail and complexity. However an appropriate mathematical description of BESs requires the development of a model framework into which robust mathematical abstractions of the elements can be properly integrated.

The development of such a system model for BESs, which are complex, nonlinear, time varying, highly coupled systems with potentially very many degrees of freedom, requires that the purpose of the model should be clearly defined. This will determine appropriate model characteristics and may avoid considerable unnecessary expenditure of effort.

BESs modelling objectives could be very diverse but a common objective is generally to describe how variables of interest (e.g. concentrations, reaction rates, flows, potentials, electric current) are affected by changes in other input variables or parameters that are accessible for direct manipulation (e.g. influent concentrations, flows, pH, temperature, external resistance). A BESs model which succeeds in doing this becomes a powerful tool for design, optimization and control because it mimics the real system behaviour under different operating conditions. More ambitious model objectives may include the generation and validation of scientific hypotheses related to more fundamental research applications, by including very detailed mechanistic descriptions of chemical, physical and biological BESs elements and processes.

A number of interesting microbial fuel cell models have recently been published, which open the route towards a comprehensive mathematical description of BESs. It is expected that models of varying levels of complexity, aggregation, and empirism will have their part to play in BESs research.

If, for example, a model is developed to predict only the response of a very specific variable over a small range of variation of other manipulable variables (e.g. current as function of substrate concentration), a simple empirical aggregated model might be sufficient and appropriate for that specific purpose. However, such a model would have a very limited scope of applicability and might not be applicable beyond the range for which it was parameterized. It will normally also be unable to provide significant insight into the modelled process.

In other circumstances, when more specific mechanisms are the subject of investigation, very detailed mechanistic approaches might be necessary to deliver a higher level of description and more importantly of interpretability of model results. Such models however are at risk of becoming overly detailed if too many processes and components are considered without justification. This may result in excessive and unnecessary complexity with a large number of parameters to be determined, some of which might be impossible to identify. The utility of such models would be very limited, not only because of its difficult management, but mainly because simulation results may be inconclusive as multiple sets of parameter values may provide indistinguishably similar outputs.

Referring to the parsimonious approach implied in the previous section, it is clear that dynamic description, segregation level and mechanistic rigour should be limited to that which is justified by the modelling objectives.

21.4 KEY ELEMENTS FOR BESs MODELLING

Figure 21.1 presents an overview of the processes and elements identified as important for mathematical description in BESs models.

A brief outline of the most common approaches used in existing models to describe these processes and elements is summarized in Table 21.1. Detailed mathematical expressions have not been included here. The reader is referred to the literature on the existing models and to general text books for more details.

21.5 EXISTING BESs MODELS

Existing BESs models are fundamentally based on mass balances. The time dependent concentrations of relevant components are generally defined as the state variables of the model and also sought as model outputs. Concentrations are coupled to the electrochemical processes through differing approaches to calculate the important electrical model outputs in the form of potentials and currents. The predominant differences between existing BESs models are considered to be in: the level of complexity; number of species and components considered; level of detail of the description of some of the BESs elements, in particular biofilms (see Figure 21.1); and different approaches to the description of the electron transfer mechanisms from bacteria to electrode. In this section a review of the main characteristics of existing BESs models is presented and a summary is including in Table 21.2.

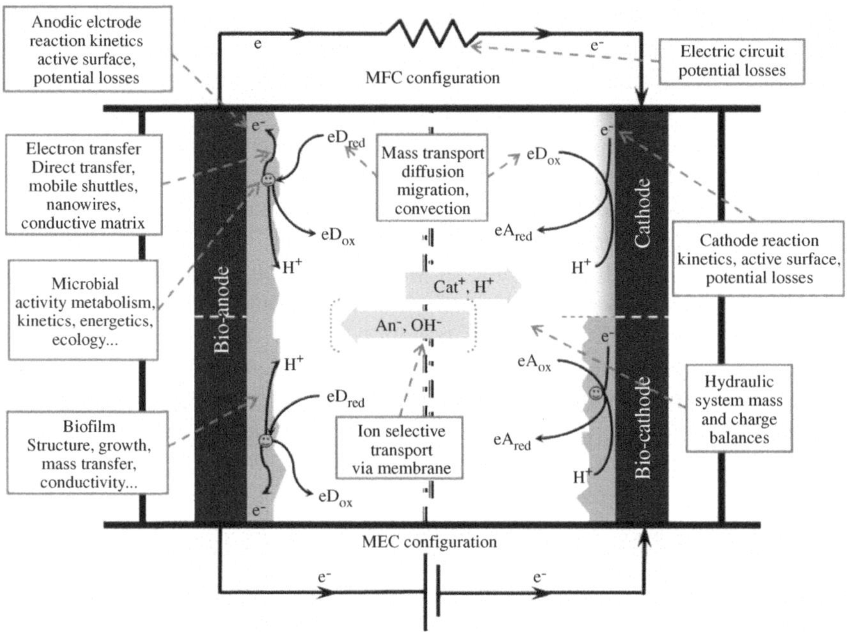

Figure 21.1. Key BESs elements and phenomena identified as relevant for modelling

Although some earlier theoretical studies have been published (Disalvo 1980; Varfolomeyev *et al.* 1993; Videla and Arvia 1975), it was to our knowledge that Zhang and Halme (1995) first published a mathematical model of a microbial fuel cell to predict current as a function of substrate concentrations in the presence of an external load. In their work, a simple model for an MFC anode with suspended microorganisms and a soluble electroactive mediator was proposed. They obtained acceptable results within the scope and purpose for which their model was originally designed, however the model is simple and hence likely to find restricted application (i.e. mass transfer and cathode kinetic limitations were neglected and no potential losses other than ohmic losses were considered). The main utility of this model would be as a basic description of behaviour in MFCs within a limited range of applicability.

Picioreanu *et al.* (2007) presented the first biofilm model of a microbial fuel cell anode based on their extensive previous work and expertise in biofilm modelling and biocorrosion (Picioreanu and van Loosdrecht 2002). Bioelectrochemical processes are integrated into a very detailed model of an anodic

Table 21.1. Summary of modelling approaches used to describe some specific BESs processes and elements of importance

Process/element	Modelling approach(es)
Electric circuit – Potential losses	– Ohm's law
Reactor system – Bulk hydraulics	– Mass balances (CSTR type, discretisation, biomass retention, etc.). – Discretisation in multiple perfectly mixed volumes.
Biofilm	– One-dimensional biofilm discretised into perfectly mixed layers. – Multidimensional biofilm description (2D and 3D).
Abiotic electrode reaction kinetics	– Butler-Volmer equation. Exponential dependence on overpotential. – Linear diffusion layer between bulk and electrode surface.
Bioelectrode reaction kinetics	– Butler-Volmer equation or Ohm's law based (depending on the electron transfer mechanism).
Mass Transport	– Diffussion by Fick's law (generally discretised). – Migration using Nernst-Planck equation.
Ion selective membrane transport	– Specific ion membrane permeabilities relative to protons.
Electron transfer mechanisms	– Mobile shuttles follow diffussion/migration transport to the electrode. – Conduction based mechanisms using Ohm's Law.
Microbial activity and metabolism	– Monod type kinetics for limiting substrate uptake. – Constant biomass yields or energy depending variable yields. – Single or multiple species considered.

biofilm. Anode-respiring bacteria (ARB) are incorporated and their metabolism stoichiometry, including biomass growth yield (assumed constant), is estimated using a thermodynamic correlation (Heijnen 1999). Detailed, mass and charge balanced chemical speciation is also included. Electrode kinetics are modelled based on the Butler–Volmer equation applied to soluble mobile electron shuttles, which are assumed to be responsible for electron transfer to the anode. The model provides a detailed and dynamic multidimensional (2D and 3D) description of the biofilm structure and current densities and is able to predict typical experimentally observed behaviour in MFC anodes, including realistic polarization curves.

Table 21.2. Summary of existing BESs models and some of their key characteristics

Existing BESs model	System modelled	Empirism vs. mechanistics	Level of aggregation	Complexity & number of parameters	Electron transfer mechanism	Key characteristics
Zhang and Halme (1995)	Anode and cathode (CSTR, No biofilm)	Mainly mechanistic (but large assumptions and approximations)	Largely aggregated (one biomass, no space dependency, CSTR)	Very simple due to the large approximations and level of aggregation. Few parameters needed	Mobile electron shuttles	– Very simple, first MFC model. – No mass transfer limitations included. – Very easy to implement and use. – Very limited applicability.
Picioreanu *et al.* (2007) Picioreanu *et al.* (2008)	Anode bulk and biofilm (Very detailed multidimensional biofilm)	Very mechanistic, based on first principles. (Some empirism used for thermodynamic estimation of biomass yields)	Largely segregated (several bacterial species and multidimensional space dependency)	High complexity; very large number of parameters (partly due to the multiple microbial species)	Mobile electron shuttles (electrons are transferred by diffusion of shuttles that react at the electrode surface)	– Very detailed multidimensional biofilm. – Detailed chemical speciation. – Multiple microbial species considered. – Difficult identification of parameters. (part due to the numerous biomass species). – Largely complex implementation.
Kato Marcus *et al.* (2007)	Anode (one dimension) biofilm	Semiempiric conduction based electron transfer model and kinetics	Aggregated (one active biomass species, no chemical speciation, 1D biofilm, lumped electron transfer mechanism)	Medium complexity, large number of parameters	Conduction (electron transfer via conduction through the biofilm matrix which has a lumped constant conductivity value)	– Lumped semiempiric approach for biofilm conduction-based electron transfer. – Integrates potential into Monod type kinetics. – Detailed 1D biofilm. – Considers endogenous respiration. – Moderately complex to implement and use.

Zeng *et al.* (2007)	Complete fuel cell: Anode, Cathode (as CSTR, no biofilm) and Membrane	Mechanistic, based on natural laws	Largely aggregated (CSTR, one biomass, no electron transfer mechanism)	Simple, low number of parameters (due to lack of biofilm and constant electrode capacitance approach)	Electrode capacitance (potential, current and reaction rate linked via electrode charge balance)	– Simple model easy to manage. – Electrode potentials related to reaction rates by charge balances with lumped constant capacitance at the electrodes. – Relatively simple to implement and use.
Oh and Sloan, (2008)	Anode chamber (EPS biofilm considered)	Mechanistic based on natural laws and electrode and electrolyte theories	Segregated (multidimensional biofilm model, detailed chemical speciation though one single biomass)	High complexity but moderate number of parameters	Electrons and protons jump through potential levels (very detailed electrolyte and electrode theories are used)	– Rigorous thermodynamics of the chemical species involved. – Integrated electrolyte and electrode theories to predict electron transfer.
Rodríguez *et al.* (2007) Rodríguez *et al.* (2009)	Complete fuel cell: Anode (biofilm and bulk) Cathode (bulk) and Membrane	Mechanistic mainly based on first principles. (Conduction-based electron transfer and energy-based estimation of biomass metabolism are semiempiric)	Segregated (1D discretised biofilm, detailed chemical speciation, detailed electron transfer mechanisms though only one active biomass)	Medium complexity, large number of parameters mainly associated to the electrode kinetics and mass transport of substrates and ions)	Mobile electron shuttles and conduction based (electrons travel both via shuttled diffusion and via conduction through nanowires or direct contact)	– Complete fuel cell with detailed chemical speciation in all compartments. – Multiple bioenergetically-linked electron transfer mechanisms. – Metabolism function of local energetics. – Configurable ion selective membrane. – Difficult identification of parameters. (electrode kinetics and electron transfer). – Complex implementation, relatively easy use.

The model has been further developed (Picioreanu *et al.* 2008) integrating it with the IWA Anaerobic Digestion Model No.1 (Batstone *et al.* 2002) in order to describe the interactions between microorganisms present in anaerobic ecosystems and bioelectroactive bacteria. The multi-biomass model version predicts kinetic-based competitive preponderance of ARB against methanogens at high load currents. The significant degree of expertise required to implement and manage the model and the large number of parameters is counterbalanced by the model's great capability to not only describe typically observed behaviour in a wide range of conditions, but also to provide insight into the biofilm structure and bacterial interactions in anodic biofilms. The strengths of this particular model can be put to very good use in the fundamental research on bioelectroactive biofilms with a very wide range of applicability.

Kato-Marcus *et al.* (2007) presented a dynamic, one-dimensional model of a biofilm anode in which electron transfer is modelled using a semi-empirical conduction-based approach. The model assumes the biofilm to be a conductive solid matrix characterized by a lumped conductivity parameter. This model is also capable of predicting typically observed data and electrode kinetic dependencies on both substrate concentration and electrical potential. The use of a semi-empirical lumped conduction-based electron transfer mechanism combined with a one dimensional model and one single active biomass species, considerably simplifies the model in terms of manageability and number of parameters. The model focuses on the anode biofilm and apparently does not provide predictions of pH and species concentrations other than the concentrations of electron donor and biomass. Ohmic losses due to ion movements and pH gradients are not considered. These unmodelled behaviours might be seen as limitations to its application in complete MFC systems and/or for studies of potential and substrate concentration in other than anode-limited current systems. However, the conduction based approach for electron transfer provides a semi-empirical way of integrating electron transfer into more comprehensive MFC models without greatly increasing model complexity. Depending on modelling objectives, the approach used can be appropriate and may even be integrated into more comprehensive and mechanistic models including all BESs elements.

Zeng *et al.* (2007) presented a simple model of a complete MFC with CSTR compartments and no biofilm. Kinetics are expressed as functions of substrate concentrations and electrode overpotentials. The model describes the electrode's overpotential as a state variable and relates it to the current and reaction rates through a charge balance, assuming a constant electrode capacitance. This lumped parameter approach results in a simple semi-empirical description of the electron transfer mechanisms. If the modelling objectives are not specifically focused on the biofilm part of the MFC, then this approach could be appropriate for some applications. The model also includes a description of ion transportation through the membrane and describes the effect of cations (as opposed to protons) being transported through the membrane. The applicability of this model is limited by the lack of a biofilm model and by a more detailed description of chemical species. However, its structure based on first principles and observed kinetics together with its ease of implementation; make it interesting for predicting general behaviour in MFCs.

Hamelers *et al.* (2008) recently presented a model to describe bioanode kinetics in which electrochemical electron transfer and substrate oxidation kinetics are combined into a so-called minimal model. Though detailed information on the model is not available yet (thus not included in Table 21.2), the authors kindly reported that an experimentally based parameter calibration was conducted and that it was possible to distinguish between the influences of biochemical and of electrochemical parameters in the anode kinetics, by using their model.

Oh and Sloan (2008) recently developed a model which specifically focused on the description of electron and proton transfer in an anode biofilm electrode. The model presents a detailed chemical and thermodynamic description of the relevant species considered and is highly focused on the electron and proton transport. The model superimposes electrolyte theory, in which electrons and protons jump through levels of potential between the electrode and the bulk media, with conventional electrode theory (Gouy-Chapman), in which an electric field governs charge distribution and electrical conductivity. The model does not explicitly consider alternative electron transfer mechanisms, such as mobile redox shuttles, but the authors were able to hypothesize a mechanistic explanation for some experimentally observed phenomena such as the uncoupling between electrode and chemical potential in some batch MFC experiments. The perceived strengths of this model are likely to find application in the detailed investigation of electron and ionic species transport mechanisms in BESs.

Rodríguez *et al.* (2007) developed a comprehensive mass balance-based mathematical model incorporating the elements and processes assumed to be of importance in BESs while aiming to constrain the model complexity down to a manageable level. The model includes both anode and cathode bulk compartments modelled as CSTRs and a one-dimensional anode biofilm modelled using discretized layers. Although one single lumped electroactive microbial species is considered, its metabolism and biomass yield are functions of the local bioenergetics (defined by local concentrations) by using a thermodynamic correlation (Heijnen 1999) evaluated at each time step. The model incorporates detailed comprehensive mass and charge balanced chemical speciation for both electrode compartments and biofilm in a similar manner to that used in the ADM1 model (Batstone *et al.* 2002). The model has been further developed (Rodríguez *et al.* 2009) and a configurable ion selective membrane, allowing for specific ion motilities and selectivity with respect to protons, has been included. Electron transfer from the microorganisms to the anode electrode is assumed to occur by two mechanisms; namely diffusion of mobile electron shuttles and conduction (through nanowires, direct contact or equivalent immobile mechanisms). The model implementation follows a recently published methodology (Rodríguez *et al.* 2008a) that allows increased flexibility and ease of structure modification through the integration of *Excel* and *Matlab* capabilities. Improvements in the current development of this model include the incorporation of additional microbial groups in the BESs biofilm. This model provides a relatively comprehensive description of the key elements and processes in MFCs while maintaining a moderate-to-high level of complexity, partly mitigated by the implementation methodology used. The model's perceived strengths could facilitate its application to control, given its complete mass balanced chemical speciation of the complete system. It's main potential may however, lie in building intuition and informing BESs research. The whole system's approach used, and its high level of detail balanced against model complexity, are seen as strengths.

Although partial BESs models based on lumped parameter equivalent circuit approaches derived from frequency domain empirical studies and electrical impedance spectroscopy (EIS) have also been published (e.g. Manohar *et al.* 2008; Ramasamy *et al.* 2008), they will not be reviewed further here. Their interpretability and applicability are limited by the fact that, a plurality of different equivalent circuits and parameters are able to represent the data used for parameterization. EIS however, remains a useful tool and such models may find utility with specific objectives suited to this technique (Barsoukov and Macdonald 2005).

There is a significant risk in any complex model with a large number of parameters, as is the case in some of the above models, that in the absence of proper parameter identification, model predictions may turn out to be inconclusive. For this reason, outputs from complex models must be always carefully analysed and discussed.

21.6 CURRENT CHALLENGES IN BESs MODELLING

Although many different aspects of BESs are currently under investigation (Rozendal *et al.* 2008), in this section, discussion of a number of specific BESs elements and processes, which are currently the subject of both research and modelling efforts, is presented. The authors have considered these aspects of interest also because they relate closely to existing models.

21.6.1 Bioelectrode kinetics

Experimentally, the bioanode kinetics, as evidenced by the electric current, display particular behaviours which may be notionally disaggregated into specific characteristics as illustrated in Figure 21.2, each generally having different response time scales:

1. The observed magnitude of the anodic current increases with an increase in the anode potential, until saturation is reached (time scale ~milliseconds).
2. The magnitude of the current increases with an increase in the concentration of electron donor, following a saturation-type function (time scale ~seconds).
3. The magnitude of the current increases with an increase in the amount of electroactive biomass attached to the electrode (timescale ~hours).
4. The magnitude of the current increases with an increase in the active area of the electrode (dependent on colonization of the surface area).

The first characteristic behaviour is observed when the activation overpotential controls the overall reaction rate, observed as current flow. The activation overpotential is essentially a potential loss, analogous to an activation energy and is directly related to the electrode potential, providing the electrochemical driving force. Saturation of the activation overpotential control is observed at the point where a mass action limitation comes to dominate. This is caused by a limiting concentration of the effective chemical species acting as electron donor for the electrode. Such a limiting concentration is coupled more or less directly with the bioelectroactive microbial reaction rate, depending on which mechanisms are involved in transferring electrons from the microorganisms to the electrode (e.g. mobile shuttling or conduction).

When soluble mobile chemical species act as electron shuttles or mediators between microbial and electrode reactions, mass limitations of shuttle species may arise because of an insufficiently high concentration of these components at the electrode surface. This insufficient concentration can be caused by slow mass transport to/from the electrode and/or a slow microbial reaction rate which is responsible for the production of their reduced speciation forms. Thus, when these soluble mobile electron shuttles are considered, separate specific kinetics need to be defined, namely (i) the microbial reaction rate that consumes the substrate and reduces the electron shuttle (controlled mainly by substrate and

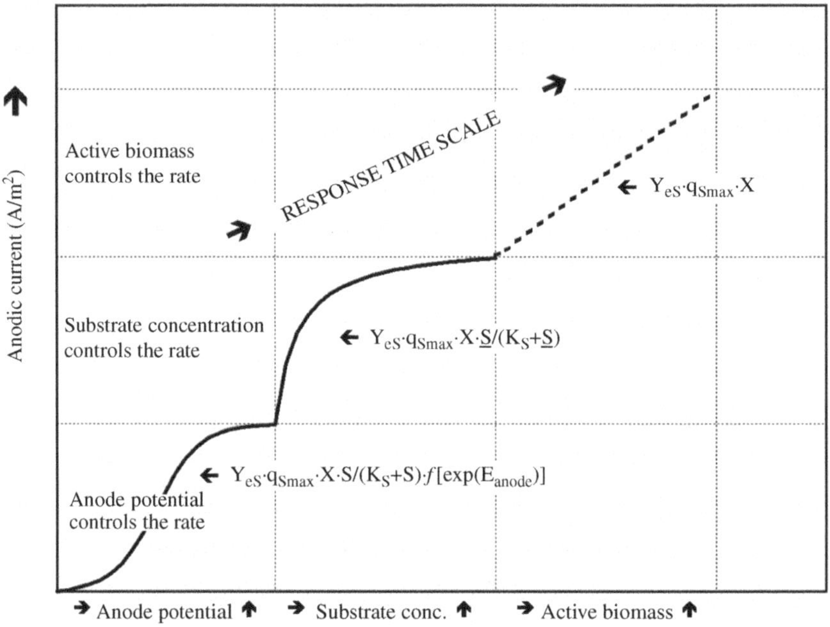

Figure 21.2. Representation of the sequence of limiting factors in observed anode kinetics. The electrode potential controls the rate until a mass limitation arises from a substrate limited microbial kinetics. The amount of active biomass becomes ultimately the limiting factor under non-limiting potential or mass transport. (Y_{eS}:number of electrons per mole of substrate [mol_e/mol_S]; q_{Smax}:maximun biomass-specific substrate uptake rate [$mol_S/mol_{CX} \cdot h$]; X:biomass concentration [mol_{CX}/L]; S:substrate concentration [mol_S/L]; K_S:substrate half saturation constant [mol_S/L]; E_{anode}:anode electrode potential [V])

oxidized electron shuttle concentrations), (ii) the mass transport rate of reduced and oxidized shuttles between microbes and electrode (controlled by concentration gradients and associated diffusion rates) and (iii) the electrode reaction rate as described above (controlled by electrode activation overpotential and reduced shuttle concentration at the surface of the electrode). The existing BESs models which assume mobile electron shuttles to be an electron transfer mechanism have considered such separate specific kinetic expressions (Picioreanu *et al.* 2007; Rodríguez *et al.* 2007; Zhang and Halme 1995).

If conductivity-based mechanisms are assumed to be responsible for the electron transfer between the microbes and the electrode, potential and substrate concentration control of the reaction rate appear much more directly coupled. This is a consequence of the much shorter response time scale of conduction-based electron transfer mechanisms, compared to those based on diffusion of mobile shuttles. The BESs models that consider conduction-based electron transfer, have both electrode and microbial reaction rates directly coupled by integrating Monod-type microbial kinetics with the electrode reaction kinetics into single kinetic expressions (Hamelers *et al.* 2008; Kato-Marcus *et al.* 2007; Zeng *et al.* 2007).

A second saturation is observed when a maximum reaction rate, associated with the microbial catalysis, is reached. This would mean that neither substrate nor electrode potential would be limiting the overall rate, rather the amount of bioelectroactive biomass (catalyst) would be the limiting factor. Because of the anisotropy (or lack of homogeneity) and the complex, time varying metabolic diversity of mixed microbial biofilms, the relationship between the specific mass of biomass and the current will follow nonlinear relationships. The current may also depend on other complicating factors such as the fraction of bioelectroactive biomass, mass transport, additional potential losses, distance, etc. and these other factors will eventually lead to saturation levels as well.

Another kinetic dependence (not represented in Figure 21.2) is that associated with the active biofilm surface area which depends largely on the morphology of the electrode and reactor design. In the absence of bulk liquid concentration gradients a more linear dependence between electrode area and current is expected, though the electrode's own heterogeneities and scale factors will disturb this *quasi* linearity at some point.

Considering the above it is clear that the validity of any empirical estimation of kinetic parameters depends entirely on the ability of the identification methods used to reveal the actual kinetics from data observed. Typically, this will be valid for a range in which the limiting rate is suitably represented by the kinetic equation which is to be calibrated. This means, for example, that determination of Monod kinetic parameters within an experimental range in which the electrode potential dominates the overall system rate, will lead to meaningless Monod parameter estimations. Calibration of kinetic equations must be therefore always accompanied by appropriate identification procedures if the specific kinetic parameter values are to be trusted.

21.6.2 Electron transfer mechanisms

Several mechanisms of electron transfer between the microorganisms and the electrodes have been reported or postulated in the literature for BESs. So far, existing models have contemplated electron transfer by two principal groups of mechanisms: (i) soluble mobile electron shuttles (mediators) and (ii) conduction through a specific medium or structure. The conduction mechanism, with only minor modifications, can include all direct contact, nanowires and conductive biofilm matrix mechanisms reported or postulated. The major difference between the two groups of mechanisms is that mobile electron shuttle mediators exhibit diffusional behaviour and conduction-based mechanisms will behave mainly in accordance with Ohm's law.

Perceived challenges for modelling in this area are not only to elucidate the dominant electron transfer mechanism(s) (and therefore which mechanisms require the modelling efforts), but also to assess to what extent detailed modelling of these mechanisms can contribute to BESs performance improvements. Another challenge lies in the design of identification experiments for both the structure and parameters used to describe electron transfer to higher levels of detail beyond solely the prediction of the electrode kinetics from electrode potential and substrate concentrations.

21.6.3 Microbial activity: bioenergetics and kinetics

The volumetric rate of formation of a product of interest in a bioprocess is typically described by the multiplication of three terms (i) a biomass specific reaction rate; (ii) a product yield and (iii) a biomass concentration. In the case of BESs, the electric current is seen as a key product formation rate of interest and can therefore be described in the same manner. Three factors therefore can describe the current production in a BESs (i) the specific reaction rate as defined by the bioelectrode kinetics; (ii) the product yield that defines how much current (product) is produced per unit of substrate metabolized and (iii) the amount of biomass (the biocatalyst), which is largely dependent on the biomass yield, in turn, determining how fast biomass grows and ultimately how much will be present in a mature biofilm. This viewpoint can serve as the underlying basis for the following discussion.

In anaerobic microbial ecosystems, like those occurring in bioelectroactive biofilms, many different types of microorganisms interact and compete with each other for survival and growth. The prevalence of one or other type of organism will depend mainly on the environmental conditions they experience, which will select for those organisms best suited. Anaerobic ecosystems are generally characterized by a limitation in the amount of energy available for maintenance and growth and organisms compete for energetic niches in which they can live and grow. The prevalence of one or other microbial group depends not only on their metabolic rate (how fast they are), but particularly under energy limited conditions, also on their growth yield (i.e. on how much they make of every unit of substrate they convert). The theoretical maximum biomass growth yield is limited by the theoretical maximum amount of energy an organism can obtain from its catabolism and this can be calculated as a function of (changing) environmental conditions using thermodynamics (Rodríguez *et al.* 2008b).

In Figure 21.3 an example of how environmental conditions (pH and electrode potential in this example) affect the energetics of three possible catabolic pathways for acetate is presented. Depending on the conditions, those organisms following the highest energy yielding pathway can theoretically obtain much more energy (and therefore grow more per unit of substrate) than others. The figure also illustrates how, if the conditions change (e.g. due to product accumulation), the energetic scenario can also change making a specific pathway more or less energetically interesting or even thermodynamically impossible.

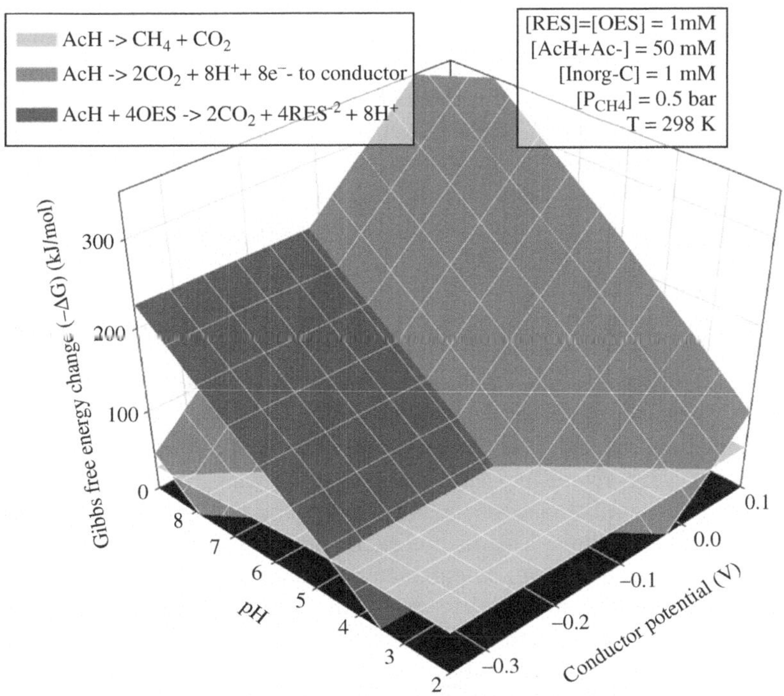

Figure 21.3. Theoretical illustration of the Gibbs free energy change (energetics) of three possible catabolic reactions from acetate at different pH and electrode potential values. This represents the theoretical maximum energy available for bacterial growth following each specific pathway. The effect of electrode potential and pH is calculated at the conditions indicated in the box. The reactions considered are acetoclastic methanogenesis (light grey), direct electron transfer to a conductor at the given potential (grey) and transfer into an oxidized electron shuttle OES (dark grey). (Note that the concentration values in the box have been chosen for illustration purposes and have no particular significance). (Adapted from Rodríguez *et al.* 2008b)

This likely effect of local concentrations (and potential) on biomass yields can have a potentially large, not only kinetic but also energetic, impact on the microbial population distribution within the biofilm. For example, abundance of specific microbial groups can differ greatly through the biofilm depth with both substrate and products concentrations, and potentials are very different so both the kinetic and energetic scenarios will differ.

Relatively easy prediction of biomass growth (and accordingly the complete metabolism) yields based on energetics are possible using existing thermodynamic correlations like that of Heijnen (1999). Picioreanu *et al.* (2007) have used this correlation in their model, as have Rodríguez *et al.* (2007) who also included the yield dependence on local concentrations as discussed above.

Considering the above, electric current and power production appear therefore interlinked with a few interesting factors. Total current is related to reaction rate, kinetics and total electroactive biomass present, which depends on biomass yield. Power, on the other hand, is a measure of energy released by the system and therefore, in a sense lost to microorganism growth. Power of course depends on both current and potential difference and the larger potential difference between electrodes (meaning lower anode potential given a consistent cathode) implies less energy for the microorganisms to grow. So there is clearly a trade-off between all these variables. These interactions in BESs can be seen as analogous to energy dissipation in any other bioprocess. In BESs, power is related to energy dissipated, which relates to available energy left for growth and to driving forces of the process kinetics.

Mathematical models which incorporate these relationships and interactions can provide very interesting information which can be complemented by detailed analyses and studies of the community and ecological interactions are present in BESs biofilms. This use of energetics is based on (thermodynamic) first principles and does not add parameters to the model.

BESs energetics can also be very closely related to electron transfer mechanisms. Depending on the conditions, the existence of any type of dominant transfer mechanism (conduction-based or electron shuttling) will ultimately lead to a decrease in the potential (energy level) at which electrons are donated when current (flux) increases (either by Ohm's law or by an increase in the concentration of reduced shuttle). This potential decrease causes also a decrease in the catabolic energy available for the microorganisms to grow, in a similar fashion to other anaerobic bioprocesses when products accumulate due to high metabolic flux. Since multiple pathways are commonly observed in anaerobic bioprocesses, leading to a spectrum of mixed products, in theory multiple electron transfer mechanism could also occur simultaneously in a BESs to overcome the same energetic limitations, thus widening the energy niche of survival.

In Figure 21.4 we have computed the potentials (energy levels) and the theoretical biomass yields for alternative mechanisms of electron transfer, namely via conduction or mobile shuttling respectively. By using realistic concentration and parameter values in BESs reactors, a maximum microbial growth rate is theoretically found at a point at which a combination of both mechanisms occurs. The existence of such a theoretical optimum distribution of electron transfer mechanisms (electron pathways) makes the hypothesis that microorganisms may follow this criterion a very plausible regulatory rule for electron transfer. This analysis serves as a good example of how modelling may inform and be informed by experimental research.

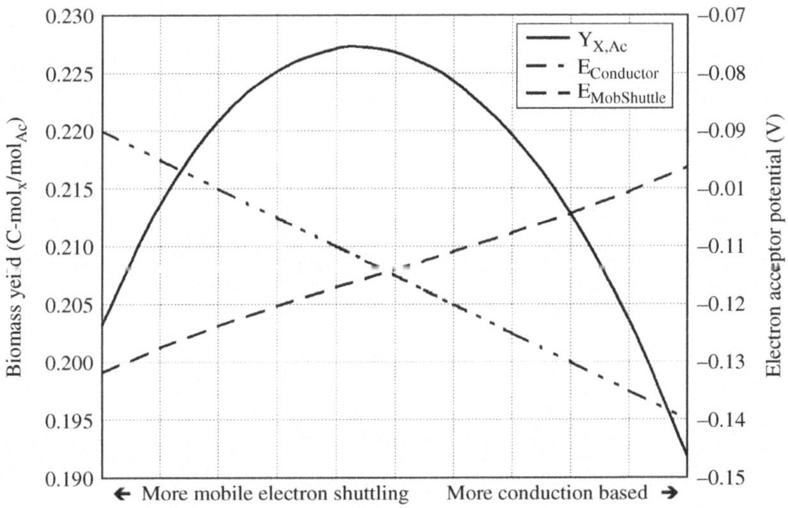

Figure 21.4. Theoretical representation of the energetic trade-off between electron transfer mechanisms and its theoretical effect on biomass growth yield. A maximum yield exists and, depending on the conditions, it may correspond to a combination of more than one electron transfer mechanism. Arbitrary, but realistic parameter values (not specified) were used just for the purpose of generating this illustrative numeric output

21.6.4 Mass transport – convection, diffusion and migration

Description of mass transport of chemical species both in the reactor bulk liquid and especially inside the biofilm is of major importance considering that mass transfer is often the limiting factor for current production in BESs, especially at high currents. It is known that chemical species can be transported by convection, diffusion and/or migration (if charged and an electric field exists).

Several of the existing models consider mass transfer of chemical species inside the biofilm through diffusion only, assuming that bulk compartments are perfectly mixed, and defining diffusion layers between bulk and biofilm phases.

Integrated diffusion and migration is usually described by the Nernst–Planck equation including a diffusion term in which concentration gradient is the transport driving force and a migration term in which electric field force on a charged species is the transport driving force added to the diffusion term.

To our knowledge, only Oh and Sloan (2008) have incorporated migration forces on charged species into an integrated transport equation in their BESs model showing that migration can be of importance, specially at high currents due to a strengthened electric field. Previously a model based evaluation of the effects of migration on the retention of charged mobile shuttles inside the biofilm had been conducted (Rabaey *et al.* 2007) using an early model version from Rodríguez *et al.* (2007). The simulations showed that migration may be important to accurately describe concentration profiles of charged electron shuttles inside the biofilm and therefore to better predict electrode kinetics and electron shuttle loss into the bulk medium.

The relative complexity of a rigorous calculation of the migration term, which requires the computation of the electrode electric field in an electrolyte, blending electrode and electrolyte theories (as conducted by Oh and Sloan (2008) in their model) may have dissuaded other models from implementing migration. An assessment of the actual impact of migration forces in BESs behaviour is necessary to determine when and whether its detailed modelling is required.

21.6.5 Biofilm and spatial modelling

There is no doubt that biofilms are a key element in BESs and they need to be included in any comprehensive BESs model. However, biofilms are very complex systems to describe given that they host a large number of processes which are affected by physical, chemical and biological variables, as well as spatial considerations. Biofilm modelling in the context of BESs must therefore be approached prudently because, while it is a crucial part of the system, it can also easily add disproportionate complexity relative to its utility for specific model applications.

Biofilm modelling has been extensively developed over the past two decades. One-dimensional and multidimensional biofilm models incorporating comprehensive descriptions of mass transport, biomass growth, death detachment and many other processes have been developed. In the context of BESs modelling some of the existing models have included simple one-dimensional biofilm implementations and also multidimensional biofilm (Picioreanu *et al.* 2008). Solution of multidimensional biofilm models will require the use of discretization techniques such as multidimensional finite element or finite difference models. Such models often require significant programming effort or are available through commercial software packages. Commercial packages largely facilitate the description of different geometries as a mesh of nodes and elements to which all transport, physical and other laws are applied. More specific components of BESs models such as microbial-related processes or electron transfer will require a degree of specific programming. Again, model complexity and time/resources needed for model development, debugging and simulation must be balanced against the objectives of a specific model and this affects particularly the level of detail in the biofilm description.

21.7 BESs MODELLING PERSPECTIVES

Alongside progress in BESs research in recent years, a number of significant BESs models have emerged. Among these models there are frequently shared characteristics while each also presents original elements. All of these models display potential for application to different purposes. The better understanding of the key regulatory mechanisms emerging from BESs research has been, or is expected to be reflected (somehow) in BESs models.

Many BESs researchers could probably put manageable BESs models to very good use, allowing them to generate and analyse simulations of their systems, describing at least the basic elements and processes. However, despite the existence of a number of such models, they remain somehow distant and inaccessible to many researchers. Accessibility could be achieved without the restraining influence of standardization, by a tool-type model which might incorporate simply but rigorous mass and charge balances of all relevant components and their ionic speciation, in a similar manner to, for example, the ADM1 (Batstone *et al.* 2002). Furthermore, a description (perhaps semiempirical) of the dependence of current on species concentrations and electrode potentials would be necessary. Such a model could have several applications including, for example, opening research paths into basic BESs control and optimization or evaluation of feasibility of BESs integration into, for example, wastewater treatment processes.

Existing models clearly demonstrate that such a model is possible and coordinated work between BESs modellers and researchers could produce a standard tool model of great benefit to the BESs research community. The development of such a model need not prevent further advancement in BESs modelling and could conversely create modelling activity while providing a useful and manageable tool model.

In order to progress in BESs model development beyond only validation and identification of existing models, there are questions that need to be addressed to drive model development efforts in some appropriate directions. The authors would like to highlight a few of them:

- What level of detail is desirable in biofilm description (1D, multi-dimensional)? Which processes should be included and what assumptions are acceptable?
- What is the optimum number of biomass species (groups) to be considered to attain a good description of the system without causing parameter identifiability problems?
- Does variable microbial metabolism offer a manageable alternative to multiple kinetically competing species in energy limited BESs?
- Is transport of charged species by migration due to electrical field forces important enough for its modelling to be required? How could it be rationalized?
- How is electron transfer into solid electrodes facilitated and how detailed do the models of these mechanisms need to be?

Addressing these questions, amongst several others, may help to focus modelling and research efforts. Despite the great achievements in BESs research in recent years, many unknowns still remain and there is much room for development in which mathematical models will certainly play a crucial role.

21.8 ACKNOWLEDGEMENTS

The European Commission under the Marie Curie Host Fellowship TOK scheme (MTKD-CT-2004-509821) and the UK EPSRC SUPERGEN Biological Fuel Cell project (EP/D047943/1-Grant 68-3A75-3-150).

REFERENCES

Barsoukov, E. and Macdonald, J. R. (2005) *Impedance Spectroscopy: Theory, Experiment, and Application*, 2nd edn. John Wiley & Sons Inc., New York.

Batstone, D.J., Keller, J., Angelidaki, I., Kalyuzhnyi, S.V., Pavlostathis, S.G., Rozzi, A., Sanders, W.T.M., Siegrist, H. and Vavilin, V.A. (2002) *Anaerobic Digestion Model No.1*. IWA Publishing, London.

Disalvo, E.A. (1980) Coupling between electrochemical and biochemical reactions. *Bioelectrochem. Bioenerg.* **7**(4), 807–815.

Dochain, D. and Vanrolleghem, P.A. (2001) *Dynamical Modelling and Estimation in Wastewater Treatment Processes*. IWA Publishing, London.

Hamelers, H.V.M., ter Heijne, A., Rozendal, R.A. and Buisman, C.J.N. (2008) Analysis of bio-anode kinetics using a minimal model. *1st International Symposium on Microbial Fuel Cells*. The Pennsylvania State University.

Heijnen, J.J. (1999) Bioenergetics of microbial growth. In *Encyclopedia of Bioprocess Technology: Fermentation, Biocatalysis and Bioseparation*. Flickinger M.C. and Drew S.W. (eds), John Wiley & Sons, New York.

Kato-Marcus, A., Torres, C.I. and Rittmann, B.E. (2007) Conduction-based modeling of the biofilm anode of a microbial fuel cell. *Biotechnol. Bioeng.* **98**(6), 1171–1182.

Manohar, A.K., Bretschger, O., Nealson, K.H. and Mansfeld, F. (2008) The use of electrochemical impedance spectroscopy (EIS) in the evaluation of the electrochemical properties of a microbial fuel cell. *Bioelectrochemistry* **72**(2), 149–154.

Oh, S.T. and Sloan, W.T. (2008). Modelling electron and proton transfer mechanism in the anode chamber microbial fuel cell (submitted for publication).

Picioreanu, C. and van Loosdrecht, M.C.M. (2002) A mathematical model for initiation of microbiologically influenced corrosion by differential aeration. *J. Electrochem. Soc.* **149**(6), B211–B223.

Picioreanu, C., Head, I.M., Katuri, K.P., van Loosdrecht, M.C.M. and Scott, K. (2007) A computational model for biofilm-based microbial fuel cells. *Water Res.* **41**(13), 2921–2940.

Picioreanu, C., Katuri, K.P., Head, I.M., van Loosdrecht, M.C.M. and Scott, K. (2008) Mathematical model for microbial fuel cells with anodic biofilms and anaerobic digestion. *Water Sci. Technol.* **57**(7), 965–971.

Rabaey, K., Rodriguez, J., Blackall, L.L., Keller, J., Gross, P., Batstone, D., Verstraete, W. and Nealson, K.H. (2007) Microbial ecology meets electrochemistry: Electricity-driven and driving communities. *ISME J.* **1**(1), 9–18.

Ramasamy, R.P., Ren, Z.Y., Mench, M.M. and Regan, J.M. (2008) Impact of initial biofilm growth on the anode impedance of microbial fuel cells. *Biotechnol. Bioeng.* **101**(1), 101–108.

Rodríguez, J., Batstone, D.J., Yuan, Z. and Rabaey, K. (2007). A comprehensive model to investigate microbial fuel cells. *11th World Congress of Anaerobic Digestion*. Brisbane, Australia.

Rodríguez, J., Guwy, A.J., Dinsdale, R. and Premier, G.C. (2008a) A implementation framework for wastewater treatment models requiring a minimum programming expertise. *Wat. Sci. Technol.* **59**(2), 367–380.

Rodríguez, J., Lema, J.M. and Kleerebezem, R. (2008b) Energy-based models for environmental biotechnology. *Trends Biotechnol.* **26**(7), 366–374.

Rodríguez, J., Batstone, D.J., Yuan, Z. and Rabaey, K. (2009) A comprehensive model for microbial fuel cells (submitted for publication).

Rozendal, R.A., Hamelers, H.V., Rabaey, K., Keller, J. and Buisman, C.J. (2008) Towards practical implementation of bioelectrochemical wastewater treatment. *Trends Biotechnol.* **26**(8), 450–459.

Varfolomeyev, S.D., Yaropolov, A.I. and Karyakin, A.A. (1993) Bioelectrocatalysis – the Electrochemical Kinetics of Hydrogenase Action. *J. Biotechnol.* **27**(3), 331–339.

Videla, H.A. and Arvia, A.J. (1975) Response of a bioelectrochemical cell with *Saccharomyces cerevisiae* metabolizing glucose under various fermentation conditions. *Biotechnol. Bioeng.* **17**(10), 1529–1543.

Zeng, Y., Choo, Y.F., Kim, B.H., Mu, S.J. and Wu, P. (2007) Modeling and simulation of microbial fuel cell. *11th World Congress of Anaerobic Digestion*. Brisbane, Australia.

Zhang, X. and Halme, A. (1995) Modelling of a microbial fuel cell process. *Biotechnol. Lett.* **17**(8), 809–814.

22

Outlook: Research directions and new applications for BES

Jürg Keller, René A. Rozendal, Largus T. Angenent, Uwe Schröder, Piet Lens and Korneel Rabaey

22.1 BES RESEARCH – FOCUS ON THE APPLICATION

As the previous chapters in this book extensively demonstrate, there is substantial scope for application of BESs in the future. However, it is also clear that there is still a major need for ongoing and, in fact, expanded research in this area. This chapter will try to highlight some of the key areas where such R&D efforts could be concentrated on. We will deliberately take a broader view than just the fundamental scientific research since the successful application of this technology will require considerably more research and development work than is currently being undertaken. As with any new or re-discovered technology field, the fundamental principles and processes need to be reassessed to deliver

 Bioelectrochemical Systems: From Extracellular Electron Transfer to Biotechnological Application. Edited by Korneel Rabaey, Largus Angenent, Uwe Schröder and Jürg Keller. ISBN: 9781843392330. Published by IWA Publishing, London, UK.

an understanding of the underlying factors affecting the performance and the behaviour, before the technology can then be used for more applied research.

In this context, we argue that the BES area is now at the stage where the emphasis should be increased on applied research and development aspects. Basic scientific research should still be continued, though, but its directions could be more influenced by the application-oriented needs and challenges. Significant breakthroughs are likely possible in the pursuit of such activities.

22.2 FUNDAMENTAL RESEARCH DIRECTIONS

22.2.1 Understanding bioelectrochemical process fundamentals

There is no doubt that the fundamental research activities into BESs will continue, and that there is good justification for this ongoing effort. For one, the underlying processes in BESs are scientifically highly novel and in many ways still not well understood. Bacteria directly transferring electrons to and from a solid matrix pose quite a unique situation that has not been investigated thoroughly until fairly recently (see Chapters 4 and 5). Such biologically mediated processes can occur at the anode or cathode of a BES, with a range of substrates and products in different phases (gaseous, liquid, solid) and under many different conditions. The closest analogy for the anodic oxidation of organics is extracellular electron transfer towards insoluble metal oxides. Here, BESs offer the opportunity to gain a much more comprehensive insight into the physiology of extracellular electron transfer, as BESs allow to: (i) control the potential of the electron acceptor; (ii) control the flow/flux of electrons within the system; and (iii) to maintain chemically consistent conditions during the experimental period. We have barely scratched the surface of microbial physiology towards extracellular electron transfer, especially on the cathode side, and are likely to see a number of unexpected processes occurring at these electrodes.

One particular question that remains to be explored is the range of microorganisms capable of direct electron transfer, and the associated versatility in electron transfer strategies. This not only concerns different phyla of the bacterial domain, but most interestingly, also representatives of other domains, such as Archaea, fungi, etc. Recently, it was suggested that Archaea could directly receive electrons from cathodes (Cheng *et al.* 2009), while mediated electron transfer from a yeast to an anode was also established (Chiao *et al.* 2006). Are these organisms capable of direct electron transfer, or maybe able to use some assisted transfer methods, such as mediators produced by other organisms within the biofilm? Their role has thus far been all but unexplored, and warrants further attention given the unique processes that some of the members of these domains are able to perform, e.g. methanogenesis. What role could they possibly play in an integrated consortium with bacterial cells? Given the highly specialized enzymes that some fungi have, they could possibly also contribute to the breakdown and electrochemical conversion of some of the more recalcitrant compounds, which is particularly attractive for BESs aiming at (lingo)cellulosic biomass as feedstock.

Furthermore, there is the equally fascinating science of the actual extracellular electron transfer mechanisms. Which roles do mediators or potentially nano-wires play and where are they actually most beneficial in an electrode-bound biofilm? How do they enable or affect the interactions between different microorganisms in the biofilm and the electrode itself? It has been suggested that bacteria can form a complex grid, electrically linking organisms over long distances (Ntarlagiannis *et al.* 2007). As is the case in many complex consortia, there is likely not one single transfer mechanism that is most efficient in all situations. In fact, there may well be a co-existence of several of these mechanisms in the electrochemically active biofilm, as indicated earlier by Rabaey and coworkers (Rabaey *et al.* 2004). We need to carefully explore the various factors that may influence the establishment of such mechanisms within the biofilm, which would most likely include a multitude of physical and chemical parameters, such as distance to the electrode, convective vs. diffusive mass transfer, electrical potential gradients within the biofilm, substrate and product gradients and their effects, such as the localized pH conditions, etc. Understanding extracellular electron transfer on a genomic and proteomic level will allow modifying organisms that were previously not capable of extracellular electron transfer, or introducing novel capabilities in electrochemically active bacteria.

Interestingly, many of these electrochemical processes and mechanisms may even occur in systems other than the BESs. One example that is potentially highly relevant is external electron transfer in fermentative anaerobic systems. It is generally accepted that fermenting organisms function best in very close proximity with hydrogen utilizing methanogens as the fermentation processes produce hydrogen (Batstone *et al.* 2006). However, given that hydrogen production and consumption is primarily a mechanism for electron transfer between these different synergistic partners, the question remains whether hydrogen is actually produced and transferred to the hydrogen-utilizing organisms, or whether they could just transfer the electrons directly from one cell to the neighbouring one. This would shift the process from 'interspecies hydrogen transfer' to 'interspecies electron transfer'. Recent work by Ishii *et al.* has focused on the interaction between *Pelotomaculum* and *Methanothermobacter* (Ishii *et al.* 2005), which coaggregate using a pilus, suggested to be conductive (Gorby *et al.* 2006). Interestingly, such syntrophic associations can be highly specific between two organisms (Shimoyama *et al.* 2009), and as yet there is no information whether such specificity is upheld within BES biofilms. Beside this example, there might be numerous other systems where the extracellular electron transfer facilitates direct interactions between different organisms in a biofilm or also within flocs or cellular conglomerates.

22.2.2 Practically inspired fundamental research areas

There is also a range of practically inspired challenges to be explored on a more fundamental level. One such example is the selection and enrichment of acidophilic or alkaliphilic electro-active organisms that could operate well in an anode under very low pH or in a cathode at a high pH. From this perspective, fungi could again be attractive as many are known to have very low or high pH optima. Bacteria may be attractive due to their genetic tractability, allowing straightforward modification to achieve a desired process, e.g. bioproduction of chemicals at the cathode. This in turn could considerably help to operate BESs more efficiently due to the inherently strong proton production or consumption processes at the anode or cathode, respectively.

Another example in this context is the specific oxidation or reduction of metal ions to extract or at least immobilize them from contaminated soils and groundwater streams. Substantial work on the removal of uranium has been performed (Gregory and Lovley 2005). Further fundamental work needs to be undertaken to investigate the often complex redox processes of these metal ions within the complex matrix they are usually found in. These findings can then be tested again in practical applications, such as *in-situ* bio-remediation (see Section 22.6).

The use of selected isolates of microorganisms, or even genetically modified organisms (GMOs), could be a further fundamental development to address specific practical limitations, as elegantly discussed in Chapter 6. The application of this strategy will mainly be viable for the cathodic processes, given that the anodic reactions primarily consist of oxidation of organic substrates to CO_2 and water. This can be efficiently achieved by a wide range of organisms and is also suitable for use with non-sterile organic feeds, such as wastewater streams. Given the separation of the cathodic compartments through an ion-permeable membrane, the use of pure cultures or GMOs is certainly a practically relevant option, even when wastewater is used as the electron-supply in the anode. There may be a surprisingly large range of cathodic reduction processes possible.

One example might be the selective reduction of glycerol (typical byproduct or waste product of bio-diesel production) to 1,2- or 1,3-propanediol, which are widely used organic base chemicals. To be cost-effective, it is likely necessary to develop highly optimized cultures of organisms capable of these selective transformations, or even the creation of GMOs specifically suited for these processes. This approach will likely need a long-term, well-funded project that requires the close collaboration between highly experienced genetic engineering researchers and more application oriented biotechnologists or process engineers to achieve the required selectivity and turnover.

22.3 APPLIED RESEARCH OPPORTUNITIES

22.3.1 Contributions and limitations of current research activities

Given the proliferation of publications on MFCs and BES in the last few years (see Chapter 1), it is valuable to evaluate what aspects are covered well and what knowledge gaps exist. This is focused mainly on the more applied research papers as much of the literature in recent times is in this area.

From a practical viewpoint there have been a number of key innovations and breakthroughs made in the last few years. Some of the major ones include:

- the use of relatively low-cost materials for electrodes and membranes, such as graphite felt (Chaudhuri and Lovley 2003) or granules (Rabaey *et al.* 2005), carbon fibre brushes (Logan *et al.* 2007) or the Ultrex cation exchange membrane (Rabaey *et al.* 2003);
- development of open-air cathodes with metal catalysts (Cheng *et al.* 2006; Zhao *et al.* 2005) or even biologically catalysed (*Clauwaert et al.* 2007b);
- development of the Microbial Electrolysis Cell, aiming at hydrogen production (Liu *et al.* 2005c; Logan *et al.* 2008; Rozendal *et al.* 2006b);
- development of stacked MFC modules to gain higher voltages (Aelterman *et al.* 2006);
- discovery of denitrification processes in the cathode of BES, and biocathodes in general (Clauwaert *et al.* 2007a; Gregory *et al.* 2004);
- demonstration of selective sulfide oxidation to sulfur on anodes (Dutta *et al.* 2009a; Dutta *et al.* 2008; Rabaey *et al.* 2006);

In terms of process optimization, there have also been substantial achievements made related to improved designs with minimized distances between anode and cathode as well as modifications in the materials and feed solutions used. On the latter, many experimental studies, however, are now employing unrealistic artificial 'wastewater' compositions, particularly in relation to the phosphate buffer concentrations (Liu *et al.* 2005a). While a high phosphate buffer concentration clearly helps in achieving maximal power outputs due to improved conductivity and high buffering capacity (Rozendal *et al.* 2008), it is also obvious that wastewater streams with phosphate-P concentrations of several hundred mg/l do not exist in any realistic situation. Addition of such levels of phosphate to actual wastewater streams would also be impossible for a number of reasons, not the least the costs and required discharge limits. Therefore, some of these 'high performance' systems have very limited, if any practical relevance.

A further main focus has been the demonstration of a range of organic substrates that could be used in the anode of a BES (Cheng and Logan 2007; Rabaey *et al.* 2005). This provides some useful contributions to determine the possible application areas of such systems. Based on these studies, it is evident now that most soluble, easily biodegradable organic substrates can be used in a BES anode. For instance, simple Volatile Fatty Acids (VFAs), such as acetate, propionate and possibly butyrate, are clearly preferable substrates (Liu *et al.* 2005b). Moreover, many other substrates would likely be converted through fermentation processes in the anode compartment or biofilm to these preferred substrates (Freguia *et al.* 2008b). Thus, they are usually also suitable, although possibly at reduced conversion rates. The use of more slowly degradable organics, including particulates, is still a major challenge for BES anodes. Typically these are either washed through the anode compartment, or may be captured and then slowly hydrolysed and fermented before being used by the electro-active bacteria.

Any of these fermentable substrates can, however, give rise to another major challenge in the BES operation: methanogenesis. Due to the long solids retention time in the anodic biofilm and the fact that hydrogen is generated in fermentation processes, it is very difficult to avoid or limit the growth of hydrogen-utilizing methanogens in these compartments. Several studies with 'realistic' wastewater sources have highlighted this problem (Freguia *et al.* 2007; He *et al.* 2005; Lee *et al.* 2008; Torres *et al.* 2007). Given the strong greenhouse gas effects of methane (approximately 21 times as strong as CO_2), this methane production must be kept to a minimum, or the methane needs to be captured. Operation at low pH (typically below pH 5) may inhibit the growth of methanogens sufficiently to avoid the problem (Kim *et al.* 2004). However, at these low pH levels, most electro-active bacteria at the anode are also severely inhibited (Rabaey *et al.* 2003), limiting the performance of the system considerably.

Due to the loss of chemical oxygen demand (COD) in the form of methane from the anodic compartment, the actual COD converted to electric current will be reduced. This COD loss, together with other, often not explicitly studied factors, such as biomass growth and loss via secondary metabolites, seems responsible for the reported low coulombic efficiencies, frequently in the range of 10–50% (Freguia *et al.* 2007). This is clearly a major challenge for any practical application since the overall energy conversion efficiency in such cases would be typically well below 20%. This is not even half the energy conversion efficiency that is achievable in anaerobic processes with methane conversion to electric power via gas engines or turbines (Ross *et al.* 1996). In order for an MFC to be seriously considered for any energy generating application, the coulombic efficiency would have to be above 90% in general, which would still likely only give an energy conversion efficiency of less than 50% given the substantial internal resistance and potential losses at the anode and cathode. Therefore, any research that is focusing on energy recovery as a main objective has to achieve a very high coulombic efficiency, while also trying to minimize the internal resistance and overpotential losses at the electrodes.

22.3.2 BESs for wastewater treatment?

While the MFC can be considered primarily as a wastewater treatment process, its performance needs to be compared to the existing main technologies for this application, the aerobic (activated sludge) and anaerobic (methanogenic) processes. Table 22.1 makes a generalized comparison between these technologies to highlight the major advantages and challenges of the MFC technology. MFC technologies share with anaerobic processes the benefits of low sludge production and net energy generation, but they require a post-treatment to achieve a better effluent quality (mainly oxidized, not reduced compounds). The possible loading rate of MFCs is higher than aerobic processes, but currently still much lower than the anaerobic technologies. The single-stage process (including energy generation) is a significant advantage of the MFC technology, while the modular scale-up approach can be advantageous for smaller applications, but will not benefit greatly from economy-of-scale advantages compared to the traditional technologies. The major hurdle though is the capital cost of a MFC, which is currently much higher than the existing technologies, mainly due to the need for expensive electrodes, membrane and current collectors within the many cells that make up an MFC. In a recent review (Rozendal *et al.* 2008), these costs and their likely future reductions are analysed in detail, leading to the conclusion that these high capital costs severely limit the commercial application of MFCs as a wastewater treatment technology.

Table 22.1. Generic comparison between main wastewater treatment technologies and MFC (Adapted from Freguia 2007)

	Aerobic (Activated sludge)	Anaerobic (UASB)	MFC
Removal rate $kg_{COD}m^{-3}d^{-1}$	1	20	3
Sludge production (kg_{COD}/kg_{COD})	0.4–0.6	0.01–0.14	0.01–0.3
Energy production (kWh/kg_{COD})	−1.2	3.8 (heat) or 1.5 (electrical)	0.8 (today) 4.4 (theor.)
Effluent quality	Good	Poor	Good
Gas treatment	No	Likely	No
Stages of treatment	Single	Multiple	Single
Reactor scale-up	Volume	Volume	Modular
Capital cost	Medium (high volume cost)	Medium (high peripheral cost)	High

22.3.3 Is power the best product from BESs?

Achieving a practically valuable energy conversion in a MFC is thus a major challenge due to the inherent constraints in the process (see previous sections). For benthic MFCs, this limitation does not impede application, as the required power densities off shore are limited and the cost per unit power is not drastically constrained. However, reactor based (terrestrial) BESs need to successfully compete with well-established wastewater treatment technologies, most notably the anaerobic digestion process that also has a net energy generation potential. The key advantage of the BES in this context is the ability to operate efficiently at room temperature while methanogenic processes work best above 30°C (Pham *et al.* 2006). Furthermore, BESs likely work better under lower influent COD concentrations (Freguia *et al.* 2008a) in contrast to anaerobic processes. Moreover, at low COD concentrations anaerobic digesters particularly suffer from significant methane losses in the treated effluent due to the relatively high solubility of methane and the overall low methane production per volume of wastewater. Certainly at lower temperatures (ambient) this methane solubility can become problematic. Given these major limitations and the severe cost constraints outlined above, it appears that alternative products to electric power may need to be generated in a BES type approach rather than a strict MFC approach. This may create an efficient, cost-competitive application successful in the 'clean-tech' market place.

22.4 POTENTIAL NEW BES APPLICATIONS

The electrochemical nature of the BES processes offers the unique ability to control two separate (bio-)chemical reactions in the two half-cells. While the oxidation at the anode may be primarily used for the oxidation of organics in wastewater, the reduction processes at the cathode offer a wide range of possible applications. These are particularly interesting since the (bio-)electrochemical reactions at the cathode can be carefully controlled to achieve selective reduction conditions, which is usually not possible in anaerobic biological processes. Furthermore, if a membrane is used to separate the anode and cathode compartments (which is typically the case) then the cathode reactions can be largely independent of the anode processes except for the electron and cation transfers. This means that, for example, biological processes utilizing pure cultures can be established in the cathode, while wastewater is fed to the anode. This creates unique opportunities for a wide range of products to be generated at the cathode.

22.4.1 Novel options for cathodic reductions

There is likely a wide range of novel biologically or chemically mediated reduction processes possible at the cathode, leading to products of far greater value than the power generated by a MFC. While many of the electrochemical processes have already been demonstrated in principle (e.g. reduction of CO_2 to form hydrocarbons (Gattrell *et al.* 2006)), the bioelectrochemical reactions are largely unexplored.

In all of these applications, power input rather than power generation is required, which slightly adds to the operating costs but also enables these processes to proceed at quite rapid rates. The latter is possible because the applied voltage over the cell can compensate for internal resistance and overpotential losses. Typically these additional voltages are only in the range of several hundred millivolts plus the thermodynamic potentials of the specific reactions (Rozendal *et al.* 2006b), but they enable considerably higher process rates compared to the MFCs. The generation of hydrogen (H_2) gas at the cathode is not further discussed in this section as it is already quite well investigated and covered in detail in Chapter 13.

One of the most straightforward cathodic processes is based in part on the fact that the selectivity of the proton/cation exchange membranes is far from perfect (Rozendal *et al.* 2006a). Due to the much higher concentration of metallic cations than protons, the former are preferably transferred. Typically sodium ions (Na^+) are most abundant in the wastewater stream, and thus are substantially transferred to the cathode. The concurrent generation of hydroxide anions (OH^-) at the cathode create suitable conditions for the formation of sodium hydroxide solutions in the cathode compartment. This cathode should be able to achieve NaOH solutions up to several weight-%, but the process is still under development at present.

A further modification of the above process can be achieved by stabilizing hydrogen peroxide, formed during oxygen reduction, at the cathode. Hydrogen peroxide (H_2O_2) is a widely used chemical for bleaching purposes in paper and pulp industries, disinfection and advanced oxidation processes (together with UV irradiation or ferric addition). The concentration that can be achieved in such a cathodic reaction is unclear at present, but needs to reach at least around 3–5% to be of practical relevance in most applications. This solution will also have a high concentration of NaOH, which helps to stabilize the H_2O_2 and in many applications is directly useful, particularly in paper and pulp bleaching (Foller and Bombard 1995).

Both of the above mentioned products are potentially far more valuable than power generated in a MFC or even hydrogen formed in a MEC. Further research will be needed to elucidate the attainable production rates, quality of the end product(s) and the stability of the process itself.

Further prospective cathode reactions may include the selective reduction of glycerol to 1,2- and 1,3-propanediol. Glycerol is a by-product of the bio-diesel production, which is generated in increasingly large quantities and has very limited value due to the contamination with salt and water. On the other hand, propanediols are major basic building blocks of the large-scale organic chemistry industry, and therefore could be readily utilized (after appropriate clean-up/separation) in the industry. However, there are several major hurdles to overcome, particularly the selective production of 1,3- or 1,2-propanediol, the possible interferences of the impurities in the glycerol from biodiesel production, the selection of suitable biological or chemical (electro) catalysts and not the least suitable reactor configurations to achieve a high concentration of the products to help facilitate the downstream separation processes. This is an application where a well-performing, pure-culture microbial catalyst could be used exclusively in the cathode instead of the typically employed mixed culture populations.

These (bio)electrochemical reductions may also be applied for the transformation of strongly oxidized functional groups in persistent chemicals. For example, the nitro-group of nitrobenzene can be selectively reduced to aniline, which can be much more readily degraded in an aerobic biological wastewater treatment process than the original compound (Mu *et al.* unpublished results). Similarly, the azo bond (–N=N–) of the widely used reactive azo-dyes is completely non-reactive in aerobic biological processes, but can be slowly reduced under anaerobic conditions. However, using an electrochemical reduction, the azo bond can be readily cleaved to form amine groups and hence break up the molecule and eliminate the colour (Mu *et al.* 2009). The further degradation of the breakdown products will need to be achieved by the (biological) oxidation processes.

22.4.2 Novel options for anodic oxidations

As yet, the use of anodic oxidation for other objectives than power generation is largely unexplored. There is significant scope for the use of anodes beyond the oxidation of wastewater 'bulk' organics. Inorganic compounds, such as sulfides, can be removed, and recovered using an anodic/cathodic switching process (Dutta *et al.* 2009b). Anodes can be used to remove xenobiotics, such as phenol (Luo *et al.* 2009), diesel constituents (Morris *et al.* 2009) and even odorous compounds, as present in, for example, swine wastewater (Kim *et al.* 2008). In the future a consecutive sequence of cathodic reductions and anodic oxidations might even be possible to provide a more comprehensive removal of xenobiotics.

De Schamphelaire and coworkers investigated the use of BESs for the removal of root exudates from rice plants. Removing these exudates may help in decreasing methane emissions of wetlands (De Schamphelaire *et al.* 2008). Likewise, sediment based BESs could even enhance the cycling of manganese, leading to improved degradation of sediment and dissolved (recalcitrant) organics (De Schamphelaire *et al.* 2007).

22.5 BES INTEGRATION INTO PRACTICAL APPLICATIONS

Given the many applications, but also constraints and particularities of the various BES concepts, the proper integration of these processes in practical applications is critically important. One of the major challenges for any novel technology, such as the BES, is to find suitable niches for the initial applications, as these will unlikely be a complete wastewater treatment system replacing existing infrastructure. Therefore, applications are most likely smaller add-on installations to generate new products that can be used on-site or for processing of a smaller stream in parallel to an existing conventional treatment facility.

It is important to note here that BESs will require pre- and post-treatment of the wastewater. The effluent is unlikely to be of discharge quality, especially when optimized for high production rates, requiring elevated COD concentrations inside the reactor. Moreover, nitrogen and/or phosphorus removal are typically not achieved in the process. Pre-treatments include solids separation (sedimentation, floatation, etc.) and buffer tanks. The composition of the feed stream and achievable effluent quality will largely define how the process is integrated into an existing installation. Generally, the wastewater feed composition should ideally have a balanced COD and buffering capacity (to be able to neutralize the generated protons in the anode directly) (Freguia *et al.* 2008a). Therefore, mainly dilute streams with COD concentrations in the range of 200–600 mg/l are usually suitable. An excellent overview of available streams and their assessment is given in Chapter 18. In case of higher COD concentrations, the use of an anode–cathode recycle could allow us to balance the COD and buffering capacity in the feed water (Freguia *et al.* 2008a), although cross-over of organics from anode to cathode may be difficult to control. The latter could give rise to the formation of a heterotrophic biofilm on the cathode, leading to decreased oxygen availability for the cathodic process. Moreover, this approach is only feasible for MFC applications where the wastewater flows to the anode and then to the air-cathode. In BES applications, aiming at production of valuable chemicals, such a recirculation is not possible, as this would contaminate the product stream. As most wastewaters will have insufficient buffering capacity to remove all the COD, either a post-treatment needs to be provided, or the buffering capacity needs to be enhanced.

Given the above constraints, BES might be integrated as a carbon and sulfide polishing step after a high-rate anaerobic process (see also Chapter 19). There is usually only limited COD remaining, in conjunction with strong buffering capacity, which enable adequate removal at the anode (see Chapter 11). Furthermore, and different from typical aerobic processes, the electrochemical oxidation of sulfide to form elemental sulfur can also be performed at the anode (Rabaey *et al.* 2006), as detailed in Chapter 12. This would allow COD polishing and sulfide removal from the effluent of the anaerobic process, with the sulfur being deposited on the anode from where it can be recovered again (Dutta *et al.* 2009b).

The application of specific cathodic processes may be easier to integrate into existing treatment processes. Nitrate reduction on a biological cathode can be used as final polishing stage of a biological nitrogen removal process that typically has 2–5 mg/L NO_3-N remaining in the effluent after treatment (Virdis *et al.* 2008). The electron supply could come from a partial (or side-stream) treatment of the influent flow, which can then be further treated in the existing process train.

For the production of hydrogen (MEC), NaOH and H_2O_2, a separate (clean) stream of water could be used in the cathode with only limited flow, while a sidestream of the wastewater influent would flow through the anode (see also Chapter 19). Since the full treatment of that stream is not the objective (it serves simply as an electron donor flow for the system), a relatively high flow rate can be applied, hence not depleting the available buffering capacity in the wastewater. One of the major challenges though for this configuration is the scaling in the cathode due to the high pH and the likely presence of cations that precipitate with hydroxide ions (such as Ca^{2+}, Mg^{2+}). Methods to avoid such scaling need to be developed.

Finally, another proposed implementation route particularly for MFC applications is the installation of complete BES modules into existing treatment tanks (Min and Angelidaki 2008). The advantage of such an approach is the low cost in structural materials required, as existing infrastructure can be retrofitted. However, current collection and the cost-effective provision of sufficient aeration for the air-cathodes are challenging aspects. While any existing aeration system could be used in this case, the energy required for this is likely considerably higher already than the energy generated from the MFC installation. Given the substantial additional costs for such installations, an overall net benefit of such a system might be difficult to achieve.

22.6 CONCLUDING THOUGHTS ON THE FUTURE OF BES

Despite the somewhat negative assessment of the future of MFCs due to the high cost/benefit ratio, the future for BES in general is brighter than ever, notably with respect to bioremediation and bioproduction.

Bioremediation Specific beneficial processes such as nitrate or sulfide removal, (bio)electrochemical oxidation or reduction of recalcitrant chemicals that are difficult to remove by traditional wastewater treatment applications pose highly exciting prospects towards application. Such implementations are typically justified due to the often considerably higher costs of alternative treatments, such as advanced oxidation processes or anaerobic removal requiring substantial amounts of co-substrates.

Bioproduction. The most promising future application in our view, however, is in the area of product formation using the wastewater stream as a free electron supply, while the product generation is performed in the cathode. There is a wide range of options available that seem to provide very attractive investment returns even in the short-medium term. Most importantly, such applications can typically also be implemented in a staged approach with no or minimal impact on the existing wastewater treatment process, hence drastically lowering the implementation barrier at any potential full-scale site.

Overall, the BESs clearly offer a highly interesting and novel range of processes, both scientifically and from an engineering/application prospect. They enable process options that have not been possible so far and new reactions and applications will certainly be developed in future as well.

REFERENCES

Aelterman, P., Rabaey, K., Pham, T.H., Boon, N. and Verstraete, W. (2006) Continuous electricity generation at high voltages and currents using stacked microbial fuel cells. *Environ. Sci. Technol.* **40**, 3388–3394.

Batstone, D.J., Picioreanu, C. and van Loosdrecht, M.C.M. (2006) Multidimensional modelling to investigate interspecies hydrogen transfer in anaerobic biofilms. *Water Res.* **40**(16), 3099–3108.

Chaudhuri, S.K. and Lovley, D.R. (2003) Electricity generation by direct oxidation of glucose in mediatorless microbial fuel cells. *Naure Biotechnol.* **21**(10), 1229–1232.

Cheng, S. and Logan, B.E. (2007) Sustainable and efficient biohydrogen production via electrohydrogenesis. *Proc. Nat. Acad. Sci. USA* **104**(47), 18871–18873.

Cheng, S., Liu, H. and Logan, B.E. (2006) Power densities using different cathode catalysts (Pt and CoTMPP) and polymer binders (Nafion and PTFE) in single chamber microbial fuel cells. *Environ. Sci. Technol.* **40**(1), 364–369.

Cheng, S., Xing, D., Call, D.F. and Logan, B.E. (2009) Direct biological conversion of electrical current into methane by electromethanogenesis. *Environ. Sci. Technol.* In press.

Chiao, M., Lam, K.B. and Lin, L.W. (2006) Micromachined microbial and photosynthetic fuel cells. *J. Micromech. Microeng.* **16**(12), 2547–2553.

Clauwaert, P., Rabaey, K., Aelterman, P., DeSchamphelaire, L., Pham, T.H., Boeckx, P., Boon, N. and Verstraete, W. (2007a) Biological denitrification in microbial fuel cells. *Environ. Sci. Technol.* **41**(9), 3354–3360.

Clauwaert, P., Van der Ha, D., Boon, N., Verbeken, K., Verhaege, M., Rabaey, K. and Verstraete, W. (2007b) Open air biocathode enables effective electricity generation with microbial fuel cells. *Environ. Sci. Technol.* **41**, 7564–7569.

De Schamphelaire, L., Rabaey, K., Boon, N., Verstraete, W. and Boeckx, P. (2007) Minireview: The potential of enhanced manganese redox cycling for sediment oxidation. *Geomicrobiol. J.* **24**(7−8), 547−558.

De Schamphelaire, L., van den Bossche, L., Dang, H.S., Hofte, M., Boon, N., Rabaey, K. and Verstraete, W. (2008) Microbial fuel cells generating electricity from rhizodeposits of rice plants. *Environ. Sci. Technol.* **42**(8), 3053–3058.

Dutta, P.K., Keller, J., Yuan, Z., Rozendal, R.A. and Rabaey, K. (2009a) Role of sulfur during acetate oxidation in biological anodes. *Environ. Sci. Technol.* **43**(10), 3839–3845.

Dutta, P.K., Rabaey, K., Yuan, Z.G. and Keller, J. (2008) Spontaneous electrochemical removal of aqueous sulfide. *Water Res.* **42**(20), 4965–4975.

Dutta, P.K., Rozendal, R.A., Keller, J., Yuan, Z. and Rabaey, K. (2009b) Electrochemical regeneration of sulfur loaded electrodes *Electrochem. Comm.* **11**(7), 1437–1440.

Foller, P.C. and Bombard, R.T. (1995) Processes for the production of mixtures of caustic soda and hydrogen peroxide via the reduction of oxygen. *J. Appl. Electrochem.* **25**(7), 613–627.

Freguia, S. (2007) Fundamental studies of anodic and cathodic processes in microbial fuel cells, PhD thesis, Advanced Water Management Centre, The University of Queensland.

Freguia, S., Rabaey, K., Yuan, Z. and Keller, J. (2007) Electron and carbon balances in microbial fuel cells reveal temporary bacterial storage behavior during electricity generation. *Environ. Sci. Technol.* **41**(8), 2915–2921.

Freguia, S., Rabaey, K., Yuan, Z. and Keller, J. (2008a) Sequential anode-cathode configuration improves cathodic oxygen reduction and effluent quality of microbial fuel cells. *Water Res.* **42**(6–7), 1387–1396.

Freguia, S., Rabaey, K., Yuan, Z. and Keller, J. (2008b) Syntrophic processes drive the conversion of glucose in microbial fuel cell anodes. *Environ. Sci. Technol.* **42**(21), 7937–7943.

Gattrell, M., Gupta, N. and Co, A. (2006) A review of the aqueous electrochemical reduction of CO2 to hydrocarbons at copper. *J. Electroanal. Chem.* **594**(1), 1–19.

Gorby, Y.A., Yanina, S., McLean, J.S., Rosso, K.M., Moyles, D., Dohnalkova, A., Beveridge, T.J., Chang, I.S., Kim, B.H., Kim, K.S., Culley, D.E., Reed, S.B., Romine, M.F., Saffarini, D.A., Hill, E.A., Shi, L., Elias, D.A., Kennedy, D.W., Pinchuk, G., Watanabe, K., Ishii, S., Logan, B., Nealson, K.H. and Fredrickson, J.K. (2006) Electrically conductive bacterial nanowires produced by Shewanella oneidensis strain MR-1 and other microorganisms. *Proc. Nat. Acad. Sci. USA* **103**(30), 11358–11363.

Gregory, K.B. and Lovley, D.R. (2005) Remediation and recovery of uranium from contaminated subsurface environments with electrodes. *Environ. Sci. Technol.* **39**(22), 8943–8947.

Gregory, K.B., Bond, D.R. and Lovley, D.R. (2004) Graphite electrodes as electron donors for anaerobic respiration. *Environ. Microbiol.* **6**(6), 596–604.

He, Z., Minteer, S.D. and Angenent, L.T. (2005) Electricity generation from artificial wastewater using an upflow microbial fuel cell. *Environ. Sci. Technol.* **39**(14), 5262–5267.

Ishii, S., Kosaka, T., Hori, K., Hotta, Y. and Watanabe, K. (2005) Coaggregation facilitates interspecies hydrogen transfer between Pelotomaculum thermopropionicum and Methanothermobacter thermautotrophicus. *Appl. Environ. Microbiol.* **71**(12), 7838–7845.

Kim, I.S., Hwang, M.H., Jang, N.J., Hyun, S.H. and Lee, S.T. (2004) Effect of low pH on the activity of hydrogen utilizing methanogen in bio-hydrogen process. *Int. J. Hydrogen Energy* **29**(11), 1133–1140.

Kim, J.R., Dec, J., Bruns, M.A. and Logan, B.E. (2008) Removal of odors from swine wastewater by using microbial fuel cells. *Appl. Environ. Microbiol.* **74**(8), 2540–2543.

Lee, H., Parameswaran, P., Kato-Marcus, A., Torres, C.I. and Rittmann, B.E. (2008) Evaluation of energy-conversion efficiencies in microbial fuel cells (MFCs) utilizing fermentable and non-fermentable substrates. *Water Res.* **42**(6–7), 1501–1510.

Liu, H., Cheng, S.A. and Logan, B.E. (2005a) Power generation in fed-batch microbial fuel cells as a function of ionic strength, temperature, and reactor configuration. *Environ. Sci. Technol.* **39**(14), 5488–5493.

Liu, H., Cheng, S.A. and Logan, B.E. (2005b) Production of electricity from acetate or butyrate using a single-chamber microbial fuel cell. *Environ. Sci. Technol.* **39**(2), 658–662.

Liu, H., Grot, S. and Logan, B.E. (2005c) Electrochemically assisted microbial production of hydrogen from acetate. *Environ. Sci. Technol.* **39**(11), 4317–4320.

Logan, B., Cheng, S., Watson, V. and Estadt, G. (2007) Graphite fiber brush anodes for increased power production in air-cathode microbial fuel cells. *Environ. Sci. Technol.* **41**(9), 3341–3346.

Logan, B.E., Call, D., Cheng, S., Hamelers, H.V.M., Sleutels, T.H.J.A., Jeremiasse, A.W. and Rozendal, R.A. (2008) Microbial electrolysis cells for high yield hydrogen gas production from organic matter. *Environ. Sci. Technol.* **42**(23), 8630–8640.

Luo, H.P., Liu, G.L., Zhang, R.D. and Jin, S. (2009) Phenol degradation in microbial fuel cells. *Chem. Eng. J.* **147**(2–3), 259–264.

Min, B. and Angelidaki, I. (2008) Innovative microbial fuel cell for electricity production from anaerobic reactors. *J. Power Sources* **180**(1), 641–647.

Morris, J.M., Jin, S., Crimi, B. and Pruden, A. (2009) Microbial fuel cell in enhancing anaerobic biodegradation of diesel. *Chem. Eng. J.* **146**(2), 161–167.

Mu, Y., Rabaey, K., Rozendal, R.A., Yuan, Z. and Keller, J. (2009) Decolourization of azo dyes in bio-electrochemical systems. Submitted.

Ntarlagiannis, D., Atekwana, E.A., Hill, E.A. and Gorby, Y. (2007) Microbial nanowires: Is the subsurface "hardwired"? *Geophys. Res. Lett.* **34**.

Pham, T.H., Rabaey, K., Aelterman, P., Clauwaert, P., De Schamphelaire, L., Boon, N. and Verstraete, W. (2006) Microbial fuel cells in relation to conventional anaerobic digestion technology. *Eng. Life Sci.* **6**(3), 285–292.

Rabaey, K., Boon, N., Siciliano, S.D., Verhaege, M. and Verstraete, W. (2004) Biofuel cells select for microbial consortia that self-mediate electron transfer. *Appl. Environ. Microbiol.* **70**(9), 5373–5382.

Rabaey, K., Lissens, G., Siciliano, S.D. and Verstraete, W. (2003) A microbial fuel cell capable of converting glucose to electricity at high rate and efficiency. *Biotechnol. Lett.* **25**(18), 1531–1535.

Rabaey, K., Ossieur, W., Verhaege, M. and Verstraete, W. (2005) Continuous microbial fuel cells convert carbohydrates to electricity. *Water Sci. Technol.* **52**(1–2), 515–523.

Rabaey, K., Van de Sompel, K., Maignien, L., Boon, N., Aelterman, P., Clauwaert, P., De Schamphelaire, L., Pham, H.T., Vermeulen, J., Verhaege, M., Lens, P. and Verstraete, W. (2006) Microbial fuel cells for sulfide removal. *Environ. Sci. Technol.* **40**(17), 5218–5224.

Ross, C., Charles, C., Drake, T.J. and Walsh, J.L. (1996) *Handbook on Biogas Utilization.* 2nd Edition, Muscle Shoals, AL: Southeastern Regional Biomass Energy Program, Tennessee Valley Authority.

Rozendal, R.A., Hamelers, H.V.M. and Buisman, C.J.N. (2006a) Effects of membrane cation transport on pH and microbial fuel cell performance. *Environ. Sci. Technol.* **40**(17), 5206–5211.

Rozendal, R.A., Hamelers, H.V.M., Euverink, G.J.W., Metz, S.J. and Buisman, C.J.N. (2006b) Principle and perspectives of hydrogen production through biocatalyzed electrolysis. *Int. J. Hydrogen Energy* **31**(12), 1632–1640.

Rozendal, R.A., Hamelers, H.V.M., Rabaey, K., Keller, J. and Buisman, C.J.N. (2008) Towards practical implementation of bioelectrochemical wastewater treatment. *Trends Biotechnol.* **26**(8), 450–459.

Shimoyama, T., Kato, S., Ishii, S. and Watanabe, K. (2009) Flagellum mediates symbiosis. *Science* **323**(5921), 1574–1574.

Torres, C.I., Kato-Marcus, A. and Rittmann, B.E. (2007) Kinetics of consumption of fermentation products by anode-respiring bacteria. *Appl. Microbiol. Biotechnol.* **77**, 689–697.

Virdis, B., Rabaey, K., Yuan, Z. and Keller, J. (2008) Microbial fuel cells for simultaneous carbon and nitrogen removal. *Water Res.* **42**(12), 3013–3024.

Zhao, F., Harnisch, F., Schröder, U., Scholz, F., Bogdanoff, P. and Herrmann, I. (2005) Application of pyrolysed iron(II) phthalocyanine and CoTMPP based oxygen reduction catalysts as cathode materials in microbial fuel cells. *Electrochem. Commun.* **7**(12), 1405–1410.

Index

 Bioelectrochemical Systems: From Extracellular Electron Transfer to Biotechnological Application. Edited by Korneel Rabaey, Largus Angenent, Uwe Schröder and Jürg Keller. ISBN: 9781843392330. Published by IWA Publishing, London, UK.

IWA Publishing's authorised EU representative for General Product Safety Regulations is Diane D'Arras, 15 rue Duret, 75116 Paris, France, e-mail: safety@iwap.co.uk.

Printed and bound by CPI Group (UK) Ltd, Croydon, CR0 4YY

12/05/2026

02108883-0002